Reef Invertebrates

An Essential Guide to Selection, Care and Compatibility

by

Anthony Calfo & Robert Fenner

Natural Marine Aquarium - Reef Invertebrates

Copyright © 2003 by Anthony Calfo & Robert Fenner
Reading Trees and Wet Web Media publications

ISBN: 0-9672630-3-4

All Rights Reserved.

No part of this book may be reproduced, transmitted or
stored in any form (electronically, mechanically or otherwise)
including, but not limited to, photocopying, Internet publishing,
storage retrieval and reprinting without explicit written permission
from WetWebMedia and Reading Trees publications.

Printed in the U.S.A.

First printing in June, 2003
Second printing in August, 2003
Third printing in January, 2005

NMA RI
PO BOX 446
Monroeville, PA 15146 U.S.A.

Reading Trees:
Voice 412.795.9461 or Fax 412.795.5702
e-mail: NMARI@readingtrees.com
web: www.ReadingTrees.com

Additional titles by the authors:
"Conscientious Marine Aquarist" (1998) Robert Fenner
"A Fishwatcher's Guide to Saltwater Fishes of the World" (2000) Robert Fenner
"Book of Coral Propagation: Reef Gardening for Aquarists V.1" (2001) Anthony Calfo

Coming soon...
Volume 2 from the Natural Marine Aquarium series:
"Reef Fishes: Selection, Care and Compatibility" by Fenner and Calfo
Volume 3 from the Natural Marine Aquarium series:
"Reef Corals: Selection, Care and Compatibility" by Calfo and Fenner
"Aquariums Alive! The Essence of Aquatic Art: Keeping Successful and Inspiring Marine, Freshwater and Brackish Aquariums" by Calfo

For more information and dealers of the above titles, please visit www.readingtrees.com

This book is dedicated to conscientious aquarists everywhere,
and the oceans and animals we steward.

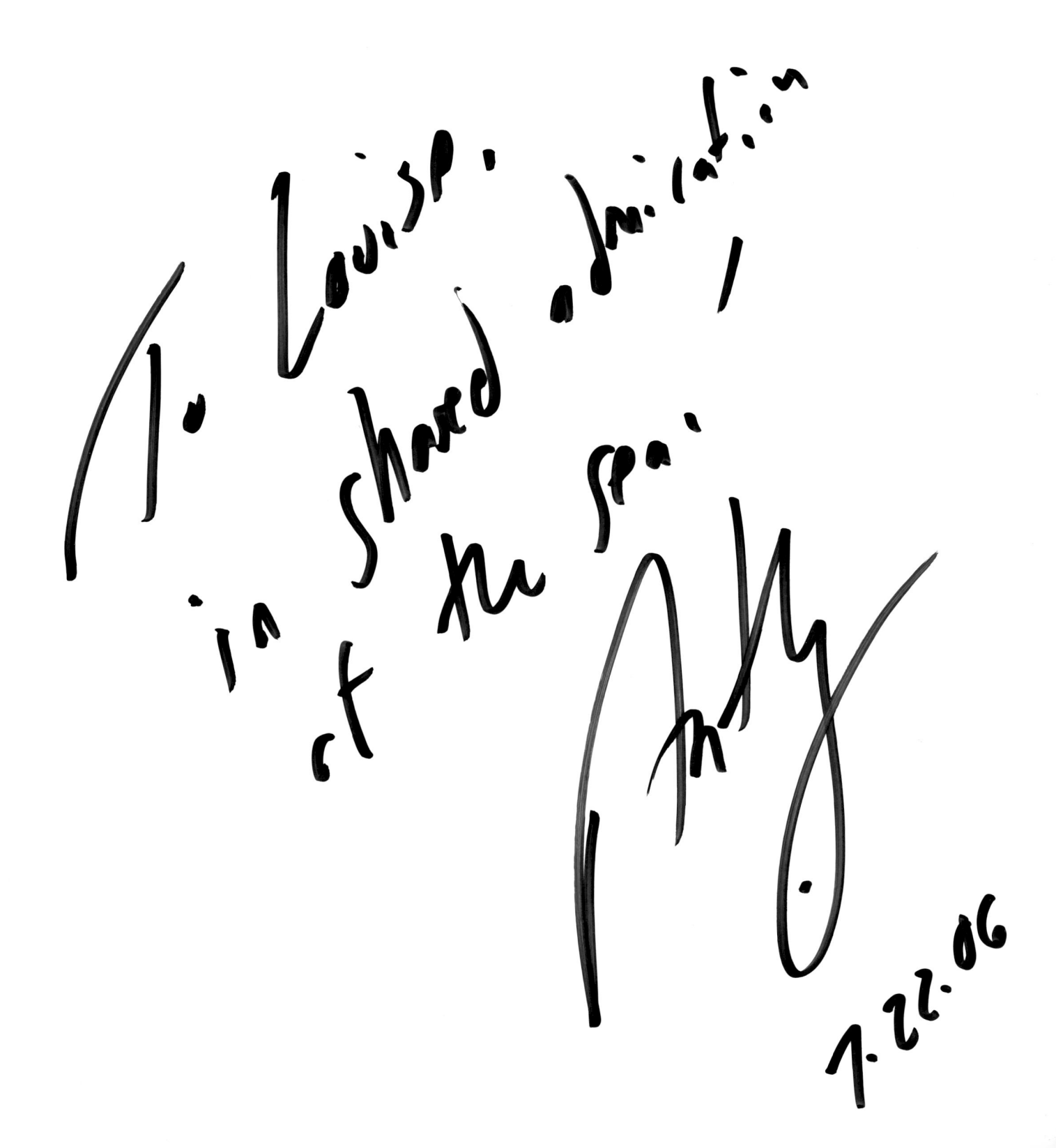

*To Louise,
in shared admiration
of the sea*

[signature]
7-22-06

ACKNOWLEDGMENTS

To Christina and Lorenzo Gonzalez our immensely gifted design & layout editors - they have inspired us all with their amazing talent in bringing art to the science of content provision with a stunning layout and finesse in presentation. Our content has been sincerely and significantly improved by such skills. The hobby and industry will enjoy and benefit ever so much more for it.

Jason Chodakowski, for great friendship and enviable skills in computer science, good humor, and dogged determination to scan, scan, scan.

A very special thanks to our friends and crew at WetWebMedia.com for supporting the web site with yeoman's duties of answering and posting daily queries for our friends in need. In doing so, we were liberated from these considerable duties to dedicate more time to creating and organizing the best possible content for you. By name, we recognize the tireless Steven Pro, Barbara Taormina, Scott Fellman, Gage Harford, David Dowless, Craig Watson, Ananda Stevens, Mike Kaechele and Peter Caterrick for keeping the machine in motion.

To Kevin Carroll for his support, talent and patience as an illustrator.

Professional kudos to Scott "Algae man" Fellman for sharing his time and contributions at large.

We sincerely thank our friend Karen Meszaros for her skill, advice and resourcefulness in helping us to arrange publication.

To Rick Preuss, Steve Oberg, and the staff of Preuss Animal House in E. Lansing, Michigan: for access to their terrific store to make many of the photographs in this book, and for their professional and personal contributions to the business and community of reef-keeping hobbyists.

Jim Nastulski of Tropical Illusions, for letting Lorenzo raid his coral farm for incredible photographic opportunity and sound advice.

Barry Neigut from Clams Direct (.com) for outstanding Tridacnid photographs and marketing support of this book.

Thanks to Walt Smith of Walt Smith International and Pacific Aquafarms for his long-standing friendship, industry involvement and penning the foreword to this volume.

Sincere personal and professional thanks to the Chua's (Edwin, Ted, Millie) and All Seas Marine, Chris Buerner of Quality Marine, Walt Smith of W.S.I. and Pacific Aquafarms, Eric Cohen and Carl Coloian of Sea Dwelling Creatures, for their support of our work in the past and the service that they provide for the aquatics trade.

Robert Fenner would like to thank the many folks who have written in to WetWebMedia.com with their concerns and input that have been instrumental in structuring the content here. This work is directed to you.

Anthony Calfo would like to offer very special thanks to friends Bob Fenner, Lorenzo and Christina Gonzalez for their priceless friendship, fellowship and inspiration in the business of our hobby, industry and life at large. I also offer my humble thanks to God and my family for their unfailing love and support.

Christina and Lorenzo owe a very big "Thank You!" to our assistants and friends: Skip Attix, Mike Bloss, Marina Harding, Timothy Rahtz, Barbara Taormina, and Jim Troeger. To each other and to our son Kieran, for simple tolerance during the final crunch-time of the book. And finally, to Bob and Anthony, for their advice and inspiration on matters far beyond this hobby.

TABLE OF CONTENTS

Foreword	6
Defining the Modern Marine Aquarium	7
Introduction	8
Living Filters	10
Live Rock	12
Live Sand	30
Refugiums	46
Plants and Algae for the Marine Aquarium	68
Selection	114
Husbandry	120
Feeding and Nutrition	126
Reproduction	132
Invertebrate Species and Family Overviews	
Sponges	142
Worms (Feathers, Fans, Bristles & Flatworms)	164
Mollusks	184
Gastropods	
Prosobranchs and Polyplacophorids (Snails and Chitons)	186
Opistobranchs (Nudibranchs, Sea slugs and Sea hares)	206
Bivalves	222
Tridacnids (Giant Clams)	230
Cephalopods (Octopus, Squid, Nautilus and Cuttlefish)	246
Arthropods **(Crustaceans)**	254
Stomatopods (Mantis)	256
Shrimp	262
Crabs	278
Lobsters	294
Microfauna	300
Echinoderms	304
Cucumbers	306
Urchins	324
Sea stars	342
Ascidians (Tunicates & Sea Squirts)	368
Ethos: Serving Life and the Living	382
Bibliography/Resources	384
Glossary	388
Index	392
Photo and Illustration credits	398
About the Authors	399

FOREWORD

Is it possible to walk into a tropical fish store these days and not have some questions regarding the care and, most of all, collection of the beautiful and diverse specimens shimmering and pulsating before you?

In all of nature's creations, is it possible to match the complexity and uncharted territory here on earth with that of our coral reefs?

Some say the coral reef ecosystem is the most threatened environment in the world. Surely, there is evidence that these claims are true. To what level does our industry inspire stewardship of this environment or contribute to its demise? After reading through the chapters ahead it became apparent that these questions will be addressed in a most compassionate way.

When my good friend Bob Fenner and my new acquaintance Anthony Calfo asked me to contribute to their new book I was honored to be included in their company. Never before has our hobby been on such a heightened "state of alert." Now, more than ever, we need intelligent offerings that allow us to ponder our own level of respect for the delicate creatures in our care. A true understanding of the role each creature plays in the wild and how it relates to our captive environment, I believe, is the goal here.

Over the past 30+ years, I have watched this hobby grow from a small handful of stores that carried exotic saltwater fish to the highly developed, state of the art systems we have today. Our hobby has not always been able to keep technology at pace with the ever-increasing demand for the exotic and rare. Somewhere in the middle of my career I can recall that live coral was limited to only a few species (mostly something everyone called "flower-pot") that were almost impossible to keep alive. This did not stop the enthusiast from wanting to know more, and a new era of amateur scientist was born. There was little to read about our mini-reef systems back then as research and authors struggled to keep up with the daily advancements of the hobbyist themselves. Although we slowly began to understand and utilize the technology being offered, from different filtering methods to advanced lighting, we still lacked a confident understanding of how our creatures related to one another. Most of the available literature at that time was outdated or simply did not exist yet. Along comes live rock and advanced technology and suddenly it was like having a newborn hobby. Keeping exotic corals and invertebrates alive was only a matter of how much you were willing to invest in your education and time. A conscientiously aware hobbyist was the new buzz and now we see would-be environmentalists growing and "fragging" their own coral in the living room next to the TV.

With overwhelming gratitude, we should all appreciate the dedication put forth in this book to make things right. Now more than ever, we realize that our captive environments are not merely centerpieces to impress our friends; they are a learning center to share the beauty and diversity of this complex creation with all who encounter it. We should be proud to have the opportunity to share this experience and pass on our humble discoveries to the scientific community eagerly watching over our shoulders.

I think, if you could say that there is a current running through this wonderful piece of work, I would like to sum it up in three words: responsibility, understanding and respect. We should never take lightly what nature has allowed us to share with others. This book will give you a better understanding of the coral reef environment, the responsibility we all share in its survival, and a profound respect for that which created it.

Read on and enjoy,
Walt Smith

Defining the Natural Marine Aquarium

The Natural Marine Aquarium method is an evolution, from the true meaning of the term... an "unfolding" of ideas and systems over the last several decades. And, of course, this is also a strategy that will continue to evolve. We can all advance by navigating its path with careful observations and the faithful exchange of information. This method proceeds based on an understanding that while we strive to replicate a small slice of the ocean, we must also accept and negotiate certain compromises due to the inherent differences between wild reef microcosms and our small closed aquariums. These differences separate and distinguish the study of marine biology from aquariology.

The fundamental components of a natural marine aquarium are living filtration dynamics. These successful reef systems often include live rock, live sand, adequate nutrient export processes, natural plankton and refugiums. But while these are the primary components, there is much more to this dynamic methodology. It is the finesse of living filters that differentiates a beautiful aquarium from an inspiring one. Due diligence of support hardware, for example, like protein skimming is pivotal for many aquarium systems. Even though one does not absolutely need to have a protein skimmer to have a successful marine aquarium, most folks will discover that it is categorically the most effective and reliable means of nutrient export (especially for beginners). And mind you, a protein skimmer is not addressed here in contrast to natural methods of nutrient export (refugiums and vegetable filters, for example), but rather as a support for all such methods. It is arguably quite natural too if you consider the abundance of sea foam washed ashore and effectively exported from the aquatic environment on a wild reef. Newer aquarists and those with modest water change schedules are strongly encouraged to employ a good protein skimmer. Systems with heavy bio-loads (fishes and/or invertebrates) will almost certainly want to employ one or two skimmers cleaned alternately to insure consistent skimmate production. One of the authors' favorite sayings is, "if you are willing to drink the collected product from a week of skimmer production, I'm willing to admit that protein skimmers are not necessary for marine aquariums." Aquarists already familiar with the extraordinary stench of exported proteinaceous matter from a skimmer are likely hacking up imaginary fur-balls at the very thought of it. There are, of course, many other subtle ways to finesse an aquarium system: tidy feeding habits, regular water changes ("dilution is the solution to pollution"), and consistency in all aspects of husbandry.

Intelligent and curious aquarists are embracing fantastic and specific biotopic presentations as well with selections of algae, plants, animals, and other components of the reef that exist together in the same natural environment or niche. It should be no surprise that there seems to be a much lower incidence of disease in captive fishes that are not forcibly exposed to other fishes and exotic diseases from entirely different oceans or locales. Atlantic species may have evolved to better contend with indigenous diseases over some period of time, but Pacific fishes may be wholly unprepared for such xenopathogenic organisms and vice versa. The same holds true as much for invertebrates. Corals collected at 25-30 meters depth (like some corallimorphs or LPS corals) and mixed in tanks with species found in one meter of shallow water cannot realistically be expected to thrive equally well under the same standardized lighting and water chemistry. The very sensation of unnatural neighbors can, and often does, illicit the ongoing secretion of allelopathic compounds (silent chemical "warfare"). Many corals may seem to tolerate this elevated level of stress for weeks or months but many will die in time from the imposition. Even if we could know that the action will not be fatal, it is clear that the expense of energies used to wage war are stolen from resources that could otherwise be used for disease resistance, growth, vigor, or reproduction. This really gets to the heart of the matter: attaining optimal health and vigor. We will maintain magnificent and inspiring displays of marine organisms largely through natural biotopic displays, considerate and adequate feeding, and proper attention to all other aspects of husbandry and maintenance of water quality in natural systems. In doing so, we open up an entirely new world of possibilities for study and enjoyment in observation of species familiar with each other and conducting amazing behaviors only possible in a truly balanced microcosm, the natural marine aquarium.

(L. Gonzalez)

Introduction

However beautiful, it's not really a "reef" aquarium without the invertebrates described herein. (*L. Gonzalez*)

In this work, the first volume of our Natural Marine Aquarium series, you hold a key for discovering the best reef invertebrates, plants and algae for aquarium use and their captive care. This reference embraces things great and small in reef aquaria outside of the popular fishes, and corals & anemones, which follow in our next volumes. Here, we offer a much needed address of the many primary and incidental creatures that make up a reef full of organisms, all found to be fascinating to reef aquarists: shrimps, crabs, worms, snails, starfish, urchins, and many until-now commonly undescribed "hitchhikers" imported with live rock and sand.

Originally we thought to call this set of books the "Modern" Marine Aquarium series, but we opted to crown our work with the title "Natural" instead. The term "modern" has a finite definition that may not seem as accurate or up-to-date years down the road when the successful methodologies and information described here remain valid and important. Indeed, the physiological profile of the fishes and the dynamics of natural microcosms within balanced ecosystems (refugiums, nutrient cycling, etcetera) will remain unchanged for more years than any of us can envision. Throughout these books, we intend to detail strategies of natural aquarium husbandry for the modern aquarist that are closer to wild, infinite "techniques" that sustain reef invertebrates in natural microcosms.

We wish to share our love of the marine aquarium hobby with you and to impart wisdom to help you in like endeavors. Whether you have been inspired to pursue a saltwater tank because of a beautiful show tank at a friend or neighbor's home, local fish store (LFS), public aquarium, or from observing the living reef while snorkeling or scuba diving, we will help you successfully recreate a small slice of the ocean. Natural Marine Aquarium methods are, in our opinion, the best ways of replicating some of the many interesting reef microcosms. You may ask, what is the "Natural Marine Aquarium" specifically? There are indeed many different and successful ways of supporting a marine aquarium: Lee Chin Eng's "nature's system," Berlin-style technology, Dr. Jaubert's plenum method, Dr. Adey's algal turf scrubbers (ATS), Leng Sy's Miracle Mud® refugiums, and more. What we have described in this series, and what many successful hobbyists have done, is a combination of many of the best attributes of the above-mentioned methods and beyond, into one that we may categorically call the "natural marine aquarium." This modern style of aquarium keeping likely utilizes live rock, live sand, protein skimming, and a refugium (sump or other). All of these various components will be discussed in detail in subsequent sections of this book and series. We also want to alleviate any concerns that this is a "hardcore" reef aquarist's book. While the first installment in this series is titled, Reef Invertebrates, many of the animals found in the tropical reef environment make their way into various types of marine aquariums. With the popularity of live rock and live sand products, aquarists across a wide spectrum of interests bare witness everyday to many fascinating and often uninvited creatures. As such, the pleasure and privilege of keeping reef invertebrates is hardly restricted to reef aquarists. With that fact in mind, successive volumes in our series on fishes and corals will be about reef animals for the marine aquarium, and not limited to marine animals for the reef aquarium. Our scope and intent in the Natural Marine Aquarium series is deliberately broad to serve all marine aquarists. We hope that you find that we have achieved our ideal with enthusiasm. The introduction to any book can be challenging to compose. In the case of this printed work, however, there were no such impediments. In your hands are our hearts and minds regarding the topic at hand: identifying and selecting the best species and individual specimens of marine invertebrates for aquarium use, and keeping them healthy.

When writing any such handbook, it is very important to summarize what the work is about: its content, purpose and demonstrable "style" to the prospective reader, to you. We are well matched to this task, as this field is well known by our assemblage. For many years we have lived in and traveled the world in the service of aquatic life, sincerely dedicated to the science and hobby of aquariology. We have vigorously lived the field as consumers, academics and industry professionals for all the modern history of captive marine husbandry, and in all of its capacities (establishing gathering stations, collections, wholesale, culture, service, and retail facilities). We have known the pioneers, studied with the pioneers, and in some instances been the pioneers. We present this with the belief that, it is essential in very few words to assert the credentials of the authors, the worthiness of the tome, and the significance of its utility. With sincere conviction, we are resolved to produce informative, useful and very practical guides for the successful care of your living reef treasures. We have attempted to accomplish this in a familiar and easy language, and in a format that we hope you will find both enjoyable and important. Indeed, many exciting paths of wonder and wisdom lay ahead for all of us in this delightfully evolving hobby.

It is our sincere desire to support you in seeking out the best available, and most appropriate marine livestock for your study and enjoyment. Knowledge from our extensive participation in the aquarium trade is freely offered every day for all who seek a shared opinion and such wisdom. Through our books, articles, photographs, video and website (WetWebMedia.com), we endeavor to communicate our love and passion for improving the success of aquarists, and the quality of life for the creatures in their care. In daily interaction, we eagerly engage aquarists in forming lists of desired additions to marine tanks. We help guide experienced aquarists and academics through their personal evolution of aquatic study and display, and we diagnose and assist those that have fallen down a rung or two through difficulties. Alas, too many other folks have made poor choices in this arena and their subsequent trials and tribulations exact a significant toll on their continued participation in the hobby. We want you to be successful in assembling your living inventory and hardware, avoiding losses with these wet pets and the heartache that goes with their passing.

Too many people miss the chance to experience the wonders of the aquatic world due to a lack of useful and timely information: either by avoiding the hobby altogether based on hearsay ("Marine aquariums are too difficult."), or by dropping out conditionally when they are unable to find good help or counsel. As "old-timers" in the hobby, science, and business of ornamental aquatics, we will earnestly assist you to keep your livestock *live* - from the very beginning, and on through your success and enjoyment of the hobby for many years to come. We endeavor to accomplish this by directing you to the best species to purchase, teaching you how to acquire the best specimens of appropriate size and number, how to determine and establish their living spaces, and provide them with appropriate care for truly long-term health and success.

The opportunity for us to co-author this reference proffers tremendous possibilities for both the reader and the writers. The dynamic convergence of our thoughts, experiences, and ideas necessarily imparts unique flavors from each author's wonderfully contrasting experiences, while reinforcing fundamental commonalities: the tried and true aspects of aquarium keeping. To that end, an intelligent consensus is formed and the reader can be assured of receiving a better version of information about the "best" (and worst!) reef fishes, invertebrates, algae and plants for the natural marine aquarium. Marvels of the sea lay ahead of you – some out in the open, some thinly veiled, some buried deeply, but all awaiting discovery for informed admirers of the ocean realm. It is our duty and desire to share with you our tools of discovery: to observe, study, experience... to learn and apply. It is our most sincere wish that you will be inspired and encouraged to share your own wisdom in kind. We can and will help you! Let us celebrate our fellowship while sharing our dreams for the future of the aquarium sciences.

In shared admiration of the sea…
Anthony Calfo & Robert Fenner

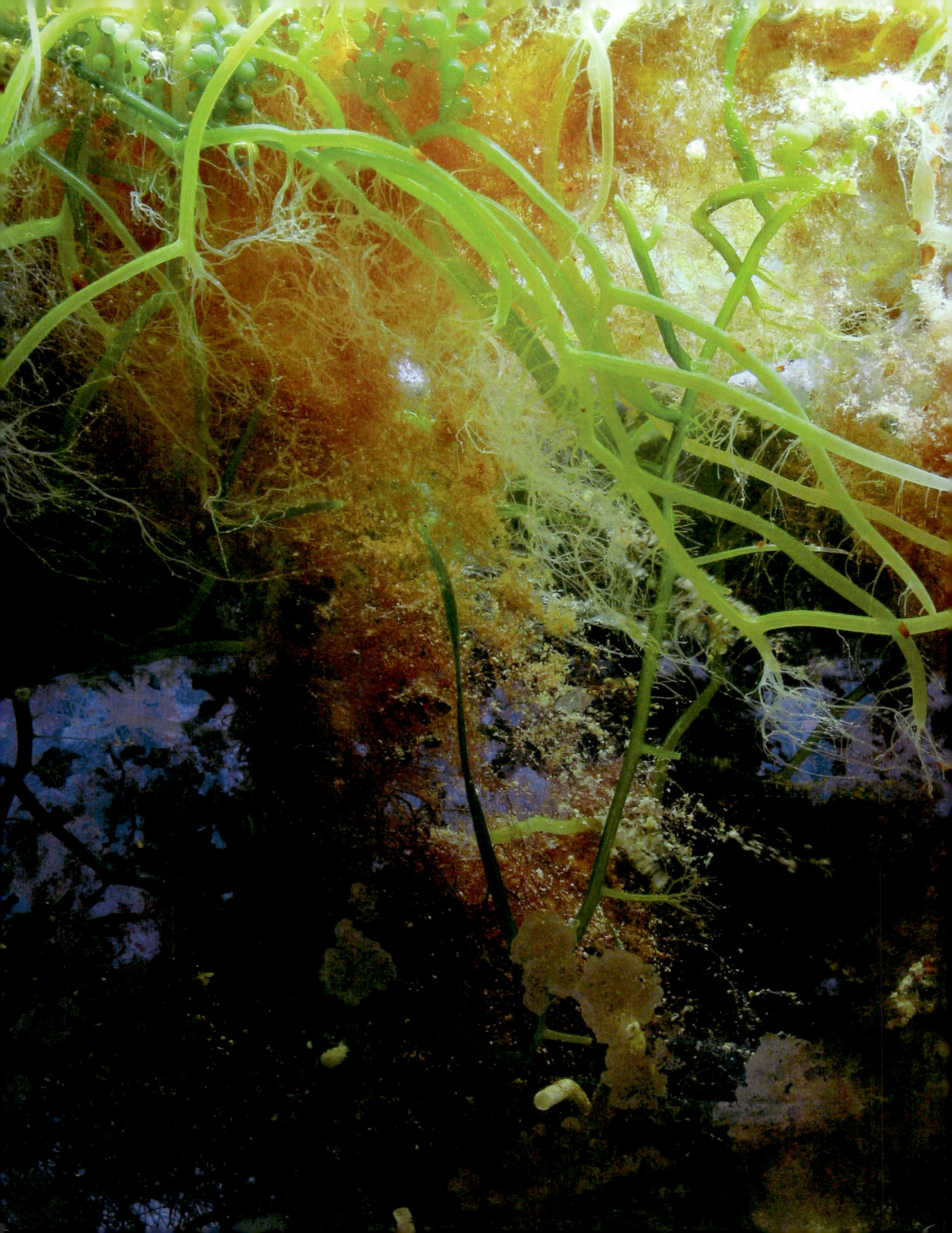

Living Filters

The term *living filter* has forged its rightful place in the lexicon of the modern aquarium hobby to the extent that many aquarists can readily explain, define or at least begin to describe this dynamic. To the uninitiated, one might wonder, "What exactly is a living filter?" In the broadest definition, we *the authors* are living filters. We take in enormous quantities of beer and filter it through our systems. Some of us are more efficient living filters than others... the best go on to chair 12-step programs and... well, you get the idea. Such is the case with marine organisms that participate as living filters; some are more efficient than others at removing various elements from the water. The microcosms of our aquariums, if successful, are teeming with living faculties over a wide scope and scale that accomplish what we may fairly describe as filtration (and very efficiently so at that). These desirable life forms keep the system in balance and harmony. Living filters may include live coral that directly consume dissolved and particulate organics from the water. Numerous bivalves seen and unseen in live rock extract ammonia and nitrogen with superb efficiency. Sponges are arguably the very best filter-feeders, processing more water per hour through their systems than bloated attendees at a salty peanut convention. A single sponge can literally process the volume of water in an aquarium through its system several times over per hour! True vascular plants and algae alike can be very effective vegetable filters. A myriad of worms, micro-crustaceans, mollusks, and other benthic animals recycle detritus and waste products into useable foods. And many thousands of bacteria and other micro-organisms hungrily scrub the system of targeted components which limit their very existence by concentration (like sources of nitrogen for the nitrifiers). It is a wonder but no surprise that reef aquariums are magnificently complex microcosms with all of this living bounty contained.

The fundamental living components of a natural marine aquarium are: **Live Rock** (LR), **Live Sand** (LS), and **Refugiums** ('fuges). The crucial role of live rock and live sand products in marine aquarium science have been time-tested and found to be indispensable for most. They can wholly supplant any man-made biological filtration and they do so with far greater efficiency and merit. There are many benefits of live substrates beyond nitrification (the mineralization of organic matter which leads to biological stability in the aquarium). These benefits are covered at length in the following chapters. In kind, the recent birth and evolution of popular refugium methodologies has inspired aquarists to completely reconsider the scope and breadth of a marine aquarium keeping and even possibilities with a self-sustaining ecosystem. Progress with deep sand beds, mangrove biotopes, plankton reactors and mud systems just to name a few applications have stimulated aquarists into experimenting with fascinating methods of alternate living filtration dynamics. A solid foundation for learning how to understand and exploit the principles of all such living filters awaits you here.

(facing & above, *L. Gonzalez*)

Live Rock

Live rock has numerous benefits in the aquarium beyond natural decor. It is an incomparable living filter that provides food and habitat for micro- and macro-organisms alike. Fiji image.

"You paid *how much*? For a live *what*?" Most of our friends think that it's at least humorous, if not grounds for committal, that we pay hefty sums for ocean rock for our aquariums. But they just don't know what we know. Live rock is remarkable from both an aesthetic and utilitarian perspective. It is a decoration that also provides a food source, filtration, habitat, and is an incomparable mediator of water quality. An aquarium supported with live rock, in contrast to an otherwise similar set-up of hardware, will fare better both stability and the vigor wise. The discovery of living substrates at large, including live rock and live sand, has improved the success of captive marine aquariology.

Live rock, by definition, is a mixture of a hard, non-living mineral matrix with an assemblage of diverse living organisms on, in, and amongst it. Nearly all living phyla of monerans, protists (including micro- and macroalgae, protozoans, and more), plants and animals make up this stony "bouillabaisse." One need only look closely to make out much of this life; scurrying amphipods, sprouting plants and algae, coral larvae, boring clams (spitting "smoke" as they mine a calcified residence), urchins, and countless other organisms.

Most live rock is formed by calcareous organisms that grow and fuse over time. It commonly includes the overgrown corallums ("skeletons") of scleractinian corals: some recent, some older, and some downright ancient. Many other macro- and micro-organisms contribute an enormous amount as mass cemented by coralline algae. Some live rock is formed through sedimentation (common in Atlantic and Caribbean waters) composed predominantly of coral sand, which has also been created in large part by calcareous algae. Other live rock is formed by volcanic activity, although this type is not common nor particularly desirable in the hobby. As a commodity in the ornamental marine aquarium trade, live rock is still of primary importance. It is a crucial resource in keeping almost all livestock. There are some challenging organisms in the trade that could not otherwise be kept without the benefits live rock provides.

Of all challenges the aquarium hobby and business has faced, the ban on collecting Florida live rock in the

12 *Natural Marine Aquarium Volume I - Reef Invertebrates*

United States, along with the fortuitous start-up of rock collections in Fiji, were benchmark events that virtually ruined, then saved the reef aquarium hobby and industry, respectively. By the time the doors completely closed on live rock collection in Florida waters (circa 1996) the hobby had developed an extraordinary dependency on this living substrate. The industry problems ran deeper than an aquaristic preference for live rock. Hobbyists could certainly maintain healthy livestock, albeit of a more limited selection, without hard living substrates. The danger and real threat to cripple both the trade and industry was the enormous dependency of merchants at all levels on profits from live rock. Indeed, live rock is more than the foundation of a healthy marine aquarium; it is the foundation of many livelihoods in the industry. The relatively abrupt cessation of its collection put dealers in many aspects of the trade in peril. Fortunately, with the opening of collections in Fiji (pronounced 'Fee-gee'), and the pioneer development of an industry there by Walt Smith Inc., the industry was buoyed. There were other significant influences to which the hobby and entire industry owes great thanks. A keystone in the development of trade in live rock was the offer of very reasonable international shipping rates by Air Pacific, which facilitated affordable export from the country. Also, great thanks are owed the local governments for their favorable treatment of new business. Kudos to the Fijians for their optimism and foresight in embracing the birth of an industry that has created thousands of jobs and funneled considerable monies into their economy. Great progress in the hobby on all levels has been realized through the popularization of successful reef-keeping with South Pacific live rock from Fiji and elsewhere, which is superior in many ways to the dense, sedimentary products of the Atlantic and Caribbean.

Live Rock Aquarium Use Benefits

Instant Karma **(Nutrient Cycling)**

Live rock that has been cleaned of sediments and is **cured** (more on this below) essentially performs as an instant biological filter. It can handle a reasonable to significant bio-load of fishes and invertebrates, and will strengthen in proportion to the load growth. In cases where cured live rock is handled and transported properly, an aquarist might begin a new tank without any of the normal cycling aquarium stresses: ammonia, nitrite, nitrate spikes. Even in these cases, though, prudence and patience is required while stocking the aquarium; even the best filtration can be unstable in the beginning. Never buy so-called "cured" live rock with the expectation that you can stock your tank within days to weeks. This is a wildly simplistic notion.

All new live rock should receive careful inspection upon arrival and at least some steps for "re-curing" need to be performed. Like corals and fishes, live rock should

> **Live Rock** (abbreviated as **LR**) fulfills some absolutely essential functions for aquarists:
>
> - establishing nutrient cycling almost instantaneously with cured product (ready bio-filtration)
> - providing food organisms of extraordinary diversity
> - utilizing nutrients ("sinking") and hence disallowing their utilization by nuisance algae
> - providing habitat and psychological benefits to captive denizens (behavioral enrichment)
> - as inoculants (seed) source of micro- and macro-invertebrates for non-living substrates (rock and sand)
> - provision of soluble carbonate and bio-mineral content for improved water quality and stability
> - natural and entertaining decorative artifacts

be quarantined for a minimum of two to four weeks for observation, stabilization (reduction of potential pest and parasitic organisms), and re-curing if necessary. After quarantine, live rock is the ultimate form of biological filtration with which to accomplish both the forward and reverse nitrifying and denitrifying reactions necessary for keeping captive marine life. Yes, to be clear here, we're stating that live rock does reduce nitrates (significant anaerobic nitrifiers exist in subsurface areas) and performs other biological faculties beyond that which any man-made filter can provide. It is an amazingly reliable and complex living filter by any definition.

Continuous, Palatable, Nutritious (Food Production)

Live rock contains hundreds to thousands of organisms, most all of which have planktonic larval stages. These life forms in adult and intermediate stages serve as prey items (plankters) to the various types of life we keep in aquariums. It is a virtual smorgasbord for your fishes, corals and other invertebrates and has tremendous regenerative possibilities if cared for well. Turf, micro- and macroalgae, sponges, hydroids, worms of many kinds, crustaceans, mollusks, echinoderms, and ascidians; these and much more make up a significant portion of fresh live rock. It's like a living restaurant for your hungry aquarium specimens. Take some time with a flashlight and magnifying glass or low-power microscope by day to observe closely what goes on both in and on your live rock. There is always some action. Aquarists have often realized while waiting for live rock to cure that they could look in their aquarium and see something new and exciting every day without ever adding a fish or coral. In fact, most hobbyists never fully appreciate the enormous diversity and bio-mass imported within live rock because of its nocturnal tendencies. Most planktonic activity occurs at night when these organisms come out to scour the tank, feed, and to release their gametes and young into the water column.

As if by magic, numerous plants and algae spring forth from seemingly plain, barren or denuded rock many months after installation when conditions are finally suitable. Many people write to us at WetWebMedia.com in a semi-panic regarding their fish and non-vertebrate livestock not feeding. Yet, with an ample stock of healthy live rock, the chances of picky feeders starving to death are vastly reduced. Newly introduced or stressed animals may not take food from your hand, yet they are eating: discreetly consuming flora and fauna found in and on live rock. While target feeding with zoo- and phytoplankton substitutes is helpful and may be necessary in some cases, no foods readily available to the aquarist can compare in nutritional quality (and sometimes quantity) to natural plankton found in live rock and live sand. Fishless refugium methods that perform as massive plankton reactors are founded on this dynamic. Live rock, like live sand and refugiums, is an invaluable and indispensable part of the natural marine aquarium.

Algae and Nuisance Organism Control (Nutrient Limitation)

One of the best ways to limit pest algae proliferation (green water, diatoms, cyanobacteria, etc.) is with the use of live rock. Good live rock (that which is rich in bio-diversity) has a wide array of primary producers (photosynthetic life) that aggressively consume available phosphate, nitrate and other potential nutrients to noxious algae. Even in systems that have established, undesirable algae populations, the addition of cured live rock goes a long way towards diminishing their presence. Aquarists have observed this many times; the stimulation of coralline algae growth staves off encroachment by less desirable algae. This is part of the dynamic of "algal succession." Chemical filtrants like activated carbon and Poly-Filter can also be used in concert with live rock to achieve this effect. The growth of nuisance organisms fundamentally boils down to limiting nutrients. A rich bio-diversity of "hungry" micro- and macro-organisms within live rock work tremendously towards achieving these ends.

Psycho-Social Effects (Behavioral Enrichment)

Imagine coming from a limitless space in which to forage, roam, and seek mates to living in a small box, like an aquarium. For motile invertebrates (and fishes) the psychological effects must be profound. What's more, think of how surprising (as in *shocking*) it must be to encounter clear walls of confinement for the first time. Whatever other novel phenomena our livestock encounters when going from wild to captive conditions, such as the lack of being able to "get away," is likely substantial. Try as we may to set up the best possible aquarium and environment, it is still unnatural and must be stressful to some organisms.

Live rock functions as a source of physical familiarity and as a natural distraction providing refuge and habitat. When investigating issues like population structure and the distribution of the many macro-organisms that we are interested in as aquarists, one finds that they do not occur randomly in the sea. They feed on a reef, live on a reef, and have evolved to survive *on* a reef. Live rock is a piece of the reef, in fact and form, which provides behavioral enrichment for aquarium specimens, while, no doubt, mediating the physical qualities of water for their collective improved health.

Seeding Non-living Substrates

There are merchants that sell what they label as "live sand" in our hobby that isn't quite as live as you'd like it to be. The pre-bagged carbonates sold as live media may contain some incidental microbial life, or at least the potential for them, but in comparative bio-diversity, these products are more like desert sands contrasted to rich forest grounds. Marketing terminology for aquarium products is in much need of definition to clarify their extraordinary and incredible claims.

Real live rock and sand have teeming multitudes of representatives of all marine phyla in and on its surfaces. For live sand, this is literally thousands of specimens of worms, crustaceans, mollusks, microbes, and more per handful. One only needs a small portion (5-10%) of live sand to inoculate the body of a bed of sterile sand with most of these desirable critters. To a great extent, healthy live rock will also do the same job. Placing purchased *previously* cured, or better still, raw wild rock (cured at home for control over life lost and saved) with a new substrate will seed the "sterile" material with more than sufficient numbers and species of **infauna**: the term used to describe interstitial life forms that live in and between substrates. Micro-crustaceans and polychaete worms are just the tip of the planktonic iceberg looking to colonize your not-yet-live rock and sand. Weakly live or even recently dry rock can be seeded very well in the same manner.

An Adjunct to Alkalinity, Biominerals and More (Mediating Water Quality)

Over time, due in large part to naturally occurring reductive activities (*as in* Reduction-Oxidation - a.k.a. **redox**), our commonly over-crowded, over-fed, under-filtered, weakly circulated and aerated systems lose alkaline reserve as well as calcium, magnesium, strontium specifically. This loss of dissolved matter (mostly of calcareous substrates like sand and live rock) results from the inevitable production of organic acids and the overall biological activity in the aquarium. Live rock contributes "buffers" (its minerals are "eaten up") in the aquarium against acidic conditions. It might surprise you, but if you were to weigh your sand and live rock when you placed them into the aquarium, and then again even just months later you'd find a good part of it missing. Often, this reduction is so dramatic that you can actually see the difference. Aragonite sand, for example, can dissolve in half in just two years time or sooner (a 6"/15 cm bed is reduced to a 3"/7.5 cm depth). Live rock will also become decidedly lighter weight over time. The physical space occupied by these substrates does decrease markedly, but please bear in mind that the dissolution of carbonaceous materials in your system is a good thing for your livestock and system at large. Liberated biominerals support water quality, and in some cases directly "feed" desirable organisms. You will want to monitor their rates of dissolution and schedule the replacement or bolstering of these non-living components as part of your long-term maintenance strategy.

Just a quick note, too, regarding other types of non-calcareous rock: by and large, you will want to avoid these types. They often lack the porosity (as with sedimentary forms) or the solubility (as with volcanic formed matter) of the more desirable types of live rock formed by calcium carbonate deposition. They will contribute very little to sustaining pH or biomineral concentrations and in some cases may contribute undesirable elements to the water.

What Better Decor? (Ornamentation and Aesthetics)

Is your house and yard filled with life-less elements: plastic plants, artificial birds outside the window, with nothing but concrete flooring inside and out of the home? Of course not, you've got aquariums for goodness sakes! Each of us seeks comforts that make our homes enriching to our lives. Even if you could live in a barren environment,

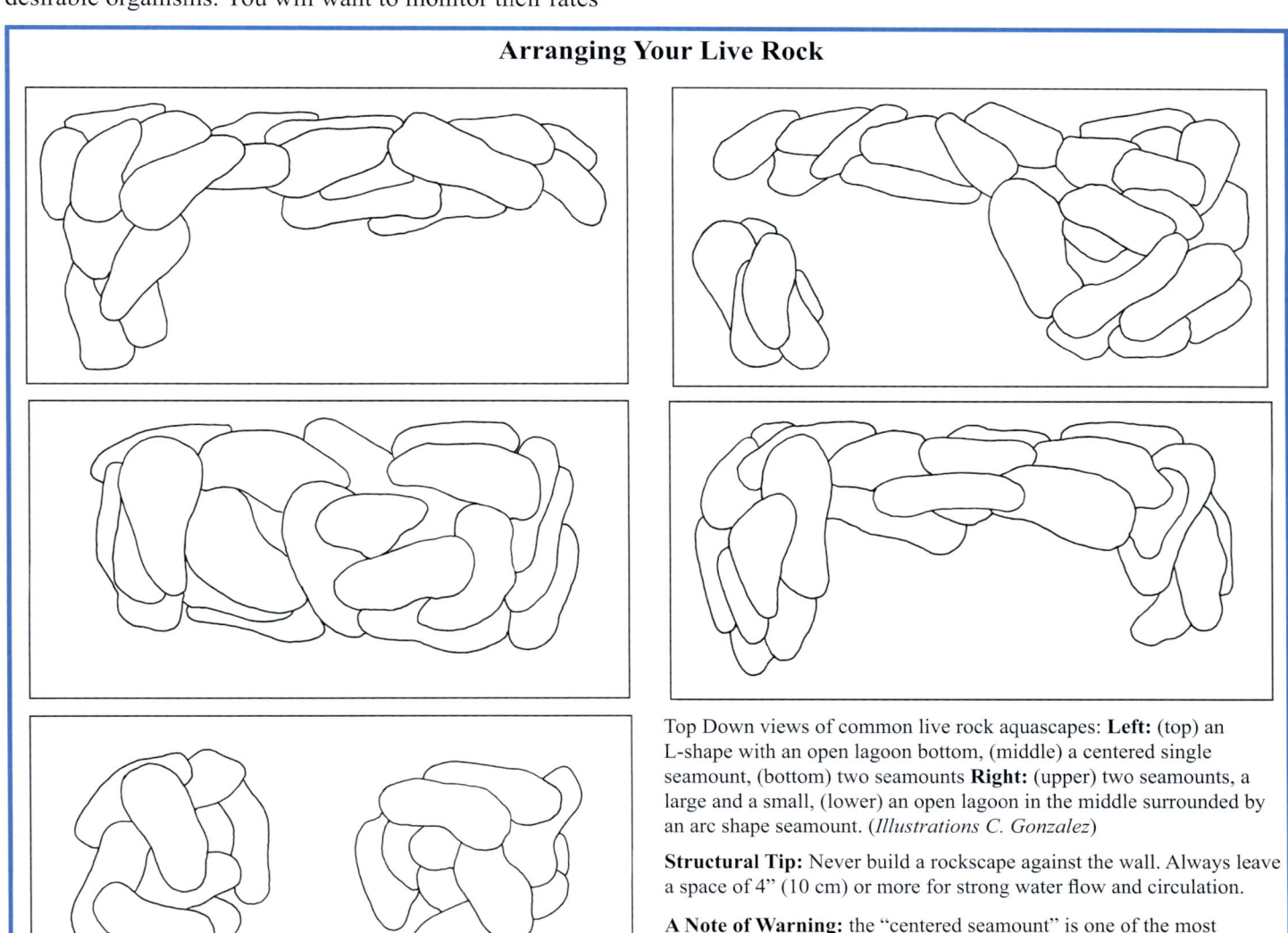

Arranging Your Live Rock

Top Down views of common live rock aquascapes: **Left:** (top) an L-shape with an open lagoon bottom, (middle) a centered single seamount, (bottom) two seamounts **Right:** (upper) two seamounts, a large and a small, (lower) an open lagoon in the middle surrounded by an arc shape seamount. (*Illustrations C. Gonzalez*)

Structural Tip: Never build a rockscape against the wall. Always leave a space of 4" (10 cm) or more for strong water flow and circulation.

A Note of Warning: the "centered seamount" is one of the most common but least successful formations as they measurably impede water flow around and through the rockscape, which often negatively effects aquarium health, detritus removal/suspension, etc.

Live Rock Collection: The First Steps

Some views of the systematic cleaning and processing of live rock in Fiji. The product is collected by locals on site by hand, then shipped on open bed trucks to their plant (here in Latouka) where it is hand-cleaned and blasted with water that has been hauled, filtered, and chilled. Every piece is handled to remove dead algae, sponges, and other undesirable growths that are likely to burden shipping and curing. The rocks are then rinsed in long tanks to remove sediments and some undesirable small organisms (worms, snails, predatory shrimp, etc.) before being boxed and exported.

would you live well, and would you even want to? The same argument holds true for keeping your aquarium pets. For the vast majority of marine systems (excepting those "biotopic" approaches to mud, seagrass, and mangrove-specific presentations, for example) the use of live rock is superior to all others in providing decor that affords natural sleeping, breeding, and feeding opportunities.

The cornucopia of live rock types, sizes, and shapes available offers untold possibilities in cave-building, stacking, layering, and overall design. Most folks find one to two pounds of live rock per gallon of display suits them aesthetically, but, due to differences in rock densities, shapes, personal preferences, and budget, you might prefer more or less rock. More can always be added later, and it is quite helpful, in fact, to add and refresh live rock periodically with new material.

SELECTION

About Live Rock Types:

A good number of descriptive names have been created to help sell rocks for many dollars per pound. "*Deco*" this, and "*Ultra*" that, bio-rubble, plant, base, premium, plating, branching, shelf, pink, purple, and a personal favorite - "pre-cured" (as if such a thing were impervious to the rigors of dry transportation).

All that you really need to know before buying live rock is:

- 1 - its general point of origin
- 2 - how cured or curable it actually is (was it held dry for days or cured underwater for weeks?)
- 3 - the nature of its composition (useful calcareous, neutral inert or other potentially troublesome matter)
- 4 - expected quality (handling on import and bio-diversity)
- 5 - the net cost to you

(1) LR Origin:

Most live rock sold in America is wild-collected from several island nations in the South Pacific. Other live rock products hail from the tropical West Atlantic, Gulf of Mexico, and Hawaii, which are essentially aquacultured. It would be difficult to say which rock is best, although some are better than others with all things considered. The "best" rock is simply that which serves your application most effectively regarding desired biological faculties, bio-diversity, availability, and price. Aquacultured rock is typically much more dense (you'll need more per gallon) with correspondingly less surface area per unit weight. It does have unique fauna, however, that one cannot easily acquire otherwise and it is the preferred source for biotopic displays of the region. It also has the added advantage of reaching the consumer faster and in healthier condition. As a result, the net survivability of incidental microfauna on and in Atlantic rock is very good.

South Pacific rock is a more mature "fossil rock," and as such has had many years to become handsomely encrusted, colonized, and richly diverse in living microfauna. These ancient stony pieces are typically less dense, are more variable in usable shape and size (with great "nooks and crannies"), and are seeded with greater number and variety of desirable organisms. For these and other reasons that you will learn, wild-harvested Pacific rock is by far the preferred product for aquarists.

(2) **"Cured" To What Degree?**:

You will want to "re-cure" your live rock, disregarding potentially superfluous claims of pre-curing. If nothing else, you must quarantine it for two to four weeks like any other wild-harvested livestock (fishes, corals, plants, other invertebrates, etc.) to screen for parasites, pests, and diseases. During this quarantine and re-curing period, you can bait for undesirable shrimps and crabs and evaluate any other creatures that become apparent. You can also stabilize weak or stressed organisms before entry into the main display. Above all, however, the isolation serves to insure that so-called "cured" rock truly *is* cured and will not foul your main display.

Unless you are buying live rock from your local merchant and have seen and smelled the product for two or more weeks, we simply cannot take the word of a seller at face value on this point. There is simply too much at risk with common misrepresentations on degrees of "curedness." If you have any doubts, ask other consumers (the Internet Bulletin Boards are of great utility here) what their experiences have been with fresh live rock.

We assure you that there is a huge difference between sources of live rock in terms of quality and consistency. Some shippers do little more than pack the rock at the beach and let it sit for untold days in the sun before exporting it (taking 4-7 days before re-immersion in water is not unheard of). This is exacerbated by the fact that some wholesalers never unpack the rock while waiting for resale. Thus, it may sit dry for another 4-7 days in a wholesaler's warehouse before it is ordered by a retailer and passed down the chain of custody.

Despite the fact that some live rock is kept out of water for a week or more since collection, an amazing number of living organisms survive the rigors of import, darkness and an extended "high tide" (air exposure). However, that does not make the rock "cured," just hardy. Below you will find details about the curing process of live rock and sand. Let us tidy up this synopsis with an exclamation point to our persistent beseech: quarantine, quarantine, quarantine!

All livestock, from rock to plants, through snails to fishes;

Healthy live rock is almost synonymous with encrusting algae. Ideally, you want to purchase rock that already has good coverage. Cultivating it is mainly a matter of providing sufficient, stable biominerals and water quality: alkalinity (8-12 dKH), high pH (over 8.2) and adequate levels of calcium (350-400 ppm). If your corallines start "bleaching" or flaking, check your water chemistry first.

everything wet with saltwater is to be kept in isolation for 30 days ideally. Two weeks is the bare minimum for quarantine. In doing so, you will have a much more enjoyable experience with the health and vigor of your aquarium guests, largely avoiding pests, parasite, and pollution.

(3) **LR Composition**:

The safest and most utilitarian composition for live rock product is calcareous in nature. Non-calcareous product is available but it is usually poor quality and we advise you to avoid it. Not only is it unlikely to buffer your pH, but it probably has no potential to add beneficial biomineral content. Some rock is even collected in cool (temperate) waters - its life suffering and your tropical livestock with it as a consequence of warm water exposure. There are people who produce legitimate aquacultured live rock or sell formulations for DIY'ers to create their own live rock. Some of this matter is very good, most are adequate at best, and some is even harmful. Once again, explore the Internet and question fellow aquarists (local aquarium societies and the like) about experiences and advice to make an intelligent decision about using these alternatives. You are strongly encouraged to avoid using volcanic, metamorphic rock as base in your system. Beyond benefits lost in contrast to real calcareous live rock, they are likely to facilitate undesirable organism growths in the aquarium and will not save you any money in the long run for the labor and expense to correct its deficiencies.

Reef Building Tips:

When evaluating live rock for use in the aquarium, seek openly structured and porous pieces to optimize bio-

Numerous plants and algae can grow from live rock but are often limited in development by the premature stocking of a new aquarium with herbivorous fishes and invertebrates. Let live rock mature in aquaria for several months without grazers for optimal "plant" growth.

diversity (acquired bio-mass and that which we hope to culture). Variable shapes and textures build a stable structure. Fluid construction with nicely articulated and open shapes affords better water flow throughout the aquarium as well as providing niches in the rocks and paths through it for display organisms.

High porosity will support a larger and more effective array of desirable fauna. Atlantic and other sedimentary rock types are less stable calcareous products in the long term. They dissolve quicker and produce more sediment over time. Avoid using sedimentary formed live rock as a foundation, especially for large tanks and those you hope to establish for more than five years. "Branch rock" consisting of thick, dense, encrusted coralliums of staghorn-type Acroporids can make a very useful foundation for small reef structures. It can suspend a full rockscape on a very small footprint (the few places where large branches bury in the sand) improving the health and bio-diversity of the sand bed below it.

Other fine structures can also be built with average varieties of Pacific live rock by strategically using "blocks" and "shelves" to form open structures. You should build your reef structure several inches off the back wall of the aquarium to support and improve the dynamic of water flow in the aquarium. Never build a reef leaning up against any wall of an aquarium. Resist the temptation to use too much live rock on startup. Although it gives the appearance of a full display immediately, consideration for animals that will be placed and grow there in the future may have been overlooked.

Most aquarists need not build a reef any higher than 6" (15 cm) below the surface of the water and much lower will be required for many. Consider the corals and fishes that will occupy the areas above your reef structure and allow room for their adult sizes. In the case of common leather corals and *Naso* tangs, for example, more than 12" (30 cm) will be needed between the surface of the water and the crest of the rockwork and even this is a gross understatement for many *Sarcophyton* leather corals. Sketch your ideal rockscape in advance and try to have a long view of the space needed for invertebrates to grow and fishes to swim. Then put your plan into action in light of these goals and expectations.

(4) **LR Quality**:

There's an old saying, "If it walks like a duck and it squawks like a duck, it's probably a duck." This analogy holds true for the quality of live rock. Cured live rock has been held in efficiently skimmed, well-aerated water for at least two weeks. It will be fully cured when all evidence of decay is absent: rotting matter has been mineralized and water chemistry does not suffer for it. If you have any doubts, let your nose be the judge. Uncured live rock has a very unpleasant smell to the point of being malodorous. It will curl your hair! Cured live rock has a mild, if not pleasant, earthy smell: a smell of the sea. As well, the dealer's water should be crystal clear and free of ammonia and other measurable nitrogenous compounds (nitrite, nitrate).

If you purchase your live rock sight-unseen, then you will definitely have to isolate the rock first. If live rock is truly cured, the holding tank water will be crystal clear, odor and ammonia free. Another thing to be mindful of is that the person responsible for curing your live rock has control over water chemistry and the ability to prevent undue distress to the rock's life forms. Perhaps the greatest motivation for an aquarist to buy live rock fresh and uncured, when possible, and curing it at home is the ability to personally oversee quality factors and their adjustment. Just because rock is bought "fully cured" does not mean it is necessarily of a high quality; it is simply cycled and stable. A previous severe ammonia spike during a bad curing process, or any other neglect of water chemistry, could have killed most of the desirable microfauna on your live rock. By curing live rock at home you can space the rock out, circulate water vigorously, protein skim aggressively and conduct water changes to be assured of retaining the greatest diversity and number of life forms.

(5) **LR Real Cost**:

Concerns regarding the cost of live rock really revolve around personal decisions - there is, or should be, little variation in the price structure of this fundamental industry staple. Live rock is generally collected by a few very large operations. It is a bulk commodity in the purest sense of the term. Prices along the chain of custody vary little. Justification for pricing is quite simple: freight is a substantially significant portion of the cost of live rock. Curing also results in very real expenses in the form of maintaining water quality. The further down the chain of custody you acquire live rock the more expensive it becomes. The final cost to the consumer is legitimately two to three times the cost of the raw rock on import.

Indeed, the cost for live rock can be high. It is possibly the most expensive component of your system. It is also one of the most fundamental and instrumental ingredients to the success of your marine aquarium. But there are ways to whittle down the cost of this investment. Curing your own rock is messy and at times labor intensive. Yet, providing your own space and labor can definitely cut costs. Finessing water quality, through aggressive protein skimming, is also a potential money saver as it reduces the amount of water changes needed.

One of the least successful ways to save on a living rockscape is the use of "base" rock or non-living rock. Know that the term "base rock" is often applied to lower-quality live rock, which are generally old pieces that haven't fared well during import. Like dry rock, it has no significant means of defending itself against nuisance algae growth, for example, as encrusted live rock does through biological means. Usually, any initial savings with "base" rock is negated many-fold by struggles with pest algae and all the laborious scraping, siphoning, water changes, skimmer adjusting, and water quality enhancing products. It is also unlikely to ever become as diverse as wild harvested live rock which has had the benefit of many more years in the ocean.

Beware of the inviting temptation with mail-order supply houses to shop 'til your mouse drops! The Internet is a blessing in many ways and yet is not wholly perfect. Buying dry goods is one matter (standardized by manufacturer), but buying livestock sight-unseen is unpredictable and potentially unwise. Due to tremendous savings afforded by not having to have so much staff and store space, e-tailers can and do offer many of the same products at substantial cost savings.

Hobbyists must also be aware that it is critical to support your local fish store (LFS) whenever possible. Livestock is highly variable in condition and health and requires inspection and observation. The local aquarium store provides an invaluable service by offering live creatures for your perusal and evaluation upon request. Even more importantly, it is critical that our local merchants continue to thrive. If they do not, advanced aquarists will not be able to enjoy their hobby without the influx of new aquarists through the local stores that actually drive the market.

Live rock Heroes! Sponges and Ascidians (Sea Squirts) are extremely important on the world's reefs and yours for cleaning up the water. They are filter-feeders par excellence, adding food (from their reproductive products and from harboring other organisms) and modifying water chemistry. Please don't dismiss these lowly categories of living organisms at a glance. True, the sponges are just "tissue-grade" life. But the ascidians are protochordates (tunicates) and are related to, of all creatures, us! They have a closed circulatory system, dorsal-oriented nervous system, a "head" as juveniles, and stiffening structural elements akin to a skeleton. They're almost us... or at least like my Uncle Al.

Sponge (left) and Sea Squirt (right) examples.

"All we are saying is give your LFS a chance." Ask them if they are able to make you a deal on bulk-ordered live rock. It may well be that they can realize additional savings by "piggy-backing" your order on theirs, and, if you're willing to place a sizable deposit, they can simply import your boxes with a small percentage mark-up. Depending on the depth of the discount and security of your order, you may be able to select from the shipment upon arrival, or you may have the understanding that they will be doing nothing more than selling you the boxed live rock "as is." Either way, your patronage serves the greater good of the hobby and the local economy. You get to see your purchase before you buy it, receive discount and savings, and your local store gets your support and, therefore, can continue to thrive and be available as a local outlet for the enjoyment of our beloved hobby. And lastly, you may want to coordinate a group purchase of box-lots (bulk-pricing) with a few other hobbyists, perhaps through an association in a local club, and barter for the real savings, which are freight discounts. This can also be done with a friendly local aquarium shop. Be resourceful when it comes to purchasing live rock, but don't scrimp. Live rock is the foundation, structurally and biologically, of your aquarium; do not short-change yourself on this aspect of the system.

Wild *versus* Cultured Live Rock

A profile of merits and shortcomings

Simply put, there is no substitute for **wild live rock**. Some aquarists consider culturing their own "base rock" (dry rock of various compositions) hoping that it will become live. To some degree it will become colonized with microbes and

> **A summary of the benefits of Wild *versus* Aquacultured live rock:**
>
> *Wild Pacific live rock...*
>
> - better collective bio-diversity (on import and/or potential for development in the matrix)
> - very mature and desirable corallines are likely
> - less dense/more porous compositionally (better infrastructure for biotic activity)
> - greater utility and arrange-ability for its varied shapes
> - produces less sediments as it degrades in time
>
> *Aquacultured Atlantic live rock...*
>
> - better ornamental algae growth
> - better sponge, mollusk and worm group representation
> - higher survivability of incidental organisms because of the shorter chain of custody on import
> - unique opportunities to acquire Atlantic endemic organisms, including corals

externally covered with various algae and some interesting invertebrates, but will never develop the diversity of life found on imported rock. Captive cultured product is simply a poor substitute as a primary source of hard substrate. At best, a mass of axenic (sterile) calcareous material is colonized by exposure to "the real thing." The end product however is limited by the original snapshot of organisms found with an imported sample of wild live rock and one's ability to support and cultivate those wild-harvested organisms. Even though one may start with wild rock from the reef, lagoon or intertidal zone, none cultured will have the impressive diversity of life forms even years later. If total bio-diversity is the objective in "rock-farming," no process can ever compare to the limitless potential of the living seas and their nurturing resources.

Aquacultured live rock may be farmed *in situ* (in place) or in captive pools, but neither is likely to be on par with wild rock. You may wonder how it could possibly be any different if it is "grown" in the sea next to indigenous growth. A glimpse into the industry of live rock-farming, however, and all will be quickly revealed. First, the readiness or "ripeness" of aquacultured rock is a matter of personal perspective. If your preference in live rock is for decorative non-calcareous algae, farmed rock can sit in the ocean for less than a year and be harvested with handsome leafy green, brown and red algae that will satisfy you every bit as much or better than Pacific wild-harvested product.

Mind you, this isn't because the Pacific live rock does not encourage or possess such attractive macroalgae, but rather because such algae is scrubbed or stripped away to improve the overall import and curing success for the longer transit of South Pacific products to the United States. Atlantic aquacultured rock is instead harvested in federal waters off the Florida coast, docked, shipped, and subsequently received by a consumer potentially in less than 24 hours with high survival of macrofauna. On this point, aquacultured live rock is likely better than Pacific sources necessarily (or at least initially, until the Pacific macroalgae grows in time).

Now, if your preference in live rock is for encrusting coralline algae, bryozoans, sponges and the like, farmed rock will need to sit in the ocean for perhaps 1-3 years to reasonably compare to wild-harvested product, although superficially it may look similar to wild native rock. A lot of aquacultured live rock is harvested within this time frame. In this range, proponents of either type of live rock will make cases for why one is better than another. With reasonably good fortune and growing conditions, we will say that the aquacultured live rock is perhaps easily on par from an aesthetic perspective with a native product here.

The real measure of live rock's merit and mettle, however, is measured by its total mass of living bio-diversity. On this point, aquacultured products may never compare with aged, native live rock, which has had the benefit of years of development and exposure to the sea. Beyond time and opporutnity for a bounty of invertebrate larvae and other growths to take up residence, many years of exposure afford a rich infauna of penetrating organisms and correlative microbes. It simply takes time for worms, bivalves, echinoids, and more to burrow into calcareous substrates and turn the previously mineral domain into a biotic haven.

Aquacultured live rock, on the contrary, is the product of a business that cannot afford to invest many years or decades in search of the ultimate product. In fact, it has been one of the leading criticisms of aquacultured live rock since its development. After the decorative plants and algae are eaten or killed, and after other superficial growths are grazed down, all that is left on most aquacultured rock is a relatively weak, biologically "barren" medium. This reality is exacerbated by the common choice of raw material that aquaculture rock farmers use which is *very dense* carbonate rock!

Why do Florida "**rock-quaculture**" merchants (we just made that word up) use such dense rock? Is it because they are trying to empty your wallet on the price per pound or the value of every box and truckload sold? We're pretty sure that's not the case. These folks are honest and hard working people who are trying to produce a good product and make a good wage.

No, instead, there is a very practical reason for why aquacultured rock rivals concrete in density: the weather. Yes, the weather in the shallow waters where most live rock is cultured can be turbulent during hurricane seasons in the Atlantic. Early "rock-quaculture" operations have literally seen hundreds of thousands of pounds of seed rock buried by mountains of sand or carried away by storm activity and with it go their investments and dreams. Current rock farmers drop seed loads of dense carbonate rock with the expectation that many thousands of pounds on the bottom of the pile will be swallowed and sacrificed. Some of the top coat at times will be covered and stifled by sand and harvest will be delayed.

Indeed, the density of Florida live rock is a necessity, although that does not make it welcome. It does, however, increase the expense to aquarists to fill the same volume of space as other less dense rock. So, what options does a rock farmer have? Lighter rock could be dropped further out in deeper and calmer waters. However, the price of such rock would be even more expensive because farmers would then have to boat further out and dive deeper to harvest it. Man-hours and fuel costs are far dearer than the savings in rock density. And that still does not change the fact that none of this rock is left in the ocean for more than a few years at best. Bio-diversity is still an issue. So why buy aquacultured rock, then, if it is less mature, denser, less diverse, and more expensive by volume? Well, if we are to believe the marketing, then we are to conclude that aquacultured live rock is the "environmentally friendly" choice.

Let's examine just how useful aquacultured live rock is. If you are willing to overlook any potential disadvantages to this product listed above because you want to make an environmentally empathetic choice, please consider the following information. Presented here are some of the marketing claims of aquacultured live rock and the facts about it:

* Aquacultured live rock provides new habitat for reef creatures in previously undeveloped areas of the wild reef: **TRUE**

The premise is that a rock pile dropped for the purpose of aquaculture will then invite the settlement of numerous reef creatures and provide habitat for organisms that are not collected when the rock is harvested and sold. This is quite obviously true and is of some benefit to the reef community.

However, critics will say that the benefit is negligible, as these rock piles are not charities and most of the generated bio-mass is taken away when the rock is harvested within just a few years. Furthermore, they argue, these small piles are weak, unreliable, and transient communities because of their sensitivity to storm activity, among many other factors, and as such provide only minimal long term benefits. As aquarists, we'll concede that any benefit by them is worthwhile and that there is no apparent disadvantage to the industry or environment for its application. If this is your impetus for buying aquacultured live rock then you will be satisfied.

The correlative "depletion" of habitat by wild rock collections, however, is a weak argument. Wild rock is collected in areas away from coral communities, and at depths where storm-tossed matter is both easy and safe to collect. Incidentally, these locales also support the growth of ideal corallines and easily assimilated growths on this rock in high nutrient and/or lower light environments with organisms more adaptable to typical aquarium conditions. Ultimately, the argument that wild rock takes away habitat is no stronger than the reality of aquacultured rock when it is harvested, assuming that it creates any substantial habitat at all.

Here is where one must look very hard at claims saying that using aquacultured rock is "saving the wild reefs." Neither product is collected from dominant coral communities. Both have the potential to create or strand fauna with their presence or collection. Storm-tossed wild live rock is also a resource that can be regenerated as the neighboring live coral communities continue to grow and produce calcareous matter, not the least of which are easily fragmented corallums. The cycle of growth and damage contributes to the rock piles below and beside the reef proper. So, there is little or no corollary fault to wild rock and a neutral or slight benefit to aquacultured live rock at

Live rock encrusted with calcereous algae. Indonesia image.

best on this point, again by the settlement of motile fish and invertebrates left behind after the rock is collected.

* Aquacultured (FL) live rock is biologically superior to native wild live rock: **FALSE**

There is a long list of sensible and practical reasons for why this not true. Coral formations in Atlantic and Caribbean waters are evolutionarily very young in comparison to Pacific coral communities. Atlantic live rock simply ships faster through the U.S. because of its shorter transit distance than Pacific products. European aquarists find Pacific live rock more diverse than Atlantic due to its point of origin just the same.

On most any biological scale, Florida waters pale in comparison to those of the South Pacific where bio-diversity is concerned. There simply are not as many possible seed organisms for aquacultured live rock in the Atlantic. Aquacultured live rock even fails when compared to native Atlantic rock if it was still lawful to collect, by virtue of its density and lack of mature infauna. There is still good reason to buy aquacultured live rock, though: the opportunity to acquire some of the unique endemics to Atlantic waters is possible within the constraints of present legislation. Indeed, live rock from the Gulf of Mexico, in particular, is easily some of the most beautiful substrate in the world, with amazing color, sponges and polychaete worms. If you favor marine plants and macroalgae, Atlantic live rock products are some of the finest available. Still, we cannot fairly say that aquacultured live rock is necessarily better.

* Aquacultured live rock is "eco-friendly" and low or zero-impact: **FALSE**

This is the most heavily promoted piece of "myth-information" about live rock collection and is used, rather ironically, as a criticism of wild products. There are numerous ways this argument is made. We will present it in its most literal form. The premise is that native (wild) rock is taken away from coral communities and that aquacultured live rock comprises no part of a living coral community (seed rock comes from mined upland or other terrestrial sources). This fact has no bearing, however, on the claim that aquacultured live rock is low-impact or eco-friendly.

Native wild rock is formed by the calcification of living reef organisms as we know: scleractinian corals, calcareous algae, and many other biomineralizing organisms as well as non-living mechanisms of deposition and accretion. The living "reef-builders" are definitely a renewable resource. That resource is admittedly not infinite. It can be stressed as we know, and as aquarists we have great admiration and empathy for it. But, it *is* a renewable resource that can be managed.

Aquacultured rock is often carbonate in nature: ancient limestone and/or old reef formations. It is a dead, *non-renewable* source of a very finite quantity that dwindles every time it is mined, or harvested on land for farming in the sea. The collection of upland and terrestrial rocks is, for all intents and purposes, a burden on this non-renewable resource, unless one's perspective is the geological time scale. In this regard, we feel that aquacultured rock is neither eco-friendly nor of lower impact on the environment. Additionally, no consideration is given for the potential impact mining or other collection processes have on land erosion or other negative possiblities. The transparent "save the reef" propaganda seen in the advertising of some aquacultured live rock does not make the product unworthy, it just makes the advertisers inaccurate.

And so, after considering the whole song and dance of praise for wild live rock, you may wonder how this can all be sorted out by empathetic aquarists. It's hard not to be biased for wild live rock products when the collective facts that should influence our buying decisions are quite real. You have probably heard, "After our perceived reality, there is what is," and will probably hear it again. And so, you may ask, are we advising you to buy wild live rock in preference? The answer may surprise you, but no, not necessarily. Each product has its merits and limitations. What we are trying to do with our quintessentially passionate style of consumer advocacy is to inform and educate you in an area of considerable ambiguity in order to help you make a good buying decision. Live rock is the very foundation of a natural marine aquarium, and may be one of the single most important decisions you will make. Make a thoughtful decision on either or both types of live rock for inclusion in your aquarium.

Care

Curing Processes and Definitions

You often see and hear of live rock described as *cured* or *uncured*. By definition, this is a valuation of its state of biological stability. In actuality, how well-cured live rock is, is more often a matter of subjective evaluation. Freshly collected live rock has a wide array of micro- and macro-organisms in varying degrees of health as the result of collection and transport duress. Some organisms on and in the live rock will be stressed but in stable condition. Others will be sound and as indifferent to the process as they are to the tides. Still others will either be struggling or dead and decaying. It is primarily the last group that determines the initial condition of newly arrived live rock upon arrival. The stressed but sound organisms are at the mercy of an aquarist's skill to handle and cure the batch for their ultimate health and survival. There are many life forms to consider: mollusks bored and buried in the rocks, worms hiding throughout, plants and algae encrusted upon surfaces, and even large animals hiding undiscovered (serpent starfish, urchins... even fishes), or even dying within the rock matrix. During the curing process organisms are stabilized and decaying organic matter is either mineralized or exported from the system altogether.

Many factors can influence the state in which fresh live rock arrives. The obvious factors that influence its arrival condition are the duration of air exposure and changes in temperature during transit. It stands to reason that the shortest possible handling time will minimize both curing time and the expected loss of life. Live rock is shipped "dry" (moist but not submerged) for two very good reasons: freight cost, and to improve the survival potential of incidental life with the rock. The expense of freight is, of course, the biggest factor. In fact, for many fishes and invertebrates that travel from afar, the cost of freight is more than the cost of the animal from the collector. The final consumer price for all marine livestock is significantly influenced by shipping expenses. Thus, it is the duty of the shipper to finesse the handling of imported livestock through the chain of custody, incurring the least freight expense (minimal water), while maximizing survivability. Since most of the organisms on live rock fare as well or better in transit by dry shipping methods, moist shipping is the protocol for this commodity.

The last statement may seem counterintuitive. "You ship the rock moist, but it's almost always collected submerged, yet more organisms survive import by being shipped just moist?" Yes! Due to long shipping times and the vagaries of handling, bouncing about, temperature fluctuations and the like, the best way to get live rock to consumers is in thermally insulated boxes, sealed shut with packing tape, in the dark, and out of water. The reason for this is that many

Live rock encrusted with life. Red Sea image. (*D. Fenner*)

organisms under duress release various waste products and noxious elements. It could be stress-induced toxins, tissue decay or simply excrement. When shipped dry, these elements are held in check as if the tide quite literally was out. However, even if airfreight were free for the industry, we still would not want to ship live rock submerged. If so, toxins and waste products would be released into the shipping water where they could contaminate, stress, and likely kill all other life forms that are part and parcel of live rock. Now you can see why dry-shipping improves survival potential for incidental life forms imported with live rock; it limits pollution and cross-contamination in transit.

Like the often disputed topic of "best reef lighting," the "*best*" way of going about curing or re-curing live rock is complicated by conflicting opinions, and rightly so. There are "many roads," as the saying goes, to get you where you want after the curing process has finished. We each have different target groups that we would like to see survive the process. Some aquarists favor macroalgae, others favor microscopic animals, while others still are simply content with strong microbial populations. These preferences raise questions and demand different handling and hardware for uncured live rock. Some of the concerns when curing live rock are:

- Can the rock be cured in the display or is it better in a remote area?
- Does the rock need lighting?
- How long will the process take?
- Is a skimmer necessary?
- How important are water changes?
- What about filtration and water movement?
- What roles do additives and water quality testing play?

The best summary of these points is that live rock is best handled in a remote vessel for curing. Water should be aggressively skimmed, frequently changed, and quality tested. All will be required to determine when the process is complete, usually between two and four weeks. Very strong water movement greatly enhances the process. Filtration beyond a skimmer may not be needed, but additives that buffer the pH, calcium and alkalinity levels will be helpful and perhaps necessary. The effect of natural acids on pH is a great burden and contributes to a harsh curing process. Wouldn't it be great if we could all just go down to the tropical beach and collect our own live rock and rush it home to our systems in minutes?

If you have the space and time to cure live rock in a separate vessel, we recommend that you seek the freshest, *uncured* live rock possible. Although it will require more work for you to make it suitable in the display, you will enjoy a much better finished product as a result of your efforts. Some folks think they are better off buying "pre-cured" rock by mail and pay a premium price for it. Too often though, the degree to which this rock is actually cured is not worth that premium if the condition of the rock is even fairly represented. Air-shipped rock is often uncured. Most local aquarium stores do sell worthwhile and reliably cured live rock. If you do not have the inclination to cure your own rock, this is usually a good source, and you can make your selections on sight. Otherwise, ask your dealer to order a fresh box for you and pick it up when it arrives at the shop.

Mail-ordered live rock, however, is a different matter altogether. It has been a disappointing and common trend to see live rock offered on the Internet as "cured" or "pre-cured," when really it is usually freshly imported product, which arrives in the original unopened boxes directly from the exporting country. Please take our advice; by and large there is no such thing as fully cured air-shipped rock. If for no other reason, the time and temperature extremes inherent in the delivery process take a toll on the product that is delivered to your door. All mail-ordered live rock should be cured or *re*-cured in a remote vessel for some time to be safe.

On arrival: *Triage*
- Inspection, Rinsing, and Cleaning Live Rock

Newly arrived live rock should be soaked in saltwater of habitable temperature and salinity. Aged tank water is better than freshly-mixed water for this application. Use the acquisition of rock as an excuse for another good water change in the display tank. A good soaking serves several purposes. The most immediate benefit is hydration of desiccated matter in order to preserve the most life possible on newly arrived live rock. This initial soak and rinse also allows the many stressed animals that are waiting for the "tide" to come back in to purge waste matter that they have been holding. It is also an opportunity to thrash the product

Life-encrusted rock in Indonesia. (*D. Fenner*)

around a bit to dislodge any coarse sand, sediment, and dead or decaying matter.

Some dead material on and in the rock appears as whitish, smelly, mucous material. During rough periods of curing, more of this matter will become apparent (often described as "foamy," like rinsed toothpaste). If solid, it should be removed with tweezers or comparable instruments; if the matter oozes or is weakly composed, simply siphon and discard the polluted water. It is very important to extract decaying matter at all stages of the handling and curing process. Necrotic matter is a severe burden on water quality and risks the spread of infectious life forms as bacteria counts run high at such times. Wearing gloves while handling live rock is recommended just as with handling live reef invertebrates.

After gross matter has been shaken off and the fresh rock has had a chance to soak for at least fifteen minutes, you will need to set to the task of coarsely "cleaning" it. Any large macroalgae, plants, and sponges should be pried off and discarded as these are very likely dead, if not dying, and will only compromise the rest of the rock's life forms. Very few of these organisms survive import, regardless of how good they look on arrival. Most are very sensitive to air exposre for even short periods of time. The risk of permitting them into the batch for curing is dangerous. No worries, though; many will "magically" reappear in the tank months later.

However, if you are willing to be very diligent during the curing process (extra water changes, over-sized skimmers, etc.), a small amount of this material can be left with hopes for survival. Plants and algae should be cropped then, instead of razed. Sponges can be identified more specifically with wisdom about the likelihood of survival by specimen: liver sponges and macro-specimens (vase, barrel, tube, tree, finger, etc.) are nearly impossible to salvage, but many encrusting species can survive and compliment the tank.

After all loose, large and obviously damaged and dying matter has been scrubbed from the rock, the receiving water is to be discarded. It is heavily laden with organics and dead organisms. You may wish to decant the sediments and solid matter to look for weak macro-organisms to rescue (shrimp, worms, starfish, etc.). Survivors can perhaps be added to a refugium. Little matter from the first rinse and scrub, though, is likely to be alive or useful. When fresh rock has been sufficiently soaked and cleaned, it is ready to deposit in a fresh tank of seawater where the curing process will begin.

Curing and Holding

Curing vats for live rock are by necessity simple vessels. Some hobbyists use aquariums, but plastic containers are likely safer and more appropriate, not to mention more durable, for the processing and handling required. Use your imagination when shopping for tanks in which to cure live rock. Plastic garbage cans are inexpensive, plastic pickle or olive barrels are available to some aquarists from the food industry and local markets, and various storage containers abound at local discount, DIY and department stores. You will need to heat the water to safe tropical temperatures, so be mindful that your heater does not melt a hole in the container. It would be advisable to place heater elements in thick-walled pipes (heavy PVC), as one does with aggressive fishes that have a flair for interior decoration (which includes smashing instruments). The curing vessel will also need very strong water flow.

Massive aeration and **water flow** is the single best thing you can do to insure a better curing process. **Aggressive protein skimming** is the next best thing. Large or sour batches of live rock will benefit immeasurably from large, efficient, or even dual skimmers cleaned alternately so as not to interrupt skimmate production. There really is no such thing as "over-skimming" when it comes to curing live rock. You will be amazed at how many times *daily* you can empty a full cup of offensive skimmate during the curing process. With the enormous influx of decaying matter, exudations from stressed animals, and the spike in dissolved organics at large, a protein skimmer is an absolutely indispensable component for curing live rock, and strong water flow supports it by keeping matter in suspension and prevents it from suffocating the substrate. Water changes will also be necessary to temper water quality parameters. Chemical filtration such as activated carbon can be very helpful, but expensive, to use in the early stages of the curing process. It may be more economical to rely on water changes heavily for the first week of curing, and then employ chemical media after one week (or the cessation of any harsh spikes in water chemistry). How one chooses exactly to finesse or manipulate vehicles of nutrient export via skimming, water changes, and chemical media is ultimately of little consequence, as long as general water quality is maintained in the process.

To exploit all of these dynamics and to insure a successful curing process, it is very important to prevent the crowding or stifling of live rock in holding. Water form and flow in the curing vessel is indeed tantamount to success. Everyone really should be in agreement on this point. The more water you have to cure your live rock in, the longer you have for changes to occur, including any stray towards undesirable ranges of water chemistry, pollution and other inevitabilities.

It's the old adage again, "*Dilution is the solution to pollution.*" As your rock cures, many things are happening (mostly bad) in the first week after import. Collection, transport from the wild, navigation to the mainland, processing by transhippers (and possibly several wholesalers or dealers) all entail some days to weeks. Success will be improved via placement of the rock in a very open vessel. **For stability, the larger the better when it comes to water volume**.

Suspension of rock off of the bare vessel floor is recommended to improve water flow all around, as well as to facilitate the siphoning of the potentially copious amounts of detritus and sediment produced. Some aquarists will employ well-spaced racks or open shelves (made of grid-like material for sediments and water to pass through). However you choose to accomplish the feat, space live rock out as much as possible and resist stacking the product on top of itself during the curing process.

You will also need to be mindful of the natural orientation of the rock and make an attempt to set each piece with the proper side up. Look for indications of light-loving algae and other photosynthetic organisms on the top side, and shadowy denizens (tubeworms and the like) on the underside. Make no mistake, however, by believing that the curing process for live rock is merely a waiting period. It is, in fact, a laborious and pro-active endeavor to rescue and protect life forms during the curing process. The benefits are known and well-defined. Who is more interested in providing you with the finest quality cured live rock, than you?

About Additives during the Curing Process:
Correct pH, alkalinity, calcium and save!

It's easy to just throw a batch of stinky live rock into a tub and forget about it for a couple of weeks. Within reason, one could literally do that and still end up with cured rock that is microbially useful, albeit stewing in water that is likely putrid. But what about all the potential life forms lost to the neglect of water quality? Failing pH and drained levels of biominerals will stress and bleach desirable coralline algae. Countless other organisms will be reduced in vitality or killed in the process. It is in your best interest to keep a very close watch on water quality parameters (daily at first, then weekly) and adjust them to save desirable life forms, diversity, and the overall finished quality of your cured live rock. If cost efficiency is a concern, simply use baking soda and kalkwasser to accomplish most or all of the buffering requirements. Test for the other fundamental nitrogenous values (ammonia, nitrite and nitrate) and make every effort to keep them near zero (definitely below 1.0 ppm for ammonia and nitrite).

Live rock in Sulawesi. (*D. Fenner*)

Live rock in the Red Sea. (*D. Fenner*)

Live rock is essentially cured after two weeks, or when spikes in water quality have ceased.

About Lighting during the curing process

There is no hard and fast rule for lighting live rock during the curing process. Either path has its advantages. Illumination during the curing process, however, does place a great burden on both the system and the aquarist, with the extraordinary levels of dissolved organics and the potential (rather likelihood) of a nuisance algae bloom occurring. In many cases, an algae bloom in a lit curing tank spells disaster as it smothers, suffocates, and out-competes other desirable forms of life imported with the live rock. The truth of the matter is that most photosynthetic organisms that commonly occur on live rock can usually weather the extended period of little or no light during curing (supported by feeding on the rich organics in the system) better than they can fend off nuisance algae while in a weakened state. The amount of light you choose to employ should reflect your knowledge and confidence in nutrient export processes created for the system. We encourage you to simply decrease either the amount of time (photoperiod) or intensity of your lighting during curing to support vital coralline algae, along with other desirable photosynthetic organisms and their welcome use of available nutrients.

Patience is a virtue in curing live rock

When in doubt, wait. It really pays to be patient while curing live rock in much the same way as it is when stocking an aquarium with fish in general. Live rock will typically take **two to four weeks** to soundly cure before it is safe for use in the display or for introducing other livestock to it. For huge systems (hundreds of gallons) some folks cure their rock in divisions of a third or a quarter at a time, adding new to the existing batch after the former is totally cured (again, no detectable ammonia or nitrites for days to weeks). Others opt for doing an entire batch at once. Both approaches have their merits, though the "bit at a time" method tends to offer more predictable water chemistry and less stress to life forms respectively. The frequent additions, however, also carry with them small nutrient spikes which can cause persistent nuisance algae blooms in the system without a proper nutrient handling processes. One's budget usually determines the rate at which live rock is added, and in this case the path is of little consequence so long as the rock is cured and water quality is maintained.

About Importing Pest Species with Live Rock

What can be done to eliminate the import of pests and predators with live rock? Quarantine!!! Curing live rock in a remote vessel for two to four weeks gives an aquarist ample time to inspect and observe for pests and predators. After a week or more, when water chemistry begins to stabilize, bait can be set to lure potential nuisance creatures. Numerous ingenious traps and techniques for accomplishing this goal abound on the Internet. One of the simplest traps for all but the craftiest of predators, like the Stomatopod mantis "shrimp," is a plain wide-mouthed glass jar (like we use for pickles or mayonnaise). A short, squat jar can be sunk in the aquarium and leaned up against the rocks. Weighted bait is placed at the bottom of the jar, like an attractive piece of smelly fish meat, krill or squid. The premise is that "pointy-toed" predators like bristleworms and crabs can slide down into the jar, but cannot easily scale the smooth glass walls to escape. Although it is hardly a comprehensive trap, it will effectively screen for some huge groups of nuisance animals. It is quite inexpensive as a first line of defense, too. Other potential pest organisms like hydroids, *Aiptasia*, and *Anemonia*, will be obvious and can be manually extracted before they spread in the display. Be resourceful in your keyword searches on the 'Net for information on how to expand and exploit the screening process while you wait for your live rock to cure. The sections of this book of reef invertebrates will also serve to help you recognize and identify potential pests and predators.

Live Rock Succession and Cycling

There are general and recognizable trends in what happens in the course of live rock navigating the curing process. During the first few days to several weeks you can expect a die-off of variable dimensions, with expected water cloudiness and rise in nutrient levels. This period is typically followed by some degree of stabilization. Your skimmer(s) wane in skimmate production (quantity and quality), and ammonia and nitrite fall to zero or nearly so, possibly with a measurable rise in nitrates (but not always). Typically what follows then is the beginning of algal succession (see more in the chapter on plants and algae). Microalgae and like organisms become established first, which often involves blue-green "algae" (Cyanobacteria), filamentous greens (e.g. *Bryopsis*), and diatoms (golden brown slime) on the substrate, rock, and viewing panes. Their presence may persist for some weeks if steps aren't taken to limit nutrient availability. With good husbandry, these detractive algal forms give rise to encrusting red (coralline) and larger green macroalgae species. In time, and often naturally, macro- brown (phaeophyte) and red (rhodophyte) algae can establish themselves in your system. Oftentimes, the soft macro species of algae are so palatable to aquarium fishes and invertebrate grazers that they may never gain an appreciable foothold without the employment of fishless refugium methodologies. Incidental invertebrates and other micro-organisms also go through successional waves in marine aquariums. Micro-crustacean populations begin to blossom. The biological faculties get stronger every day with a stable bio-load. The very microscopic foundation of the living reef reveals itself in splendor with time and due course. Be sure to look closely and make use of a magnifying glass or jeweler's loupe in observance of the miniature drama that unfolds in your aquarium system every day.

When is Live Rock "Fully" Cured?

Like concrete, live rock is never really finished curing; it is an ongoing process. Organisms are never really done changing their population densities upon and within the matrix. Initially, new rock has the most biodiversity for coming from a much more dynamic setting than our aquariums can afford: the ocean. Over time, this diversity is lost by number although not necessarily by abundance. Indeed, the bio-mass of well-tended live rock can and will increase, however limited in speciation it becomes. What we are interested in as aquarists is that live rock be cured "enough" for our purposes in a closed aquarium system. This demands that succession, the dying off, out-competition and re-establishment of organisms, wanes enough to afford adequate mineralization and stability of water quality via biological faculties. A true climax community is never really attained in captive systems, but rather the tides of succession turn over slower, revealing a loss of overall bio-diversity as a hallmark of maturing systems.

Age and Senescence in Live Rock

Live rock has a "life of its own" - a distinctly active period of time where it optimally affects an aquarium's physical and chemical make-up and supports a large diversity of micro- and macro- life forms. There comes a time, however, when these processes sharply drop off in utility and action. After a year or two in tanks stocked well with fishes, most types of live rock will "run out of gas." The ready sites for the dissolution of carbonate are lost; populations of some organisms become extinct while others are noticeably reduced. This is an example of gross over-simplification, but in the majority of reef systems there is a general trend for the stated effect: the appearance of "old-age" in live rock. You certainly can extend the effective lifespan of your live rock and sand by having a larger system and a greater quantity of product per capita of predatory fish and invertebrates. This is the very dynamic that helps large public aquarium displays thrive longer than most small private home aquariums. Being diligent concerning your husbandry, including the mindful dispense of foods, supplements, etc., helps to support a strong bio-diversity. The use of a fishless refugium to supply and bolster populations of plankters in the display is also of tremendous benefit. One of the very best ways, however, to preserve diversity and useful surface area of your substrates (rock and sand) is the simple augmentation or rotation of live rock with fresh product periodically. An influx of fresh live rock and sand once or twice yearly will provide noticeable and immeasurable benefits to any marine aquarium.

Conservation Concerns: Be Concerned But Don't

There are voiced concerns outside the trade that live rock collection is deleterious to the world's reefs. Are we really removing the reef? Are we really going to run out of it? The answer is no. Live rock is a living, growing and regenerative resource whose collection is quite unlike the mining of finite limestone deposits, or the dredging of reefs for airports and ship harbors. Collection of live rock is usually storm debris and other portions removed from the mass of the reef by natural causes.

Is Live Rock Indispensable?

All marine systems benefit from the use of live rock, from tropical to temperate systems. Fish-only and fully invertebrate displays alike will derive nutrition and sustenance from the living fauna beyond the mediation of water quality by mineral content and microbial activity. Indeed, the use of live rock, along with vast improvements

Coralline-encrusted live rock in a home aquarium. (*L. Gonzalez*)

in filtration (live sand, refugiums) and lighting applications, has revolutionized marine aquarium keeping. Using live rock won't guarantee success, but it will greatly increase the chances of it. There are many types of successful aquarium systems and, subsequently, ways to work around using live rock; deep sand beds and mud methodologies with vegetable filters (incorporating macroalgae or true vascular plants like seagrasses) are among the many possibilities. Still, many more species of fishes and invertebrates have been, and can be, kept successfully with live rock in the aquarium. Please do include at least some in your systems, and add new stock periodically as funds allow, ensuring optimum populations of diverse and desirable organisms in the wondrous living matrix that we call live rock.

Examples of desirable life forms with live rock

Macroalgae are favored strongly by aquarists for their utility and aesthetic impact (as opposed to the generally less desirable micro-algal forms). These large "almost plants" are denoted by color, Green, Red, or Brown, but are classified in science by their types of photosynthetic pigments, storage foods and reproductive biology. Most macroalgae is removed as live rock is processed, since it will likely die in transit adding waste and malodor to the shipment. Nevertheless, there are "seeds" and starter colonies on healthy live rock, and the growth of these simple photosynthetic organisms should be encouraged. They are used for nutrient export, as both direct and incidental sources of food, and, obviously, decor. Seeding tanks and refugiums with deliberately acquired specimens is also recommended when a good understanding of their sometimes demanding husbandry is acquired.

Sponges and Ascidians

These are two of the most under-appreciated but key groups of incidental species that make successful reef-type aquariums possible and practical. They are often the bulk (more than half) of the living part of hitchhikers on live rock and they serve the aquarium and water quality as superb filter-feeders. Don't despair if you don't see many of them on your live rock initially. The larger specimens are likely to desiccate or be crushed in transit and are deliberately removed in processing. Rest assured though, that much of the interstitial parts of live rock contain these organisms; most are only visible by breaking the rock open and inspecting the infrastructure and crevices.

Live Sand

*An **Archaster** "Sand Star" surfaces from a deep bed of sand where it spends most of its time buried. These voracious eaters can sterilize a captive sandbed in short order - and subsequently starve. Six square feet of sand per individual is a fair minimum. Aquarium image. (L. Gonzalez)*

The application of "live" sand is a commonly misunderstood aspect of marine aquarium husbandry. The aquarium industry at large has had tumultuous experimental relationships with various substrates that have waxed and waned through the years. Recommendations and moods on the matter (composition, depth, size, etc) have oscillated from thick to thin, coarse to fine, stirred not shaken... can I get an olive with that magical mud martini?!?

Looking back on the young history of reef-keeping, we have seen a shift from far right (bare-bottomed) to far left (deeper than quicksand!), to the current state of confusion and misinformation that we currently find ourselves in. Rest assured, though, that here we can and will distill the facts from the myths and guide you with safe and reliable direction in all areas of this topic.

The Evolution of Live Sand Applications

In the early 1980's, announcing the popularization if not "birth" of formal reef keeping, there was a looming shadow in the hobby: nitrate accumulation. High nitrates were a serious impediment to keeping some interesting fishes and many invertebrates. Issues about it were at the forefront of all things salty: magazine articles, new products and claims, species selection (awareness of sensitivities), etc. The hobby by this point had long since evolved beyond the dark days of box filters and was growing out of undergravel filters with the arrival of wet-dry/trickle filtration, some of which incorporated purposeful denitrators. As such, issues of nitrification were of less concern for keeping marine species. Good canister filters, experimental Dutch tray filters, and the blossoming trickle filter styles were exceeding the productivity of the now tried-and-true undergravel filters. The accumulation of nitrates, however, from these very efficient filtration strategies had to keep pace with the new super-charged faculties of mineralization. Denitrification became the new focus.

Thus, necessity inspired ever more creative invention and experimentation... this time in the form of a race for nitrate control. After a brief diversion in the industry with weakly or wholly ineffective nitrate absorbing products (some of which simply masked nitrate readings, but absorbed no nitrate at all!), the hobby quickly set its sights on intelligent address and solutions to the problem. The focus turned quickly and sensibly to exporting organic waste before it was processed (fully) by nitrification. Filter pads and pre-filters were rinsed daily (actually a great, albeit tedious, strategy to this day for large bio-load systems), water change protocol became more aggressive (25% weekly water changes or better were common... another very fine idea) and protein skimmers were forever thrust into the limelight as now-indispensable components of saltwater aquariums for nutrient export. Despite these great efforts, nitrates still persisted problematically in some systems. As an aside, part of the reason for this dilemma was the grossly underestimated need for water flow in the aquarium. Indeed, reef environments are very dynamic, and most aquarists to this day still don't have enough proper water flow. At best, some folks had excessive laminar flow with powerheads blasting the tissue off corals and body-slamming fishes into the walls of their aquariums. Even then, dead spots and accumulating detritus were likely realities somewhere in the tank. The quest for nitrate control trudged on.

Logic, in time, led aquarists toward "Berlin-style" reef keeping and variations thereof (most commonly... bare-bottomed tanks and aggressive protein skimming). Reefscapes were built without sand or other substrates, and the rockwork was propped atop grates and shelves (or stilted on pegged founding rocks). This allowed aquarists to quickly see and siphon out the inevitable accumulating detritus on the glass bottom of the aquarium and from under the "suspended reef." For the most diligent aquarists, siphoning their tank more than once weekly, nitrate control was realized... but at a dear cost in labor. There had to be a better way! And so, the hobby took a harder look at natural substrates like newly imported live rock, and embraced it fervently for the wondrous and incomparable natural filtration that it could provide, as well as for its merits as a natural decor and as a food source. Large amounts of porous rock did provide some denitrification, but at last... it still wasn't enough!

The recognition of natural nitrate reduction (NNR) dynamics, seen in some tanks stocked heavily with live rock and experimental depths of substrate, led some aquarists to reconsider the possibilities with deeper substrates. Until the early 1990's, using any significant amount of sand or gravel was shunned by most marine hobbyists. "It's a mess... a source of nutrient accumulation, and a possible death-trap by anaerobiosis," the critics raved. Fortunately, with the pioneering and popularized work of a wide range of people in the aquatic sciences abroad, including Dr. Jean Jaubert, a sensible appreciation for the tremendous benefits of various (natural) substrate methodologies has been embraced.

What is *"Live"* Sand?

Live sand is essentially a combination of non-living substrate (usually calcareous in composition but sometimes silica-based) with a myriad of tiny beneficial life forms

Some challenging marine fishes can fair well in properly designed marine systems that include fishless refugiums and rich, live sand bed strategies. Arguably, an aquarist might not want to attempt to keep such animals without the use of live sand.

The surface of the seafloor is an enormous source of activity and attraction for reef animals feeding on, in and with it! (top row, left to right) *Istigobius ornatus*, *Ophioderma* sp., *Valenciennea muralis* (bottom, left to right) *Archaster typicus*, *Holothuria floridana* Yellow Sea Cucumber, *Nassarius* sp.

infused throughout it: organisms living on (meiofauna) and between (infauna) the substrate. Creatures found in this medium range from visible zooplankton down to a wide range of microbes dominated by bacteria. Indeed, live sand is much more than microbial colonies battling it out for space and nutrients. All phyla of marine life have representation in sand on the living reef. Some of the most commonly encountered organisms are segmented worms (annelids), roundworms (nematodes), micro-crustaceans (amphipods, copepods, mysids and the like), and bivalves (mollusks), but there are many, many more. Hence the need to speedily and carefully handle wild-collected live sand; to delay or ignore this plethora of life forms in handling and transit is to reduce its biodiversity.

The extraordinary biodiversity and depth of living substrates in the salted seas is almost inconceivable. The role of this medium as a fundamental component of so many dynamic processes is inalienable in the ocean and aquarium alike. They import many nutrients, export others, and serve as an extremely efficient living "filter," housing both nitrifying and denitrifying biological faculties (deeper down). Space and food are exploited by the colonization and proliferation of these crucial micro-organisms. The by-products of mineralization and other biological processes in live sand also liberates various nutrients and bio-minerals, like calcium and carbonates notably, for some macro-organisms - many of which are listed here in this book! Other undesirable elements from the water are precipitated and bound into the substrate. Live sand certainly is a complicated and fascinating microscopic world of its own, and quite worthy of a closer look... so get that magnifying glass or microscope out!

The Lower End of the Food Web

Just think about it... what does the marine life that we study do most of the day long? Seek or make food! So what could be better than having some food available at all times? Who wouldn't want to live inside of a large aquarium with a comfortable couch, an entertainment system and a cloud overhead that always rained cheeseburgers and beer ... and donuts (sugar coated, of course)? In the aquarium, a properly designed and mature live sand bed can be a significant source of food for many marine animals (fishes, corals, and numerous other filter-feeding invertebrates). For some creatures like seahorses and dragonets, a deep sand bed is almost necessary and can make the difference between success and starvation.

Many popular marine animals have evolved to have an intimate and vital association with specific substrates in the wild and will not fare well in captivity without it. Some of these creatures feed on the sand coatings (deposit feeding sea stars) while others feed in it (various gobies sifting for micro-crustaceans). Certain creatures consume sand to digest infused particulate organics (Holothuroids/sea cucumbers) while others ingest it as a mechanical tool to aid digestion (like the crop of a bird, as with surgeonfish and the Siganids/rabbitfishes).

Live sand is a veritable microscopic zoo. A healthy live sand bed can provide a habitat for bacteria, algae, diatoms, bristle worms, round worms, flat worms, various snails (*Cerith*, *Stomatella*, *Strombus*, *Nassarius*, etc) and other mollusks, sea stars, brittle stars, and numerous other organisms. Any of these can serve as a food wholly in various trophic levels, but equally important is their participation reproductively with the release of large amounts of gametes and larvae. This planktonic material can be a tremendous source of nutrition for corals and other filter-feeding invertebrates. There is no artificial methodology that can simulate this dynamic with even a remotely similar degree of bio-diversity and efficiency. The manual application, otherwise, of target-cultured organisms (phytoplankton, rotifers, brine shrimp, etc) is tedious at best. Live sand and refugium methodologies are beginning to simulate this dynamic. If you cannot provide a valid reason for not using live sand in your aquarium, perhaps you should be knee deep in it... or at least wrist deep?

The Impact of Live Sand on Aquarium Water Quality...

In creating your own slice of the ocean, surely you have wondered about the differences between a home grown reef and the wild environment? Certainly our aquariums are smaller, more densely stocked (if not outright overcrowded) and due to these features the chemical and physical influences on the captives therein is more mercurial. The ocean environment is, of course, a vast and nearly inconceivable volume of water. Parameters of water quality are remarkably homeostatic or easily reconstituted when it is not so (seasonal and sudden influences of weather). A crucial component of this remarkably stable environment is the enormous depth and breadth of the living seafloor. Among many desirable benefits of this largely calcareous matter, the substrate (loose aggregate and the hard reef proper) serves as an alkaline "bank," dissolving in resistance to reductive/acidifying influences and buffering the pH of of the system upward. This stabilizing action is accomplished by the replenishment of cations of importance, like calcium, magnesium and strontium from carbonate mass, to natural and captive reefs alike. We aspire to replicate this bastion of bio-mineral stability and sustenance, in part, with the incorporation of live sand methodologies in aquariums.

What Type of Live Sand is Best?

Composition

When shopping for sand, the type you see most often in the hobby nowadays is aragonite. There are also calcite and silica-based products, but most aquarists will opt for one of aragonitic composition. Depending on there being other ready sources of biomineral and alkaline reserve, there may be no significant disadvantage to calcite or silica, but aragonite has a distinct advantage in seawater. It is able to dissolve at a significantly higher pH than calcite (and silica does not impart any buffering minerals by dissolution at all), which makes it a much better buffer in the marine aquarium. Since this source of calcium carbonate is comprised of old processed products of the reef (chomped coral passed through parrotfish, storm broken and pulverized shells, etc), it stands to reason that it will be an ideal supplement of bio-minerals to form new coral and shells! Aragonite will begin to dissolve, in fact, at a pH over 8.0 (a still safe level for marine life by far), while calcite does not readily dissolve until a pH below 8.0! This means that calcite is not likely to impart any significant measure of useful minerals (buffer/alkalinity) into the water until the pH falls to a dangerous level for most marine life. In this regard, the old argument of dolomite & calcitic crushed coral versus non-calcareous freshwater "gravel" for marine aquariums in the early days was a moot point. Again, calcite is not harmful in the marine aquarium... it simply is not as helpful regarding its buffering ability. This distinction between calcite and aragonite ultimately is not all that significant in healthy systems with a calcium reactor or diligent supplementation to maintain stable calcium and alkalinity.

Aragonite and calcite are forms of calcium carbonate. The difference between them is the arrangement of their crystalline structures. This is analogous to how diamonds and graphite are both made of pure carbon, but differ on the atomic arrangement level. Diamonds are slow formed and very dense (the most concentrated form of carbon known), while graphite is considerably less dense and faster to form. Both, however, have the same fundamental chemical composition (carbon)... just like the calcareous media aragonite and calcite.

If you ultimately do consider calcite or silica-based forms of substrate, simply be aware that you cannot rely much or at all on such products for supporting water quality with a buffering reserve. Nor will you enjoy the significant liberation of bio-minerals as with aragonite. Either situation can be corrected or compensated for with due diligence in water testing and supplementation. Still... given to choose, aragonite will serve most aquarists best.

Size of Grains:

Indeed, there is no best or ideal grain size for sand in marine aquariums. The selected substrate must serve the needs of the system and its inhabitants. If the livestock have no specific demands or limitations (as with free-living corals or sand-sifting fishes requiring fine sand... never coarse media), then the aquarist must decide what benefit(s) by grain size are valued most. Be sure to keep a bed of consistent size grains... mixing sizes casually can result in

Contrary to popular misconception, the discoloration of sand visible substratum through the glass is not dangerous anaerobic activity or even indicative of any other organisms spread throughout the rest of the sand bed. The "colors" most often are simply algae and other micro-organisms flourishing in a very thin layer between the glass and sand from direct or refracted light through the aquarium. (*H. Schultz*)

the limitation (through compaction and channeling) of any grade's merit by design.

Sugar-fine grade (0.2~1.0 mm) - excellent for denitrification and deep sand bed (DSB) strategies. Requires less product by volume to reach an anoxic state to accomplish natural nitrate reduction (NNR). For optimum nitrate control, always maintain at depth over 3" (7-8 cm) with 4" to 6" recommended. Ideal at depth for culturing seagrasses (*Zostera*, *Syringodium* and *Thalassia* species). Very supportive of capillary root structures in red mangroves (*Rhizophora mangle*). Encourages the finest zooplankton (copepods) to develop in fishless refugia. Ideal substrate for free-living corals like Fungiids (Fungia, Plate anemone, slipper, tongue and helmet corals), *Trachyphyllia* (Open brain coral), *Goniopora stokesii* (Green Flowerpot), and *Catalaphyllia jardinei* (Elegance). This is the ideal grain size for most detritivores and sand sand-sifting reef animals.

Medium grade sand (1.0~2.0 mm) is similar in form and function to sugar-fine media. Requires slightly greater depth to accomplish denitrification (4-6" minimum recommended). Very shallow beds are recommended instead if NNR is not required (maintain at just 1/2"/ 12 mm or less). Coarser grain size affords a nice mix of micro-crustacean species to proliferate (e.g. amphipods and copepods). Calcareous plants flourish here including *Halimeda*, *Udotea* and *Penicillus* (baby's bows, fan algae and shaving brush respectively). Medium fine grains may still support some free-living corals like *Cynarina* (flower, button, donut corals). This sand is borderline inappropriate (too large) for sand-sifting detritivores like sea cucumbers and gobies.

Coarse grade sand (2.0~4.0 mm) is a challenging grain size to employ in deep sand bed strategies. Rather too coarse to support anoxic faculties. Requires fairly strong water movement to prevent the buildup of detritus. Supports larger micro-crustacean populations, like amphipods, very well. Easy to service with a gravel siphon, not as messy or easily disturbed like a sugar fine grade. Too coarse for most macroalgae and stony free-living corals. An acceptable substrate for culturing most soft corals in grow-out trays. Recommended depth of 1/2" or less (<12 mm) for low maintenance with high water flow.

Intermediate depths of 1-3" (25-75 mm) require regular (weekly) stirring or siphoning for optimum success. Maintenance neglect of this size sand at depths of 1-3" can lead to dangerous accumulations of detritus known as a "nutrient sink" condition. Mismanaged coarse sand at this depth has been the single greatest burden on the Deep Sand Bed methodology's reputation... an entirely misinformed application of the strategy (fine sand is recommended for NNR).

Very coarse media (4.0 mm and larger) of any kind over 4 mm requires special considerations as a substrate in marine aquariums. For the casual aquarist seeking the greatest "bang for the buck" (benefits & ease of maintenance)... it isn't here. Rubble, gravel, and coarse shell demand extra service in the course of routine maintenance (weekly siphoning is strongly recommended) and greater attention to flow dynamics and the performance of system hardware (clogging pumps, waning flow and skimmer performance to prevent detritus from accumulating) will be required. For the least burden on your husbandry, shallow beds work best (1/2" or less, <12 mm) in aquariums with very strong water flow and aggressive nutrient export mechanisms. Large fishes, messy feeding animals, and overfeeding is to be avoided on very coarse substrates. Some reef animals do enjoy at least a little coarse media (like jawfishes and octopuses), but others do not fare well at all on it (like wrasses). Others still will suffer terribly if forcibly kept on coarse media (like some eels, sharks and especially rays) as evidenced by developing sores and lesions. Few true plants or macroalgae short of nuisance species will grow here. However, shallow beds of coarse rubble, gravel or shell can work quite well for aquarists settling cultured coral in propagation endeavors. The largest micro-crustaceans also flourish in coarse substrates and can be set up in a fishless refugium full of such media to perform as a zooplankton "reactor."

Shape:

The "ideal" shape of marine substrates is irregular to spherical, not flat. Of the desired shapes, small, round forms settle more densely and afford better environments

Live sand methodologies have exorcized their perceived demons and have found a deserved place as an effective strategy for natural nitrate reduction (NNR), nutrient exchange, and as sources of bio-minerals. For nitrate control, the key is maintaining depths over 3" (75 mm) while ensuring adequate water movement above and turnover of the substrate by either manual means or incidentally by the livestock. A deep sand bed (DSB) application pictured here. (*L. Gonzalez*)

for desired microbial activity. Irregular and angular shapes create considerable spaces for colonization by larger micro-organisms like worms and crustaceans. Such shapes, however, are more likely to catch and trap detritus, leading to a higher local concentration of dissolved and available nutrients, and ultimately growing nuisance algae species faster or stronger. Some say this is the substance behind criticisms of silica-based sands growing algae faster than calcareous media (sharp/irregular grain shapes settle and grow diatoms faster). Fine, calcareous media (spherical) will serve most keepers of marine invertebrates best.

Depth of Bed:

A recommended depth of substrate is perhaps the most controversial aspect of live sand methodologies. The truth of the matter is that any depth of sand maintained properly can be safely enjoyed as time-tested and proven by thousands of aquarists. Likewise, any depth of sand neglected can become a burden or danger to life in the system. In brief summary of the key points of sand depth dynamics, we can say the following:

- Very **shallow** beds of 1/2" or less (<12 mm) can easily be maintained in an aerobic state. They present little help or harm to the aquarium system with reasonably good water flow and agitation (mechanical or natural). With very coarse substrates like gravel and shell (crushed oyster shell, large Puka, scallop shells, etc), a shallow bed is the only practical option for enjoying such media. Any appreciable depth beyond 1/2" requires regular maintenance (weekly stirring and/or siphoning for all but some sugar-fine substrates) and fine-tuned nutrient export mechanisms in the system.

- **Intermediate** beds of 1" to 3" (25-75 mm) are the most popular and yet precarious. When maintained properly, they offer swell advantages with little trouble. If neglected, however, they can quickly be problematic (largely by prolonged neglect and nutrient accumulation). Some say beds of sand in this range of depth is neither shallow enough for optimum nitrifying faculties (aerobic), nor deep enough for optimum denitrifying faculties (anoxic). Aquarists with a decided microbial preference should commit to a more shallow or deeper bed of sand as necessary. Casual aquarists in the middle here are simply encouraged to maintain good water flow and stir or gravel siphon the substrate regularly in good husbandry.

- **Deep** Sand Bed strategies require sufficient depth of substrate to support denitrifying faculties. With increasing grain size, the penetration of oxygen rich water increases and so does the need for more sand at depth. Sugar-fine sand (.2-1 mm) should be applied at a depth of more than 3" (75 mm). You must be mindful too of the wonderful, fast rate of dissolution of aragonite. In most healthy systems, aragonite has a "half-life" of 18-24 months. That means that after two years, a 3" sand bed will have been reduced to 1.5" and possibly failing in natural nitrate reduction. For this reason, aquarists seeking optimum nitrate control are advised to resist being frugal and apply honestly deep sand beds of more than 3," with 4"-6" (100-150 mm) as a good start. Of course, you will need to prevent the bed from waning below a 3" minimum in time.

Substrate Varieties

Coral Sand (0.2~1.5 mm)

Advantages: Deservedly the most popular type of media for marine aquariums. Relatively easy to maintain, provides numerous benefits with few disadvantages. Deep sand beds (DSB) in excess of 8 cm (~3") can provide fast, efficient and reliable denitrification (nitrate control) through natural nitrate reduction (NNR). Numerous desirable micro-organisms fare well and proliferate in these substrates including worms, copepods and other large plankters. The top-most layers of DSB and shallow beds also offer significant nitrifying faculties. Many fishes require fine substrates for health and survival in captivity (wrasses, some sharks and rays, eels, etc). Some free-living corals derive sustenance directly from the microclimate surrounding their residence in placement on the sand floor of a display. Corals and other reef invertebrates at large benefit significantly from nutrients imported to and exported from the DSB. Aragonitic material is especially useful for traditional reef aquariums for its fast and significant dissolution... releasing bio-minerals and lending support to the buffering pool of the system.

Disadvantages: Easily disturbed by active fishes and larger motile invertebrates, if allowed. Coral sand is too fine to use as a flow-through substrate/sub-stratum filter and packs easily if neglected.

Special notes on coral sand as a substrate: Fine coral sand appears in the aquarium industry as calcite or aragonite. Calcite is more common compositionally on wild reefs, but aragonite has dominated the hobby trade of substrates for aquarium use. The collection of aragonite for use by all industries (casting glass, construction, and other industrial applications, recreation, aquarium use, etc) is a very simple matter endeavored commercially by few entities. The source for all is quite similar or the same due to the limited geographic distribution and availability of aragonite (unlike calcite and silica based products). As such, aquarists can rest assured that there is very little difference

Who'd have thought there could be so many names and modifying descriptions for substrates?!? Aragonite, oolitic, calcite, dolomite, crushed coral, bio-active, alive(!), special grade, reef grade... and the list goes on. What's the difference? Marketing mostly. Decide what your primary purpose for using sand is. Start simple first: denitrification versus plankton culture, or aesthetics versus bio-minerals, for example. Then determine which composition and grain will serve you best. Pictured is 0.2 mm grain. (A. Calfo)

compositionally between the various brands available... certainly not enough of a difference to warrant great disparities in price! Packaging, marketing and advertising entail real and certain costs... pay for as much or as little as you like when you choose a substrate for your aquarium.

Coral gravel and rubble (2.0~15.0 mm)

Advantages: Not easily disturbed by fishes or other physical actions like dynamic water flow. Does not pack or settle easily in neglected displays. Coarse grain encourages the growth and proliferation of some zooplankton species. Excellent for settling new divisions of soft coral in culturing troughs. Ideal for flow-through substrate/substratum filtration.

Disadvantages: As with all coarse media, requires greater attention to proper husbandry: very strong water flow and gravel siphoning may be necessary to prevent the bed from trapping detritus and becoming a burdensome nutrient sink. Requires much greater depth for any chance at denitrification (over 4"/10 cm likely). Often calcitic in nature and is a poor media for buffering and dissolution of bio-minerals when so. It is not as successful in scope and size of bio-diversity compared to finer media as a "live" substrate.

Special notes on coral gravel and rubble as a substrate: Like coral shells (below) and all coarse media, aquarists are advised that gravel and rubble requires significant maintenance (water flow and servicing) and attention to the health of the bed. Such media is best suited for rubble troughs and refugiums dedicated to the culture of target organisms (like zooplankton and coral divisions). It will also serve well in displays with larger motile invertebrates and active fishes inclined to disturb the substrate and keep detritus in suspension. Casual aquarists seeking a traditional reef display will likely be served better by a finer substrate at any depth.

Coral Shells (4.0~25 mm)

Advantages: Attractive shapes and colors. Coarse nature provides great haven for "critters" (amphipods and other micro-organisms). May be ideal for a fishless refugium, with adequate water flow, focused on generating zooplankton.

Disadvantages: Offers little buffering ability as calcite (dominantly), compounded by its coarse physical nature (large chunks are slow to dissolve and provide bio-minerals). Traps excessive detritus, offers little or no support of denitrification without extraordinary depth. It is difficult to maintain mechanically (gravel siphoning is tedious and strong water flow may not sufficiently purge it). Few "reef safe" detritivores have the muscle or means to service this media. Shells should not be mixed with fine sand as they will separate and may impede other desirable actions. Few substantive benefits of this substrate beyond aesthetics... some of the least useful media for marine aquariums.

Special notes on coral reef shells as a substrate: such coarse media is extremely challenging to good husbandry and water quality... the irregular shapes and coarse nature of even the smallest grade quickly accumulates unhealthy amounts of detritus. The large, crushed, oyster shell-type substrates can just be a plain nightmare to maintain even at shallow depth! For all shell substrates, maintain very strong water flow in the aquarium and know that manual siphoning of the media with regularity (weekly) will likely be necessary. For ease of maintenance, apply at no more than 12 mm (~1/2") depth.

Non-coral Substrates

Advantages: Many novel, aesthetically attractive and unique options. Can be instrumental for replicating specific niche-environments, educational microcosms and geographically accurate biotope displays.

Disadvantages: Highly variable and sometimes undefined compositions. Terrestrial aggregates and volcanic media

A popular and attractive pink beach sand. Approx. 1-2 mm size. (*A. Calfo*)

may be a somewhat greater risks for imparting undesirable or harmful elements into aquatic systems.

Special notes on non-coral media as a substrate: Non-traditional substrates are best suited for specialists with the need and knowledge to incorporate them into deliberate systems. Some will require an adjustment to system design or maintenance... others may dissolve or "react" in an unfavorable manner (imparting deleterious elements into seawater). Casual aquarists are encouraged to use one of the many popular and predictable calcareous substrates instead. All of the limitations mentioned herein for the various grades and shapes of calcareous substrates apply.

So you've picked a substrate... *now what?*

There are many more decisions to make regarding live sand methodologies even after you've picked a type, grain size and depth. What color sand should you pick... seriously?!? To reefkeepers: have you ever wondered why some of your mature and grown corals have pale tissue (without zooxanthellae or other pigmentation) on their undersides? It isn't merely an artifact from the shadow of growth... its much more than that. In the open ocean, corals do not receive light from a static posterior source (like an aquarium lamp), but they are bathed in lumination from the sun that travels in a wide and seasonally variable arc across the sky. The path of the sun irradiates corals at sometimes severe angles and this is something that we usually do not enjoy in the aquarium unless we move our lamps on tracks and rotating timers. One dynamic of this process that we can exploit to compensate for our lack of balanced/traveling illumination is the refraction of light off the sand bottom. Light on a reef is refracted from many places like glitter lines from the dancing waves and off of a light colored seafloor. Bright white sand can be an enormous reflector of radiance for zooxanthellate-symbiotic corals and reef invertebrates. For aquariums that are not littered with coral and rubble on the substrate and have somewhat open rock structures (hopefully, for good water flow as well as aesthetics!), an open bed of light colored sand will be a significant benefit to many plants and animals in the display.

And once you have installed a functional sand bed... be sure to exploit it! Many species of free-living coral must live on the seafloor niche or suffer, among other things, from the loss of micronutrients from the sand. Beyond attrition, these corals may also suffer injury and die from abrasion to

A medium grade coral sand. 1-3 mm size. (*A. Calfo*)

Very coarse sand, small rubble and gravel can be practical in large marine displays and many ancillary components of a marine aquarium system such as refugia and grow-out troughs for cultured corals and other reef invertebrates. Diligent husbandry techniques must be practiced if coarse media is to be used as a substrate. Frequent stirring and gravel siphoning combined with vigorous water flow will be needed to prevent the buildup of detritus in the media bed. Approximately 15 mm size. (*A. Calfo*)

Natural Marine Aquarium Volume I - Reef Invertebrates

Fine, broken shells are marketed in mixed sizes from ¼" (6 mm) to fully 1" (25 mm) or more. Shells are categorically poor choices as marine aquarium substrates. They require greater maintenance, offer fewer benefits and can easily become quite a long-term burden on husbandry and aquarium success overall. (*A. Calfo*)

soft tissues by unnatural and repetitive polyp cycles against hard rocks (and gravel or coarse rubble/shell too). Some corals that must be kept on sand include (but are not limited to): Fungiids (*Fungia*, Plate anemone, slipper, tongue and helmet corals), *Trachyphyllia* (Open brain coral), *Goniopora stokesii* (Green Flowerpot), and *Cataphyllia jardinei* (Elegance), and *Cynarina* (Donut coral).

Purchasing *Live* (Wet/Wild) versus Dry Sand

One of the first questions new aquarists ask merchants of sand products is, "What's the difference between live and dry sand?" The answer depends on what you regard as "live" and which target organisms you seek to exploit in culturing a sand bed. For aquarists seeking denitrifying faculties to reduce nitrates, the difference is only a few weeks! Yes, in just a few weeks a bed of dry sand, accelerated by a handful of mature live sand or live rock for inoculation, can actually begin to measurably and significantly reduce nitrates. It is an amazing feat to discover... very fast development of microbial activity.

Aquarists seeking the development of larger zooplankton like worms and crustaceans, however, will have to wait weeks or months for a sand bed to develop significant populations from "fresh" live rock or live sand inoculation. If the system is stocked with fishes (or overstocked), such plankton may never develop. This is why we strongly recommend that refugiums be maintained fishless and display aquariums stocked slowly for optimum plankton culture.

As a rule, wild-harvested live sand is only necessary in small amounts to inoculate beds of dry sand. The bio-diversity will develop nicely and naturally in time if spared the burden of premature predation by fishes and other hungry invertebrates. When large quantities of live sand are used, standard procedure for quarantine and curing on import are required as with live rock.

* Note: A word about packaged "live" and "bio-active" sand products...

In our hobby, we have a remarkable situation of extraordinary commerce with little consumer advocacy. There are few if any watchdogs outspoken about the practical truths of products and claims in our trade (a reality that the volunteers at www.WetWebMedia.com live everyday, consulting and consoling fellow aquarists in need). One of the latest trends in our hobby has been the marketing and promotion of packaged, so-called "live" sand products. To generalize: these are bags of moist or wet sand marketed in an attempt to profit from the popularity of wild-harvested live rock and live sand products of the sea. The problem is that aquarists are allowed, if not instigated, to believe that they are getting a richly diverse and "power-packed" living substrate that just sits full-strength (and hungry apparently!) on a merchant's shelf indefinitely waiting to be purchased and called into action. Sure. Funny thing, though... when you look at the sand in the bag- it doesn't look alive. No worms, crustaceans, sea-monkeys... nothing of size. But then again, the manufacturers do not claim to have any of these creatures in their product: just biological activity. So... the next question you should ask is, "what does biologically active mean?" If we soak some dry sand with salted water and then bag it unsterilized... is that biologically active? Sure. If we add some cultured bacteria to the sand and water mixture is that better? Perhaps. And how is it again that these living "power-

Just a sampling of the many non-traditional substrates available to the marine aquarist. Most require special considerations, reasonable as they may be, to incorporate. The unique sizes and nature of alternative substrates can impart a significant practical or aesthetic component to special biotope and niche displays. Perhaps the exotic black sands are reminiscent of a mysterious Kona, Hawaii or Sulawesi Indonesian reef! 2-3 mm size. (*A. Calfo*)

Living Filters: Live Sand

In the muck and algae off a North Sulawesi coast we find a wonderful mix of life! True plants, calcareous algae, worms, epiphytes and bacteria unseen. A snapshot of living filtration.

stations" of microbial activity are remaining in stasis on a dealers shelf... in a sealed bag... without food... or climate control... no oxygenation... or expiration date on the bag? Some sort of alleged dormancy?!? It must be torpor like when the batteries die on your TV remote control and you just sit frozen on the couch waiting for a miracle to make the batteries work again. Even if any and all of the claims by merchants of such sand were true, do you really want to pay a premium for sand that has no benefit that cannot be achieved by putting the same amount of dry sand in an aquarium and waiting two weeks? Even better... add a single handful of live sand from another healthy aquarium system to your dry bed. You can otherwise pay for advertising, sexy packaging and (dubious?) research if you like... the rest of us will be content to spend that money elsewhere - like on a good book or another invertebrate.

Curing live sand

As with "live rock," wild-harvested live sand is composed of a vital mix of sessile, benthic and free-living organisms amidst solid matter and organic detritus. And like newly collected live rock, fresh live sand may have a lot of stressed living matter on and in it struggling to live through the chain of custody and rigors of import. It cannot be trusted for direct placement into a thriving aquarium system. There must be an acclimation period known as "curing" to allow dead and decaying matter to be processed naturally (mineralization) or with support from an aquarist (water changes & filtration). Uncured and fouling live substrates produce ammonia and other deleterious by-products of decay in excess that can stress or infect other biological components of the system. Conservatively, you must cure live sand in a separate vessel. Never cure live rock or live sand in a stocked display aquarium; the risk of fouling healthy living components is too great. While curing live substrates, provide very strong water movement and aggressive vehicles of nutrient export including dedicated protein skimming, frequent water changes, water quality testing and quality chemical filtration media.

Plenum Or No?

Since the popularization (and misapplication) of the Jaubert-style plenum for deep sand bed methodologies in the early 1990's, much has been written, debated and revealed about the use of this feature. For those of you new to the issue, rest assured that there is little you truly need to know as a casual aquarist. A fundamental summation of its benefits are included here. For all others interested in the history and subtleties of the process, we suggest that you explore the references in the back of this book and use the Internet to acquire a broader consensus on the matter.

A plenum is a physical water space underneath a deep bed of sand. It is used to create a hidden, reserved body of static water to facilitate the diffusion of nutrients and other vital components of biological processes through the substrate. The premise here is that the biological faculties we seek to harness for natural filtration in live sand can be supported and encouraged by this feature. In the bigger picture, this is indeed true. Some of the challenges of employing this technology in the past were understanding and adapting it from the original recommendations that trickled into popular aquarium literature. It seems that at least some of the early systems incorporating this strategy were very large semi-closed or open systems (fresh flowing seawater) with extraordinary depths of sand not easily replicated by home aquarists. Without getting too involved in the "how's" and "why's" of the matter, let us summarize that the plenum methodology has not been demonstrated to be exceedingly useful nor harmful at all. Most aquarists find that there is little difference with or without a plenum for a deep static bed of sand in a healthy aquarium system. This should bear no reflection on the validity of the methodology, but rather illuminates that the adaptation for home aquarists, especially with smaller aquaria (let's say under 200 gallons), may have little impact. There are no hard and fast rules here. You may have an interest to experiment with the strategy and are encouraged if so. Just know that having a plenum is not critical to success with live sand methodologies. If you are inclined to attempt a plenum system, please seek the academic work of Dr. Jaubert specifically and regard the writings of other authors that have or have had a vested interest (sales of the technology and related services) with due consideration.

One last mention of the improper implementation of plenum strategies. We should like to dispel the most common corruption of the application for those interested to know or try it. Severe criticism of plenum use has faulted it for becoming a "nutrient sink": trapping and accumulating detritus to levels that cripple water quality and fuel nuisance algae growths. The reality of the matter is that incorrect applications of the technology caused the rift. As aquarists, we too often have inadequate water flow which prevents detritus and organic particulates from being exported by protein skimming and filtration dynamics. In turn, excess detritus can settle in pockets and migrate down into the substrate. Furthermore, coarse sand and gravel is still quite popular to this day and allows particulates to settle and accumulate rather easily. The killing blow to a flawed application with coarse substrates in weakly circulated aquariums is the aforementioned popular employment of intermediate depths of sand at 1"-3" (25-75 mm). In this mid range, the sand is often too deep to be wholly aerobic and yet not deep enough for efficient denitrifying faculties. As such, the two dominant (and desired!) biological populations are restricted if not excluded and the sand bed may become a dead zone... a nutrient sink. Intermediate sand depths can be maintained successfully, but require due diligence with vigorous sifting naturally or mechanically, strong water flow in the tank, reasonable bio-loads, etc. And as far as plenum methodologies are concerned, these are some logical reasons for an undeserved bad reputation.

Maintaining Live Sand

"To stir or not to stir?" That is entirely dependent on the grain size, depth and husbandry in the system. There are benefits both to stirring and leaving a sand bed undisturbed.

If you are a casual reef aquarist, should you periodically stir your live sand, or even vacuum it ("gravel siphon")?

Live substrates are home to many unheralded macro-organisms like the true vascular plants (an encrusted blade of *Thalassia* turtle grass pictured here) and even some corals (a solitary stony cup coral top center).

Living Filters: Live Sand 41

Perhaps not. If you install and maintain your sand bed sensibly, you won't have to service it much or at all. As mentioned above - dynamic water flow, aggressive nutrient export (protein skimming, water changes and regular chemical filtration) and appropriate stocking levels will support a thriving, low maintenance and productive substrate. If you want denitrification, install a deep sand bed of sugar fine grains, maintain very strong water flow and good water quality and little else will be required. If you have no need for nitrate control, a shallow bed with the same considerations could likewise be fine with little service. You may want to seed the bed with appropriate types and limits of creatures (a starter culture of live sand, worms, snails, 'pods, starfish or other detritivores) and allow time for them to establish. There are many such organisms that will do most of the necessary "stirring" for you. A balanced and reasonable crew of sand-dwellers ensures that the bed is well-aerated at the surface and scrubbed of detritus that would otherwise accumulate and challenge water quality. Please note, however, that some popular sand-sifters are quite voracious and may be a burden that far outweighs their merits in the bigger picture. Some of the more dubious "clean-up crew" candidates include: hermit crabs, sleeper gobies, goatfish, most conchs, large sea stars and very large sea cucumbers. Brittle and serpent starfish are popular, hardy and generally helpful instead. There are numerous gastropods that are also quite useful and innocuous for sand stirring. See more about these creatures below.

In contrast to a low maintenance strategy of undisturbed live sand keeping, there are some realities and unrealized benefits to consider by stirring sand. Again, most aquarists do not have as much water flow in their aquariums as we would like to see (by volume or distribution). Most reef tanks lack an appropriate dynamic turnover of water (more than 10 times a tank's volume per hour), and many that have it lack a proper delivery of it instead. Severe laminar flow (one direction, as with powerheads) at the surface of the aquarium can leave a significant deficit near the bottom resulting in poor oxygenation of the lower tank region, accumulated detritus and poor sand bed health. Overstocking and overfeeding are also common problems that can burden live sand methodologies. Stirring sand not only supports a stimulating aeration of the surface of the substrate (no need to stir deeply) but also provides food for the many filter-feeding animals in reef aquaria by dissolved and particulate organic matter. It also makes solid matter more likely to be exported by good protein skimming and other aspects of filtration. Several aquarists have had

"Live" Sand-in-a-plastic-bag will never look like this! While it may not be possible (or desirable) to recreate the muck of the Indonesian reef-floor shown here, we've seen some refugiums that come close. (*L. Gonzalez*)

It may look funky to some, but for the owner of this richly populated deep sandbed, it's the basis of a beautiful, thriving system. This 8 inch (20 cm) bed is also required by the Banded Jawfish just visible in the top-right corner. (*L. Gonzalez*)

tremendous success with challenging aposymbiotic species like Scleronephthya (colored Cauliflower soft corals) by stirring the sand bed weekly, if not daily. One could make the argument that the liberation of organics and nutrients in this manner is not unlike storm activity that cycles and moves vital nutrients about on the living reef. The concern here with stirring would be the liberation of excessive nutrients that have been allowed to accumulate, causing blooms of nuisance algae and other troublesome organisms. Weigh the merits and possible disadvantages of stirring live sand with the needs of your tanks inhabitants. If you keep a typical reef tank with photosynthetic small polyped stony and soft corals that feed more by zooxanthellate partnership than by absorption or organismally, you may not need to stir your sand much with proper maintenance.

Keeping up with Live Sand... Replenishment

Another very important aspect of live sand maintenance that should be mentioned again is replacement or augmentation. In time, you'll notice that your live sand slowly dissolves. It may not seem to have quite the same diversity of organisms living in it as it once did. You may even notice that it's buffering ability (maintaining pH, Calcium and alkalinity levels) is waning. You're not "seeing things." There's no doubt about it... the mineral part of your substrate is dissolving. You will need to replace or supplement your live sand on an as-needed basis. There are several different methods for doing this. One common way to do this is to add thin portions (less than 1/4") of new sand atop the old. We need to be mindful, though, not to bury or stifle the aerobic fauna that have developed. It may be better to gently push the existing live sand aside temporarily and add new material to the void before gently recovering this new media with the mature substrate. Some folks even siphon out a portion of the existing live sand periodically and replace it altogether. All methods are attempts to maintain targeted depths and overall vigor of the living substrate. Whichever method you choose, you will likely find it helpful to presoak the sand in seawater first to reduce clouding from fine carbonate matter. Adding sand to an empty tank is always

Living Filters: Live Sand

easiest, but supplementing a water-filled aquarium can be as simple as sinking plastic bags of soaked sand to the bottom and very gently cutting them open to minimize turbidity. It should be obvious that pouring fine grains of sand openly into an aquarium will make a dreadful mess of turbidity that will stress corals and aquarists alike! Beyond physically sustaining a bed of sand at depth, we must also be concerned about bio-diversity over time. As mentioned before, there are numerous fishes and invertebrates that will exploit the living creatures on and in the sand. As such, many desirable living components can be exhausted in time. Popular fishes like wrasses, gobies, anemonefish, damsels, pseudochromids and many more will quickly reduce a large population of micro-crustaceans in live rock and sand to irreconcilable low numbers. For this reason if no other, fishless refugiums plumbed in-line to an aquarium system can be invaluable for producing and replenishing such micro- and macro-organisms. At any length, you will occasionally need to reseed most substrates to maintain and improve bio-diversity. There are many specific reasons for why populations of some plants and animals have peaks and valleys or suffer local extinction. Beyond predation, there are issues of microscopic competition and succession. Certain husbandry factors also play a role. Perhaps the temperature of your tank will favor some species over others, or available foods support specific populations. Whatever the reason, all sand beds lose diversity in time and can benefit from fresh inoculations. Adding fresh live sand, live rock, or some of the commercially available detritivore/flora & fauna/"whatever they call them" kits will help tremendously. Many aquarium societies often have "sand swaps" where everyone brings a sample of their sand to a club meeting, it all gets mixed, and everyone takes a cup of sand back home (a great idea, but its best to quarantine such sand for a few weeks first, of course). No one source of inoculation is necessarily better than another. You are simply advised to use several or as many of these different donors as possible to diversify your live sand bed over time.

Vulnerable fishes living and feeding in the open sand flats like this blenny (left of center) in Sulawesi often evolve remarkable cryptic coloration... or they learn to swim *really* fast.

Natural Marine Aquarium Volume I - Reef Invertebrates

Summary of Live Sand

Ultimately, the advantages to keeping live sand in the marine aquarium far outweigh the easily prevented dangers. With live sand, the following benefits can be enjoyed:

- Natural source of bio-minerals to support invertebrate growth and buffering of the system
- Natural and incomparable source of live food for animals at all levels
- Required to keep certain niche-specific invertebrate species including free-living corals and anemones
- Natural habitat for many fishes and invertebrates
- Reduces burden on water quality with efficient natural nitrate reduction (NNR)

Some of the potential problems with live sand are:

1) Stress or death of livestock from anaerobiosis in a neglected substrate

2) Release or accumulation of nutrients within the substrate that could lead to the growth or proliferation of nuisance organisms such as microalgae or pest anemones.

Both conditions are easily avoided.

How are anaerobic by-products (methane and sulfide) produced in the sandbed and imparted to the water?

- Excess food and waste accumulating in the substrate by overfeeding and/or weak water movement causing the rapid proliferation of microbes and hypoxic conditions that subsequently lead to anaerobic activity.
- Inadequate natural or mechanical aeration of the sand as well as mixed or packed sand (like calcium spikes causing cementing in aragonite).

How are unfavorable levels of nutrients pooled and released by the substrate?

- Undesirable elements in the water like phosphate will precipitate (as with kalkwasser use) or are deposited otherwise (detritus and organic matter in general). Sudden changes in water chemistry, mineralization/acidification by microbial activity, and aggressive disturbance of the substrate can liberate various nutrients into the system.
- Abrupt changes in water quality may stress and kill significant populations of the fauna in the substrate. A correlative proliferation of bacteria and other decomposers quickly change the dynamic of the sand biota, environment and byproducts leading to a change in water chemistry and composition. These changes may be instigated by power outages (drop in temperature, oxygen), hardware failures or simply poor maintenance (lack of water changes, sand aeration, etc).

The flaw and solution for each event listed above is obvious and should be easy enough to recognize and avoid.

At last, controversy over most issues concerning live sand applications in our hobby is truly unnecessary. There are aquarists and industry professionals who eschew the use of live sand, preferring bare-bottomed displays (or nearly so) or rock structures raised on grids/shelves to facilitate easy cleaning. And there are numerous proponents in contrast discussing at length which type, grade, arrangement and husbandry to apply to deeper living substrates. Here, we offer suggestions on what we have found to work best with the knowledge that there are aquarists who have succeeded with just about every possible combination imaginable on this issue. We have seen successful sand beds ranging from less than one inch (25 mm) thick to more than a foot deep (300 mm). Excise the tidbits of aesthetic and functional appeal from the matter at large that suits you best. Use the information here to make an informed decision, and know that you can succeed with live sand no matter which strategy you choose.

Defining Refugium Culture

(L. Gonzalez)

What's in a name? Refugia… refugiums, refugium… by any other name, it would smell as… well, it would just smell the same. If you have "aquariums" in your home, then you have "refugiums" too. If you are a keeper of "aquaria," then you may very well have "refugia." They are plurals (refugia or refugiums) of the singular (refugium).

Defined in aquaristic terms, a **refugium** (L.- *noun*) is about as literal as it's translation - *refuge*: a place of shelter or protection. We could go on *ad nauseum*, *ad infinitum* or even *ad me* (to my house?!?!) spouting Latin and waxing philosophically about the broadest definition of a refugium. Technically, an abandoned algae scrubbing pad in the aquarium will become a refugium from which some organisms are protected from predation. The recesses of live rock in the display are a safe-haven just the same. For practical purposes however, we are talking here about separate and dedicated containers for the protection and cultivation of targeted plants, algae and animals. In doing so, we accept a generic understanding of the fundamentals of refugium applications and then make distinctions about what makes a *better* refugium.

Most refugiums are simple aquariums made of glass, acrylic or another plastic, which are plumbed in-line (in series) with the display or main tank to benefit from and contribute to the overall health and vigor of the system. A successful refugium will establish itself as a distinct but integral part of the aquarium dynamic- a synergistic process that makes the system better than the mere sum total of its two or more parts.

The benefits of refugiums and their many interpretations in practice are wonderful and complex. Ultimately, they can be as low maintenance or labor intensive as you choose. They can be as productive or purely aesthetic, just the same. Some folks prefer the benefits of an algal "scrubber" or vegetable filtration with algae and plants. This refugium would be filled with fast-growing and (hopefully) stable and weakly noxious "greenery," consistently and systematically harvested as a means of nutrient export. Other aquarists instead may seek great zooplankton production from their refugiums to overflow and feed populations of corals and fishes nightly. For this, a well-fed vessel filled with rocky rubble or a dense artificial matrix is in order ('pod condominiums like bonded filter pads). For still other aquarists, an ornamental display of rare flowering marine plants may be the goal: seagrasses and mangrove tree seedlings growing right in the

A Summary of Some Refugium Benefits:

- refugiums will passively or actively facilitate the culture of desirable organisms as stable bio-mass, free-living plankton and/or shed epiphytic matter

- they can participate as living filters to limit the growth of undesirable organisms elsewhere in the system:

    ~ as a vegetable filter with easily harvested plants or algae for nutrient export

    ~ as a settling chamber for the collection and handling of solid particulates

    ~ as parasite control (limited) with established communities of filter-feeding, predatory micro-organisms

- refugiums can be used to contribute nutritive dissolved elements to filter-feeders by shedding metabolites (proteins, vitamins, etc) and other elements like CO_2 from photosynthetic activity

- cultured micro-organisms produced in the 'fuge (slang for refugium) can also be recycled whole as food to higher animals in the display (like algae to herbivores and zooplankton to organismal feeders like fishes and corals)

- the vessel at large supports or mediates dynamics of water quality in the system (like pH stability from respiration/photosynthesis of organisms on a reverse photoperiod from the illuminated display)

- biotic faculties can specifically limit or consume undesirable elements like nitrate and phosphate

- a refugium increases water volume and overall stability of the system

- it provides ornamentation, entertainment and education... a diverse microscopic world that is itself a microcosm of the home reef microcosm

- is an option for biotopes and specialized organisms that are not convenient or possible in the display proper (like a mangrove community, or a baby fish nursery amidst the spines of *Diadema* urchins)

- a handy emergency space for sequestering plants and animals for observation or recovery

- a successful and artistic refugium installation will make you look and feel younger. It can grow hair too (hair *algae*, that is)

home! Before deciding on the "how" and "where" for your refugium, consider the "what." Contemplate the many possible benefits of refugium applications and focus on accomplishing the top one or two most important to you Most other benefits will likely follow to some degree.

There are many desirable organisms that will develop in your refugium in a short time. Some will be carried in with living substrates like live rock and sand. Others may be imported as epiphytic matter on introduced plants and algae. Still others seem to appear from unknown locations as juveniles or even larvae. In one of the authors' systems, a series of coral-culturing aquariums were essentially functioning as fishless refugia with a deliberately "sterile" set-up (to avoid the introduction of a pest, predator or disease). Tanks were filled with dry sand, only synthetic seawater and coral fragments were later introduced only as fleshy cuttings. Again, the purpose was to have a controlled culture without any wild substrates (live rock or live sand) that could introduce an exotic pest or predator to the corals. In due time, the DSB (deep sand bed) became biologically active on a microbial level, but no larger micro-organisms were ever expected (flatworms or micro-crustaceans). Imagine the surprise when Atlantic *Cassiopeia* jellyfish developed into full-grown adults (over 6"/15 cm) in the coming months! This was especially amazing when the only display organisms in the tank were a monospecific stand of Pacific coral. Ultimately, it was realized that some innocuous (and improperly quarantined) *Astraea* snails had carried "larval" jellyfish medusae into the aquariums on their shells (settled) and perhaps from a small amount of introduced shipping water. Many wonderful organisms will develop in your refugium with time and they will delight you with all of the magic and wonder as that first batch of sea monkey *Artemia* you hatched as a child... or child-like adult.

Undesirable guests can also develop without invitation in your refugium. Pests are often fragments of nuisance algae that gain a foothold in the typically higher nutrient, slower flow environment of a refugium. Few undesirables are ever outside of your control by limitation of nutrients, light or other critical parameters. In a worst-case scenario, the refugium can be temporarily turned off-line or bypassed to rectify the problem. All told, far more desirable organisms appear than do nuisance organisms.

Living Filters: Refugiums

Caulerpa is only one of the many possible organisms for a refugium functioning as a "vegetable filter." Be aware of its many benefits and dangers when cultured *en masse*. (C. Gonzalez)

Size of Refugia

It's a short story- "bigger is better" in matters of refugium size. The popularity of plastic hang-on refugiums and other vessels of mere gallons/liters in capacity are beneficial adjuncts to systems that can't be fitted with larger dedicated refugiums over, under or to the side of main display systems. However even a 2, 3 or 5-gallon refugium can do wonders in making a reef system stable, and better fed overall. From the standpoint of plankton generation, of course, production is all about refugium size and it doesn't matter if your refugium's cubic footage is nestled within the rockscape or hanging on back of your aquarium- the micro crustacean population is a function of sheer volume. A large remote refugium will serve your goals best in ease of installation, productivity, upgrade options and more. By virtue of size, small refugiums (less than 20% of the tank's total volume) are unlikely to make as much an impact on water quality as any number of correlative dynamics can in the display proper (deep sand bed, fresh and copious live rock supplies, water changes, etc). Let's be clear, however, that we are not discouraging the use of small refugiums! Even the smallest can impart worthwhile benefits to a system. If you have a choice, install the largest refugium that you have space for. You are only limited in the design phase by your expectations for its production; engineer your refugium vessel accordingly.

Refugium Flavors

Above all, it is important to remember that the word "refugium" should not to be identified with any one specific strategy, group of organisms, or application. Many different combinations of livestock and hardware can be assembled to work as refugiums for a wide array of purposes. For the new marine aquarist beginning to read hobby literature and advertising abroad, it is a disappointing reality that some sources lead them to believe that *Caulerpa* macroalgae, for example, is synonymous with refugiums. Such advice is misleading and falls woefully short of the tremendous potential of refugium applications. *Caulerpa* is just one of the many possible genera worthy of target culture in refugiums. However, it should not be promoted as the best or only organism available. On the contrary, it is especially challenging and potentially dangerous to employ by inexperienced or ill-advised aquarists. The specifics of various merits and challenges with culturing *Caulerpa* are outlined in the dedicated chapter on plants and algae. To summarize, however, it is a potent, noxious (toxic exudations) algae that can be a wonderful boon or a fearsome scourge depending on how it is kept. There is no one best plant or algae for nutrient export or habitat for plankton culture. Any plant or animal worthy of inclusion in a refugium application has its own unique attributes and limitations. Aspire to discover what is best for your particular system and do not subscribe to any one method.

Placement of refugiums

The placement of a refugium can be succinctly categorized as internal or external (remote). External refugiums are either upstream or downstream. See the accompanying images and illustrations for visual cues to typical installations on the following pages.

Refugium Functions

Refugiums are typically designed for one or more of the following purposes:

- food culture
- nutrient export
- mediation of water quality
- ornamentation
- complimentary culture of certain organisms to support or limit the growth of others in the display
- target culture of organisms (necessary isolation or a concerted farming effort)

Most of the refugium types listed below will accomplish all of the above purposes with varying degrees of success. Consider which style serves your system best.

Refugium Types

- Vegetable filters: algal "scrubber" (microalgae or macroalgae), true plants
- Animal filters
- Plankton generators (zoo- and phyto-)
- Natural nitrate reducers- deep sand beds
- Mud systems
- Ornamental and Alternative

Vegetable filters

Some of the earliest refugiums were lagoon-style aquariums connected to a main display where fishes and other macro-organisms could travel freely through tunnels between a seagrass or turf algae exhibit and the reef proper. In public aquaria, this offered a very clear picture of how nutrient fixing and cycling and recycling worked between niches in the overall reef environment. These multi-tank mesocosms were able to illustrate some very simple but important relationships on the reef between various organisms like the surgeonfishes that eat algae, pass waste, which is then dissolved in the water and feeds new algae, which is eaten again in a circle of life. It is a crude but poignant example of how some of the processes work on a nutrient *concentrated* reef (recall that reefs are not nutrient poor). Indeed, healthy reefs are teeming with nutrients, but they are fixed and concentrated within the dense and diverse biomass. In a natural marine aquarium, we are trying to replicate the balance and movement of various elements from the products of photosynthesis, with the burden of imported nutrients from food and mineral-rich water into the aquarium. Management of a variety of forms of algae is natural and necessary in this dynamic.

In the broadest definition of a refugium employed as a vegetable filter, we have the deliberate culture of a plant or algae for the purpose of concentrating or exporting nutrients from the system via the biomass of a growing target organism. Nutrients taken in by a plant or algae may simply be fixed in the growth of a stable organism like a mangrove tree, or they may be volatile as in the fast-growing but precariously short-lived cycle of the single-celled algae *Caulerpa*. Mangrove trees do not grow quickly, and do not die back or wane easily like *Caulerpa*. Thus, the nutrients that they draw from the aquarium are fairly stable and secure. In this manner, mangroves are slow but steady bastions used to out-compete lower order plants and algae that are often regarded as nuisance organisms. They have other benefits in culture too, as ornamentation and as a fairly low maintenance organism. For their slow growth and sensitivity to pruning however, they are not ideal species for systems with a heavy bio-load in need of a faster growing medium that can be harvested as a vehicle for aggressive nutrient export.

One of the dangers with faster growing but less stable algae in an aquarium is the risk of a "vegetative" event where pigments (notice green tinge to the water) and numerous other compounds are released suddenly and can impact water quality. (*S. Attix*)

Some common guests in refugiums (invited and uninvited)

- Masses of **micro-crustaceans** - amphipods, mysids and everyone's *cuckoo* for copepods!
- Gaggles of **gastropods** - shelled and shell-less snails, limpets, chitons, etc. *Stomatella* and *Cerith* are fine incidentals, *Strombus* and *Nassarius* are helpful intentionals, sessile Vermetid snails may be inevitable, they are helpful filter-feeders, but harsh on one's hands for their sharp, sturdy calcareous spiral tubes.
- "Plagues" of **polychaetes** - the much-maligned (and quite unfairly so!) errantiate polychaetes (bristle worms) are wonderful additions and beneficial for aerating and purging sediments from the substrate much like earthworms in soil. Numerous sedentariate polychaetes Sabellid (fanworms & featherdusters) and Serpulid worms (calcareous tubeworms) can be expected too... occasionally we can acquire the delightfully utilitarian Sipunculid "peanut worms" from imported substrates. Spaghetti worms are another prolific and welcome worm group in refugium methodologies (*Loimia*- the Medusa worm, and like genera).
- **Sponges** - better than a third kidney. A great many sponges will inevitably grow in your refugium, sumps, plumbing, and tank recesses. They are efficient filter-feeders and able to process extraordinary amounts of water per hour. Higher forms take some months to develop. Little *Sycon sp.* sponges are one of the most prolific (free-living) and misidentified sponges in refugium culture- they look like dingy, white-yellow, pillowy capsules.
- **Ascidians** (Sea squirts) - populations are more fickle and less prevalent than sponges in casual culture, but various colonial and free-living species are known to flourish in some aquaria.
- **Bivalves** occasionally appear incidentally and are usually welcome additions if their strict nutritional needs can be met. Like sponges, they are extraordinary filter feeders, able to process a lot of water per hour. Deliberate Tridacnid clam refugiums can be very effective at consuming nutrients.
- **Forams** (Foraminiferans) are exceedingly common and most always unnoticed or incorrectly identified. They are filter-feeding protozoans that generally prey on bacteria and fine organic matter. Some are photosynthetic and occur in brightly lit areas including species that are epiphytic on seagrasses. Other are calcareous and flourish in shaded, high flow areas of the reef. The small spiny red (calcareous) foram, *Homotrema rubrum* is perhaps the most widely recognized.
- **Echinoids** aplenty: *Ophiuroids* - Brittle and serpent starfish are welcome and little burden, even when less than ideal. Some tiny species are live bearing, prolific and a great boon to refugium methodologies. They are good detritivores overall and adaptable to a variety of food sources. *Echinoids* - Sea urchins are generally harmless to useful in refugiums. Favored species are well-behaved herbivores that can be quite helpful for concentrating nutrients by active grazing of microalgae and the passing of compact fecal pellets. *Holothuroids* - the utility of sea cucumbers in a refugium runs the gamut from innocuous to precarious. "When in doubt- leave them out."
- **Hydroids** - are rather like sea cucumbers, "When in doubt- leave these out as well." Some species can be useful and prolific filter-feeders, but way too many can easily become unmanageable nuisances. Make your decision for inclusion or extraction by species with research.
- **Algae** - are the very heart and soul of most refugiums. Various nuisance varieties are inevitable with the slow flow and high nutrients of many applications. Even then, they can easily be managed and used as indicators of overall system health by their presence in the refugium but not in the display proper.
- Other **Macro-organisms** - like fishes, shrimps and crabs are sometimes deliberately added to refugiums and usually ill-advised. Most take far more than they give to the overall system when placed here and are a burden. If their presence is desired and can be supported, so be it. Some popular shrimp species are added for their frequent reproductive activities and release of eggs (food for filter-feeders). Even in such cases their value is generally overstated. These organisms are often predatory on other valuable and precarious life forms (polychaetes and micro-crustaceans especially), and their participation should be measured carefully. Crabs are perhaps the worst of the three large groups mentioned here to place in refugia, as they are voracious and indiscriminate opportunistic omnivores.

Natural Marine Aquarium Volume I - Reef Invertebrates

A Brief Summary of Refugium Placement Strategies

Internal Refugiums (inside the display):

- Easiest to incorporate for most systems. Deliberate places of refugia are created by submerging cages or hang-on internal miniature aquariums to block out predatory macro-organisms

- Often limited in production by virtue of size

- Generally challenging to efficiently service

- May be the least expensive installation... utilizes display lighting and flow

- Severely limited if at all possible to upgrade

- Indiscreet or inconvenient from an aesthetic perspective

Pictured here is an interesting internal refugium created by a sealed partition in the display tank. Holes in the divider at the top prevent the largest fishes and most predatory crustaceans from entering. Although not ideal, it is a useful application.

A productive vegetable filter will instead utilize a plant, like seagrasses, or algae that is fast growing but stable, easily harvested and maintained, weakly noxious (if at all) and efficient at nutrient uptake.

Vegetable filters typically employ microalgae, macroalgae or true marine plants (read more about all in their dedicated chapter). Mixing groups is possible but generally challenging and less successful in the long term. A brief summary of some popular groups and genera of plants and algae for a refugium follows:

Microalgae - general

- excellent and efficient for nutrient export

- moderately easy to grow (sometimes slow to start but reliable once established)

- easy to maintain and harvest (palatable to grazers)

- low to moderately noxious or inhibitive to other life forms

- generally stable in culture

- high impact on water quality (desirable if harnessed, dangerous if unattended)

- useful for limiting competitive organisms

- very good for food culture- (excellent for co-culturing microfauna, very good as a whole food, grazed)

Macroalgae (non-calcereous) - *Caulerpa sp.*

- moderate to very good for nutrient export

- easy to grow

- can be difficult to harvest and maintain stably long-term (as a single-celled algae, pruning can be harsh... pull whole fronds instead)

- moderately to very inhibitive to other life forms (severe noxious exudations)

- precarious stability in culture (3-6 month life cycles, sensitive to nutrient levels in step with growth, e.g.)

- have high impact on water quality (desirable if harnessed, dangerous if unattended)

- can be effective for limiting competitive organisms

- moderately effective for food culture (weak as habitat for co-culturing microfauna, variable to good as a whole food)

Macroalgae (non-calcereous) Other: *Gracilaria sp., Chaetomorpha sp.*, etc.

- moderate to very good for nutrient export

- easy to grow

- moderate to good stability (life cycles, suffrage of pruning, luxury storing of nutrients)

- mild to moderately inhibitive to other life forms (variable noxious exudations)

- have high and generally safe impact on water quality (desirable if harnessed)

- variably effective for limiting competitive organisms (generally good)

Living Filters: Refugiums 51

Upstream Refugiums (next to or above the display):

- Easy to install- can be operated with a small pump (on a dedicated loop) or simply fed by return from sump before overflowing into the display (see illustration)

- Easy to maintain- may be plumbed in remote location (next room, next floor above, shelf on wall, alongside display, etc)

- Tremendous freedom for dynamic displays and species selections (the only practical application for mangroves, for example)

- Gravity overflow to display spares impeller sheer of plankton (generally an overblown concern but nonetheless...)

- Most forgiving style for modifications: emergency bypass, upgrade, etc

- Conspicuous placement may be an advantage or disadvantage. Proponents of this application exploit it as a focal point with specimens that have a high educational or aesthetic value (seagrass or mangrove communities, commensal relationships like *Diadema* urchins as fish nurseries, or motile anemones that shouldn't be kept with sessile corals). Unattractive as a vegetable filter in most cases.

- Near-display installations may be able to take advantage of indirect light from the full-reef lighting (radiant/pendant halides).

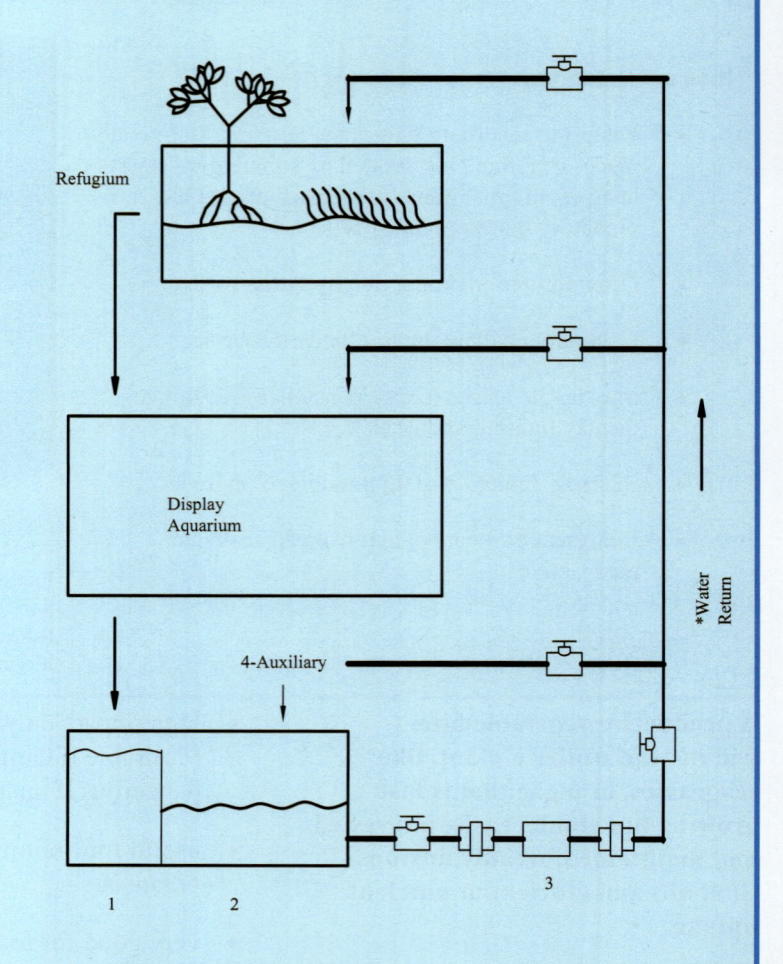

The above diagram illustrates the fundamentals for plumbing an upstream refugium (the topmost vessel here). All raw overflowing water feeds into the skimmer compartment (1) for maximum skimmer efficiency. True unions flank the pump (3) for ease of removal for maintenance and replacement. Gate or ball valves are used to finesse the restriction or diversion of water flow throughout the system. The auxiliary line (4) may be used as a bleeder for oversized pumps, to power an ancillary feature, or simply for emergencies and the temporary diversion of all water flow to avoid shutting the pump/system down (as with water changes on the display). Section (2) is the sump proper. * NOTE: an upstream refugium may also be fed on a dedicated loop by a pump in the display and overflow just the same. (*K. Carroll*)

- moderate to very effective for food culture (moderate habitat for co-culturing microfauna, very good as a whole food)

Macro algae (calcereous) - *Halimeda sp.*, *Udotea sp.*, *Codium sp.*, etc.

- weak to moderate nutrient export abilities

- moderate to challenging to grow

- variable growth, generally stable

- moderately inhibitive to other life forms (variable noxious exudations)

- sensitive to mineral and nutrient levels in step with growth

- moderate impact on water quality, consistent growth is a good indicator of mineral levels

- moderately effective for limiting the growth of competitive organisms

- moderately effective for food culture (good as habitat for co-culturing microfauna, weak as a whole food)

True Flowering Plants: Seagrasses and Mangroves

- weak to moderate nutrient export abilities without large populations in aquaria

- moderately challenging to grow

- slow to moderate growth but very stable

- weakly inhibitive to other life forms (mildly noxious exudations if at all)

- moderate to very good impact

Downstream Refugiums (under the display):

- Moderately challenging to install - conveniently capitalizes on raw overflowing water from the display above

- Often severe limitations in size of refugia here by space or structural artifacts under the aquarium (posts, pillars, other hardware and the like)

- May be a separate vessel that catches gravity overflowing water from above before continuing downstream to sump or may be integrated into the sump proper

- Sump-integrated refugiums are especially limiting: can be challenging to plumb, reduces sump volume and safe running margins for water levels in the event of a power outage, usually awkward to service

- Separate downstream refugiums between the display and sump are usually more convenient than sump models to install and maintain

- Aesthetically discreet if the aquarist does not want any focus on the refugium

- Restricted possibilities for upgrading

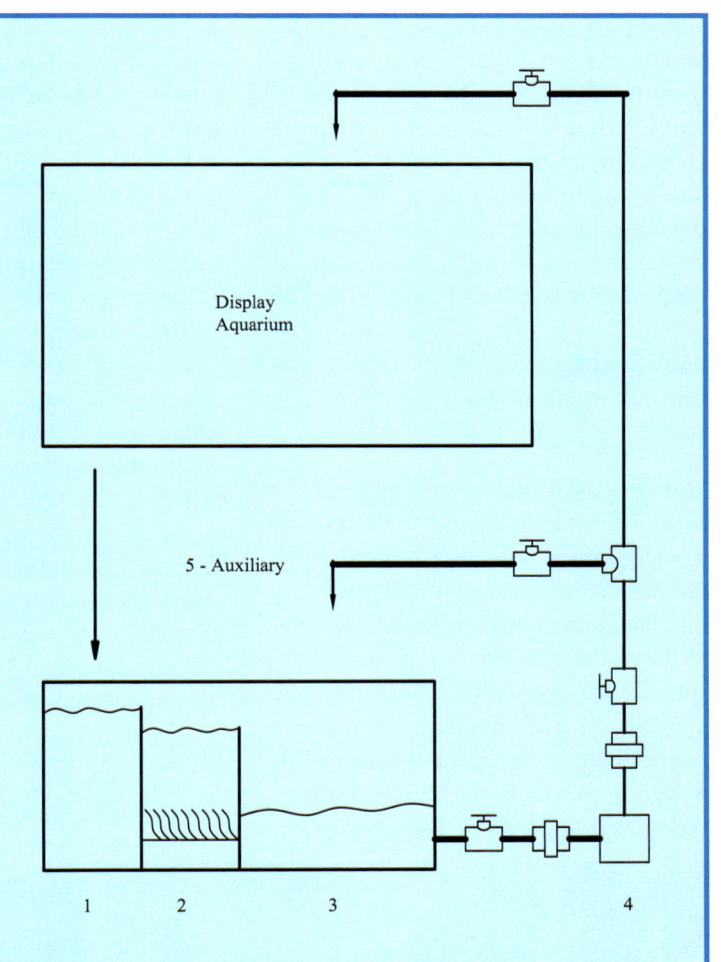

The above diagram illustrates the fundamentals for plumbing a downstream refugium (2). All raw overflowing water feeds into the skimmer compartment (1) for maximum skimmer efficiency. True unions flank the pump (4) for ease of removal for maintenance and replacement. Gate or ball valves are used to finesse the restriction or diversion of water flow throughout the system. The auxiliary line (5) may be used as a bleeder for oversized pumps, to power an ancillary feature, or simply for emergencies and the temporary diversion of all water flow to avoid shutting pump/system down (as with water changes on the display). Section (3) is the sump proper. *NOTE: a downstream refugium may also be installed as a separate vessel in-line between the display and the sump. In this application it can be fed by overflowing water, by a separate pump in the sump on a dedicated loop, or by the auxiliary line (5). (*K. Carroll*)

on water quality (mediating, stabilizing)

- moderately effective for limiting the growth of competitive organisms

- moderate to very effective for food culture (very good as habitat for co-culturing micro fauna, moderate to very good as a whole food with grasses)

Some interesting aspects and applications of vegetable filters:

Epiphytic matter - Numerous organisms including other algae live as *epiphytes*: freeloading, piggie-backing, hanging-on organisms residing on the backs and bodies of some plants and algae. Various groups, families and phyla of organisms exist as epiphytes and many are very important in the food web for filter-feeders. They grow, over-grow, break away and drift, or are liberated by rasping grazers or water movement to become part of the natural plankton that we seek to encourage in aquaria. Little is known about the nature and extent of this dynamic in the aquarium, but intuitively we know it to be good, and anecdotally we believe that we are starting to see results from experimentation. Dense refugiums of seagrasses, for example, have contributed to the successful keeping of some challenging filter-feeding corals for the shed epiphytic matter.

Reverse daylight illumination of the refugium has a clear advantage of pH stabilization for those that seek this benefit. Also known as RDP (reverse daylight period), colonies of photosynthetic plants or algae in a remote vessel can be illuminated on a photoperiod opposite of the main display. Most aquariums stocked with photosynthetic corals, anemones and algae experience a mild to significant drop in pH at night. A healthy aquarium is likely to resist any change of more than two tenths of a point (8.4 to 8.2 for example). If the biomass of photosynthetic organisms (or the impact of their potential) is

Living Filters: Refugiums 53

adequately balanced between refugium and display, the effects of a pH drop from the respiration of both can be tempered in opposition. As a rule, we should be desirous of very stable water quality including the range of pH in the aquarium. There is some merit, however, to allowing a slight natural drop, if consistent and mild, of the pH at night to dissolve oolitic material in the aquarium to provide bio-minerals for reef invertebrates and system stability.

Harvesting biomass in a vegetal filter is crucial husbandry for this style of refugium. There are numerous, dangerous implications to neglecting unchecked growth of plants and algae in the aquarium. At issue for some species of algae is the maturation of a colony in a life cycle that brings about an act of **sexual reproduction**. Sexual reproduction of algae in the confines of all but the very largest home aquariums can be catastrophic. Noxious exudations, the sudden and sum total release of nutrients absorbed over weeks or months, coupled with the sheer volume of gametes (sex cells) can overwhelm most aquaria. Beyond any issues of toxicity, the enormous influx of dead or dying organic matter (sex cells and the husk of the donor) stimulates a sudden bloom in microbial activity to degrade the organic matter and there is a correlative drop in oxygen levels from the biotic (bacterial) bloom. Many fishes in the aquarium die simply from oxygen deprivation during these events. Ironically, sexual events are fairly predictable, and they are easily avoided even when they are not clearly defined. Most popular macro algae in the trade have known life cycles. For those and all others, proper and consistent pruning to interrupt the process of maturation to sexual maturity can stave off acts of sexual reproduction.

Aside from reproductive issues, **proper and consistent pruning** facilitates the primary purpose of a vegetable filter: nutrient uptake and export. To accomplish this is very simple. Either the plant or algae is very stable and fixes nutrients in a slow-to-change (die) mass or vitality of the plant, or algae is fast and fickle but frequent pruning spurs a constant re-growth and serves as a vehicle to export absorbed nutrients. As an aquatic horticulturist, you should determine a reasonable schedule of pruning for your cultured species to

Pictured here is a nicely varied vegetable refugium. Mangroves are true marine plants providing some filtration and handsome decor, the macroalgae can be farmed systematically as a means of nutrient export, and the plastic bottle culture of unicellular algae can provide a constant slow drip of nutritive phytoplankton to various organisms. (*L. Gonzalez*)

maintain them in a controlled portion that does not overgrow the boundaries of the system, minimizes the theft of nutrients from desirable organisms, and does not risk maturation and reproductive events. For volatile species like *Caulerpa,* pruning may be required weekly (and specifically, *Caulerpa* is to be thinned and not "pruned.".. read more about this in the "Plants and Algae" chapter). For more stable species like the seagrasses and mangroves, pruning may only be required monthly or less. Actual success and growth of targeted organisms in the refugium will dictate necessary control measures.

With or without pruning, the deliberate culture of large masses of plants and algae in any system will put particularly strict demands on an aquarist to maintain water clarity in the system due to the sometimes copious amounts of **water discoloring compounds** that they shed. Even a mere week or two of neglect can create an environment where light penetrating the water is measurably impeded. The discoloration of seawater in closed systems with plants and algae is generally a significant concern. If not addressed, it can become a *seriously* stressful problem for photosynthetic organisms in the display within months. For the high cost of owning and operating reef lighting systems, aquarists should be proactive in preventing any light-inhibiting discoloration of tankwater. Carbon and ozone are two of the best chemical agents for maintaining water clarity. Ozone is to be metered by a redox (ORP) controller and all treated water and air should be passed over carbon to remove residual ozone. Carbon alone simply and inexpensively improves this and many other aspects of water quality. We recommend changing small amounts of carbon weekly rather than large amounts monthly to reduce any possible shock from a sudden increase in water clarity.

Animal filters are an interesting and novel vehicle for the export of nutrients. The very notion is a testimony to how far and fast the hobby has evolved in a relatively short period of time. Not quite twenty years ago most people would tell you that it was very difficult if not impossible to keep corals and other reef invertebrates alive in an aquarium. Nowadays, we are so successful in keeping reef animals that we're farming some attractive and ornamental species as living biomass to be harvested as nutrient export mechanisms! The premise is very simple though we must abide by essentially the same limitations that one would with the above-mentioned plants and algae in a vegetable filter. Ideal animals will be fast growing but stable, easily harvested and maintained, weakly noxious (if at all) and efficient in nutrient uptake (feeding aggressively by absorption, suspension and organismally in this case). Popular species for this application include anemones like the nuisance species *Aptasia* and *Anemonia* as well as the magnificent *Entacmaea quadricolor* (Bubble Tip), which is currently being farmed aggressively by imposed division (cutting!). Many polyps are used as well like zoantharians and the so-called "*Parazoanthus*" yellow polyps. True corals work well just the same with *Xenia* being one of the most popular for its fast absorption, weak aggression and high resale value abroad (seek wholesalers in big cities like LA, Miami and NY if your local market is flooded). All of the animals mentioned thus far have been listed for their efficient feeding strategies organismally and by absorption. Several other animals, however, are even more effective: sponges and clams in particular are aggressive filter-feeders able to process many gallons of nutrient-rich seawater per hour through their bodies.

Beyond their merits as a living filtration dynamic, the use of an animal filter allows an aquarist to work with more appealing and aesthetically satisfying living filter mediums. The harvest of animal biomass is desirable to other aquarists, stores and wholesalers (gift, resale and trade). The very activity of producing more reef animals captively in these refugiums is empathetic to the reefs we admire, and is supportive of the hobby and trade at large by displacing animals that would otherwise be wild-collected on demand.

Living Filters: Refugiums 55

Getting creative, this aquarist has a refugium overflowing to another refugium before overflowing to the sump. The separation affords better control over farming different targeted organisms. Here, competitive species are kept in the same system, but not the same vessel. (*G. Rothschild*)

Refugiums as Plankton Generators

Also known as a plankton "reactor," the hobby and trade have created several fancy names for refugiums designed specifically to produce plankton. Few are more than simple, remote aquariums devoid of predators that would otherwise eat that which will bloom and grow naturally in your system. However, some are literally empty or nearly so as described below.

As we move into this discussion of various types of micro-organisms that we call plankton, a short list of handy definitions is proffered for clarity:

- **planktonic** is a "lifestyle," living in the water column, temporarily or permanently, unable to move greater than ocean currents

- **plankton** is a group of pelagic organisms (heterotrophs and autotrophs)

- **plankter** is an individual *plankton*

Not all of what we perceive to be zipping around the ocean *is* plankton. There are several types of matter adrift: plankton, nekton, and Vinny. **Plankton**, as stated, are pelagic organisms that drift in the ocean (various larvae and sex cells, for example). **Nekton** are pelagic *swimmers* like fishes. **Vinny** is a guy that owed us a lot of money who is still floating in the ocean somewhere. A very large planker. Are you keeping up with us? Good, most home aquariums are only acquainted with plankton and nekton.

We then break the category of plankton down to **Holoplankton** or **Meroplankton** (seriously... we're not making these words up). Holoplankton are organisms that remain free-swimming through all stages of its life like copepods (a zooplankter). Meroplankton includes organisms that spend only part of their life cycle as plankton before settling out, like coral larvae.

We also distinguish plankton by Kingdom: Zooplankton, Phytoplankton and Protoplankton.

- **Zooplankton** includes animal plankters of which more than half of all in the ocean are *copepods* (the number is perhaps closer to 70%). These are heterotrophic species that cannot manufacture their own food (photosynthesis).

- **Phytoplankton** includes algae plankters and is defined by autotrophs (photosynthetic species).

- **Protoplankton** includes the ever-important but oft-overlooked bacteria.

These various plankters are described at times by depth.

- **Epiplankton** includes plankters found in less than 200 meters from the surface

Natural Marine Aquarium Volume I - Reef Invertebrates

- **Bathyplankton** includes plankters found in waters greater than 200 meters from the surface

- **Hypoplankton** includes plankters found near the bottom of the ocean

- **Mesoplankton** includes plankters specifically found in the deep sea

And lastly, plankton is sorted by size ranging from microscopic all the way up to plankters the size of small dogs. Listed incrementally by increases in size, we have the following: ultaplankton (femto-, pico, nano-), microplankton, mesoplankton, macroplankton and megaplankton.

Most plankton currently produced in marine aquariums falls into one of two categories. The most conspicuous organisms in evidence to us are **macroplankton** and include most of the micro-crustaceans that we see with the naked eye (like copepods) and fish larvae. Ever more common nowadays with successful reef aquarium husbandry is the production of **microplankton**, which largely includes invertebrate larvae.

In the current state of "enlightenment," the hobby at large is making attempts at strategies that will produce more **nanoplankton** with the hope of being able to keep more of the challenging reef animals like magnificently colored azooxanthellate Nepthids (cauliflower corals) and *Goniopora*. Live-cultured and commercially prepared phytoplankton (*Nanochloropsis* and the like) are steps taken to deliberately produce target species of nanoplankton.

Zooplankton "reactors" (micro- and macroplankton generators): One of the easiest and most useful refugiums to employ is a simple zooplankton refugium. Their purpose is to produce large populations of micro-crustaceans ... AKA 'pods, bugs,

> **Merits of using corals, anemones and polyps and other animals as living filters for nutrient export:**
>
> - harvested bio-mass is usually very saleable
>
> - cultured animals sold, traded or given away displace demand for wild-harvested counterparts, and are hardier, more adaptable to aquarium conditions
>
> - target organisms are usually aesthetically more pleasing to culture than algae (!)
>
> - there is little or no concern for shed discolorants in the water or catastrophic events of sexual reproduction as with plants and algae
>
> - most animals will be more forgiving to fluctuating or inconsistent available nutrients as they usually have mechanisms of "luxury" food storage, and many can be target fed easily
>
> - benefits to water quality like pH stabilization (with RDP) can be attained with photosynthetic animals as with plants and algae

"creepy crawlies." The process is achieved simply by having any dense physical matrix to support the creeping and crawling flurry of amphipods, copepods, mysid shrimp and the like. The matrix can be wire algae, hair algae... or low-maintenance spun, plastic fiber pads! Of course, if you use living algae for habitat to cultivate micro-crustaceans, you will have all of the merits, challenges and limitations of the living medium (lighting, water flow, nutrient base, etc). If you choose an inert and artificial media like fibrous pads instead, there is almost no maintenance to speak of. The refugium can simply be an unlit and empty vessel full of

openly stacked or threaded (hung on strands like a clothesline) pads that are bathed in reasonably good water flow. For this purpose aquarists commonly utilize batting material (Dacron) from yardage stores, blue bonded filter pad, stiff algae or dish scrubbing pads, pond filter pads or pre-filters, and even very course foam blocks. The functional excellence of fibrous padding is witnessed every day by aquarists with live rock or sand in a mature aquarium with a trickle or canister filter. A cleaning of the mechanical pre-filter (floss, pad, etc) will often send chills down your spine when a scene from a bad horror movie unfolds- swarms of amphipods and like-fauna teem and crawl all over your hands as they are disturbed from the dark and food-laden pre-filter from which they were quietly residing. About all one has to do to encourage the production of these wonderful crustaceans is to offer the colony a source of food. If the colony is not supported adequately by particulates in the raw water that feeds the refugium, supplementation will be necessary. Different grades of media will encourage different organisms, of course. Various natural substrates can be employed with the similar results instead of algae or floss. A tank full of rubble (crushed live rock, course calcareous gravel, etc) will easily provide a good habitat for larger micro-crustaceans. Finer gravel and sand may be necessary for smaller micro-crustaceans like mysids and copepods. A zooplankton refugium is regarded as highly useful for aquaristics because most popular fish and coral in the hobby are zooplankton feeders.

Phytoplankton "reactors" are vessels dedicated to the culture of "greenwater" (unicellular algae of many possible genera). Most *phyto* species fall into the category of nanoplankton by size. They have great appeal and potential in the hobby as food for some challenging corals and other reef invertebrates that until recently have been too discriminating in their feeding habits to keep (compared to large-mouthed tank mates that feed on zooplankton). Phytoplankton is also a fundamental foodstuff for many micro-organisms in the reef like copepods, which are in turn crucial in the web of life (accounting for more than half of the zooplankton in the sea). There is always some concern that too much "phyto" introduced into an aquarium (or just enough into the wrong aquarium, as with nutrient-rich systems in bright light), can lead to a stressful "greenwater" bloom in the display. In most healthy systems with a judicious application of phytoplankton dosing, however, there is little cause for concern. Refugiums for phytoplankton range from simple pop-bottle cultures (manual dosing) to elaborate metered systems with UV sterilizing light filters on the effluent drip feed (automated dosing) to prevent the viable proliferation of unicellular algae in the aquarium.

Phytoplankton in the sea are mostly single-celled algae of the following divisions: diatoms, dinoflagellates, cyanobacteria, and to a much lesser degree- green algae. Aquarists are familiar with these four principal groups: 1) the **diatoms** for their golden brown films on aquarium walls and substrates, 2) **dinoflagellates** as endosymbiotic zooxanthellae in many invertebrates (yes... they are found floating around the sea waiting to inoculate or become palate fodder) as well as the causative organisms (*Oodinium*, *Amyloodinium*) in freshwater and marine Velvet disease, 3) **cyanobacteria** - "slime algae," what's not to hate about this nuisance from an aesthetic point of view?, and 4) **green algae** as suspended algae blooms. There are many non-parasitic species (thank goodness!) that serve as food for zooplankters, suspension and filter-feeders at large. As a fundamental food for higher order micro-organisms, greenwater is vital. Some cnidarians like Neptheids and Gorgonians, we believe, also depend heavily on phytoplankton as a measurable portion of their diet. A variety of other reef invertebrates like bivalves, sponges and tunicates perhaps with varying dependencies will also benefit from "phyto" supplementation in the aquarium. Please note, however, that for the sole purpose of target feeding average reef corals, phytoplankton is not likely to be a satisfactory staple in light of the preferences of corals to feed on zooplankters. This mention is an admonition to aquarists with huge swimming pools full of greenwater cultures, or those with stock in the bottled phyto-supplement products: too much of any good thing is bad! In fact, there is another name for excess phytoplankton supplements in the carnivorous reef aquarium- *fertilizer*! Algae goes to algae- rotting excess phyto is simply food for another nuisance organism to grow from. Aquarists with pest algae growths in the aquarium should be very judicious when dispensing phytoplankton and any organic foods or supplements.

Herbivorous grazing fishes like the blenny hiding in this image are some of the few macro-organisms permissable in food-culture refugia. Most others are predatory on plankton, like coral in fact, and are not ideal when plankton culture is a priority. (*A. Calfo*)

The protocol for culturing live phytoplankton is well-documented and very straightforward, although strict. *Nanochloropsis* and *Isochrysis* species have been quite popular commercially but numerous viable genera exist. Some aquarists regard the effort to grow and maintain greenwater tedious, while others find it simple enough. For those that fall in the former category, many live, semi-live, liquid and concentrated paste products are available commercially to spare you much or all of the work of building and operating a live phyto refugium or food station. Most aquarists will be satisfied and well served to experiment intitially with these products. More ambitious hobbyists can search for live cultures, kits and information from aquaculture supply houses like the pioneering organization, Florida Aqua Farms and the reference, "Plankton Culture Manual" by Frank H. Hoff and Terry W. Snell.

Summary of plankton at large...

The term "plankton" in gross terms essentially refers to any and all animals, algae, and other organisms living in the water column whose locomotion is insufficient to determine their motion; in other words, all life forms that are carried about on the currents by more than their own volition. This is an enormous and important source of food for autotrophic (photosynthetic), filter-feeding, and predaceous species alike. Do understand though, that their motion is not only guided adrift by latitudinal movements (with currents), but many of these life forms travel great distances vertically... some moving several hundred feet daily toward and away from the surface!

As alluded to previously, organisms that can move faster than the currents are called *nekton*. A good deal of nektonic life starts as plankton, such as most embryonic and larval fishes, crustaceans, echinoderms, mollusks, worms, etc., floating about in tides and currents as they develop, which leads to a wider distribution of the species (dispersion) at the risk of predation. Zooplankton does occur in aquariums commonly and passively. With the advent of refugium methodologies, natural plankton occurrence is surging fantastically. Zooplankters are particularly evident during the nocturnal cycles of our aquariums. At night, various micro-crustaceans, shrimp, worms and much more rise up from sand, rock, and from amid living substrates wherein they had been hiding all day away from hungry planktivorous fishes. Phytoplankton occurs too, although it does so more sporadically and less passively (it requires more deliberate cultivation than zooplankters). Most often it appears as single-celled nuisance algae blooms (AKA "greenwater") due to lack of filtration and excess nutrients. In such cases, the best long-term prevention is the use of protein skimmers or other means of aggressive nutrient export. Ozone or UV sterilization of the water will also quickly remedy an outbreak, although they only treat the symptom and not the problem (again, excess nutrients).

Alas, in the wild, planktonic flora and fauna are far more varied, constant and numerous. However, even the vacillating, depauperate motley crew of what we call "aquarium plankton" is of great use and consequence. Micro- and macroscopic planktonic life forms, as adults and larval forms, serve as food for all manner of life in our systems. Improved aquarium husbandry and refugium technologies are stimulating an ever-increasing number of organisms (worms, snails, corals, fishes, etc) to breed with regularity in the home aquarium and provide invaluable and nutritious plankton in the process. Surveying the shallow-water marine environments, one is taken with how much of the life depends on plankton for vital sustenance. In all aquariums, let us aspire to cultivate natural plankters that will better sustain our cherished captive charges.

Deep sand bed (DSB) Refugia and Natural Nitrate Reduction (NNR): The essential mechanics for accomplishing nitrate control with sand at depth, with its merits and limitations of various grain sizes and amounts, were detailed in the chapter on "Live Sand." Here we wish to remind you that NNR is not restricted to large, dense rocks in the display or enormous beds of sand. What's that you say? You don't need to have

Macroalge in the display can quickly grow out of control and requires due maintenance (pruning manually or by natural predation) if it is not a featured organism. (L. Gonzalez)

A conspicuous upstream refugium is one of the simplest and most effective styles of refugia. This newly adapted and empty refugium may be stocked with ornamental species like mangroves, or be seascaped with a biotopic micro-habitat to house special display-species that are unsuitable for the main aquarium. (*S. Boyer*)

a small beach lifted and transported to your house for nitrate control in your display? Why no, not at all! Very efficient nitrate control can be maintained in DSB refugiums with a variety of options for complimentary culture. The size of the refugium that you need to accomplish this depends entirely on your particular system's propensity to accumulate nitrogenous matter. With that said, we will offer a very basic guideline that a DSB refugium should be at least 20% of the displays volume in size while closer to 40% would be ideal. Sand is to be maintained at more than 4" depth (10 cm) with 6" or more (>15 cm) ideally.

The simplest refugium for NNR are filled with deep sugar-fine sand and kept unlit. Here, the primary purpose is nitrate reduction. The absence of (dedicated) light reduces the ability for autotrophic nuisance algae to gain a foothold. It also spares the need for much or any support from detritivores (sand-stirring creatures such as hermit crabs, sea cucumbers, starfish and shrimp). Good water flow and occasional sand stirring (manually) will work well if detritivores are not employed. Indeed, one of the primary reasons for including detritivores in illuminated displays is to reduce the growth of diatoms and nuisance algae on the surface of the illuminated sand.

The heavy microbial activity of the unlit sand bed, however, will still compete heavily and aggressively in the system for nutrients like nitrate, which could otherwise have an undesireable impact on water quality and the health of livestock. A deep sand bed can also sequester precipitated phosphates, which are therefore effectively removed from participating in the system as bound matter if not wholly used as a nutrient by the biological faculties flourishing on and in the substrate.

A deep sand bed will also facilitate the development of numerous polychaete worms and micro-crustaceans that become priceless natural plankton. To exploit this medium for plankton culture to the fullest extent, this vessel, like most refugiums, will need to be

Nearly Unseen...

Various micro-crustaceans like amphipods, copepods and mysids are welcome additions to your refugium. They are quite small and often nearly unseen. They also provide an excellent and crucial food source for fish like mandarin dragonets, wrasses and just about any other denizen of your reef aquarium system. (*C. Gonzalez*)

kept fishless and without any other carnivorous predators (Arthropods, Cnidarians, etc). Coarse sand is better suited to fostering the development of macroplankton but may trap detritus more easily without regular sand stirring by detritivores or manually with a tool (at least occasionally). When nitrate reduction is the primary goal, sugar-fine sand (again) is recommended instead. And as previously mentioned, course grades require greater depth to accomplish NNR.

Mud-based substrates in refugia:
Methodologies for employing "mud" and soil-like products in refugium substrates has been a very interesting and progressive area of development in the marine hobby. The application has many potential benefits and limitations as well. Newer aquarists will benefit from the evolution of the methodology into widely varied styles for the natural marine aquarium. Tenured aquarists will remember the early years of this strategy when it was strictly associated with *Caulerpa* species used to exclusion on a 24/7 photoperiod (lights always-on) in the vessel. Some extraordinary, although perhaps dubious, claims have been made as well by the proponents of "mud" systems. Marketing of the application has literally stated that there is no need for carbon, protein skimming or mechanical aeration with a mud-filled refugium. There is some truth to these claims, but then again, the confluence of all strategies (mud refugiums *with* skimmers, carbon use and mechanical aeration) is actually an even better mode of husbandry with a synergistic effect on water quality. Indeed, with decades of proven support, it would be hard to argue that skimmers, carbon use and extra agitation of seawater are harmful: quite the contrary! Thus, newer aquarists are especially encouraged to resist the hype of avoiding carbon and protein skimming, but advised instead to employ such methods to enhance water quality with mud refugia until

(A. Calfo)

skilled finesse of these parameters is learned in time.

Other questionable claims about mud system merit fall very, very short of unadulterated marketing propaganda, with statements like a tank will never need a water change again (or more recently with concession, that the need is dramatically "reduced"). Such claims, quite frankly, are potentially dangerous when aquarists do not appreciate the dynamics required to support such a style of husbandry. It seems very unlikely and even impossible that mud refugium methodology has reckoned how to successfully manage the neutralization or export of all deleterious elements for aquatic life while pooling and preserving the life-supporting ones. There are even more confusing claims still about mud substrates like, "prevents HLLE disease in fishes," and "no more balancing of chemicals." Claims that the calcium supplementation of reef aquaria is not necessary with non- or weakly calcareous mud products are simply bizarre. Without scientific data to support incredulous claims, we recommend that you simply embrace the many possible benefits of mud substrates in refugia with little regard for advertising claims.

Some of the benefits of using mud (with or without fine sand) in refugium substrates include:

- unique nutrients provided by mud are not readily furnished by other marine substrates

- medium of mud supports the production of unique plankton not readily cultured in other media

- can be physically and nutritively beneficial for plants and animals in the system

- supports various plants and algae for display and growth as vegetable filters for vigorous nutrient export

Some of the concerns with using mud in refugium substrates include:

- compositional nutrients are not clearly defined, easily monitored or standardized... thus success for casual aquarists with the methodology is difficult to promise or predict (relative to claims made about the process)

- Extra attention paid to water clarity may be necessary via carbon, ozone, water changes, skimmers, etc.

- expense of commercially prepared mud can be dear and hard to reckon without clearly defined benefits

- it should not exclude or be exchanged wholly for aragonite material in the system. Both have useful benefits and limitations

- unused nutrients imparted by the substrate can fuel the growth of nuisance organisms

Ultimately, mud-based substrates have much to offer in refugium methodologies. Any limitations that we currently recognize are generally easy to reckon by good aquarium husbandry. It is a methodology well worth judicious experimentation. Aquarists interested in true marine plants like seagrasses and mangroves are especially encouraged to experiment here. Mud-based substrates enhance vegetable filter style refugiums and cultivars in reverse daylight photoperiod applications.

Ornamental & alternative refugia

Refugium methodologies are still a highly experimental aspect of aquariology. Even the styles already listed above are only a beginning realization of the many benefits and potentials of these ancillary vessels. Casual aquarists can be satisfied to employ these. Progressive aquarists may wish to consider some of the below-listed ideas and should certainly cultivate and share new refugium concepts of their own.

Batch treatments in refugia: While most refugia will be installed as flow-through components of a system, a batch-fed refugium can have some very unique benefits. The premise is simply that water fed to such an aquarium is metered by a timer and not fed continuously as flow-through. Thus, for a predetermined period (a setting on the timer), water is pumped from the system into the refugium, which displaces "aged" water that has been sitting since the last cycle for return to the system. For intertidal simulation, an undersized drain can be fitted to slowly empty the vessel in step with the tides and cause exposure of some rocks and organisms (calculation and experimentation is required to accomplish this, of course). Otherwise, the refugium is simply fitted with a properly sized high-water overflow bulkhead. The most common purpose of batch-treatments is denitrification. The stale intermittent period affords denitrifying faculties the time and matter for nitrate reduction. A marine aquarist here can replicate a mangrove niche, an intertidal zone, or some other marine microcosm where oxygenated water is not assured. In this anoxic environment, biotic faculties not readily available in the oxygen-rich environment of the display can convert various organic matters. The replication of a muddy lagoon or deep sand environment with mangroves works very well here and makes a fantastic natural and functional display. If nitrate reduction is not desired, perhaps stronger populations of certain zooplankton can be cultivated by the intermittent overflow of water (without the fear of excessive overflow of brood population by flow-through applications). At any rate, batch treatments in refugia can be an interesting and challenging vehicle

A unique downstream refugium catches all raw overflown water from the display above into a lagoon-style vessel. Sealed in the back of this forward refugium is a glass dam and hidden sump. (*A. Calfo*)

for creating aspects of the marine ecosystem not commonly replicated to date in traditional installations.

Intertidal refugia: Without batch treatments, an intertidal display can still be created with a flow-through application by having aspects of the installation mimic a tidal environment. Incoming water can be delivered across exposed shore-like rocks and other hard substrates for the purpose of culturing immersed tidal species that are not inclined to grow in fully submerged environments. Different algae and motile organisms of the waters edge can be kept here. Some species like fiddler and mangrove crabs are fascinating "air breathers" that will require such an intertidal environment. Many beautiful hermit crab species also cannot live fully submerged but would be very interesting in an intertidal collection. Gobiid Mudskippers would of course be a very logical choice here too as denizens of brackish, mangrove and mudflat regions. The creative freedom to create a tide pool with sandy puddles, rock waterfalls, and numerous cracks and crevices alive is very exciting even if only from an aesthetic perspective. One might even be inclined to meter and dose all evaporation top-off freshwater into this vessel to further emulate the interface between land and sea. If you have never had the pleasure of searching through tide pools and various intertidal regions, you simply don't know the wondrous life forms that you've been missing. Aspire to know it, and consider replicating it.

Multi-tank systems: It's also a pleasure to say that there are no rules with refugium applications. You are limited only by your imagination. Multi-tiered systems can display a remarkable collection of otherwise incompatible species- each in refuge from another. For example, an aquarist that is enamored by large angelfish, plants & algae, and mandarin fish is unlikely to succeed in keeping all three in the same 200-gallon aquarium. But, a 100-gallon display with an angel that supplies nutrient-rich water to a 50 gallon *Chaetomorpha* vegetable filter, which overflows to a 50-gallon lagoon-style haven for the mandarin fish, can work very well. The microalgae grows largely from the heavy nutrients by the angel... the matrix of algae provides habitat for zooplankton which overflows to feed the mandarin... and the mandarin lives without aggressive competition from the angel (if not fear for its life!). This is but one of numerous scenarios of refugiums in series and multi-tank systems. Another popular expression of this methodology is the replication of specific niches on a reef (back, fore, crests of the reef, etc) or even the evolution of fauna with a progression down the reef slope with each tank representing, say... a 20 foot increment downwards. Again, you are only limited by your imagination in alternative refugiums. Of course, intuition and common sense help here too.

Twilight & cryptic fauna: One of the most exciting developments in recent years for marine aquarium enthusiasts has been the realization that some "delicate" twilight and cryptic fauna can in fact be kept successfully in captivity. Refugium methodologies have been instrumental in making this happen. There are many fascinating fishes and invertebrates that occur at depth or merely in the dark crevices and caves of the shallows that simply do not fair well in direct light or in unnatural competition with autotrophic "nuisance" organisms like microalgae. Indeed, many delightful heterotrophic animals (organisms that seek food but cannot produce it like photosynthetic species) are not inhibited by light at all. For many, they simply have not evolved mechanisms to easily defend biologically against the encroachment of autotrophic species. For a filter-feeder that only occurs at depths or niches where microalgae rarely if ever occurs, a battle with nuisance algae in all-too-common nutrient-rich aquariums has been a tremendous obstacle. Nowadays, we readily embrace the in-line installation of a dimly lit refugium that affords the benefits of water quality from the system while being spared competition from the autotrophs (corals, algae, etc). A small twilight refugium also facilitates the practical advantage of concentrated target feeding for such heterotrophs. Otherwise, the keeping of hungry filter-feeders and aposymbiotic corals in displays dominated by zooxanthellate species usually means that at least one group will be underfed or overfed. Maintaining good water quality for such large mixed-group displays is generally an extraordinary challenge. The cryptic refugium, however, lets one focus on organismal feeders in an isolated area to prevent broadcast slurry of polluting food in the reef display proper. The twilight refugium is a perfect place for keeping the wondrous and photogenic *Tubastrea* corals, for example. However, this refugium style is not only for organismal feeders, or those that require target feeding. Filter-feeders on nanoplankton and dissolved organic matter can also reside peacefully here

Many species of Sabellid fanworms will flourish in protected refugium vessels. (A. Calfo)

It doesn't have to be pretty... just practical! Refugiums are fertile grounds for aquarists with DIY imaginations. (L. Gonzalez)

provided with ample nutrition from the influx of system water. Some of these "shadowy" creatures may be deliberately collected and placed herein, others will develop naturally. Many aquarists have enjoyed this dynamic in the recesses of dark sumps and trickle filters where fantastic cryptic fauna sometimes develop. The care and study of tunicates and sponges will be greatly supported by refugiums styles. We cannot forget the shy twilight fishes either. There are more than few magnificent species of fish collected in the depths and shadows that will never acclimate to a brightly lit aquarium. Improvements in the safe collection and delivery of deepwater fishes and invertebrates at large will fuel the development of twilight and cryptic refugiums for the natural marine aquarium.

Species-specific refugia: At last, refugia can serve in the most practical and distilled version of their namesake as a place of shelter for species in need. Aquarists are sure to desire some organisms that simply cannot be kept in the display proper for any one of a number of reasons. There may be issues of aggression, competition, feeding preferences, habitat requirements, or other parameters that are best satisfied by a species-specific set-up. For this, a refugium can afford a place to include special animals in a collection without having to establish a separate and dedicated system. One likely situation for such a refugium is the keeping of a tiny fish or shrimp that would get lost in the display, rarely to be seen, even if it could otherwise survive there. A refugium isolates these minute guests nicely for dedicated care, study and observation. Perhaps you've discovered a wonderful pair of clown shrimp, instead, that eats the easily cultivated and prolific *Asterina* starfish. These *Hymenocera* shrimp could wreak havoc on the desirable echinoderm population in the display proper, but can be appropriately isolated, fed and enjoyed in the confines of a species-specific refugium. Another common, albeit less-than-ideal, application of this type of refugium is in employ as a makeshift quarantine for new or stressed animals. When a fish or invertebrate needs to be isolated, a separate quarantine tank is the first course of action. If a separate QT vessel cannot be procured, however, isolation in a refugium is better than putting any such animal into the main collection. Although the risk of introducing a parasite, pest or disease is still a danger to the system, the animal in duress is spared extra stress from immediate and direct competition from the established denizens. In a worst-case scenario, the aquarist is also spared the risk of needing to find a dead or dying animal in an inaccessible region of the display rockscape. There are numerous other and apparent benefits to species-specific refugia. Have an open mind about the many possibilities of such ancillary vessels.

Lighting and Water Flow

Any discussion about water circulation in refugia needs to specifically

Natural Marine Aquarium Volume I - Reef Invertebrates

address the needs of the organisms in residence. It is not fair or possible to proffer advice that will serve all applications. Here, we can serve you better by dispelling some popular myth-information regarding light and water flow in refugia.

Lighting is a fairly straightforward matter with refugiums. Photosynthetic organisms have specific needs that most often are easily discovered. One of the most common problems with keeping plants and algae in refugia is inadequate light (wattage or installation of lamps). Some of the most popular algae for vegetable filters occur in very shallow tropical waters. The amount of natural light required by these species is extraordinary and in some cases very difficult to replicate captively with fluorescent lamps. Does this mean that metal halide lamps are necessary for your refugium? Likely not. But compromises will need to be made. Even with higher intensity fluorescents (VHO, PC, T5), lamps will need to be kept close to the water (not more than 3" off the surface) and water clarity (yellowing agents) will need to be quite clear by carbon or ozone, for example. The depth of plant or algae-filled aquariums should also be shallow (ideally less than 18"/45 cm but certainly less than 24"/60 cm) to facilitate faster growth. In this manner, some of the best algae for vegetable filters like *Gracilaria* and *Chaetomorpha* can be cultivated in a brisk and proper manner for nutrient export. One will need to be very mindful, however, of good aquarium husbandry in a brightly lighted refugium. A slow flow, high nutrient, brightly lit environment is a recipe for nuisance algae growth. Due considerations will need to be made here regarding the finesse of lamps and photoperiods on refugia.

There are, of course, variations of refugiums that impose special circumstances with regards for lighting. Cryptic and twilight creatures will favor wavelengths on the blue side of the spectrum. Red lights may be employed to observe some shadow dwellers that do seem to be bothered by it (but will shy from white daylight). Very shallow water organisms on the contrary may flourish under very warm colored lamps emphasizing the red, orange and yellow wavelengths of the spectrum (like terrestrial "plant" type bulbs from the hardware store). Do exercise some caution here as warm light tends to favor undesirable algae (diatoms and brown algae commonly).

At length, all lighting over photosynthetic creatures needs to be of a stimulating and useful composition. For zooxanthellate corals and reef invertebrates, most fluorescent lamps are only good for 6-10 months before the aged bulb strays in useful spectrum. Interestingly, these lamps tend to migrate to the warm end of the spectrum and may be useful to you still over refugia if you keep shallow water algae. Else, light your refugium with the same care and consideration as the reef proper. For economy and a longer view, consider that some metal halide lamps often last 2-3 years and put out more light per watt than most other bulbs.

Water circulation in any aquarium is one of the most underestimated physical parameters of successful aquarium keeping. Much attention is given to lights, feeding and filtration but usually, water circulation is overlooked for its significant influence on coral and invertebrate growth, nutrient export, the limitation of algae, and water quality overall. It is equally important in refugia and, like lighting, is usually misapplied. So often, aquarists are advised that refugia are vessels in need of very slow water flow. Alas, this is often not true and too often misinterpreted as nearly stagnant! We can agree that most refugiums will require less water flow than the reef display. However, the correct amount will depend on the needs of the organisms being kept and cultured in refugia. Informed aquarists will realize that a slow flow vessel being fed raw aquarium water will act as a settling chamber. Settling chambers can be a boon or scourge depending on how efficiently sediments are processed. If your refugium has high light, slow flow and the accumulation of sediments in a stagnant pool... you have a recipe for nuisance organisms to grow, if not bigger problems. We must be sensible

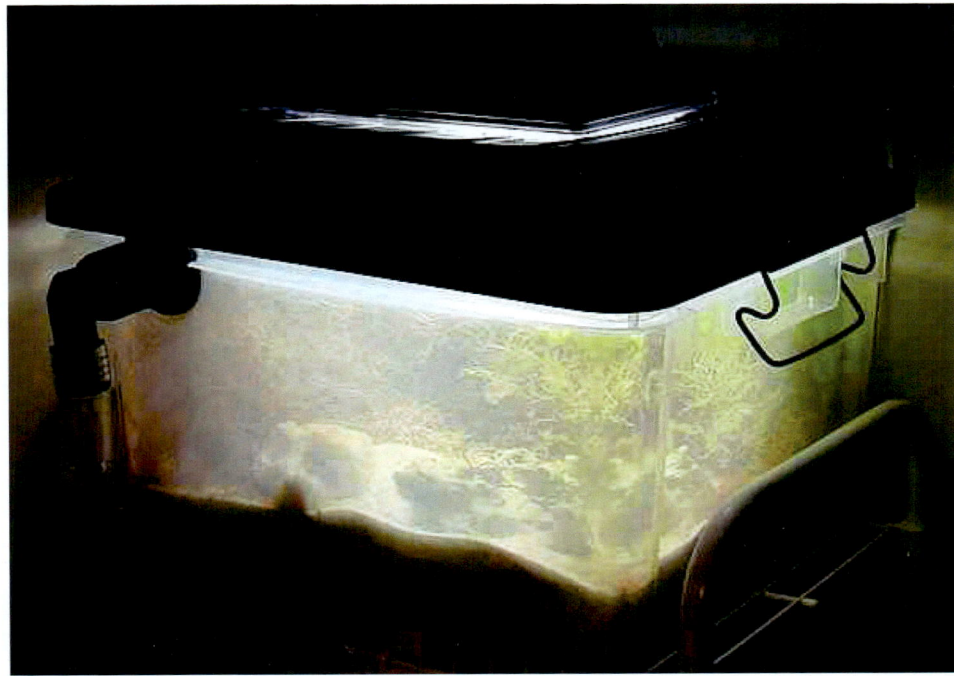

Refugiums have been associated with weaker light and water flow than the main display too often. The true application of these parameters is to be determined by the target species kept in refugia. (*S. Attix*)

about water flow with regard for the fundamentals of aquarium keeping.

We recommend moderate to strong but diffused water flow for most refugia. Tremendous amounts of water flow can be significantly tempered by the application of a spray bar or manifold. A vertical or horizontal spray bar is fairly self-explanatory. The velocity of incoming water is diffused through a pipe that has been drilled with many holes for a gentle dissipation. A manifold can be similarly constructed by tapping a solid pipe with an assortment of directional tees. In this manner, a large volume of water can be run through the refugium with little physical disruption from an otherwise dynamic laminar nozzle. Aquarists keeping the popular *Gracilaria* species and other free-floaters will find this method quite necessary. *Gracilaria* is one of the finest algae for vegetable filters and has the potential to be a very fast-growing and effective vehicle for nutrient export. It requires very bright light and a brisk water flow to keep the free-floater tumbling in suspension. Many aquarists fail to receive or appreciate the importance of these requirements, however, with these delightfully useful algae.

When keeping animals like corals and anemones in refugia, an even stronger flow may be required. For refugium creatures that desire stronger water flow, converging effluents that produce random turbulent flow will be simple and effective. Surge-style flow is generally difficult to produce in smaller aquaria, but a larger refugium with seagrasses will be very well served by a surge application. This water flow stimulates and "cleanses" the blades of this plant while liberating metabolites and epiphytic matter that is quite nutritious to filter-feeders in the system.

There is no rule of thumb, unfortunately, for water flow in refugia. It should be dictated by the needs of the captives and target organisms. For a very gross frame of reference, most refugia will require a turnover of at least five to ten times the tank volume per hour. Many refugiums with *Chaatomorpha*, *Gracilaria* or turf algae, for example, can take the full volume of water flow from the system's recirculating pump if it is diffused (spray bar or teed manifold). Refugiums designed to culture fine zooplankton, on the contrary, will require decidedly slower flow.

Feeding a refugium

Perhaps it would be a better headline for this small section to say, "*Finessing* food in a refugium." Most refugium applications are net *exporters* of nutrients; they are repositories or catch basins for dissolved and particulate matter to be consumed by target organisms. True plants and some animals used as living filters consume organics from raw system water and fix it in their stable growing masses (like corals and mangroves). Other refugium guests grow quickly, like algae, and it is intended that they should be harvested regularly to export their less-stable fixed resources. The food required by a refugium will depend, of course (sounding like a broken record by this point!), on the needs of the organisms in residence and the incidental matter imported by the feed of raw water from the system. One can imagine that a reef aquarium with a very large population of fishes that are fed daily and heavily will be a tremendous source of dissolved and particulate organics for a refugium. Depending on the size and nature of the refugium population, it may even be too much to handle (requiring extra water changes, an additional protein skimmer, etc). On the contrary, a thick and healthy seagrass refugium may require a solid mix of nutritive sediment and mud in the substrate and perhaps supplemental fertilization just to maintain vigor. In the case of a refugium dedicated primarily to the cultivation of macroplankton (microcrustaceans specifically), it may be necessary to literally feed dry prepared or thawed

A simple, "hang-on" refugium packed with sand, live-rock rubble, and macroalgaes. (*C. Gonzalez*)

frozen meaty foods. Identify the needs of your aquarium life forms and then evaluate how best to serve them.

A fox in the henhouse… keeping refugiums "safe": Many hobbyists are still trying to develop a clearer picture of just what exactly is occurring in the broadest scope of a refugium dynamic. At the most basic level, a refugium by definition must give refuge to something. Most refugiums are intended to produce "food" for the system in the form of plankters. However, if you put fishes or corals or other organismal feeding invertebrates (shrimp, crabs, anemones, etc) into a refugium, the refugium then ceases to be just that. At best, your vessel is a refugium for planktivorous predators in the form of the hungry fishes and invertebrates you just put in there! Thus, macro-organisms have no place in refugia when they are not the primary organisms being cultured. This is an area that passionate aquarists show weakness in; they see a new fish or invertebrate that they want but know they have no compatible place for it in the display. At that point, the refugium ceases to exist and becomes "my second fish tank." Corals are not to be underestimated as predators either. Anemones even more so. Most of these popular creatures feed on a wide variety of plankton. For a successful refugium, you must address your targeted guests and protect them with sensible stocking and husbandry.

Co-culture of life forms and species succession: In much the same way that hobbyists are tempted to mix predators with prey in refugia, the temptation to mix refugium cultivars is strong. Most refugia for home aquaria are already undersized. To then attempt to mix two or more species of plant, algae or animal in a small ancillary aquarium is unrealistic and challenging to success if not an invitation for failure. Does this mean that all refugiums should be monospecific? No… but many indeed

The science experiment behind the facade. This is a peek at the refugium support of a beautiful mini-reef aquarium by Ken Uy. (G. Rothschild)

would perform better for it. Refugia designed for nutrient export (control of nuisance algae, nitrate reduction, etc) are definitely best suited for monospecific culture. With only one species, the target organism for your animal or vegetable filter is relieved of wasting resources (shared or defensive) in competition. Its specific needs can be addressed and maximum growth can be enjoyed in this manner. When co-culture is possible, make a concerted effort not to mix directly competitive species.

At any rate, despite our best efforts, the natural order will impose its will through species succession. We find that some organisms mixed unnaturally close will simply dwindle in ranks while others flourish. We see this in reef aquariums crowded with corals especially after a year or two. You may wish to avoid the cruel beauty of species succession by deliberate and conservative stocking of refugia.

Summary

We would like to strongly encourage you to consider any of the many possibilities with refugium methodologies for the modern marine aquarium. We have no doubt that in time, refugia will be as indispensable as live rock and protein skimmers are for the casual aquarist. Just as the introduction of live rock dramatically improved the quality of life and longevity for our aquarium guests, refugiums are exponentially expanding the diversity and vitality of reef life in captivity.

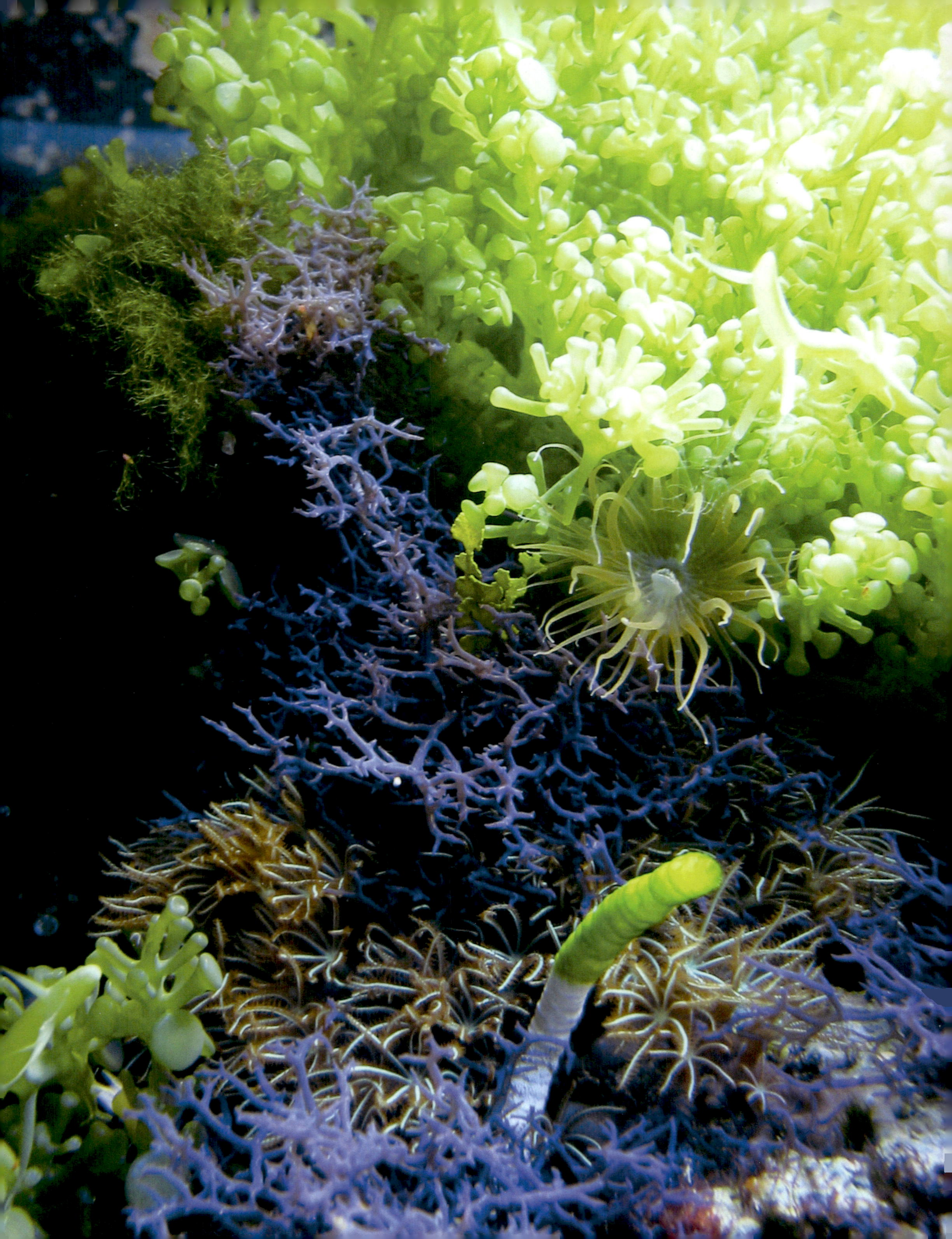

Marine Plants and Algae

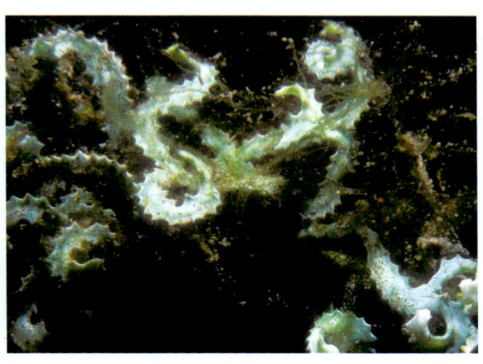

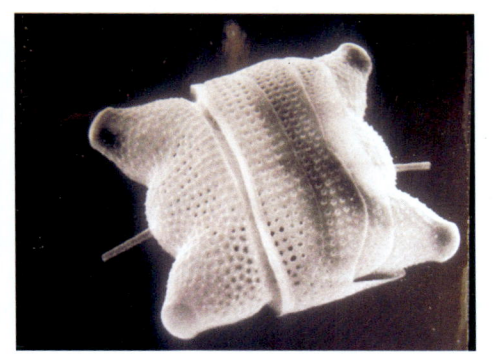

Above left the red algae *Meristiella* (turns red in shade, or creamy yellow in full sun), at center Sawblade *Caulerpa serrulata* (green algae), and at right a centrate diatom under an electron microscope.

Algae are the fundamental food and nutritive foundation of marine ecosystems. True vascular plants are also tremendously pivotal where they occur in shallow water reef environments. These plants, however, have not been as readily incorporated into captive reef systems until recently with the exploration of refugiums, multi-tank displays and other ancillary vessels. Any discussion of a natural marine aquarium must address at least some of the many instrumental types of algae and true plants in reef ecosystems.

For the purpose of this reference, we will spare you a dissertation on the scientific aspects of algae classes, divisions and other taxonomic issues. You may also be relieved to know that we will not be offering any painfully detailed illustrations of cell structure, function or similar biology except when specifically practical and pertinent to aquarium husbandry. Instead, we wish to provide lucid and useful information for aquarists that can be readily applied in the everyday enjoyment of the hobby. If you read it, enjoy it and easily understand it, then we have done our jobs. We believe that fellow hobbyists will appreciate and benefit from this approach while the more academic-minded aquarists can explore the bibliography and beyond to discover why blue-green algae lack *plasmids* or how many freckles a mangrove tree has.

Putting Algae *IN* Your Aquarium?

From an aesthetic perspective, it's easy to see why people are not especially partial to algae in the aquarium. Slimy, stringy, splotchy, toxic and pathogenic, very few algae occur in seasonably fashionable colors, and some have bad habits that run the gamut. A cursory glance at some of the best-selling products in the aquarium trade is a clear indication of their "popularity." Algae scrapers, algaecides and algae-grazing livestock abound. One must understand, however, that algae are essential for good aquarium health, and life at large.

Prior to World War I, the study of marine plants was little more than an academic curiosity. However, when the Germans cut off the supply of potash (as a source of elemental potassium for making fertilizer and gun powder), during World War I, and the Japanese later cut off the supply of agar (extracted from *Gelidium*, a red algae), during World War II, research in earnest on the processing and industrialization of seaweeds was expanded greatly. The roles these algae and plants play in the great circle of life extend far beyond simple aquarium ornamentation. They are fundamental in food, fodder, fertilizer, medicinal, pharmaceutical, and industrial applications. Oxygen and nitrogen production as well as the fixing of carbon make algae the foremost primary producers in the sea. In laymen's terms, they produce food and they *are* food. Most are photosynthetic, some are heterotrophic, and some are phagotrophic (Yikes! Algae that eat things!? Well, tiny particles at least). Most algae are unicellular, though some are multicellular, including kelp species that can be very large. The majority of the unicellular varieties are microscopic like "greenwater" phytoplankton. Others, such as *Caulerpa* fronds, are huge single cells. The multicellular varieties range from millimeter sized scraps to the massive kelp forests as tall as buildings! By any measure, algae are vital components of the wild and captive reef.

Doesn't it seem logical then that an aquarist should strive for a realistic "slice" of the ocean environment in the captive reef? Well, maybe not. Suppose an aquarist were to have a huge private aquarium of around 1,000 gallons that would be set up to resemble the wild seas. Most of the space would be taken up by water, with very little live rock or coral, and almost no macrolife. It is difficult to try to conceive of the vastness and enormity of the marine

Left: A diverse assemblage of life: single-celled algae, multicellular algae, coral, and anemone... all in a beautiful but sometimes fierce battle for space in a hobbyist's refugium. (*L. Gonzalez*)

environment from the confines of one's living room. Instead, most "captive seas" are purposely overcrowded, and thus overfed and infused with nutrients. As such, aquariums are inherently challenging to keep balanced in the aquarist's favor, and support from even seemingly innocuous and unassuming microalgae can be a tremendous benefit. Higher forms are not difficult to keep or culture, and there is no need for a tremendously sophisticated filtration system or university degree in botany to succeed with them.

For aquarium use, there are three purposeful groups of algae: Brown (Phaeophytes), Green (Chlorophytes) and Red (Rhodophytes). Of further interest and concern to aquarists are the Diatoms, Dinoflagellates, Cyanobacteria and True Vascular Plants. Members of all of these groups are covered here with regard for their mettle and merit in the aquarium.

Aquarists generally make distinctions between "microalgae" and "macroalgae," although these terms have no taxonomic base. In gross terms, macroalgae are plantlike and usually "desirable" while microalgae tend to be microscopic, and often regarded as aesthetically or functionally detractive to the system. Either type can be a boon or scourge in the aquarium under various circumstances. Some algae are less than attractive, grow to excess, compete with desirable organisms in the display for space and nutrients, burden water quality, and can literally be toxic or pathogenic to other display organisms. These "nuisance varieties" get most of the attention in popular literature.

Avoid direct contact between invasive algae and sessile invertebrates. There can be long-term complications. Instead, utilize refugium methodologies for a more vigorous culture of plants and algae not typically found on a reef. (*L. Gonzalez*)

Know that successful aquariums often have any number of potentially unfavorable algae lingering in spots. It is quite natural and acceptable; good aquarium husbandry will prevent any such risk from becoming a liability. Rest assured that nuisance algae are imported with live rock or livestock, and algae spores in shipping water. Have faith that they will not flourish if proper water quality is maintained. This is a reality with almost any (excess) nutrient-driven organism in an aquarium. As an example, an aquarist can buy a new coral or piece of live rock covered with pest anemones (*Aiptasia* or *Anemonia* sp.) or nasty brown diatom algae without fear if their aquarium is well-designed and operated. These nuisances will only flourish in the presence of excess nutrients and a lack of competition or predation. Conversely, weak water flow, lingering particles of excess food, poor protein skimming, etc., could easily lead to a pest anemone farm covered in brown algae. Most nuisance algae are easily limited by nutrients and can be stalled or starved into submission within weeks given proper aquarium husbandry.

On a brighter note, even nuisance species of algae can be helpful in moderation. They are the simplest oxygen-producing life forms on earth and they translate their duties

---

A summary of the potential benefits of algae in the aquarium:

• Most all fishes and invertebrates augment their diets with algal material or its by-products (epiphytes, metabolites and other things that rhyme with "-ites")

• They serve very well as bio-indicators, showing signs of degrading water quality before many other livestock by their vigor or duress as the case may be

• Algae stabilize the captive environment by introducing beneficial microbes, absorbing nutrients, and producing matter that mediates systems biologically

• Some macroalgae can control undesirable microalgae by utilizing nutrients and light that would otherwise be available for nuisance growth forms, and producing chemicals that limit other algal metabolism

• Algae work well as both habitat and ornamentation; they provide hiding spaces, natural beauty, and culturing mediums for organisms such as plankton

• They can be harnessed to have desirable effects on water chemistry as with pH stabilization in Reverse Daylight Photosynthesis (RDP) systems from the dynamics of photosynthesis and respiration

in aquaria readily. Most algae are decidedly autotrophic (self-feeding) and have no complex organization in the form of leaves, roots, stems, or a vascular network. They do, however, contain chlorophyll and other photosynthetic pigments. Beyond their key role in the food chain, their presence provides a medium for the culture of many other desirable organisms such as microcrustaceans and various epiphytic matter. Some algae also shed reproductive units, such as gametes and spores in the form of phytoplankton, and nutritive elements such as sugars and fixed nitrogen/proteins for use by other reef denizens. For successful marine aquarium keeping, one must discover which algae are most useful and which are too problematic to collect or encourage.

A summary of the main groups of algae

Brown Algae (Phaeophytes): They are macroscopic for the most part with no unicellular or colonial forms. They have no roots, leaves or flowers, however they look very much like true (vascular) plants. The whole body of a brown alga is termed a **thallus,** meaning "all about the same body," and the erect stem-like stalk is specifically called the **stipe**. The portion that attaches the thallus to the substrate is simply called a **holdfast**. Though they may resemble the roots of terrestrial plants, holdfasts do not absorb and transport nutrients to the rest of the thallus. Almost all brown algae are attached, with a notable exception of the classic genus *Sargassum*, which also occurs free-floating and makes up the bulk of the magnificent floating rafts teeming with life that make up the Sargasso Sea. There are a few other distinctions of note to aquarists regarding Phaeophytes. Although brown algae contain the requisite photosynthetic pigments, chlorophyll A & C, they are characteristically colored brown due to the accessory carotenoid pigment fucoxanthin. Unlike green algae, they do not produce starch and most are limited to marine environments. These brown algae also endure handling quite well, in part due to the production of a jelly-like compound named algin that they use to retain moisture and contend with tidal exposure. Many species of brown algae are intertidal and have evolved to be quite robust to endure this changeable environment. Some varieties form little air sacs known as pneumatocysts to aid in the buoyancy of their necessarily dense structure. Specimens in the aquarium trade are

It might surprise you to learn that this aquarium image depicts one green, and two red algae species! At left, *Caulerpa*... and the Red algae *Ochtodes* (center) and *Botryocladia* (at right). (*L. Gonzalez*)

Living Filters: Algae & Plants

A diverse grouping of algae in this Red Sea photo by Diana Fenner.

generally collected along both coasts of the United States in cooler waters. They are present in tropical waters as well, although to a lesser degree than red and green algae. Brown algae are generally beneficial in the marine aquarium. The brown water scum and spots commonly thought to be brown algae are actually colonies of diatoms. There are few nuisance brown algae species and these are easily controlled when they occur.

Green Algae (Chlorophytes): These are the most commonly observed algae by species and number in the tropical marine environment. Taxonomists recognize over 7,000 species occurring in a tremendous range of forms and functions. Calcareous species contribute significant measures of calcium carbonate to the reef ecosystem; in some areas they are responsible for much of the gorgeous white coral sand that is seen. Most aquarists are familiar with huge single-celled species such as *Caulerpa* fronds, as well as the microscopic single cells that make "greenwater" (phytoplankton) if found in high concentration. Many aquarists cherish decorative macroalgae yet chastise the ubiquitous turf algae. Part of the challenge of nuisance green algae species is their remarkable adaptability to a wide range of physical parameters in their environment. Many are hardier than the featured animals kept in display, and as such demand special attention and due diligence to control. Aquarists should study and recognize the incidental species that develop in their aquarium and develop a long-term view with intentionally acquired specimens. Green algae can be either a boon or a scourge to marine aquariums.

Red Algae (Rhodophytes): These algae are highly variable in form. Non-specialists rarely identify these organisms correctly for what they actually are: true algae. The larger red species are either plant- or kelp-like with leaf-like fronds that attach to the bottom with holdfasts. The encrusting and branched calcareous varieties are often mistaken for encrusting scleractinian coral formations. Other red algae are more expectedly soft, flat sheets and flexible in texture. Some are decidedly filamentous and traditionally "algae-like" altogether. Although most Rhodophytes appear red-hued in color, they also contain green chlorophyll. This pigment is usually masked by other pigments; particularly phycoerythrin. Depending on growing conditions, a given species might appear to be orange, brown, pink, purple, burgundy, dark red, or even or blackened. Most red algae live attached to rock, or epiphytic on other algae, shells, and marine plants. None of the red species are free-floating like the brown alga, *Sargassum*. Some red algae are microscopic, while a few grow eight to ten feet tall. Most species offered in the aquarium trade grow 2-10 inches (5-25 cm). Rhodophytes occur over a wider range than most any other algae, with species found at extraordinary depths to 600 feet due to the light-capturing capacity of their photosynthetic pigments. Beautiful calcifying coralline species are the very foundation of living coral reefs, producing more carbonate material than the stony corals themselves. They serve as the "glue" holding living and non-living materials together that make up reefs. To a large extent, it is the calcareous red algae that make reefs hospitable to stony coral proliferation. The few non-calcareous nuisance species encountered by aquarists can usually be controlled by herbivores and nutrient limitation. Rhodophytes may reproduce by spores, sexually, or simply by asexual division. Most can be enjoyed both aesthetically and functionally in the aquarium.

Diatoms (Bacillariophytes), are a group with enormous importance not only in the marine environment, but globally, for their food production, principal place at the base of food webs and oxygen production. These unique algae produce siliceous skeletons that are called, in fossilized form, diatomaceous earth. In practical applications of marine aquariology, however, one common type or variety is most familiar: nuisance golden-brown or slime algae. Unlike the widely represented and highly variable red and green algae, some generalizations can be made about diatoms at large. On the whole, they are:

- Limited in growth by nutrient levels
- Significant producers of oxygen
- Crucially important for banking food/nutrients in the environment through photosynthetic carbon fixing

In aquariums, diatoms are necessary, inevitable, readily cultivated when desired, and easily controlled in terms of massive overpopulation when they are not welcome. Keeping brown diatom algae growth in control is simple; a modest growth that remains manageable on the interior of the glass is tolerable if not helpful.

Dinoflagellates: These are unicellular life forms that have struggled to earn respect from aquarists despite their tremendous importance in the marine ecosystem and beyond. Aside from the mysterious and fascinating symbiotic **zooxanthellae**, most organisms in this group are regarded as either a nuisance or dangerous. Some of the bad boys of the "family" include: toxic "Red Tide" species (the neurotoxin-producing plankton that kills masses of fish life and more), the pathogenic Velvet diseases of fishes (*Oodinium, Amyloodinium*) and the toxic slime algae *Gambierdiscus toxicus*. Most disdain in the hobby surrounding dinoflagellates is due to the fact that they are not easily controlled and must be allowed to run their course which often includes harming or killing display animals. Prevention is the best way to avoid complications from unwanted dinoflagellates. Problems with dinoflagellates encountered in display aquariums may be tempered by improved water quality, including high, stable pH and alkalinity, as well as aggressive nutrient export mechanisms. Beyond zooxanthellae, few species of dinoflagellates are deliberately cultured or welcomed by aquarists.

Cyanobacteria: They do not appear to have any redeeming qualities at first glance from an aquarist's perspective. It should be noted, however, that they are of tremendous importance to aquatic ecosystems as a rare vehicle for fixing nitrogen and providing ammonia and nitrite in nutrient-poor environments. This is not easily harnessed in a closed aquarium system, however, and the deleterious qualities of most Cyanobacteria far outweigh their potential merits. The most commonly encountered varieties are the so-called "slime algae" and may occur in colors of red, maroon/burgundy, blue/green or almost black. They are not readily grazed by herbivores and some species are toxic. There is at least one Cyanobacteria, spirulina, that is of commercial importance and value due to its high protein content and nutritive value. This densely nutritious alga is marketed as health food for human consumption and is an important ingredient in many foods for aquatic animals. Control of nuisance Cyanobacteria is generally accomplished by aggressive nutrient control. Improved water flow and water chemistry can also temper growths of slime algae. The presence of significant amounts of Cyanobacteria in aquariums is generally indicative of a flaw or deficiency in the system.

True Vascular Plants: Angiosperms, the true or flowering plants, are becoming more familiar and recognized by aquarists with the advent of refugium technologies and multi-tank systems. There are but a few species of true vascular plants encountered in the trade, but their utility and novelty will forge a place in the market in time. At present, we enjoy only a handful of species of seagrasses and mangrove trees, with many more to come. Marine plants offer many of the benefits of macroalgae including

Microscopic diatoms (Baccillariophyta) are responsible for the the golden-brown "scum" we commonly see coating the inside of our aquariums.

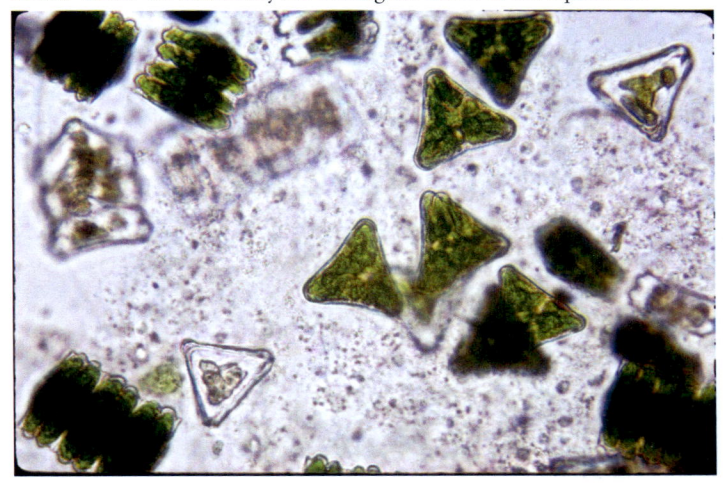

aesthetic value, nutrient export capabilities, shedding of nutritive metabolites & epiphytic matter, and the culture of incidental or ancillary desirable organisms. Fortunately, they do not share the majority of the risks that popular macroalgae species impart such as precarious survival and dependence on nutrient levels, severe noxious exudations, and potential to quickly overgrow or out-compete desirable display organisms. True marine plants are overwhelmingly useful, albeit slow-growing, and valued additions to modern marine aquariums.

Algae Applications

In popular aquarium literature, algae are most often discussed with disdain. However, in modern aquariology algae must be recognized for their vital importance and fundamental role in the biggest picture: the "macrocosm," if you will, of the marine environment. Good hobbyists will translate the pivotal role of algae into the microcosms of their home aquariums.

Algae play many roles in various applications. In simplest terms, they serve as bio-indicators, providing a good sign that the system is biologically viable. After all, at least algae can grow in it! An aquarist can expect to watch and see a familiar series of algal succession from brown diatoms through soft greens to calcareous reds for most maturing systems. Even more generalized is the transition of "power" with microalgae seceding to macroalgae after an aquarium becomes established. Unassuming sheens of brown and green films give way to handsome plating forms and plant-like structures. We may choose to preserve,

Not all *Valonia*-type bubble algae are nuisance growths. This "vasicularia" type forms very dense, tight weaves close to the rock. In healthy systems with adequate nutrient control, it may not be invasive and can be ejoyed ornametally for its metallic green sheen and complex ecomorphology.

or introduce if necessary, any number of these algae for deliberate culture in refugiums and vegetable filters as described in the chapter on refugia.

At the present time, the most popular varieties in the trade are green algae (Chlorophyta), especially of the genus *Caulerpa*. It is unfortunate how much emphasis this genus gets, but have no doubt that this will soon change. Numerous brown and red algae, such as *Sargassum* and *Gracilaria*, are being cultured, explored and promoted for their merits in aquarium keeping. The competitive culture of some of these varieties can be used to limit the growth of other undesirable species, provide food and stabilize captive systems.

As both natural habitat and ornamentation, algae serve as a food and as a medium for culturing food. The relationship is so intimate that numerous fishes have evolved to graze algae incidentally just to derive their required sustenance in zooplankton from within the algal mass. The evolution of algal species in an established aquarium is also more aesthetically pleasing than a well-scrubbed aquarium.

Aside from refugium strategies and participation as vegetable filtration, a popular application for desired marine algae is employing them for their stabilizing effect on water chemistry. There are some tremendous benefits to be harnessed in a significant mass of algal matter from the products of photosynthesis and respiration. Useful amounts of oxygen and CO_2 are shed from photosynthetic activity. In a practical twist on algal scrubber and refugium strategies, some aquarists like to illuminate their ancillary vessels for algae culture on a reverse photoperiod (RDP- reverse daylight photosynthesis) from the display, which has a direct and promptly stabilizing effect on pH by preventing the typical nighttime drop.

Certainly not to be forgotten among the popular implications of algae is the extraordinary measure of cardiovascular exercise that nuisance algae in particular imposes upon the aquarist. Lifting heavy buckets of water for water changes and feeling the "burn" in tired muscles from the redundant scraping of microalgae; is that not the most significant application of all?!?

Controlling Nuisance Algae

Algae have been on the planet for time untold and they are likely to be here much longer. Given adequate water, light, and a nutrient base, they have an extraordinary ability to occupy almost any niche in the global environment. They occur in fresh and saltwater habitats, in soil, hot springs, snow, and even on or *in* living plants and animals. Along with some fungi, there are even algae that live on bare rock as lichens in such forbidding areas as the Arctic! Talking to aquarists abroad about algae reminds one of the broadest

Dictyota, also known as Y-branch algae, is a potentially nuisance variety that can grow fast by fragmentation and is not easily erradicated. There are many beautiful species and colors of it though that can be enjoyed if controlled diligently. (*L. Gonzalez*)

definitions of what people call "weeds": things that we have yet to find a useful purpose for. However, a weed is defined by the beholder and not by the organism itself. Thus, what one man considers a weed may be a useful species to another. Is that turf algae for a vegetable filter, or is it an unsightly growth of hair algae in the display? Alas, most aquarists are more interested in avoiding algae altogether than understanding them better. It's too bad because many perceived "nuisance" organisms could be quite beneficial to the ecology of an aquarium if tempered and harnessed. *Chaetomorpha* "hair algae" is one of several so-called nuisance species finally getting due recognition for its utility in a healthy marine aquarium as an effective means of nutrient export. Understanding the potential benefits and risks of keeping algae in the aquarium is a worthwhile and necessary endeavor.

In the aquarium, some nuisance algae are noxious, others are invasive, and others still are simply unattractive. Most can be considered worthwhile in some capacity in moderation though. At any length, they can generally be manipulated by control of nutrients or light, with most species being limited by the former.

The focus here is describing the prevention and control of undesirable forms of algae. The fundamentals of doing so are as follows:

- Impose strict control on nutrient import through mindful application of foods, supplements, and control of source water for aquarium use

- Maintain strong water flow to keep detritus in suspension for nutrient export mechanisms to process

- Apply vigorous aeration to water to limit or prevent excess accumulation of CO_2

- Maintain good water quality: high pH, high Alkalinity, high Redox (ORP)

- Be mindful of light and heat: avoid using old lights that have strayed spectrally or new lamps designed to provide warm colored radiance under 6,000K. Aspire to provide strong daylight or cooler colored light (>6500K). Avoid warmer water temperatures as well

Living Filters: Algae & Plants

Gracilaria is one of the finest Red algae for vegetable filters, as a vehicle for nutrient export, and as food for fishes. Provide bright light and strong water flow to keep floating masses tumbling. (*L. Gonzalez*)

- Conduct aggressive nutrient export: fine-tuned skimmers producing dark skimmate daily, regular water changes, frequent exchanges of chemical media, and routine harvesting of bio-mass

These control methods can be summarized as: prevention, controlling physical parameters, or controlling chemical parameters. Success with controlling nuisance algae really need not be complicated. A strict application of the described methods can reduce or eliminate most frustrating plagues in a tank in a matter of just a few weeks. By any measure, however, prevention is the very best treatment to curb nuisance organisms before they become established.

Maintain strict control of imported nutrients: As the sayings go, "An ounce of prevention is worth a pound of cure," and, "The best offense is a good defense." Preventing algae is much easier than battling it once it has established. Since most nuisance species are instigated or fueled by excess nutrients, one must examine closely how a nutrient base is formed. Let there be no question that the growth of algae, like any other organisms in the aquarium, is entirely at the aquarist's mercy; all nutrients needed for life are imported by the aquarist.

Source Water: Algae thrive in nutrient rich water. Insuring the quality of source water used for evaporation top-off and the manufacture of synthetic seawater should therefore be a matter of great importance. However, in popular literature and especially in commercial marketing propaganda, the influence of trace impurities imparted by tap water and various brands of sea salt are vastly overstated. There are far greater contributors to the nutrient base that feeds nuisance algae in the aquarium. Nevertheless, any source water that admits a measurable amount of phosphate, nitrate, or various other nutrients must be some point of contention that warrants address. We do not disagree with any recommendation to begin with high quality purified water (De-ionized, Reverse Osmosis, or otherwise filtered) to be reconstituted for aquarium use. Aerating and buffering such water and using a proven quality brand of sea salt is a great habit for overall beginning water quality and consistency. It allows you to start with clean water of a very safe and known composition rather than using variable unfiltered tap water or variable quality natural seawater.

Liquid "Supplements": With the deluge of advertising and propaganda that hobbyists are subjected to regarding liquid food supplements, it is a wonder that more aquarists aren't buried up to their armpits in nuisance algae. This is a bit of an exaggeration, admittedly, but the truth is that even the "best" supplements are easily abused and safer alternatives exist for nutritive supplementation. There are, of course, some very good liquid supplements on the market, but most are little more than pollution in a bottle. Liquefied food (in contrast to liquid trace-elements & bio-minerals) for reef invertebrates are particularly bad habits if not outright harmful to water quality even when the assay asserts they are competent nutritionally. Many such products are plagued with inevitable clotting as the product ages, questionable or absent expiration dates, cumbersome particle sizes and in some cases, unknown or unstated ingredients! Without getting into a raging debate about actual nutritive value, let us cut to the chase and reiterate: most liquid supplements are easily abused, and of little use. Little, if anything, that they provide cannot be supplied with greater integrity though effective fishless refugium technologies and water changes. The use of liquid supplements should be treated as experimental and applied judiciously. During events of nuisance algae outbreak it is advisable to reduce or stop the application of such products to see if they are a contributing factor. The nutrient levels in reef supplements are usually concentrated and potent, and few if any are necessary. Algae are almost entirely composed of water; just a speck of solids from the above dense sources can produce several orders of magnitude of weight in unwanted algae growth.

Particulate Food: If liquid foods are bad for an aquarium at large, then solid foods are too, right? Perhaps not; there are several significant differences between liquid food for invertebrates and solid foods for the same purpose. With solid foods, an aquarist has greater control over the particle size offered organismally or prepared for target feeding as a suspension. The prey is always of a known composition and can be verified fresh and nutritive as such. Fresh and frozen matter also generally have a much higher moisture content and are not very dense. The main concern for aquarists feeding any foods, whether dry, live or thawed frozen, is that it is metered with prudence. Simply be sure that most all solid food used for target feeding is promptly consumed and not left to rot. This really gets to the heart of the matter: feeding the *animals* and not the *water column* or the substrate. **Note**: One of the most common misapplications of food to marine animals with thawed frozen matter is the conveyance of nutritive pack water with solids into the aquarium. Pack juice from thawed meaty foods should be drained and discarded since it is a tremendous source of dissolved nutrients for nuisance algae. New aquarists have a tendency to feed either too much or too fast. Very small and frequent feedings are much better for animals and aquarium health than occasional large feedings. Some marine fishes, such as Anthias or Butterflies, arguably need three small feedings daily rather than one very large meal. Be mindful of water flow and direction to see if it serves the needs of the inhabitants by keeping food in suspension in a useful manner, or if it is simply carrying particles into locations where they will get trapped, degrade and contribute to excessive dissolved nutrient levels. Target the animals that feed organismally rather than offering liquid slurry to the water column. For animals that need a suspension and prolonged feeding opportunities, explore refugium technologies and live food drips.

Water Changes: "Dilution is the solution to pollution" is sound advice for all aquarists to live by. Regular and significant water changes are perhaps the single best thing one can do to improve water quality and system vitality. Water changes bring in fresh bio-minerals, reduce DOC levels, bolster alkalinity and stability, support higher Redox and oxygen levels, dilute noxious exudations and so much more. With specific regard for accumulating nutrients from imported food, source water contaminants, and decomposed organic matter, water changes are the foundation for delaying and diluting the fuel that can flame outbreaks of nuisance algae. Case in point, imagine if a tank accumulates 2 ppm of "magic algae food" every month and the threshold for an undesirable algae bloom is 5 ppm. Even a 50% water change monthly only staves off the inevitable accumulation of nutrients for some time as 50% of the dissolved and accumulating nutrients are left behind each month! Since no one has invented a filtration system that identifies every desirable component of seawater to leave behind while removing every undesirable component, aquarists have to rely on good water quality with water changes to support other good habits of nutrient control. How much and how often for water changes is a matter of some debate. Studies have shown on paper that one large monthly water change can dilute given components slightly better than cumulatively equal weekly exchanges (20% monthly instead of 5% weekly, for example). However, what such studies do not show is the effects of prolonged exposure of aquarium inhabitants to higher levels of deleterious components before the large water change reduces them. Small, weekly water changes of at least 10% for most aquariums are recommended, and more if the bio-load and water quality dictates it. Reef invertebrates demand dilution of their various and sometimes potently noxious exudations more than marine fishes or most other aquatics.

Maintain strong water flow: Water flow is one of the simplest, yet most commonly misapplied influences on algae in the aquarium. The sum and substance of the matter is a battle to keep detritus and organic debris in suspension. Without settling, organics in suspension may be consumed organismally or physically exported by protein skimming. The other possible outlet for channeled organic matter is, of course, man-made filtration. However, it is desirable to minimize this in natural aquariums. It is far better to have dissolved and particulate matter consumed

Various *Caulerpa* spp.

Living Filters: Algae & Plants

directly by living animals, plants and algae, or exported entirely from the system as skimmate rather than permitted to degrade and be converted in mechanical filter media. However organics are handled, know that the amount allowed accumulating in the cracks and crevices of slow flow areas must be tempered. If it is allowed to accumulate appreciably in the display due to inadequate water flow, an aquarist will be challenged to prevent a nuisance algae bloom. This is a far more likely contribution to the nutrient base than any trace contaminants possibly trickling in from source water. Many harsh judgments about the deep beds of sand or gravel have been leveled unfairly due to the commonly inadequate dynamic of water flow in modern aquariums. Aspire to create strong, random-turbulent water flow in the aquarium when a surging pattern is not convenient or possible. Resist creating strictly laminar flow, as from powerheads, in the display. Most nuisance algae fare better in low flow environments. In an otherwise optimized aquarium, an increase in water flow can sometimes be the cure for a plague or nuisance algae.

Vigorous aeration of water: Concerns about the accumulation of CO_2 in the aquarium are valid, however secondary, to more prevalent issues of nuisance algae control such as inadequate water flow, poor nutrient export and overfeeding. CO_2 is an essential component of photosynthesis and is a limiting factor for plant and algae growth. If allowed to accumulate, it is like rocket fuel for accelerating the process! It is also easily dissipated in the aquarium with good aeration. When trying to diagnose the occurrence of unwanted algae in the aquarium and CO_2 accumulation is suspected, simply aerate a glass of aquarium water to discover the truth. Test the pH before, and then again several hours after aerating the water sample vigorously. Well-circulated aquariums with adequate gas exchange will not reveal any significant increase in pH. There are other implications to an aquarium with accumulated CO_2 and poor aeration. Inadequate water flow, low dissolved oxygen and low Redox levels are also likely and indirectly supportive of an environment that encourages nuisance algae. Aggressive protein skimming also serves to aerate a system well beyond the primary benefit as a vehicle for nutrient export.

Microscopic photo of a pinnate diatom.

Water Chemistry: Various aspects of water quality are at least indirectly inhibiting to the growth of undesirable algae. However, these guidelines are to serve not as a primary course of treatment for controlling nuisance species, but as a reminder to maintain high water quality as a means to a successful aquarium overall. The aquarist must make a choice to replicate an environment similar to that of the distant pristine reefs, or one that is more akin to the lagoon that serves as a catch-basin for a sewage-spewing coastal resort!

Key parameters that influence algae growth include water temperature, pH & alkalinity, and Redox. The relationship between the proliferation of algae and pH/alkalinity is somewhat complex. The short story is that accumulated CO_2, carbonic acid and a low pH are part of one big happy family that is conducive to pest algae growth. A high pH and strong alkalinity are sometimes less hospitable to the successful growth of nuisance species in part by virtue of the fact that the condition inherently neutralizes and limits carbonic acid/CO_2. Kalkwasser use has been demonstrated to be indirectly helpful in a similar vein. It is naturally caustic, it tempers acids and supports alkalinity, and may precipitate the "super algae fertilizer" phosphate. A favorably organic-rich environment as a rule will nourish algae. A high oxidative state (high redox) also limits the availability of micronutrients. Redox can safely be increased and maintained with the use of an ORP controller to meter and dose ozone. Ozone will specifically limit the growth of algae by oxidizing crucial micronutrients as well as increase dissolved oxygen. The manipulation of salinity has also been implicated in the control of nuisance algae, however it has not been recognized as a safe or measurably effective strategy. Algae can tolerate a wider range of salinity than most any other desirable organism. While it is true that algae can be shocked or killed by a sudden change in salinity, especially to a lower specific gravity, it is also true that most of the favored microfauna and display animals will shock and die first. Focus instead on creating an environment of high, and most importantly, consistent water quality. Stable and moderate levels are better than occasionally spiked ideals. Target: pH 8.3-8.6, temperature 76-82°F, calcium 350-425ppm, alkalinity 8-12 dKH, and Redox of 325-450 mv.

Light and Heat: The manipulation of light to temper nuisance algae can be a tricky endeavor for some aquariums with zooxanthellate-symbiotic organisms such as corals, anemones and other desirable plants and algae. If the aquarium houses fish only, or only possesses aposymbiotic (non-photosynthetic) invertebrates, the photoperiod of light can possibly be reduced in an effort to slow or stop nuisance algae. Light is a limiting factor for many undesirable species, but keep in mind that nutrients have the same or greater impact. When lighting

A beautiful juxtaposition of "Red Grape" algae (*Botryocladia*) and tentacles of a *Macrodactyla doreensis* anemone. (*L. Gonzalez*)

seems to be a significant factor in nuisance blooms, it is more often the quality of light that is the problem and not the quantity. Please keep this in mind if choosing to restrict light over a system. Also know that the treatment is secondary, and really a matter of addressing the symptom and not the problem. Consider instead if the aquarium isn't suffering from a bigger issue of excessive nutrient presence or inadequate nutrient export as detailed above. As a guideline, employ lamp colors 6500K and cooler, and avoid warm lights under 6000K; change fluorescent lamps every six to ten months and metal halides approximately every eighteen to twenty-four months.

Higher water temperatures, positively associated with high light intensity, have been implicated in nuisance algae blooms. There is more merit with warmer water in general stimulating algae than the issue of brighter light doing the same. At the risk of making a gross generalization, most algae offered in the trade are collected from shallow water where the water is warm in temperature and color, hence the concerns with aging bulbs straying to the warm end of the spectrum and stimulating growth. When nuisance algae are of concern, avoid temperatures above 82° F (28° C).

Aggressive nutrient export

The various options available can be categorized as mechanical, biological, or chemical control methods.

Mechanical nutrient export

For aquarists seeking the quick fix, a fine-tuned and well-designed protein skimmer is the best mechanical nutrient exporting tool. It is strongly recommended for all beginners and most advanced aquarists; be sure to pick a proven style and model. There are very few nuisance algae populations in the aquarium that cannot be starved into submission within weeks by a skimmer that produces significant dark skimmate daily. A good skimmer will easily produce skimmate with consistency and to see it happen while curing an algae-plagued tank simply underscores the tremendous dependency of most nuisance growths on nutrients. Skimmate not only contains micronutrients and organic matter that would otherwise feed an algal bloom, but it also contains algal spores, cellular phytoplankton ("greenwater") and filamentous algae fragments adrift.

Rhizophora mangle, AKA the Red Mangrove: This is the most commonly utilized species by aquarists due to its adaptability to emersion in fully undiluted seawater. It has been cited as reaching twenty-two meters tall in the wild, although that is quite extraordinary for this categorically slow growing plant. In aquariums, it is capable of breaking the strongest materials of containment, so be sure to plant and pot them spaciously with a long-term view. Loss of magnesium ions, along with sodium pumping, is a common complaint in the aquarium with heavy mangrove populations. Test Mg concentrations and supplement as necessary. At large, they are not useful for aggressive nutrient export, but they have many other benefits beyond their unique and naturally beautiful inclusion in a balanced reef microcosm.

Product removed by a skimmer has the distinct advantage over other forms of mechanical filtration such as filter floss, pads and media-based filters, including trickle filters and fluidized bed filters, in that it wholly isolates collected organic matter from the system. Collected skimmate cannot contribute deleteriously to water quality as it degrades in time, unlike rotting matter trapped in filter media. This is the primary criticism of man-made bio-filters in tanks with only light or moderate bio-loads: nitrate production from the mineralization of nitrogenous matter. Live substrates and protein skimming are preferred in such scenarios. Life forms on the live rock and sand can directly consume any organics that the skimmer fails to remove. Refugium technologies with alternate biological nutrient export abilities have also evolved into good choices for natural aquarium strategies.

Biological nutrient export

Biological means of nutrient export have been earning recognition for their importance. One of the most useful applications of a biological mechanism for nutrient export is the vegetable filter which includes algal "scrubber" or uptake technology. Farming plants and algae can be an extremely effective means of nutrient export and an aesthetically fascinating dimension of the system. Refugiums also have numerous ancillary benefits such as the incidental culture of micro-crustaceans and other small plankters. They occur in a wide range of styles and applications and must be considered indispensable for a natural aquarium.

Beyond the traditional use of a refugium to farm "plant matter," animals can also be used as biological vehicles for nutrient export. The proliferation of feather dusters, syconoid sponges or even reasonably inoffensive soft corals such as Xeniids (pulse corals) may be enough to offset an offensive nuisance algae bloom. Although a bloom of undesirable organisms may seem mysterious at first, it is important to understand that they exist because of nutrient levels within the aquarist's control! Given to choose between a plague of hypnotic pulse corals and a plague of slime algae, most people would choose the Xeniids; they're more attractive and easier to trade and sell. Better still, Tridacnid clams are one of the finest living filters for marine aquariums. For displays where they might be at risk with predatory residents, they may be kept in a shallow downstream raceway or refugium that catches all raw overflowing water for "clam scrubbing." Over shallow water less than 20" (50 cm) in depth, lighting can be provided inexpensively with standard output fluorescent lamps. The possibilities with employing living organisms as nutrient export mechanisms are numerous in modern reef keeping. To be effective, they simply must be stable and accumulate mass, or be routinely harvested.

Chemical nutrient export

Chemical filtration is generally an unrealistic means of nutrient export by virtue of its expense. It can be recommended, however, as a supplement to primary vehicles such as protein skimming and biological mechanisms, provided it is used in small amounts and changed frequently. Chemical filtration is also quite important as a method of maintaining water clarity for photosynthetic organisms. Activated carbon and exchange resins such as beads and other media are the most common and effective choices. Some media, such as the Poly-Filter by Poly-Bio Marine, changes colors to indicate contaminant adsorption. Others must simply be exchanged routinely as a blind media. The advantage of chemical filtrants over biological and strictly mechanical products such as floss, bio-balls and fluidized bed filters is that at least some of the organics trapped by the media cannot interact fully with water quality as they degrade; instead, they are chemically adsorbed rather than absorbed. Such organics, however, can undergo biological reduction (nitrification) if the media is ignored long enough and becomes inoculated microbially. This is one of the reasons for changing carbon weekly instead of monthly. Other chemical filter media tries to be more specific and target nitrate or phosphate extraction. Know that most of these products, however, are weak; either they don't work at all under practical conditions, or they perform minimally. Some such products in the past have been accused of imparting elements that simply corrupted test kit readings of the targeted nutrient! Rest assured that activated carbon is tried and true and several notable exchange media products have very well established and easily confirmed reputations.

Other Controls

Elbow Grease: Another obvious method of algae control not yet mentioned is manual extraction. Numerous commercial and do-it-yourself (DIY) products have been devised to rasp, scour and scrape algae from the various surfaces of the aquarium interior. Razor blades and plastic edges, spun fiber pads and a variety of stiff brushes, palm-fitted tools and devices on sticks all grace the merchant and aquarist's shelves. All treat the symptom and not the problem. For minor control of casual algae growth, these tools are necessary and useful. For plague and nuisance populations, however, an address of the above-outlined causative agents is required. It is exceedingly labor intensive to control undesirable populations by hand and in some cases the very activity of cropping stimulates new and excessive regrowth.

Slime algae (cyanobacteria, diatoms and dinoflagellates) should be siphoned clear and away into a bucket during water changes. Never agitate or disturb particles of slime algae in the aquarium unless the desired goal is to spread it. Tough and turf species that require vigorous scrubbing should be treated similarly for fear of reproductive cells spreading. One of the best mechanical means for extracting tenacious encrusting forms is to scrub and siphon simultaneously with a stiff toothbrush tied to and slightly forward of a running flexible siphon hose. Alternatively, rocks with large colonies can be removed and scrubbed under running water in a sink. Most all wanted livestock on the rock will endure this temporary insult.

Other popular and inexpensive algae scraping tools include plain, spun plastic fiber dishwashing pads, credit cards (for acrylic tanks), single-edge razor blades (for glass aquariums), and recycled toupees (still paying attention?).

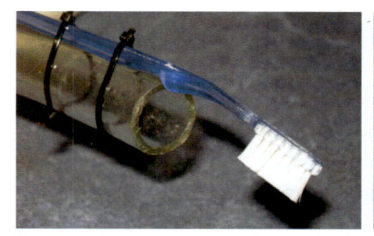

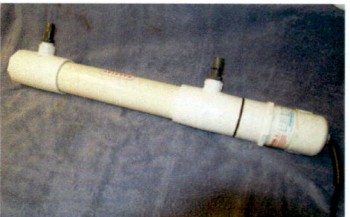

Possible algae control methods. Left: a toothbrush connected to a siphon hose is particularly useful for tackling BGA's and cyanobacteria without spreading it to other parts of your tank. Ultraviolet Sterilization is also promoted and used for controlling unicellular algae.

Algal Predators: Algae grazers must also be given consideration. Various fishes, crustaceans, mollusks and other creatures naturally and eagerly graze algae. Although they cannot directly export nutrients, their activity reduces or controls the proliferation of algae. Digested algae are excreted in convenient fecal pellets that drift through the aquarium and may be directly consumed by corals or other invertebrates, or simply handled physically by filtration aspects such as protein skimmers. Disturbed bits of algae that have been broken, cropped or otherwise freed may be liberated from the system by the same methods. *Turbo* and *Astraea* snails, some blennies, tangs and rabbitfishes are among good grazers. Snails are the most widely used herbivores and generally the most innocuous choice. Crabs, sea slugs and urchins can also be very effective but often have special challenges or risks. Amphipods and other micro-crustaceans are some of the most effective and important yet underestimated herbivores in marine aquariums!

Herbivorous Menagerie: The Algae Grazers

As the fundamental food staple on a reef, most organisms eat algae, or they eat something else that eats algae! Aquarists keeping fish-only displays realize this quickly with some predatory fishes that wane in color or health if they are not fed whole prey foods. The specific problem is that fish fillets, prepared shrimp, or otherwise incomplete (gutted) prey lack the nutritive value that gut-loaded prey

Chlorodesmis, Turtleweed, is a rather attractive multicellular algae that often occurs in bright green manageable tufts. (*L. Gonzalez*)

Living Filters: Algae & Plants

has. It is for this reason recommendations are frequently seen for feeding whole foods. The lionfish and grouper that eat small fishes that eat algae all day long are sustained in part by the algae in the fishes' gut. There are many other such examples on the reef too. Triggers, damsels and gobies will feed on algae for the microcrustacean content, however they still derive significant nutrition from the greenstuffs. There are also numerous dedicated herbivores such as the popular surgeon fishes that mostly eat algae as a primary staple. Gut analysis of many common herbivorous fishes found in the aquarium trade has revealed some wild diets of nearly 100% algal matter. It is a wonder that any adapt to the dry and pelleted foods offered routinely in aquaria. Skilled and empathetic aquarists will want to research and deliver appropriate foods to their charges with all of these considerations.

Accompanying this section is a list of some of the more common, popular or noteworthy herbivorous organisms. Please be sure that their demand for algae can be satisfied if one is taken into collection. For example: ever wonder what those faint and seemingly harmless fine red streaks are on the rostrum of long-nosed tangs? Symptoms of blunt force trauma! They get a bloody nose from days, weeks or months of smashing their poor little snout against bald and barren rocks and glass every waking moment of the day. Even when fed well, foraging is an inalienable behavior among these fish. This is not rocket science. Algae eaters eat algae and they will suffer without it regardless of how sexy the packaging is on pelleted and flake food containers.

A conclusion about controlling algae

Is such strong emotion as fear or enmity deserving of such simple life? Are algae problems in your aquarium inevitable? Are they to be avoided at all costs? The answer to all is a resounding no! After all of the considerations, it ought to be pointed out that sometimes it really is better to just let it be. The growth of algae in moderation on a reef and in aquariums alike is an indication of a normal, healthy state that can usually be ignored within reason. Recognize key species and movements that help keep a system balanced and stable. Well-groomed and regularly harvested algae from a controlled area will benefit any system. By purposely cultivating desirable forms, the incidence of undesirable varieties can be reduced. A proper refugium or vegetable filter is one of the best measures of insurance against the proliferation of nuisance organisms. Aggressive nutrient export with tuned protein skimmers can correct most any other problem.

Cultivating desirable species of marine plants and algae…

Marine botany in the aquarium is an endeavor still very much in its infancy. This field will hopefully gain favor in North America faster than its freshwater equivalent has. Marine algae are readily available and capable of culture in aquaria and they perform several important aquaristic functions beyond their natural aesthetic beauty.

The declaration so often heard in the hobby about what "nutrient poor" conditions exist in tropical reefs is at least somewhat misleading. In the context of an admonition to new aquarists about being mindful of nutrient control, it is an appropriate statement. Specifically, though, the reefs are not at all nutrient poor, but rather nutrient *concentrated*. Obviously there is a tremendous amount of "fixed carbon" and flow of energy from everything eating and being eaten.

Padina scroll algae is a beautiful species that is prevented from growing in many tanks with herbivores stocked too early. It commonly sprouts from Fiji live rock.

Why then would the water be considered "nutrient poor?" Simply, it is because the wealth of nutrients are jealously bound up by the very life forms themselves and not merely dissolved and waiting invisibly in the water surrounding the reef. When a natural reef accumulates high levels of dissolved organics, it suffers and often becomes a dead zone with exception to the murderous nuisance algae and like organisms that flourish to exploit the influx of nutrients and space from the dead, dying organisms it can settle on. This very same thing happens in the home aquarium. Feed invertebrates well, feed plants and algae also… but please do not feed the water! So much of successful aquarium husbandry is founded upon nutrient control.

Selection

Understanding what to look for in health and vigor is a different matter entirely from knowing how to find suitable specimens. It will be some time before a wide selection of plants and algae are available in many merchants' displays with regularity. There are more than a few obstacles limiting selection and viability of such marine "greens." To summarize, algae have never really enjoyed a collective reputation as favorable organisms. When recognized as a harmless or beneficial candidate for the aquarium, previous aquarium methodologies using harshly herbivorous organisms have been unfavorable to the establishment of plants and algae. Fortunately, modern refugiums and remote ancillary vessels such as algal filtration technologies are changing that. Finally, when all other obstacles have been reckoned, there is still a great challenge to locate and acclimate specimens to a waiting system. Despite the general tolerance of most algae to suffer extremes of water quality, they do not take such exposure abruptly and they do not ship well by and large. Changes in salinity can be stressful or fatal to marine plants and algae. Other aspects of water quality require strict attention as well.

No matter where in the world you live, there is assuredly an aquatic environment within hours travel, if not minutes, by plane or car to a collection site with marine algae. Despite this marvel of modern transportation, it can be quite a challenge for an aquarist to find viable starter specimens of algae. To begin with, one must seek species that are suitable for the biotope that is being replicated. For most aquarists, that means utilizing tropical species only. Be sure to resist temperate plants and algae that will burden an aquarium with their struggle and decay during transition to a warmer environment. In most of America and some of Europe, numerous species of tropical plants and algae from the Atlantic and Caribbean are most common. These include seagrasses and mangrove tree seedlings (true vascular plants), *Halimeda* and other calcareous algae, numerous soft greens, some reds and some brown algae. The Internet should be used as a research tool and catalog reference

> **Chemical Control Note**:
>
> A brief mention of the use of chemicals or antibiotics to control nuisance algae: don't! Using any such agent to control algae is the least desirable route with regard for overall biotic system health and vitality. Categorically, it is a weak treatment of the symptom (nuisance algae) and not the cause (nutrients). Ask instead, "What are the factors that are contributing to this system being out of balance?" There are obvious downsides to altering the natural evolution of a system with toxic anti-microbial agents. Do not forget that antibiotic means "against life." There are several brands of antibiotics and algaecides on the market. Erythromycin and synthetic derivatives are commonly employed, as is copper (usually in some format of copper sulfate solution). All should be avoided for algae control as one can be quite certain that desirable microfauna, including biotic filtration faculties, will be harmed or destroyed. In tanks with symbiotic invertebrates it is also important to remember that the life-giving dinoflagellates inside of cnidarian tissue are algae too.

when determining candidates for an aquarium. If necessary, ordering specimens by mail is an option via online portals, although the risk of getting inferior product is greater for shopping sight-unseen and for the rigors of postal carriage. The best advice is to find a local merchant that stocks specimens, which can be closely examined; if necessary, special order species through the merchant's professional divers and experienced channels of shipping aquatics. One way to ferret out a viable supplier is to look for other species offered in their selection including common Atlantic reef denizens such as small hermit crabs, *Astraea* snails and *Diadema* urchins. If they carry these animals at affordable prices, they may have an able-bodied collector or resource to get a variety of plants and algae collected.

Once a source for marine algae has been identified, the integrity of available specimens needs to be evaluated. Are they healthy, or stressed and perhaps even dying? General appearance will indicate a specimen's vigor and viability to some extent. Are there apparent damaged areas, pale or discolored spots, or any necrosis? Is there a staggeringly offensive odor to the plant? Is the species understood well enough to recognize ideal color and expected texture and firmness? The entire process is rather "hands-on" and demands a close inspection. Plants and algae are easily stressed in transit and often acclimate poorly. For species that occur attached to rock, pay special attention to the holdfasts and anchoring aspects for evidence of damage. Always acquire plants and algae with natural substrate whenever possible.

Herbivorous Menagerie: The Algae Grazers

Surgeonfishes, Family Acanthuridae:

The two common *Zebrasoma* "Sailfin" tangs are now united taxonomically as *Z. veliferum*. The trade still makes a distinction between the Pacific and the Red Sea (formerly *Z. desjardini* - at left)

Naso and *Acanthurus* tangs are more specialized algae controllers (*Naso* with bubble algae and *Acanthurus* with turf species). Because of special challenges with these genera in contrast to the more hardy and manageable *Zebrasoma* and *Ctenochaetus* species, these specimens are not recommended for casual aquarists. Please research thoroughly to learn the special needs such as large and mature aquariums with high water flow for *Naso* and *Acanthurus*. Adult Nasos grow to well over 12" (30 cm) and need very long and large aquariums over 6 feet. And though the "Powder Blue Tang" is an understandably popular surgeonfish, this species is best kept by advanced aquarists with mature tanks over one year old.

There are some truly magnificent *Zebrasoma* tangs if your patience and pocketbook can bear the rare acquisitions. Black tangs (right) are stunning as adults with an icy blue sheen at the base of the dorsal fin while Gem tangs (spotted) are striking and uncommon at any size.

(J. Cross)

The extended rostrum of long-snout *Zebrasoma* species is good for tearing hairy types of algae away from hard substrates. They are not "perfect" grazers though in so much that their activity merely crops and essentially farms the turf (like cows munching grass). This genus is best for controlling growths on hard irregular substrates such as rock. They do not fare as well for grazing glass and sand.

Ctenochaetus species are the best tangs for aggressive control of nuisance growths. Their specialized comb-tooth mouthparts are very efficient at rasping algae clean and low to the surface. They forage better on glass and sand than any other surgeonfish and are equally competent on hard rocky substrates. It should be noted that they will also feed on some unpalatable growths, such as Cyanobacteria, not readily grazed by other organisms.

Rabbitfishes, Family Siganidae:

On the whole, Rabbitfishes (AKA Foxfaces) as a group are some of the very best algae grazers. They are more active and less discriminating grazers than many popular surgeonfishes by comparison. They do, however, possess venomous spines and require larger aquariums (over 100 US gallons ideally). This group is distinguished as including members that will feed upon some of the most unpalatable or sinewy microalgae such as mature hair algae that has become too long for even *Zebrasoma* to graze, and bubble algae.

Angelfishes, Family Pomacanthidae:

Although angelfishes consume algae as a significant part of their diet, this group is not recognized as primary controllers of nuisance algae in aquariums. Still, they are fishes of extraordinary beauty and grace and may be considered for grazing modest populations of algae or simply enjoyed as a living marvel. Dwarf angels of the genus *Centropyge* are more likely to eat a staple diet of algae more reliably than other genera of Angelfish.

Blennies, mainly Family Blenniidae

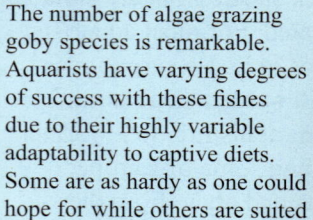

The number of algae grazing goby species is remarkable. Aquarists have varying degrees of success with these fishes due to their highly variable adaptability to captive diets. Some are as hardy as one could hope for while others are suited only to a species tank. Several that are of great utility and durability in the aquarium are *Salarias* "Lawnmower" blennies (150 species in this genus!), *Atrosalarias* spp., and *Amblygobius phalaena* "Dragon," or "Bullet" gobies. *Salarias* control algae very well on hard rocky substrates and glass, while *Amblygobius* are outstanding sand-sifters.

Natural Marine Aquarium Volume I - Reef Invertebrates

Echinoids, Sea Urchins:

Despite their formidable appearance and inaccurate reputations for destruction, sea urchins are outstanding algae grazers for most any marine aquarium. Although some may be clumsy or cumbersome, most are categorically safe, useful and harmless scavengers of algae and other organic matter. Two of the very best specimens for the marine aquarium are *Diadema*, a surprisingly nimble and "reef-safe" long-spine variety, and *Mespilia*, a handsome, fascinating and short-spine decorator species.

Arthropods, Real and False (hermit) Crabs:

There are numerous hermit crab species that are useful in aquariums because they scavenge on all kinds of debris. However, they should not be added with the expectation that they can support themselves. It is common for hermit crabs to starve when preferred foods become depleted. Most crabs will attack and kill desirable organisms in displays, including snails, fishes and corals. Please refer to the section on hermit crabs and other sources to gather data about desired species. Make an informed decision about hermit crabs' needs and your ability to meet them.

Percnon gibbesi, AKA the Sally Lightfoot crab, has developed a reputation for being an effective algae grazer but there is just cause for concern in keeping this arthropod in marine aquariums with small fishes, corals and other weakly defensible invertebrates. True to crab form, they are opportunistic predators. They eat algae, polyps and sleeping or stressed fishes. They are hardy and fascinating but should only be kept with the sturdiest tank mates.

The Majid crabs of the genus *Mithraculus*, AKA Emerald or "Mithrax" crabs, have earned an excellent reputation in the aquarium trade. Emerald crabs are good herbivores and will consume nuisance growths, including bubble algae. They eat other varieties of algae and take prepared foods when algal masses have waned provided the aquarist remembers to feed them. However, they have large and able claws and grow larger than most "reef-safe" janitors. They occasionally attack fishes and can be disruptive to rockwork and precariously seated sessile invertebrates. Generally though, *Mithraculus* is a good scavenger.

Micro-crustaceans as a group are some of the most under-rated yet significant herbivores and detritivores on the reef. They can consume enormous amounts of bio-mass quickly. Dead fish, algae and invertebrates lost in the rockwork can be processed by copepods and amphipods in days, if not hours. They also consume a remarkable amount of greenstuffs and are found in multitudes within algal masses. The best way to cultivate these organisms is to have at least one inline fishless vessel that serves as a refugium.

Asteroids, Seastars:

Mentioned here as a warning to aquarists, most sea stars have limited use in the reef invertebrate aquarium as herbivores. The safest species are generally delicate and require very large habitats (one starfish per 100 US gallon minimum) while the worst species are ravenous and predatory and may digest just about every bit of living sessile organic matter in their path. Please reference the Asteroid and starfish sections in this text for tips on species selection.

Mollusks:

The list of safe and effective snails for reef aquariums is delightfully long. Be mindful not to add more individuals to a display than is needed for the long-term. Stocking densities can range wildly with most coin-sized species weighing in at around one per ten gallons of display water (a very gross generalization). The most popular group of snails is the spiral shaped turbans: *Astraea*, *Turbo* and *Trochus* species. *Stomatella varia*, AKA the Paper-shell snail, from Indonesia is a very useful and prolific incidental organism as is the *Cerith* species from the Atlantic. *Strombus alatus* the Fighting Conch), are large but effective cultured mollusks. Nerites, Chitons and Limpets may be collected deliberately, or acquired as hitchhikers, and are innocuous. Many of these snails at large will reproduce in the aquarium with regularity and provide the added benefit of nutritious plankton for other captive reef invertebrates.

Some shell-less snails such as the *Elysia viridis*, a green sea slug, will actively graze on various species of algae and assimilate plastids to use as ongoing, food-producing symbionts. This particular species is a wonderful utilitarian grazer that favors *Cladophora*, *Chaetomorpha*, *Codium* and *Bryopsis* algae. This slug's color is amazingly dependant on the color of the algae that it is feeding upon.

Other shell-less snails, such as sea hares, sea slugs, and nudibranchs, may prove useful to feed upon even a specific alga. Unfortunately, the difficulty with finding, collecting and effectively shipping many of these snails in good health exceeds the value for importing them *en masse* from both an economical and ethical standpoint. Casual aquarists are discouraged from pursuing such species experimentally.

Living Filters: Algae & Plants

Care

Maintaining a tank with plants and macroalgae is no different than maintaining other types of marine systems. Some arrangements are easier to conduct than others. Obtaining good, healthy specimens and providing a suitable habitat is the first step and a key to any system's success.

Acclimating new acquisitions to the tank must be done slowly and gently. Plants and invertebrates are much more sensitive to osmotic shock and differences in water quality than fishes. Although algae grow in almost any conceivable salinity of water, few are tolerant of sudden changes in specific gravity or other fundamental parameters of water quality, excluding most intertidal specimens. Tolerances will vary by species, however all will appreciate a slow drip acclimation over fifteen to thirty or more minutes in a temperature-insulated vessel. Avoid transporting algae out of water only to be literally tossed in a new system for lack of pack water to use for dilution.

A formal two to four week **quarantine (QT)** is recommended for all aquatic life, not excluding plants and algae. Beyond the tremendous investment most any marine aquarium has in hardware, time and livestock, aquarists have a moral and ethical obligation to all living organisms that are taken into their charge. New specimens of plant,

Numerous invertebrates, plants and algae like this Rhodophyte, *Laurencia* (perhaps *L. poitei*), are never given a chance to flourish or even begin life when aquariums are stocked too quickly with fishes. (*J. Troeger*)

algae, live rock, coral, etc., are to be propped up on PVC fittings or similar shelf. Small bits of meaty food bait the tank at night in search of predatory hitchhiking worms, crabs, etc. Stabilization of a new specimen can then occur in the confines of a quiet and fishless system. Medications have no place here and are categorically harmful or fatal to most plants and algae. With patience, the four week protocol will also reduce most fish and invertebrate parasites in the absence of a host. The procedure is not foolproof, but it is time-tested and proven to save lives and a tremendous amount of grief and hardship for aquarists. Quarantine, quarantine, quarantine!

Before screened algae is acclimated to the main display, the system should be biologically stable. Although macroalgae can feed significantly on nitrogenous matter, many species will not acclimate to unstable water with already elevated levels of ammonia and nitrite. Some residual nitrate of just a few ppm (parts per million) is helpful for growing macroalgae as well as the endosymbiotic zooxanthellae in coral tissue; simply avoid a flat reading of zero. A steady pH higher than 8.2 and full-strength seawater (1.024-1.026 S.G.) are also recommended for most plants, algae and invertebrates for their intimate dependency on dissolved matter. Therefore, it stands to reason that unnaturally dilute seawater is less favorable. Consideration must also be given to the fact that most reef organisms which aquarists keep are **stenohaline**: narrow in their tolerance to change in specific gravity, and appreciate constancy. Those that are not will still benefit from the stability.

It makes little difference incidentally, whether the aquarist chooses to start with synthetic seawater or natural seawater. However, the "real thing" is far more work to prepare safely. Aquarists living on coasts should not be tempted to use natural seawater due to serious concerns regarding

Halimeda are heavily calcareous species and contribute significantly to the production of reef sand when their calcium rich matrix dissolves.

water quality. For the breadth and depth of civilization near any marine coast, issues of pollution are pervasive. There are extraordinary challenges in securing and conditioning natural seawater for safe aquarium use. The process is quite inconsistent and tedious, and potentially infectious or toxic for those living in populous areas. Synthetic sea salts from reputable companies are time-tested and true, with many marine organisms being raised entirely and successfully with them for well over thirty years! Industry and hobbyist chatter regarding synthetic sea salt toxicity is by and large ludicrous, and rather irresponsible. Both the hobby, and science have validated the use of synthetic salts for consistency and reliability.

Adequate nutrient levels are a matter of great controversy among aquarists. There is a wide range of opinion on the method and matter of nutrient supplementation for plants and algae. Likewise, there is concern surrounding features such as skimmers and carbon filters removing excessive nutrients. Rest assured that skimmers and activated carbon do far more good than harm and are very helpful, if not necessary, components of a modern marine aquarium. If, for example, carbon use was to be excluded on the basis of its aggressive nutrient removal, aquarists would need to avoid keeping corals and *Caulerpa* too! The various merits of carbon use and protein skimming have been covered in great detail elsewhere, however it is important to reiterate their contribution to maintaining water clarity to enhance light penetration for the benefit of photosynthetic inhabitants.

Fundamentally, aquatic algae require the same sixteen or so essential nutrients that terrestrial plants do. It is for this reason, in part, that popular marine mud substrates are believed to be effective since they contain these fundamental nutritive elements. A significant difference, however, between terrestrial plants and marine algae is that aquatic species absorb nutrients through all parts of their structure and not just their "roots." Thus, it is not critical for aquarists to fertilize their aquarium substrates. Algae should, however, be purchased attached to solid matter if naturally occurring on hard substrates. Burrowing forms require either anaerobic or oxygen-limited substrates and must be promptly and gently buried in deep soft media. This trick is crucial for establishing seagrasses such as *Thalassia* and *Syringodium*; plant them at a minimum depth of 2-3" (5-8 cm) in a deep sand bed greater than 6" (15 cm) in depth.

The truth of the matter is that most aquariums have excessive nutrient levels after just a few months of maturation. As such, few systems need any significant supplementation. Regular water changes and daily feedings to fish and invertebrates can be more than satisfactory. Without adding metal, biomineral, alkalinity or other types

Porolithon sp. in Cozumel. Photo by Diana Fenner.

of supplements, most algae will not die, they just will not grow as fast. There is legitimate cause for concern against both excessive supplementation and excessive proliferation of macroalgae in home aquariums. Uncontrolled and random supplementation is more likely to boost the growth of undesirable species of nuisance organisms and create other unwanted consequences. Algal metabolism can sway pH and other parameters significantly. Additionally, the biomass of a large population of algae ripe with metals, phosphate and other potentially harmful nutrients can instigate a system-wide catastrophe in the event of a sudden algal stress or reproductive event that results in the sudden release of the slowly accumulated compounds within the bio-mass. The point of this dissertation is to practice *moderation*! Dose supplements known to be helpful such as iodine, iron, magnesium and other various elements in judicious moderation. Most are to be maintained in very low measures of one or two tenths of a part per million (ppm) and sometimes less depending on other factors and indicators such as nuisance algae growth. It is a fair rule to only dose products that can be tested for and monitored. For all else, rely on water changes for dilution and replenishment.

Beyond the fundamentals of nutrient supplementation, some algal varieties are calcareous and require special attention to stable and consistent calcium and alkalinity levels. The calcareous species such as *Halimeda*, in fact, are good indicators of these minerals evidenced by consistent growth (AKA "Baby's bows" at one coin per day). Aspire to maintain calcium levels near 350-425 ppm, alkalinity levels near 8-12 dKH, and magnesium between 1200-1400 ppm. In contrast to nuisance species, desirable plants and algae will appreciate a high Redox potential (ORP over 300 mv). The key to sound and stable growth in the marine aquarium is consistency with these parameters even at moderate levels.

It is also helpful to remember that good and bad things should happen slowly in a successful aquarium. Dose

Ochtodes is a beautiful macroalgae that commonly occurs in shallow waters with an irridescent blue or purple color to it. It requires bright light and turbulent water flow to flourish but has proven to be quite adaptable. It is useful for vegetable filtration like **Chaetomorpha** and **Gracilaria** and should be treated similarly. (*L. Gonzalez*)

experimentally in moderation and with patience when elements cannot be tested for and metered. And always rely on water changes to bring the system back to center when necessary: "**when in doubt, do water changes!**" There is perhaps no better supplementation for plants, algae and invertebrates than fresh, constituted seawater. As an aquarist becomes more involved with the culture of photosynthetic plants and algae, there are further issues of micronutrients, CO_2 infusion, and manipulations of the photoperiod that can be finessed. Please seek references in the bibliography of this text and elsewhere for a better understanding of these dynamics that cannot fairly be discussed in the brief format of this comprehensive title.

Lighting for photosynthetic organisms is the keystone of life. Where nuisance algae are limited (or energized) by nutrients, desirable plants and algae are more intimately dependent on the quality of light for growth and vigor. Undesirable algae can usually fare very well in poor light with high nutrients, but many desirable species often fare poorly under the same conditions. With the wide range of niches, depths and oceans that the many popular algae hail from, one cannot fairly cite an ideal light or lamp to serve all targeted species. For guidance, however, lighting that favors the warm daylight end of the spectrum is favorable for algae and plant growth. Higher algae will usually tolerate or thrive with lighting that leans towards 10,000K. At any rate, lamps in the 6,500-10,000K range will serve most photosynthetic reef applications. Fluorescent lamps over shallow water can keep most algae, while metal halides are recommended for true vascular plants and aquariums with water deeper than 24" (60 cm). A photoperiod of around ten to twelve hours is reasonable for most systems while a shorter period may be needed on intense halide-lit systems. It is difficult to render a rule of thumb for the amount of light by measure. It is easy to understand with the range of species, locales and tolerances in this category, any given watts per gallon rule will favor some, suffer others, and barely maintain others still. Nonetheless, a suggested minimum of 10,000-15,000 lux will work as a starting point for most algae. Many red algae will tolerate less light, and most green will demand or utilize far brighter. This is likely to translate into at least four watts per gallon over a 24" depth of water or less (<60 cm). Be sure to also use light timers for daily and consistent delivery of light to your photosynthetic organisms. Other than *Caulerpa*, which can be kept illuminated twenty-four hours a day and remain in stasis, all marine plants and most algae require a day/night cycle for respiration. For dynamic growth, increasing the duration of the photoperiod is not recommended, however an increase in luminary intensity may be helpful. It is difficult to over-illuminate most algae, which hail from shallow tropical waters, in an aquarium.

Water temperature is an uncomplicated matter with higher algae. Maintain a stable temperature from 76° F to 82° F (26°-28° C) as would be done for the growth of corals and fishes. Avoid warmer temperatures (above 82° F), which favor nuisance algae blooms. Water flow is also a straightforward although more highly variable parameter. Research the niche and habitat of a given species with the intent of replicating its specific needs. Please note, however, that moderate to strong random turbulent or surging flow is stimulating for most desirable algae while weak water movement encourages nuisance growths. There has also been a tendency in the infancy of refugium applications to use inadequate flow; please know that even lagoonal and nearshore environments can be dynamic with an often greater turnover of water volume than most home refugiums enjoy. *The bottom line:* don't be shy about **strong water flow** in most areas of the modern marine aquarium. Good water flow will carry nutritive metabolites and epiphytic matter away for filter-feeders in the aquarium and prevent detritus and debris from stifling good growth.

Pruning and predation are necessary and logical catalysts for the vitality and growth of true plants and algae alike. The world's reefs are vivacious environments, with webs of ever-changing, inter-cooperative relations at play, and organisms competing for space, light, and nutrients. Natural predation and mechanical pruning will need to be variously imposed to thin out old and dying matter to promote colony health and vigor. In aquariums, the process can be an excellent means of nutrient export as well. In other situations, it is necessary to interrupt the natural life cycle of a colony to prevent a potentially noxious event of sexual reproduction. Special considerations must be heeded for some algae. The single-celled varieties such as *Caulerpa* are very sensitive. A frond, if cut, will sap contents at the risk of the entire cell dying and releasing significantly noxious material. The arena of chemical interactions amongst sea life, particularly the **allelopathogenic** effects of some algae and other photosynthetic organisms against each other, is not well-known amongst hobbyists. Culturing a noxious species or groups of species in a system can have profound effects on the growth, reproduction and very life of other organisms. Some of these compounds discourage the encroachment of other organisms, germination, or growth. All can be exuded *en masse* with excessive pruning and by instigating a "crash" of the colony. This can potentially stress or kill any number of living organisms in the tank. When thinning algae colonies, it is preferable to remove whole and individual strands in contrast to cutting out sections or randomly breaking fronds. Multicellular colonies are generally much more forgiving to both grazing and manual removal by hand. When trying to establish a new colony, be mindful of protecting the mass from early or vigorous grazing by tangs, angels, butterflies, crabs, snails and others creatures that relish destroying and devouring "greenery."

Reproduction in algae occurs in several ways: asexually by simple division or fragmentation, by spores, and sexually by gametes (the potentially devastating acts of sexual reproduction that aquarists fear). Unlike true plants, they do not produce flowers, seeds, or embryos. Propagation in the vascular marine plants (flowering Angiosperms) is a much more complicated process with large, mature specimens as a rule, and is beyond the scope of this reference or home aquarium keeping. In most cases, algae have complex life histories including reproduction by alternating asexual and sexual generations of a gamete-producing generation followed by a sexual spore-producing generation. This is a most fascinating aspect of the dynamic and physiology of reproduction in algae known as the "alternation of generations" and it's a challenging concept to grasp. Essentially, some algae have life cycles with mature specimens that reproduce differently than we'd expect in a sexual event. Imagine the reproduction of an alga, but instead of participation by only two types of individuals, one male and one female, the species that is encountered can be one of three possibilities: male, female or a sporophyte. The sporophyte produces spores that develop into the gametophyte generation, which in turn reproduce via motile, flagellated gametes. In some cases, this unique "alteration" (the gametophyte generation) has been mistaken for a different species or even genus altogether!

Parthenogenesis is another amazing reproductive strategy, form and life cycle that occurs in familiar hair algae species such as *Derbesia* and some higher seaweed species with regularity. The halicystis stage of *Derbesia*, in fact, is a large green bubble that resembles *Valonia* but gives rise to the filamentous form of the genus. Other interesting and complex modes of reproduction do occur with fascinating reproductive structures including *sori* and *conceptacles* (the conspicuous raised bumps). Beyond hobby applications, further detailed explanations are outside the scope of this reference as a practical guide.

What an aquarist will want to remember about sexual reproduction in algae is that it can be very dangerous in the confines of a closed aquarium. This is but one important reason to prune algae often and routinely in order to interrupt or forestall the life cycle of sexual maturation of a given colony. Algae can and do modify their environment profoundly, poisoning themselves and other life forms at times with events of sexual reproduction and in defense of predation. As mechanisms for distribution and survival, this makes sense considering their lives in the open sea. In closed aquariums, however, it can be real trouble. Be on the look out for sudden milky exudations or cells turning clear. The best response to such an event in the confines of smaller home aquariums should be an immediate and significant water change. Utilize aggressive nutrient export tactics (protein skimming, carbon filtration) and preparedness for more water changes as necessary during events of sexual reproduction in algae.

More often, algae simply reproduce by fragmentation or simple division. Due diligence with nutrient control and active herbivores will render this strategy relatively harmless. At large, algae reproduction and cultivation is a matter of balance, choice and due diligence in maintenance; make peace with it.

Thalassia testudinum - **Turtle Grass**: Growing up to 2 ft. tall (60 cm) but usually found shorter. Forms flat leaves ½" (12 mm) wide of a handsome dark green color with adequate light. It typically occurs from the shallows to depths of sixty-five feet, but has been recorded deeper. Photographed here in the Tropical West Atlantic in muddy sand beds in which it is deeply anchored. (*D. Fenner*)

Compatibility & Algal succession

Regarding the compatibility of marine plants and algae, there is little to discuss. They don't like each other and they don't like other groups of organisms altogether! Marine algae are very competitive and have been documented at length to produce numerous and copious noxious strategies to inhibit or kill competitive biotic growth, including plants and animals alike. Some corals and reef invertebrates are literally out-competed for food, space and light if not outright killed by toxic anticompetitive exudations from algae. Marine algae are unltimately necessary and helpful for the aquarium. Keep them in moderation in the display proper, or enjoy larger masses in a proper refugium or remote culturing vessel under a watchful eye. The real dynamic of interest with algae presence and compatibilities is algal succession.

The notion and nature of algal succession in the marine aquarium is a matter of great predictability and should be of interest to aquarists. By definition, there are many possible scenarios and participants in a given order of algal succession depending on the available components for life. In layman's terms, there are various species of algae that are expected to appear, and then be displaced by other organisms in due process and time. The framework of the process might be the change of seasons. A freshwater lake, for example, is likely to see a diatom algae bloom in the spring due to warmer waters and the accumulated nutrients from winter. By summer, proper green microalgae will become established and strip nutrients and positions from the diatoms. By late summer, however, the increased biomass from a good season's growth of green algae and plants will begin to break down and overwhelm the environment with decay and overgrown matter which invites the blue-green algae (Cyanobacteria) to gain a foothold. At last, by late fall, the diatoms regain their footing before dormancy and winter. The direction of a succession in evidence can speak volumes about the health of a system. In a healthy marine aquarium, one can expect a diatom bloom on start-up from the availability of fresh and nutrient rich seawater. These diatoms may persist beyond the first couple of months if an aquarist struggles with control of nutrients. There may be a

There are many attractive species of *Caulerpa*, most of which can be used and useful if maintained diligently. (*L. Gonzalez*)

Acetabularia sp. - Mermaid's Cup

Actinotrichia sp. - Spikeweed

Maroon BGA slime algae (*L. Gonzalez*)

learning curve with adjustments to filtration dynamics, or from excessive feeding and stocking issues. By four months established, however, we can realistically expect the diatoms to give way to a clear presence of green microalgae. Around then, and soon after, classically higher forms of greens and reds should appear including macroalgae and plant-like structures. With adequate and consistent levels of calcium and alkalinity, the calcareous species should begin to be expressed appreciably. It is at this point that a healthy marine aquarium might plateau in the order of algal succession: calcareous greens and reds with varying amounts of non-calcareous higher reds and greens. If at some point the multicellular browns take a dominant position, examine nutrient export processes and the quality of light on the system. Warmer temperatures, warmer lamp colors, and accumulating nutrients will all lend themselves to this less desirable direction of algal succession. If such problems get worse, the system may suffer the sudden expression of blanketing growths of slime algae (Cyanobacteria or dinoflagellates) or diatoms again. Rest assured that each stage is telling and predictable and is a very good indicator of system health. Take heed and comfort in the order or algal succession for good marine aquarium husbandry.

Marine Plants and Algae Species Overview

Images and essentials for ready identification of their use in the aquarium

Please notice that the following plants and algae are not grouped taxonomically, but instead listed in *alphabetical order* with scientific and common names provided, and many of them cross-referenced. This provides a casual break from tradition with a different educational approach for aquarists that, by and large, are hobbyists without an advanced education of the life sciences necessarily fresh in their memory. Therefore, rather than require the reader to either know or learn that "blue-green algae" are not algae in the purest definition, that diatoms *are* algae, or that brown kelp is also not a true plant, identification and information is organized by the first thing an aquarist needs need to know or learn about an alga: its name!

• *Acetabularia* - **Mermaid's Cup** or **Wine Glass**: This delicate macroalgae is a real treasure in the marine aquarium. Few aquarists get the pleasure of keeping a large mass of this alga. They are very sensitive to shipping and handling and rarely survive import in large numbers. Most are discovered in aquariums incidentally as hitchhikers on live rock or substrate with an invertebrate. This tiny organism resembles a finely ribbed and sculpted wine glass with a "cup" (cap) diameter of a mere 1/4-1/2" across (6-12 mm). The color of this alga is naturally a pale green-white that can fluoresce magnificently under actinic light! By virtue of its sensitivity and demure presence, it serves little purpose as a vehicle for nutrient export. In turn, it has little mass to provide any appreciable habitat for microcrustaceans and other desirable fauna. Essentially, they are to be appreciated for their rare grace of the captive environment and regarded for a modest but lovely contribution to the overall diversity of an aquarium. *Acetabularia* is found in shallow water less than twenty feet in depth in protected areas. Provide bright, warm light and moderate water flow in the aquarium while avoiding strong or linear movement. Mermaid's cup will grow solitarily or in clusters and prefers hard substrates.

• *Actinotrichia* - **Spikeweed**: A densely branching and intricate, calcified red algae. This bushy coralline shares lineage and traits of captive husbandry with *Galaxaura* (Thicket Algae or Boneweed).

• Blue-green algae (BGA) - see Cyanobacteria

• *Bossiella* - Twig Algae: A common branched calcareous genus. Some members of this family form very attractive shapes, including palm-like fronds. This genus and *Corallina* are notably tolerant of human pollution, often growing luxuriously near outfall "boils."

• *Boodlea* - "Crispy Green Algae": This unusual species of algae is sometimes called "Hair Algae" although one

Living Filters: Algae & Plants

Botryocladia sp. Red "Grape Caulerpa" (*LG*)

Bryopsis sp. - a type of Hair algae.

Valonia sp. - Bubble Algae. Clear cell gone sexual. (*AC*)

might not agree without a closer look. It usually forms short crunchy, brittle growths that fragment and spread easily. It can quickly become a serious nuisance with high light and high nutrients as in aquaculture facilities such as coral farms. It responds well to most measures of control (predation, nutrient control and manual extraction).

• *Botryocladia*- "**Red Grape *Caulerpa***": This is a true red alga and not at all related to the genus *Caulerpa*. Imports often travel with live rock and are moderately durable on acclimation at best. Colonies, like the name suggests, resemble clusters or strands of grapes rather like *Caulerpa racemosa* in gross form. It is typically too sensitive and too palatable to herbivores to ever grow nuisance levels in most aquariums. Little has been reported in hobby literature about its merits and limitations in culture. It is very attractive and unique ornamentally. It is strongly recommended that cultures are started in a fishless refugium in order to establish colonies without predation prior to attempting introduction of "Red Grapes" into the display aquarium.

• *Bryopsis*… see "Hair Algae"

• **Bubble Algae**: There are numerous species and more than a few common genera of "Bubble Algae" represented in the hobby. Some are prolific and bona-fide nuisance organisms with little redeeming qualities, while others are innocuous and can be tolerated if not enjoyed. A few are even considered to be beautiful by some people (usually non-aquarists looking at a tank!) for their glossy marble-like shine. Regardless, they serve little purpose in aquarium systems due to their potential to infiltrate and pollute most any illuminated place in the system through tiny daughter cells. Bubble algae can grow in high or low flow, high or low light, and it regrettably is not limited by excess nutrients like most other nuisance algae. More often, though, it is found in protected areas that are not necessarily polluted or high in nutrients. Bubble algae can be found naturally on reef flats, back reefs and lagoons with corals and invertebrates familiar to aquarists.

It is indeed possible for a reasonably healthy aquarium to sustain a bothersome population of bubble algae despite good husbandry. Control is best accomplished by persistent manual extraction and herbivory. Siphon them off rocks with a toothbrush tied to the end and slightly forward of the mouth of the tubing. In this manner, you can scrub and siphon ruptured cells simultaneously to control the spread of daughter cells. *Naso* and *Zebrasoma* tang species (*Z. veliferum* Sailfin types) may work well for large tanks. *Mithraculus* crabs have earned quite a good reputation for control in smaller tanks. Be warned, however, of their large and aggressive adult stage and know that like most crabs, they can be opportunistic feeders and predators. There are other herbivores, including gastropods (shelled and shell-less snails) and echinoids (*Diadema* urchins), that also consume bubble algae. If you see your bubble algae waning clear as depicted in the image above, pack up the family and the dog and run screaming from the house! Well, perhaps it's not that bad, but bubble algae can bloom rapidly by vegetative growth (budding) or by this wickedly successful strategy with numerous daughter cells formed within a reproductive donor (a mature bubble). The cell will become soft and clear then rupture to release a bounty of "baby bubbles"…. Aieee! The following genera are similar in form and/or function: *Valonia, Boergesenia, Ventricaria, Dictyosphaeria*.

• ***Caulerpa***: What a paradox for use in the marine aquarium! Species in this genus share traits that make them some of the most useful and most undesirable, even dangerous macroalgae for aquarium use. Like so many aspects of aquariology (and life), success with *Caulerpa* species is critically dependent on having a clear understanding of their use, implications, and complications. Suffice it to say that if you do not know precisely why you might want to keep *Caulerpa* in your aquarium, you shouldn't keep it! Informed aquarists instead can enjoy some wonderful benefits from this macroalgae.

Caulerpa racemosa (D. Fenner),    *Caulerpa* sp.    *Caulerpa prolifera*

Benefits to *Caulerpa* Use

- Its vigorous growth can be harnessed for aggressive nutrient export

- *Caulerpa* species are some of the most attractive algae available and found in a wide array of forms

- The potential dangers of using *Caulerpa* are well understood and easily tempered or avoided with due diligence

- Most species are eagerly grazed by herbivores (although one must be careful not to abuse this due to the cumulative toxicity of some *Caulerpa* species)

- *Caulerpa* has the unique ability to be cultured with 24/7 illumination and remain in stasis. It has significant merit in refugiums and vegetable filters

- Exudes some nutritious organic matter into aquatic systems, including vitamins and proteins

- Requires no special substrates and grows anywhere, attached or unattached, and can grow under a wide range of physical parameters (various light & water flows).

- Some species are rather innocuous and quite temperate for aquarium use in comparison to other dominantly noxious or dangerous species in this genus such as the C. racemosa varieties).

Concerns with *Caulerpa* Use

- *Caulerpa* has been well-studied and documented to leach numerous <u>noxious or toxic exudations</u> above and beyond the scope and potency of most other popular algae

- Toxic to some fishes forced to eat it regularly over time (as with captive aquarium fishes)

- Toxic to some grazing invertebrates forced to eat it regularly over time (as with snails and echinoderms such as sea urchins)

- Toxic exudations can directly inhibit the growth of stony and soft corals. Some *Caulerpa* appear to chemically "burn" SPS tissue, for example, with contact.

- Non-toxic exudations, including phenols, that can significantly discolor the water and reduce light penetration in aquariums without ozone or aggressive chemical filtration like small weekly changes of carbon. This dynamic is mitigated by occasional fast growth or large masses of the colony.

Aquarium Use of *Caulerpa*:

- *Caulerpa* may grow explosively and become invasive. As such, it is best kept in controlled refugiums and should be avoided in casual reef displays. Species in this genus can overgrow and harm or kill other sessile reef organisms including coral

- Some *Caulerpa* contain remarkably potent anti-predation compounds that reduce natural herbivory to the extent that aquarists must control growth manually.

- Acts of sexual reproduction can cause a sudden and potentially dangerous release of accumulated and manufactured compounds that can harm or kill other life forms in the system, up to and including notorious whole-tank "wipe-outs." This can be avoided by using constant lighting and consistent pruning or predation to skirt the three to six month life cycle of most species

- Natural or manual control should be done systematically and in moderation. *Caulerpa* are large single-celled organisms that can "sap" or leach if pruned excessively. Complete failure of a colony is

Living Filters: Algae & Plants

- possible in severe cases, which imparts a large and sudden amount of nutrients that can be overwhelming

- Colonies can be very sensitive and demand consistent and increasing levels of nutrients, that match their growth in stride. Without a stable "luxury consumption" of nutrients these algae may crash and die when the critical mass fails to be sustained by necessary elements.

Caulerpa has enjoyed a popular history of usage with marine aquarists. Until the recent advent of refugium technologies, however, it was rarely grown in great quantities, as most tanks have inevitably had an effective herbivore of some kind that would control or eliminate such "greenery" in the display. Indeed, most marine fish aquariums through the years have had at least one angel or tang in them, which would readily graze *Caulerpa*. Because of these realities, most aquarists had not realized some of the severe challenges and dangers to keeping this genus *en masse* by virtue of traditional colony sizes, which historically have been small, until recently. Ironically, any arguments about *Caulerpa* being a boon or the devil incarnate may soon become moot. Love it or hate is it, *Caulerpa* has drawn the attention of the "powers that be." Legislative bodies have recently begun to make the possession of these potentially useful algae illegal! Ultimately, it will be the vigor and adaptability of this macroalgae that will be its undoing in the trade. Legislators have legitimate fear of *Caulerpa* being released to wild environments where it can establish and displace native species and become a formidable physical impediment to pumps, intakes, watercraft, etc. It is of utmost importance that no exotic organism be released into a novel environment where it might persist, displace native life, introduce pathogens or other natural competitors, or become a burden in any appreciable way.

For those aquarists that wish to take the opportunity now, or in the future to keep this macroalgae, a bit of history and husbandry follows:

Fauchea sp.

There are over seventy known species of *Caulerpa*. Most look like creeping vines with alternating leaf-like projections emanating from their two-dimensional profiles. Their holdfast aspects (rhizoids) provide anchorage in the sand, gravel and upon hard substrates. These structures are also part of the more notorious concerns with keeping *Caulerpa* due to their chemical exudations that kill other organisms in an effort to clear organic matter from hard substrates allowing for expansion of the colony. Odd as it may seem, the biggest problem with this genus is its success, in the wild and in aquariums. Not only must there be concern surrounding the confiscation of nutrients, light and space that are needed for other livestock such as corals and reef invertebrates, but there are real possibilities of **Caulerpa** altering water chemistry unfavorably in its exuberance. There are more than a few noxious or toxic exudations in this genus produced as anti-biotic and anti-predatory deterrents. Some have been shown to concentrate in fishes and invertebrates that graze and kill them. This may or may not occur commonly in the wild where a hungry grazer has options, but theoretically can happen in home aquaria when an herbivore is confined to live in a field of *Caulerpa* unnaturally. It would be like forcing people to live in a room full of french-fries for the whole of their abbreviated, heart-diseased lives. Oh, sure, it's cool at first, but after the beer runs out and the ketchup comes from somewhere other than Pittsburgh, all that is left are greasy fries to poison the body every day. This bizarre, if not poignant, analogy should serve to remind aquarists of several aspects of good aquarium husbandry underscoring the possibility of "too much of a good thing is bad."

Wipeout! One of the most feared aspects of *Caulerpa* maintenance is the aforementioned risk of a colony going "sexual," resulting in a wipeout that would take the fish, some coral and the VCR with it! It is a state whereby this large single-celled organism gives up the ghost and releases the tremendous lot of its contents and composition, including the good, the bad and the deadly. Beyond any issue of toxicity there is the simple dilemma of a massive and sudden influx of various elements that had been imported into the biomass slowly and in small quantities over time. Many aquarium systems simply cannot handle the burden of the abrupt degradation of water quality. Colonies may crash because their growth and needs have exceeded limits of available nutrients. They are also notoriously sensitive to sudden physical or chemical changes in water quality. Changes in salinity are also quite stressful on *Caulerpa*, as with the pouring of less saline water for water exchanges or evaporation top off into the system, and it can be a possible catalyst for a colony crash. At last, the absolute worst way to induce a wipeout of a colony is to simply do nothing at all. Huh?!? Well, with good water, ample nutrients and time without disturbing the colony, *Caulerpa* will "go sexual" and reproduce:

Chaetomorpha sp. (*LG*)   *Chlorodesmis* sp. - Turtle Weed (*A. Calfo*)   *Codium* sp. - "Dead Man's Finger" (*LG*)

self-inducing a wipeout of the colony with the expulsion of gametes. If an aquarist has a strong growing colony of *Caulerpa* that hasn't been pruned for at least three months, there exists a possibility for an event of sexual reproduction that could be overwhelming to the system.

With all of the many concerns detailed above, one can understand and possibly agree with the recently popular admonition to aquarists for keeping macroalgae: "Anything but *Caulerpa*!" Whether aquarists choose to subscribe to this ideology or not depends on how hard they are willing to work in order to enjoy the many benefits of this macroalgae. In defense of the genus, its demands are not extensive so much as they are strict. Aquarists need to be mindful of the sensitive needs of this macroalgae, which requires stable and increasing supplies of vital nutrients. In keeping with good husbandry, the regular and generous use of ozone or activated carbon (small portions changed weekly) will be necessary to reduce discolorants in the water and improve overall water quality. Regular water changes of a small amount weekly will also work towards this end as well as dilute some noxious exudations and replenish vital trace elements and minerals. As with so many things in life, please apply only in moderation. And no, a garbage can sized refugium full of this algae plumbed in-line to a nanoreef does not constitute moderation.

• **Chaetomorpha** (also see "Hair Algae" for a broad overview): *Chaetomorpha* is a highly variable genus of macroalgae whose members can be observed in a variety of forms, ranging from simple hair-like colonies to intricately woven masses which might resemble, of all things, a kitchen dish-scrubbing pad. It has also been described as resembling a green mass of steel wool or tangled fishing line. *Chaetomorpha* is often found in high nutrient or polluted areas of the marine environment. It occurs fixed to hard substrates or unattached. Unlike most other genera of "hair algae," this macroalgae is less of a nuisance due to its temperate impact on water chemistry, vigorous nutrient export capabilities, innocuous carbon-fixing, and overall stability in culture. When properly controlled or harnessed, *Chaetomorpha* is not only tolerable, but also beneficial for marine aquariums.

In aquariums, hobbyists will often encounter the aptly named, "Spaghetti" variety of this macroalgae. This dark green form is prolific, easy to culture and quite useful for a variety of applications. A cursory glance on the various Internet message boards reveals that *Chaetomorpha* is rapidly gaining favor with hobbyists all over the world as an ideal macroalgae for refugiums and vegetable filters. It is weakly noxious in comparison to *Caulerpa* and other popular macroalagae when cultured *en masse* and colonies do not crash easily. In essence, it can be used aggressively with fast growth and harvest as a vehicle for nutrient export, especially in systems with high bioloads. The dense matrix of its structure makes it an ideal home for the development of desirable plankton, including numerous microcrustaceans such as mysids and amphipods. *Chaetomorpha* can be utilized in sumps, refugiums, and possibly in the display proper. The popular strategy for keeping *Chaetomorpha* is in separate vessels as a free-floating colony, where it can be grown and harvested with minimal complications.

Requirements for culturing this macroalgae are easily met. Nutrient-rich seawater and bright daylight illumination are the foundation for this species to thrive. Harvesting *Chaetomorpha* consists of a gentle portioning and separation of the colony. Farmed product may be recycled as food to herbivorous fishes in the system or exported to a fellow hobbyist to help them start their own culture. For aquarists fearful of or frustrated with keeping precarious *Caulerpa* colonies, *Chaetomorpha* is an excellent choice of macroalgae for cultivation in the marine aquarium.

• *Chlorodesmis* (*C. fastigiata* and others) AKA **Turtle Weed** or **Maiden's Hair**: This green "hairy" alga is found widely in the South Pacific and enjoys a strong demand in the marine aquarium industry. It occurs in gorgeous bright-green clusters of very fine filaments. The dense structure of turtle weed provides an outstanding medium for the culture of microcrustaceans while being somewhat

Coraline algae - *Peyssonnelia* sp.

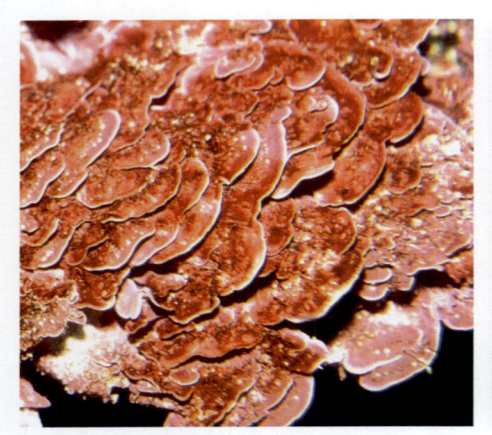

The Colors of Scum: No, not political affiliations, but varieties of slime algae! Often referred to as blue-green algae ("BGA"), these mat-forming growths may be black, brown, green, blue, purple, burgundy, red or yellow.

resistant to herbivory due to its noxious inhibitory exudations. Turtle weed is unfortunately rather difficult to culture in the home aquarium. Inclusion of this alga in the aquarium/reef display proper is challenging. *Chlorodesmis* prefers extraordinarily brisk water movement and very bright light in the form of metal halides over shallow water. Its extreme needs for husbandry are incompatible with most coral commonly kept. *Chlorodesmis* can be a pest if too prevalent, but is otherwise useful for nutrient uptake/consolidation and as food. Colonies of turtle weed generally wane within months in a casual aquarium display due to its specialized needs.

• *Cladophora* AKA **Wire** or **Hair Algae**: This genus usually forms thick, course "segmented" (multi-nucleated) filaments that may be attached or form free-formed clusters. Fine-threaded species are also observed. These firm tubular clumps are sometimes mistaken by aquarists as a variety of nuisance bubble algae such as *Valonia* and other similar species. Unfortunately, this alga too can become a troublesome nuisance that blooms with spikes in common nutrients including nitrate & phosphate. Regrettably, by virtue of its rough texture and large filament size, fewer natural predators are available to control it in aquariums. Control measures should focus on maximizing long-term nutrient export processes and may have to embrace some methods of manual extraction. As a medium for plankton culture, especially 'pod growth, it is very good with its cushion-like masses. At large, however, it may be best to avoid the deliberate culture of *Cladophora* in vegetable filters and refugiums when so many better options are available. *Cladophoropsis* is another "Wire" or "Hair Algae" that is similar to *Cladophora*. This genus, however, is less inclined to be fast-growing and is easily removed manually.

• *Codium*- AKA **Dead Man's Fingers**: *Codium* is a fascinating green calcifying alga that forms thick, fuzzy, velvet-like green fingers. These digitate extensions are composed of a matrix of finely woven filaments and are actually buoyant! This alga is recognized as a very hardy but slow growing organism in the aquarium. Although it generally cannot participate in any significant dynamic of nutrient export, it does make a very fine addition to the natural appearance of a display. It is a very unique alga in many ways. *Codium* is found on both coasts of America and beyond, ranging from Baja California all the way to Alaska in the Pacific! Species in this genus are found over a wide range of depth as well, although it is collected as an intertidal species. Its harsh life in the tidal zone makes it a reasonably good candidate for aquarium life. Despite its residence in areas of dynamic light and flow, tempered physical parameters are tolerable for many specimens collected under ledges and in protected areas of the reef. Sexual reproduction has been observed in captivity with this dioecious (separate sex) genus. *Codium* presents no significant challenges, benefits or dangers in the marine aquarium and can grow toward one foot in height and diameter (30 cm), tethered by a single holdfast. Maintain fundamental water quality for marine life and be mindful of calcium and alkalinity levels to support growth in this calcareous genus. *Codium* is also known by several other common names including: Green Finger "sponge" and Green Fleece algae. It is also called "food" by *Elysia viridis*, a green sea slug, which actively grazes this algae and assimilates plastids to use as on-going, food-producing symbionts. Ahhh, photosynthesis: the gift that keeps on giving!

• **Coralline algae**: Its difficult to decide where to even begin discussing the monumentally important group of calcifying algae known as corallines. Articulated and encrusting species of calcareous red algae are of extreme importance to shallow marine environments. To a large extent, they contribute mightily to reef building and nourishing of all life found there. They are the cement that binds the living reef together and are some of the most important organisms in the entire marine ecosystem, in large part because they make up a good deal of the biomass and productivity. In aquariums, it is most common to see

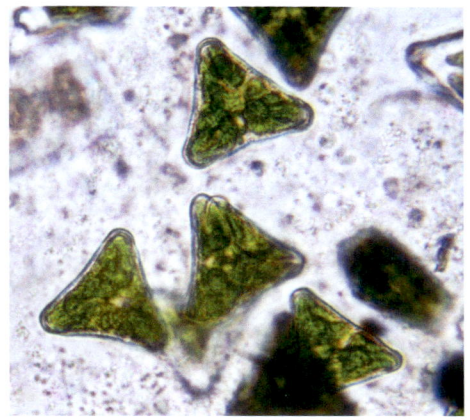

Diatoms of the centrate variety.    *Dictyota* **sp.**, an unusual blue variety. (*LG*)    *Dictyosphaeria cavernosa* - Buttonweed.

encrusting forms that plate objects in a most attractive fashion in time given good water quality. Needless to say, the taxonomic classification of this enormous group of red algae is the subject of much debate in many areas.

Delicate branching forms are observed less often in the hobby but are a real treasure to behold (see Twig Algae). These segmented forms are mainly members of the subfamily, Corallinoideae, comprising the jointed, or articulated, corallines. Both groups are found worldwide and common in tropical to sub-tropical seas with colors ranging but dominated by red, pink, orange, burgundy and purple.

Regarding their importance in marine aquarium keeping, corallines play more than a few crucial roles. Some of the significant benefits of this group include the following:

- They are a desirable and high ranking species in the order of algal succession. Very few nuisance algal species can displace corallines once established in healthy systems

- Their considerable mass and surface area utilize nutrients in the water and contribute significantly to water quality by shedding oxygen and "nutritious" compounds into the water: vitamins, proteins and nitrogenous "food matter" of use to other organisms including corals in reef aquaria

- Their consistent growth and vigor are meaningful indicators of water quality for keeping other calcifying plants, corals and animals

- There seems to be no significant deleterious exudations from corallines that would impede coral and reef invertebrate health in captivity. Conversely, numerous species of reef invertebrates will only settle larval forms on specific coralline algae

- Many corallines can provide all of these benefits in low-light environments that very few other photosynthetic algae can

Most corallines can adapt to a wide range of physical parameters. Bright light, however, should only be imposed slowly and with due diligence of acclimation over time as one would do with coral. As with cnidarians, corallines are much better suited to adapt to lower light environments if compensated with adequate nutrients. Adequate levels of biominerals are also crucial for successful cultivation of calcareous algae. Stable levels of Calcium and Alkalinity will be necessary for optimum growth and survival of coralline species. Consistency is far more important than trying to maintain unrealistically high ideals. Various other trace elements and compounds, including iron, magnesium and strontium, are stimulating to vigorous coralline growth. Anecdotally, the use of organic calcium gluconate has been demonstrated to nicely accelerate coralline algae growth safely and effectively. Sugar-based calcium should not, however, be used as the primary vehicle of calcium supplementation for stony coral. Control of coralline growth, if ever necessary, is best employed by various mollusks and echinoids (snails/limpets/chitons and urchins respectively) with sturdy grazing mouthparts.

Summary of Coralline genera in the aquarium:

- High Current, High Light, Tidal Exposure: *Lithophyllum, Porolithon, Titanoderma*

- Medium Light, moderate current, low intertidal or subtidal niche: *Hydrolithon, Mesophyllum,* and *Sporolithon*

- Low Light, moderate current, subtidal niche: *Peyssonnelia*

- **Cyanobacteria** - AKA **Slime Algae**: Despised, disgusting, slimy, and offensive: must be a politician! Hmmm… often this would be true. However, this description is in reference to nuisance blue-green and red slime algae. So what's the story? Are they algae? Are they bacteria? Does anyone care? Cyanobacteria are basically photosynthetic bacteria that are barely, but importantly, related to algae. Like some other large groups of algae,

Living Filters: Algae & Plants

these organisms run the gamut in form and function. They may occur as single cells, clusters, threads and chains. Regardless of general disinterest for them in aquariums, they are crucial for the global environment.

By texture, some slime algae are stranded or filamentous, but most common aquarium varieties are mat-forming. They often contain trapped air bubbles similar to some diatom growths and many dinoflagellates. Direct siphoning of Cyanobacteria may be helpful, but avoid other disturbances, which can spread the "problem." Most Cyanobacteria are toxic or unpalatable to herbivores. Pathogenic species exist as well, including some coral diseases. Most all can be tempered and reduced by improved water quality, circulation, and lighting that encourages other photosynthetic life.

How did it get here and how do we get rid of it? The type of lighting used might influence the color of slime algae, but nutrient levels predominantly limit their presence. Cyanobacteria are most often nuisance growths and their presence in significant quantity is an indication of a flaw in water quality or system dynamics. Aquarists will typically find BGA in aquariums that lack sufficient water flow, nutrient export mechanisms, or have poor aeration. To inhibit slime algae, a system will benefit by high alkalinity (10-12 dKH), high Redox potential/ORP (over 350 mv), low nutrient levels (nitrates less than 10 ppm usually) and fully saturated levels of dissolved oxygen. The correct use of a calcium reactor and an ozonizer can help to accomplish these goals.

Bad habits that can nourish Cyanobacteria include pouring thawed pack juice from frozen foods in the aquarium, overfeeding or overstocking of fishes, and abuse of bottled food supplements such as phyto- and zooplankton substitutes for invertebrates. Always strain or decant meaty foods away from pack juice, and refrain from overfeeding or overstocking a system.

Good habits that can prevent or eliminate Cyanobacteria include maintaining aggressive water flow and protein skimming, keeping aquarium lamps fresh and of a sharp, cool color, and using kalkwasser regularly. Kalkwasser precipitates phosphate, which is a nutrient for slime algae, improves protein skimming and indirectly supports high alkalinity by tempering organic acids due to its causticity. Changing small amounts of carbon and other similar filtration media weekly instead of monthly will also improve water quality.

- *Derbesia*… see **Hair Algae**

- **Diatoms**: There are many faces of these brown algae and most are familiar to aquarists. Most prominently, the fast and furious golden-brown slime that coats aquarium glass so quickly is, as one might guess, a diatom. It is also important to note that they fix about 25% of earth's carbon and singly produce a measurable amount of the oxygen in the atmosphere. Yikes! Put down the napalm and flamethrower, these algae are important! Occurring as unicellular plankton or colonial benthic fauna, diatoms are fundamental food products in the big salty picture. They also occur in freshwater, brackish water and even moist terrestrial niches.

The first diatom exposure by most aquarists is the aforementioned slimy, scummy golden-brown algae. It is one of the first algae to appear on any high nutrient scene. You might even notice tiny, trapped bubbles in the blanketing film as it advances. This is quite common in new aquariums loaded with fresh nutrients, a lack of competitors and an unrefined schedule of husbandry. Stressed or neglected systems that have been established may also see resurgence in diatoms. Diatoms are one of the first varieties to occur in the order of algal succession. In time, they are supplanted by green algae and ultimately calcareous species such as corallines. A strong presence of brown diatom algae after three or four months in a new system may be an indicator of less than perfect water quality.

While all aquarium systems must inevitably include some diatom growth, excessive blooms are often the sign of a spanner in the works. High concentrations of organic compounds, low ORP, weak water flow, the accumulation of detritus, and spiked nutrients all contribute to excessive diatom growth. An improved water change schedule and aggressive protein skimming alone can easily temper or eradicate even the worst blooms of diatoms in aquariums. A well-designed protein skimmer is perhaps the single most effective method of control for most aquarists when keeping a casual reef display. If there is a problem with persistent diatom growth, do examine if the skimmer has been running at full efficiency and producing dark skimmate daily. Many other viable and effective alternatives to diatom control exist beyond a good protein skimmer. Vegetable filters and refugiums can compete with diatoms for growth-limiting nutrients. Corals in a reef display can do the same. Dilution of micronutrients through frequent water changes may also be of use. If frequent water changes do not offer some relief, an examination of the quality of incoming water may be in order. There are also numerous natural herbivores of these largely palatable algae. Among invertebrates, the ever-popular *Astraea* and *Turbo* snail species do a fine job. Do remember to avoid keeping *Astraea* snails in displays with large expanses of fine sand as these "rock-dwellers" do not right themselves well on soft substrates. *Strombus* and *Nassarius* snails are better suited for disturbing or dissuading the growth of diatoms on sand. *Ctenochaetus* tangs, such as the *C. hawaiiensis* and *C. stiratus* (Kole tang), are magnificent

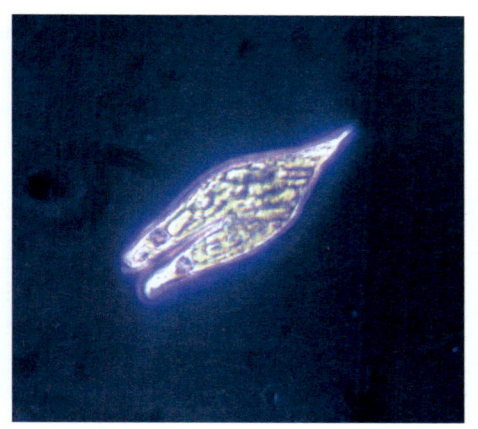

Dinoflagellate - *Euglena*  *Gelidium sp.* - a nuisance Red Hair Algae (*AC*)  Red Kelp - a desirable Rhodophyte (*A. Calfo*)

grazers of this and many other nuisance algae.

• *Dictyosphaeria* (AKA Buttonweed) - see **Bubble Algae**

• *Dictyota* the brown algae known as "**Y-Branch Algae**": *Dictyota* is commonly acquired incidentally on rock and can be quite beautiful to observe as some species have a remarkable iridescent blue color. There are yellow-brown, purple and green-hued varieties as well. Aquarists beware, however! This uniquely formed beauty has all of the potential makings of a dreadful plague or nuisance in the aquarium. It may bloom and grow quickly, and it fragments and spreads easily if attempts are made to extract it manually. This algae naturally grows in most marine environments over a wide range of physical parameters including fine sand beds. *Dictyota* seems to have a preference for hard substrates including coral skeletons (corallums), and once established, it can beat living cnidarian tissue into submission and smother living colonies. Few herbivores will eat this seemingly unpalatable algae and its excessive growth can shade or out-compete many desirable plants and invertebrates in the display. It does not weaken or crash easily and seems to pose no remarkable threat to water quality when compared to other algae. Aesthetically, it is rather attractive to some people. If *Dictyota* is observed in an aquarium, keep a very close eye on the size of the colony and control growth swiftly if necessary. Although any algae might arguably be useful as a vegetable filter, deliberate and excessive cultivation of this species is not recommended when so many others are available and better suited with regard to nutrient export efficiency and convenience in culture. For all practical purposes, *Dictyopteris* and *Stylopodium* are handled with the same care and regard in the aquarium.

• "**Red *Dictyota*** " AKA "Red Y-branch" (*Nitophyllum*): This alga is a very flat, forked red variety with curled edges and resembles the brown alga, *Dictyota*, in form. It has a color range and texture that may also resemble *Halymenia* (Red Leaf Algae). Likewise, this delicate and decorative genus is readily eaten by many herbivores and is easily controlled. It may or may not be fair to call this alga a nuisance.

• **Dinoflagellates** are another large and diverse group of organisms that have tremendous importance in the world. The many ways that these Protists impact people's lives, economies, and environment at large is really fascinating for even non-science types. They truly run the gamut of form and function, ranging from photosynthetic to non-photosynthetic, endosymbiotic, free-living, nutritious to toxic, and even parasitic. With a couple thousand species and more than one hundred genera in the group, dinoflagellates are characterized as single-celled, with two flagellar processes: one circular, and one distal. Ultimately, aquarists can rest assured that most species are safe and manageable.

In aquariology, there are several very prominent dinoflagellates of interest. Most notably, reef aquarists are likely desirous of learning more about the endosymbiotic, photosynthetic zooxanthellae that provide most of the sustenance, in the form of fixed carbon as sugars and oxygen, in symbiosis with many popular corals, anemones and Tridacnid clams. Beneficial dinoflagellate symbiosis can also be observed in various flatworms and foraminiferans.

There are also some dangerous members of this group. For example, dinoflagellates are responsible for the infamous "Red Tide" conditions that kill masses of fish species and can afflict humans that ingest the dangerous neurotoxin produced by the algae. The event is uncommon in aquariums, although it has been observed. More prevalent toxic dinoflagellates can be found in the common brown slime algae (*Gambierdiscus toxicus*) that is seen in aquariums. This nuisance growth is categorized by having trapped bubbles within its matrix and is decidedly toxic to many herbivores including some snails and fishes! Improved water circulation can temper growths indirectly by keeping detritus and other nutritive organics in suspension and providing a better chance at export.

Living Filters: Algae & Plants

Gracilaria sp.   *Galaxaura* sp. - Boneweed   *Halimeda* sp. (*LG*)

Deliberately stirring or agitating slime algae is never recommended. Slurp siphon this alga out directly with straight tubing, not a tapered gravel siphon, and focus on more aggressive means of nutrient export in order to starve the nuisance into submission.

• ***Enteromorpha***… see "Hair Algae"

• *Galaxaura* AKA **Thicket Algae** or **Boneweed**: This unique and attractive red alga is a heavily calcified Rhodophyte with distinctly dichotomous segments of pink to red color. These "branches" (segments) may be tubular or flat, smooth or fuzzy. Colonies remind one of a coarsely structured bird's nest with stiff and profuse "crunchy" branches. *Galaxaura* forms small bushes up to 6" (15 cm) in diameter and attaches by a single holdfast. Generally found in shallow water, this alga can be quite adaptable in captivity if given stable and supportive water quality. Moderate to bright light and tempered water flow will be appreciated. This alga is often found in protected areas by patch reefs and attached to hard substrates. Its purpose and participation in the aquarium is almost purely aesthetic.

• *Gelidium* (and *Gelidiella*) AKA **Red hair algae & Red "Kelp"**: An immediate distinction must be made between the desirable forms of this family which are rather plant-like, and several other nuisance species, such as *Gelidium pusillum*, which are wiry and problematic. Desirable forms of red kelp are deliberately collected for the aquarium trade and resemble fern-like fronds ranging in color from pink to red to orange, with some yellow shades. They can be quite leafy and decorative. The genera at large are highly variable in form, however many are intricately branching. In aquariums, they generally do not grow fast or fare well. In part, this may be due to their need for very bright light and extremes of water flow. Most species occur in very shallow water and many are intertidal. Some species in this group, particularly the nuisance varieties, occur interspersed with various algal turfs on hard substrates. If an aquarist manages to succeed in culturing members of Gelidiaceae, it is best done in a dedicated, remote vessel to prevent encroachment on scleractinians in the reef display proper.

Desirable species are prized as ornaments, nuisance species should be promptly extricated, and none in this family are remarkably useful for aquarium ecology or filtration dynamics. Wiry varieties are most always short-napped less than 1" (25 mm), while arborescent forms aspire to 10" (2.5 cm) or more. *Gelidium* is a commercially significant global commodity that is harvested for its agar base. Several other Gelidiales, including *Onikusa*, *Pterocladia*, and *Pterocladiella* share gross similarities.

• *Gelidiopsis* AKA **Burgundy-Red Wire Algae**: The sparse and wiry filaments of this alga are neither attractive nor desirable. Colonies rarely form any appreciable mass to be of use for nutrient export. Attempts at mechanical harvest by hand, or natural harvest by grazing herbivores, often serve to fragment the colony and mitigate potential problems with its proliferation. Although generally slow growing, *Gelidiopsis* can establish itself upon coral skeletons and encroach upon living tissue. Natural herbivory is challenging for all but the most robust grazers due to this alga's integrity. This genus has few, if any, redeeming qualities to marine aquarists and control is recommended early.

• *Gracilaria* are some of the finest macroalgae for marine aquarium use. They are not only decorative, but also remarkably friendly and safe for deliberate culture *en masse* to aid in nutrient export. Additionally, this alga is simply a fabulous food genus for both hobbyists and fishes. They go by several common names and some are marketed as an absolutely heavenly food for Tangs. Numerous invertebrate and vertebrate grazers alike covet *Gracilaria* and it is farmed extensively for both uses. Long before humans ever thought to keep *Gracilaria* in aquariums this algae was known by the name "Ogo." It has been farmed for centuries by the Hawaiian people for human consumption and is utilized in a variety of Hawaiian and Asian cuisines. Much of the cultured algae in the aquarium trade still come from Hawaii, where they are an economically important product. The tremendous demand for Ogo and resulting over-collection in recent decades has forced Hawaiian seaweed

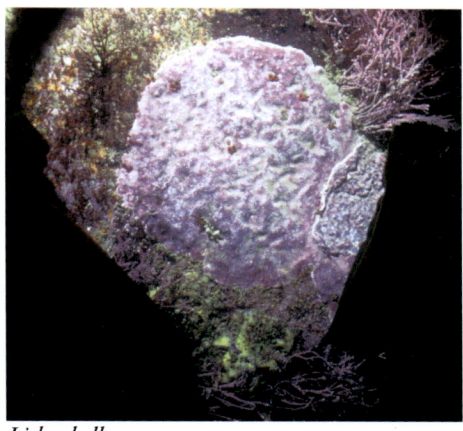

Halymenia sp.　　　　Kelp - the *Sargassum* variety (*LG*)　　　　*Lithophyllum* sp.

farmers to import a Floridian species, *Gracilaria tikvaheae*. *Gracilaria* has considerable global commercial appeal for various uses, including high quality agar, and it is highly regarded for its ease of management, high productivity, and minimal nuisance epiphytic associations.

The genus *Gracilaria* at a glance encompasses approximately forty species. Its scientific name means "slender" or "delicate" and certainly describes this macroalgae well. One of the most common species encountered in the aquarium trade is *Gracilaria parvispora*, a deep red variety that grows in a floating mass of fronds known as thalli. As an aquarium specimen, *G. parvispora* grows rapidly and is a highly efficient consumer of micronutrients. Cultivation of *Gracilaria* is a relatively simple matter: provide bright light in shallow water with brisk water flow. Nutrient-rich water overflowing from the display aquarium into a *Gracilaria*-filled vessel makes for a productive refugium. This algae will perform as an efficient vegetable filter with diligent harvest and its structure will provide a productive environment for culturing desirable plankton species including microcrustaceans and epiphytic matter. One of the keys to success in growing *Gracilaria* is to keep the algae in constant motion within the water column. Words such as "rolling" or "tumbling" are often used to describe the proper movement of Ogo in culture. This helps to prevent detritus, debris, and epiphytic materials from settling upon and stifling the fronds. Be mindful not to apply excessive flow, however, which can slow growth or damage the colony altogether.

By any definition, *Gracilaria* is one of the best algae for display and culture in marine aquariums. It is nutritious, functional, friendly, easily controlled and innocuous with very little of the "baggage" associated with some other popular macroalgae such as the potentially invasive and noxious genus, *Caulerpa*.

• **Hair algae** (Green) including *Bryopsis, Chaetomorpha, Derbesia,* and *Enteromorpha* commonly: Algae in this group are an interesting paradox with great potential to be a scourge and/or a boon for marine aquariums.

Managed properly in refugia or ancillary vessels, hair algae are tremendously effective and natural substrates for desirable microfauna including worms and so much more. Many popular fishes, like Rainford gobies and mandarin dragonets, can forage a healthy lifetime in beds of hair algae. Young growth in hair algae is also an excellent food source for herbivorous grazing fishes and invertebrates, although even the best of them have difficulty mowing down long, mature and sinewy aged growths. Hint: help tangs along in problem tanks by cropping old hair algae down to a length shorter than 1" (25 mm). This will allow them to more easily control and manage new, short growth. It's generally a simple pest to control when it blooms, these species are easily limited by nutrient control. Aggressive nutrient export and address of water quality at large is the primary course of action. Use high alkalinity (10-12 dKH) and various natural predators in support of control measures. Hair algae will be readily consumed by Echinoids, Gastropods, some fishes, and microcrustaceans.

~ *Bryopsis* is a distinct hair algae often found in feathery tufts with a rich green or blue tint of color. It is perhaps the most difficult hair algae to control as a nuisance. *Bryopsis* maintains a tenacious grip on hard substrates, is slow to respond to nutrient control and releases distinct noxious exudations to reduce predation. Manual removal is often necessary with extended diligence in follow-up to water quality and nutrient export processes. By any definition, this is an unfavorable "hair algae" for the marine aquarium, requiring labor-intensive supervision at best to harness.

~ *Chaetomorpha* is the darling of "hair algae" if there is such a thing to reef aquarists. It is a coarse and wiry green algae that can occur in free-formed clumps or fixed to hard substrates. It can easily be removed manually and is eaten by some popular herbivores, although the coarse texture of some forms in this genus make it difficult to graze for others. *Chaetomorpha* is fast growing and a most convenient vehicle for nutrient export in refugiums as a living vegetable filter. It is an equally ideal substrate for cultivating natural plankton. Utilize bright daylight in shallow water with good water flow for deliberate culture.

Living Filters: Algae & Plants

~ *Derbesia* is a very fine-threaded hair algae that is sometimes described as silky or slimy to the touch and a greater nuisance than most other genera in this group. It is distinguished as more invasive and slower to concede to nutrient and herbivory control methods.

~ *Enteromorpha* is a medium coarse, light-green hair algae that is tubular in structure. *Enteromorpha* tends to grow in tufts or clusters as an epiphyte (growing on organisms like other plants and algae). This genus is remarkably cosmopolitan, adaptable, and found in brackish and marine environments alike. Although still regarded as a nuisance, it may be less suffocating than *Derbesia* and more responsive to control measures.

~ Red Hair Algae... see *Polysiphonia*

• **Halimeda**, AKA Cactus algae, Baby's Bows, and Money plant: This green algae was one of the earliest macroalgae to be cultured in aquariums (circa 1960s). There are more than a few species from this group commonly found in the trade including: *H. incrassata*, *H. discoidea* and *H. opuntia*. *Halimeda* is one of the toughest calcareous genera available. Most are dark green, but a few occur in hues of light green or even bright yellow. There are quite a few species and varieties among *Halimeda*. The genus has been found over a very wide range with colonies observed below 200 feet! As such, the range of needs and tolerances among the many specimens that may be found in the trade will vary remarkably. Nonetheless, good water quality, nutrients and minerals are necessary to succeed with these algae. In the aquarium, they can be planted in rock crevices, however they fare better planted in sand at a depth of 2" (5 cm) or more. Aquarists should know that newly acquired specimens will often die back entirely, then grow anew months later from the sand or from settled colonies on the rockscape. Acts of sexual reproduction with *Halimeda* in the aquarium have been observed commonly and demand prompt address of water quality with aggressive nutrient export and water changes. In such events, the parent colony appears bleached and bespeckled afterwards and will die. In the wild, it is known to degrade and be a significant and extraordinary component of coral sand by weight and volume. In part due to its inclusion of calcium in growth, this algae is unpalatable to many grazers for its gritty texture and apparent "taste." From a practical standpoint, diminished degrees of herbivory are of little consequence because *Halimeda* is remarkably easy to control otherwise and poses no more threat, by noxious exudations or excessive growth, than any other macroalgae commonly kept. If anything, *Halimeda* has been demonstrated to be more useful and less noxious than most others from several perspectives. Consistent calcium and alkalinity levels may be evidenced by steady growth in *Halimeda* species. As such, some aquarists like to watch this macroalgae as a water quality indicator for a regular "coin" or "half-coin" growth per day to have faith that they can support other desirable calcifying organisms such as hermatypic corals. Fears of it competing too aggressively with such corals for available minerals are largely unfounded. As one might assume, limiting calcium is one of the easiest ways to control growth of this algae. Again, manual extraction is not very difficult or inconvenient. Regarding noxious exudations and discolorants in the water, *Halimeda* is also only weakly contributory. Overall, this genus of calcareous algae is well-suited for culture and display in the aquarium.

Note: "Catch me if you can!" **Chloroplast migration and Halimeda changing colors:** Chloroplast migration is a sometimes unnoticed and oftentimes misunderstood phenomenon of *Halimeda* biology. Some aquarists may have noticed that *Halimeda* changes color between night and day, as well as on stress of import or neglect. They simply pale in color from dark, rich green pigmentation to a weak yellow during spells of darkness and deficient illumination. With proper illumination, they resume dense green pigmentation. The mechanism for the magic is a redistribution of chloroplasts in their cells. By day, these chloroplasts are packed tightly against the outer wall of the structure like Bob Fenner with his nose to the passenger-door window of an automobile when we take him on a drive for ice cream. At night, these chloroplasts dive deep into the segment and impart a pale, if not chalky, color to the body of the calcareous genus. The color change is remarkable and can be quite dramatic. As one might assume, photosynthetic activity wanes at this time too. The process surely demands considerable energy to conduct, and the reason for it doing this is not entirely clear. One of the theories for this phenomenon is to minimize damage from predation as concentrated chloroplasts are protected from rasping grazers at night. Another theory is that new segments of this fast-growing calcareous alga are supplied quickly from these chloroplasts which are stored nightly. With a growth of one-half to one new segment daily for many species, chloroplast migration imparts the means for a new division to begin photosynthesis and calcification almost immediately.

• *Halymenia* AKA **Red Leaf Algae**: *Halymenia* species are some of the most beautiful and decorative marine algae available. Occurring commonly in watery colors of red, pink, orange and sometimes yellow, this algae has a welcome presence in most systems. The leafy structure is delicate, almost gelatinous to the touch, and is very palatable to many herbivores. Blooms and fast growth are generally fueled by excess nutrients and can be controlled by limiting a primary food source such as nitrates. It is sometimes collected deliberately and oftentimes develops from unnoticed origins on live rock. Growing from a single, central holdfast, transplantation and import of this flimsy

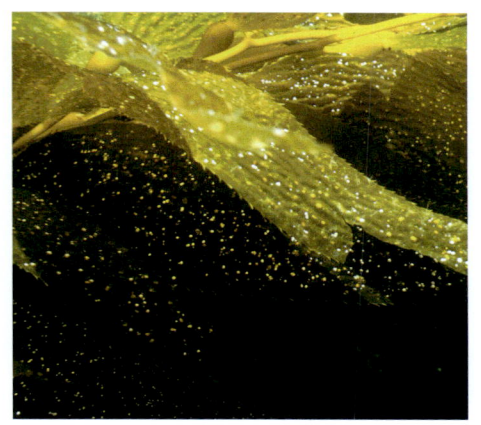

Lobophora sp.   *Macrocystis pyrifera* - Giant Kelp   *Rhizophora mangle* - Mangrove

alga are challenging at best. Nonetheless, there are more than a few exceptionally attractive species in this genus to attract aquarists. They do not seem to be remarkably noxious, nor are they difficult to extract if they become a nuisance. *Halymenia* may be targeted for ornamentation and utilitarian uses in the aquarium.

• *Hypnea* AKA **Hookweed:** This is one of the most unrecognized red algae found in aquariums. The word "unrecognized" is chosen because the alga generally occurs in most any color but red! Specifically, it can be found in brown, gray, purple, pink and red; sometimes with a magnificent blue iridescent sheen to it. Most often, it is an orange or watery tan-brown color. Many *Hypnea* have a unique and attractive morphology with forked branches that finish with pointed tips. Not much has been written in popular literature about experience with this algae in aquariums. Due caution in its culture must be exercised though with regard to nutrient control and proper lighting. In the wild environment, *Hypnea* has gained global attention as an invasive species with many thousands of tons of this alga washing up on shores. In some areas, such as Hawaii, *Hypnea musciformis* is a dominant form of algae, found in patches with nearly 100% coverage. The impact of this sort of biomass accumulation is grave: loss of biodiversity, habitat loss, and all of the implications from these effects higher up the food web. The potential to be a nuisance in aquariums exists just the same. Hookweed can grow attached to hard substrates or as an epiphyte and can be spread by the tiniest fragment. Lighter colors of yellow, tan and orange occur in brightly lit specimens. Darker brown or red colors may be induced in lower light environments. Most herbivores inclined to sample this weakly structured algae have little trouble consuming it. Control excessive growth in the aquarium as necessary, either naturally or mechanically, in order to reduce competition with slower growing photosynthetic organisms in the display. *Hypnea* has been considered commercially for a variety of reasons including potential as a source of carrageenan. In aquaculture, it is recognized for its high productivity and nutrient export capabilities, however it is disdained for its difficulties to manage and high rates of incidental nuisance epiphytes.

• *Jania*... see **Twig Algae**

• **Kelp** (various Brown and Red Algae)... see *Macrocystis* or *Sargassum* for more information

• *Lithophyllum* is one of the most widespread genera of marine algae in the world, reaching both poles of the globe and everywhere in between. It can be found at depths of 500 feet, but is commonly observed on reef crests with amazing surf. *Lithophyllum* forms range from thin calcareous growths to stout, dense and knobby masses that are commonly mistaken for stony coral. The importance of this calcifying alga in the marine environment cannot be overstated. For additional information, see "coralline algae" in this list.

• *Lobophora* AKA **Brown Wafer Algae** or **Encrusting Fan-Leaf Algae**: This algae resembles several other common genera including *Padina, Mesophora* and *Pseudolithoderma*. As its common names plainly suggest, *Lobophora* species are fan-shaped algae that encrust upon rocks and glass in the aquarium. It is found in both the tropical Pacific and the Caribbean/Atlantic and occurs in a wide range of colors and shades including brown, red, and green, although curled brown species are most common. At first, the prostrate blades of this algae are often mistaken for brown coralline algae. One touch, however, and the gelatinous and almost rubbery feel of this algae lets one know that it is not calcareous. It is not a significant algae to aquarists with regard for useful inclusion in living filtration dynamics, nor is it especially noxious or inclined to become a nuisance. *Lobophora* may be enjoyed in the marine aquarium for its unique appearance, and it can be controlled by limiting nutrients or using harsh grazers, such as *Diadema*, as necessary.

• *Macrocystis pyrifera* AKA **Giant Kelp**: Perhaps the largest, fastest growing organism on the planet, this species grows up to two hundred or so feet long and more than a

Living Filters: Algae & Plants   103

Nemastoma sp.   *Neomeris* sp. - Fuzzy Tip Algae (*LG*)   *Ostreobium* sp., Boring Green Algae (*A. Calfo*)

foot (30 cm) per day. Aquarists are not likely to encounter or manage this crucial species at home, though it is commonly displayed in huge public aquariums. Its obvious needs for temperate waters and vast amounts of space and nutrients are impractical. Nonetheless, it is a wonder of the natural world and thought-provoking for aquarists that consider its pivotal role in a specific biotope with hopes of understanding and replicating other unique biotopes in the captive marine aquarium.

• **Mangroves**: Much has been written about the ecological enormity of the importance of mangrove trees in the marine environment. At first glance, it is easy to notice that they provide habitat for countless life forms above and below the surface as well as at the very water's edge. Birds, reptiles, mammals, fishes and invertebrates exploit mangrove communities for food, shelter and reproductive activities. The utilization of these communities as a nursery environment for larval species has extraordinary ramifications far up the web of life. The very structure of these tangled trees and lesser plants, including no less than fifty-three families of plants with species called "mangroves," is crucial to coastlines for protection from erosion and storms, and in the stabilization sediments from run-off that could otherwise pollute the reef community and subsequently destroy the fish and invertebrate life forms dependant on all. The protection of mangrove habitats is crucial for the survival of coral reef ecosystems and all that depend on them, from the fisherman to the fished and down to the living substrate.

Three genera of mangrove trees are commonly recognized, although aquarists are predominantly interested in the most aquatic variety: the Red Mangrove (*Rhizophora mangle*). The Black Mangrove (*Avicenia germinans*) and the White Mangrove (*Laguncularia racemosa*) are not readily tolerant of full submersion in seawater.

A summary of each mangrove genera:

Red Mangrove (*Rhizophora mangle*)

- can live submersed, emersed or fully terrestrial (if well-hydrated)
- tolerates fresh, brackish or full seawater, but cannot be freely moved between gradients
- favors fine sand or muddy substrates, although it can be grown in coarse substrates or none at all (hydroponics)
- is the most temperature sensitive of three genera listed here and requires warmer temperatures
- is the most sensitive to pruning. Immature or improper cuts can harm or kill some trees

Black Mangrove (*Avicenia germinans*)

- can live as emersed or fully terrestrial
- tolerates some salted air and water but favors fresh environments
- favors substrates with a decidedly significant soil or muddy component, and tolerates sand
- is reasonably tolerant of pruning and tolerates mild frost conditions

White Mangrove (*Laguncularia racemosa*)

- lives as a fully terrestrial plant
- naturally must tolerate salted air (coastal) but suffers in excess salted water
- is very tolerant of pruning
- is moderately to very tolerant of occasional frost conditions

For aquaristic purposes, the red mangrove is specifically addressed. Indeed, they are the first species likely to be encountered by aquarists. They are the most commonly photographed for their magnificent aerial prop roots (the arched and exposed knobby knees plunging into the coastline and shallows). Also, they are by far the most important genera of this family to marine environments.

It is their very elaborate and extensive root system that must be given due consideration in an aquarium. Even a seedling mangrove can develop a formidable root system that can stress or damage glass or acrylic refugiums in as little as three years. Therefore, it is recommended that mangroves are potted in containers and vessels that are as large as possible, which will minimize future disturbances of the tree without making it overly difficult to service for transplantation if necessary. Rest assured that growth is so slow that these fascinating Angiosperms can be enjoyed perhaps indefinitely in an aquarium system.

Although the collection of mangrove trees is outright forbidden in many areas, the harvest of their abundant seeds (known as **propagules**) is fairly unrestricted on the whole. Un-sprouted propagules look like long green cigars with a narrowly tapered end, where the leaves and branching canopy will sprout from, and a thickened, blunt end that is often tinged brown. The blunt end of a propagule is appropriately weighted by design to increase the likelihood of finding its way into a substrate when cast or carried adrift. **Note**: un-sprouted seedlings can survive out of water in temperate conditions for up to a year. Propagules can be sprouted in fresh, brackish or saltwater and will do so even fully submerged, although this is not recommended for aesthetics and the cultivation of prop roots. One thing, however, is certain: a mangrove cannot be moved between saline gradients quickly, if at all! Aquarists are strongly advised to only acquire un-sprouted seedlings. If there are any roots or leaves in evidence on arrival, one must be told what salinity the propagules were sprouted in. Failure to abide by this advice is likely to be fatal for the seedling, evidenced by a shriveled desiccation and demise within weeks. After determining the nature and need of the seed or seedling, a serious planting decision must be made. The matter really boils down to long-term plans versus short-term residence, and the encouragement of aerial roots versus profuse establishment of capillary roots in a substrate. When mangroves are used primarily for their aesthetic function, the encouragement of decorative aerial roots is a large factor. One will find that the cultivation of arched prop roots on mangroves is very easy to encourage, or fail to encourage, despite the lack of tidal cycles in the aquarium.

In the absence of tidal cycles, red mangrove propagules can be trained to develop noble aerial roots by beginning their life tied gently with flexible gardener's tape to a thin plastic pipe or rod. Be sure to use flexible tape, as a rigid tie will otherwise cut into the plant as it grows. An un-sprouted seedling can then be tethered at a depth where only the lower 1/3 of the propagule, the thick, blunt discolored end, is initially submerged in water. Roots will sprout, incidentally, before leaves will. As roots begin to grow and develop, the body of the plant is slowly moved upwards on the stake. In this teasing manner, strong roots will grow thickened and extensive to support the weight of the body above the water. It will take many months before the body of the propagule can grow above the water with an arched and anchored root system. The plant will likely need to be potted in the future, but only after satisfactory roots have developed above the water surface.

If instead the aquarist simply sticks a propagule into a bed of sand like a dart, root development will occur fast and profusely. Note that prop roots are unlikely at all in this situation without a replication of tides and the exposure of some roots to air. Red mangroves will grow in a wide range of substrates, however they prefer fine sand and muddy sediments ideally. Fertilizing the substrate may be helpful, but it is unnecessary and potentially dangerous in average aquariums where levels of dissolved organics are typically high. One exception to the matter may be elemental magnesium; mangroves have been implicated in aquariums as depleting Mg to the point where it skews the balance of minerals. Trace elements may be supplemented deliberately if tested for and monitored. Alternatively, one may simply rely on regular partial water changes for this and contributions to overall water quality. As with most vegetable filters and refugiums, a mangrove basin should be fed raw overflowing water from the display for opportunities to exploit dissolved and particulate matter. It would be counter-intuitive to feed mangroves clean-filtered water given their natural habitat and abilities.

Lighting is a simple matter with mangroves. They are quite adaptable to a wide range of light but prefer bright illumination. Expensive reef aquarium fixtures are not necessary however. Common plant bulbs from the local hardware store are quite fine. Many aquarists have grown nice mangroves under incandescent plant-growth spectrum floodlights or spotlights, including mercury vapor and metal halide lamps. Fluorescent lamps are found in useful spectrums, however they lack intensity in all but the closest placement applications with mangroves.

Pruning red mangroves is a sensitive matter and rather a moot point for most with consideration for their categorical slow growth in captivity. If one must trim a tree, be sure to resist any pruning until after the axial tip has branched. Damage to the lead tip before splitting can be fatal for young specimens While discussing the topic of growth, it

Padina sp. - Funnelweed or Scroll Algae.

Penicillus dumentosus - Neptune's Shaving Brush

Porolithon sp. - Purple Coralline Algae

should be interesting to note that irrigation of the leaves is nearly as much of a limiting factor of growth in *Rhizophora* as light and nutrients! Misting the leaves daily, or at least several times weekly with purified, mineral-free water helps to rinse away the salt crystals exported through the leaves of this marine plant: a fascinating adaptation! With other aspects of good mangrove husbandry in order, a lack of leaf irrigation will still significantly reduce growth.

It is surprising to hear aquarists debate the efficacy of mangroves in the marine aquarium as vehicles for nutrient export when their functional abilities can be weighed clearly against their growth, which is dreadfully slow. In fact, numerous governments recognize their naturally sluggish growth on native coastlines where legislation controls or forbids pruning. At large, even occasional storm damage can be devastating. In wild habitats, they are outstanding vehicles for nutrient export, fixing nutrients in their enormous and collective biomass. In the aquarium, however, there is not a forest of twenty or thirty feet tall mangroves! The scrappy little seedlings that aquarists do have, instead, demonstrate leaf growth concurrent with leaf drop at times. The proof is in the pudding, as they say: **they are weak nutrient export mechanisms** in the aquarium because they do not produce stable or harvestable mass quickly. For the aquarist looking for a vegetable filter, there is a long list of algae, plants, and seagrasses that can provide greater harvestable mass for nutrient export. Mangroves are simply marvelous to look at and a pleasure to include for aesthetic, although minor, biotic advantages in the home reef ecosystem.

Summary of Husbandry for the Red Mangrove in the aquarium:

- Acquire un-sprouted seedlings (propagules) whenever possible

- Discover and match the salinity of seedlings purchased if they have already sprouted

- Fine sand and muddy sediments are preferable to coarse media for rooting

- There is no such thing as "too bright" when lighting properly acclimated mangroves

- Leaves must be irrigated several times weekly or better to shed exuded salt crystals

There are many aesthetic and mildly utilitarian applications for mangroves in aquaria and few notable disadvantages. One of the best and most natural ways to enjoy this rare flowering marine plant is in an upstream, fishless, deep sand bed refugium. Beyond significant natural nitrate reduction (NNR), the copious production of natural plankton and epiphytic matter in a mock mangrove microcosm can be enjoyed. If the fishless requirement is compromised, consider a school of cardinal fishes living in a magnificent commensal display with the remarkably agile *Diadema* urchin among the roots. An aquarist is limited almost only by his or her imagination in the application of mangrove trees in the captive marine aquarium.

• *Nemastoma* is a unique, if not bizarre, red alga that forms bulbous, translucent, branching and club-fingered growths. The body of the mass is usually quite uniform in thin, watery colors of red, maroon, orange and yellow hues. The tips are often light or contrasting in color. Little is published about this fascinating genus regarding its use in aquariums. It has a wide distribution in the world, but is commonly acquired by aquarists as a hitchhiker with live products such as coral, live rock, etc., from reefs in Fiji and Indonesia.

• *Neomeris,* AKA **Fuzzy Tip algae** or **Spindleweed**: aquarists most often discover this innocuous, calcareous green alga as an incidental organism imported with live rock. It is more likely to be found on rocky substrates than are other "plant-like" calcareous algae. They are not inclined to spread prolifically in the aquarium but can be maintained with stable water quality Although unique and attractive to many aquarists, this alga is not actively cultivated and serves little purpose beyond ornamentation.

Sargassum hystrix - a variety of Kelp.   *Thalassia* spp. - Turtle Grass   *Syringodium* sp. - Manatee Grass

They are quite adaptable to a wide range of light including modest levels. *Neomeris* can be found toward 100 feet at depth, in low light or shade, and often away from the reef.

• *Nitophyllum,* the red alga known as "Red *Dictyota*" or "Red Y-branch algae" (see *Dictyota*)

• *Ostreobium* sp., **Staining Green Algae** (AKA Boring green algae): A dreadful but slow growing organism. Exactly as its name suggests, this alga appears like a green stain on the exposed corallum of stony coral into which it bores. This nuisance is generally stimulated or limited by nutrients such as nitrate and phosphate, which ironically limit calcification in the victimized scleractinian coral. *Ostreobium* may be tempered by some echinoderms such as the harsh grazing *Diadema* species, but herbivorous fishes are unlikely to make any significant impact on this alga. Nutrient control is recommended to temper this organism.

• *Padina,* **Scroll Algae** (AKA Funnelweed) is an attractive and uniquely-formed brown alga that occurs in hues of gray, white, yellow and pale brown. A few especially handsome varieties also occur with iridescent blue/green highlights. Once acclimated, they do quite well in captivity, but they are notoriously difficult to ship and transplant. Subsequently, they are best acquired while attached to a hard substrate. Most aquarists get scroll algae from live rock. It can grow vigorously and may be regarded as a nuisance, but it is rather easy to control mechanically and has a history or likelihood of volatility and fragility. Aquarists have little to worry about with this alga as a long-term problem since it is not believed to be especially noxious and it is readily grazed by a variety of herbivores. Deliberate culture should emphasize a warm lighting spectrum as most *Padina* hail from very shallow waters less than two meters in depth. Strong water flow and stable chemistry are crucial for long-term success with this weakly calcareous alga.

• *Penicillus,* AKA Neptune's **Shaving Brush** is an attractive calcareous algae so-named for its uprising tufts of dense, bristled filaments on the apex of a stalk (reports are still unconfirmed if Neptune actually uses these living artifacts to shave with). Sand-anchored by a rhizoidal ball, most species for the aquarium trade are collected in shallow waters from protected sand flats or from within meadows of seagrass. It is important to plant *Penicillus* at depths greater than 2" (5 cm) in fine sand or a somewhat muddy substrate. Old growth will often die and degrade on transplant only to give rise to new buds weeks later; do not throw out seemingly dead specimens! Like most calcareous algae, Shaving Brushes are used more for their decorative appearance. *Penicillus* does not grow quick enough to be used in vegetable filters, however it can serve admirably in fishless refugiums to provide habitat and nutrients for natural plankton.

• *Peyssonnelia,* AKA **Maroon Coralline** algae: This calcified red alga is one of the most welcome, desired and sought-after corallines by aquarists. It is very attractive in color, prefers low light, and is forgiving of low water flow. *Peyssonnelia* on the reef occurs in shaded environments when shallow, and can also be found at amazing depths (200m). The color is rich and typically striking in shades of maroon, red and burgundy. Its structure is weak and rather crustose in plating forms which aids in dispersion with the slightest damage and distribution of fragments. *Peyssonnelia* requires good water quality and very stable mineral levels; consistently moderate calcium and alkalinity levels will grow this alga faster than random or occasionally high levels. Organic, sugar-based calcium (gluconate) seems to be stimulating to grow this and many coralline algae.

• *Polysiphonia,* AKA **Red Turf/Hair Algae**: We suggest aquarists resist the casual cultivation of *Polysiphonia*. This filamentous alga may occur in long strands, however it gets more attention as a tough little turf species. It can be used and is productive as a means of nutrient export in vegetable filters. Left unchecked in the display proper, however, *Polysiphonia* can be a wicked foe to corals and other sessile reef invertebrates. Tenacious settlements of this turf alga may encroach and overtake other desirable

Living Filters: Algae & Plants

Zostera sp. - Eel Grass (*D. Fenner*)    *Halophila* sp. - Paddle Weed (*D. Fenner*)    *Turbinaria* sp. - Saucer Leaf Algae

species. It is especially harmful to scleractinians once established upon the corallum (skeleton) of a living animal. Natural herbivory is randomly successful and manual extraction is tedious at best. Limiting nutrients may be the best long-term means of control for this and most other nuisance algae. *Centroceras*, *Ceramium* and sporophytic *Asparagopsis* are recognized and treated similarly.

• *Porolithon*, AKA **Purple Coralline algae:** Little can be said about this desirable calcareous red algae for aquarium use that hasn't been mentioned in the section on corallines or with similar genera elsewhere. It is noteworthy to mention that *Porolithon* is perhaps the single most important reef builder, beyond stony corals, on many reefs. It is also categorically one of the most durable calcareous algae, tolerating extremely bright light, staggeringly powerful water flow and long periods of desiccation. It is the diametric opposite of *Peyssonnelia* maroon coralline algae on the reef and in aquariums with regard to living requirements.

• **Red Hair Algae**… see *Polysiphonia*

• *Rhipocephalus*, AKA **Pinecone Algae:** This unique calcareous green alga grows 5" (125 mm) in height and resembles a stalked pinecone by growing overlapping thalli, which emanate from a single, attached stalk. It has similar requirements in husbandry to other calcareous algae such as *Halimeda* and *Penicillus*. Pinecone algae is found mostly in shallow waters on rock and in silt-laden or muddy soft sand substrates, but some specimens can be adapted or cultured in low light. Stable water quality with adequate calcium and alkalinity are necessary for success. They grow too slow to be of use as a vehicle for nutrient export, but can provide useful habitat for micro-organisms in fishless refugiums. This algae is collected deliberately for aquarium use in the Atlantic, although old growth on newly acquired specimens does not readily acclimate to captive parameters. Rest assured that after appearing to die, many of these algae give rise to new buds within months. Plant specimens in fine sand at a depth of 2" (5 cm) or greater and leave them alone to thrive or recede with hope that they will sprout anew.

* *Rhizophora mangle*… see **Mangroves** (the Red Mangrove)

* *Rhodymenia* species are leafy red algae that have a historical and extraordinary popularity outside of aquariology. To non-aquarists, they are known by many names including the popular title, Dulse Leaf. One of the many distinctions it carries is that of a "famine food" eaten raw, or cooked historically, by peoples in tough times. Today, it has numerous places in culinary preparations worldwide from simple coastal cuisines to students of organic living. It also has numerous uses in homeopathic medicinal preparations. They are categorically attractive in popular forms with flat, branching dichotomous leaves in handsome deep red and burgundy colors. Their presence in the aquarium is more ornamental than utilitarian.

• *Sargassum*, AKA Sargasso weed, *Sargassum* Seaweed, Gulf weed, Brown Kelp: This alga is more plant-like in form than most any other that is commonly seen in aquariums. It is often called "kelp" although most encrusting varieties will not grow more than 20" (50 cm) high in the aquarium, which is miniscule compared to its relative brown alga, the Giant Kelp (*Macrocystis pyrifera*). It occurs in very large colonies attached as well as free-floating and provides the diver and aquarist alike with endless possibilities for observing many forms of micro and macro life associated with it such as fishes, plankton, shrimp, crabs, and so much more. *Sargassum* species occur in fragments and colonies mere inches in length to sections up to 5' (150 cm) long, most often in shades of brown, but some with yellow, orange or green hues. The large masses of this brown alga are one of the crucial nursery niches for larval species in marine habitats. Attached forms produce marvelous little air sacs for the buoyancy of their arborescent structures. *Sargassum* is deliberately collected in the Atlantic and commonly navigates import with small beginnings encrusted upon live rock. It grows

108 *Natural Marine Aquarium Volume I - Reef Invertebrates*

quite large and can easily shade other photosynthetic organisms in aquariums. Although it is not regarded as being especially noxious, it is inappropriate for use in most smaller aquaria if only for its sheer potential in mass and demands for pruning. It may be farmed in refugiums and used as a vehicle for nutrient export, but categorically it is less productive than turf algae or other soft green species in a proper algal scrubber. Because of its tough texture and apparent lack of palatability otherwise, few but the toughest grazers can control this weed and it can become a nuisance in some aquariums. Ironically, as a "nuisance" algae, it does not only prosper by nutrients and poor water quality; it fares as well or better in well-maintained aquariums with very good water quality and bright light. Manual extraction may be necessary to control this alga. Otherwise, employ large Rabbitfish or Surgeonfish such as the *Naso* species if tank size allows.

• **Seagrasses** (*Halophila, Syringodium, Thalassia,* and *Zostera* - the Angiosperms): The pursuit of cultivating marine plants like the seagrasses is a very exciting new dimension of the hobby. There are many varieties that can and will be utilized by marine aquarists in the future. The evolution of aquarium keeping with refugiums, surge devices, and multi-tank displays is forging a greater desire for, and understanding of, marine vascular plants. For the aquarist looking for a real biotopic challenge, consider them true vascular plants.

Seagrasses are vitally important in some natural marine reef environments, serving in numerous pivotal capacities. They are nurseries for countless juvenile reef denizens and function as enormous food banks, not only in composition, but also in ability to impart nutritive elements including metabolites and epiphytic matter to the reef for filter-feeding organisms. Most grasses occur in the shallows up to the first fathom of water, though some, such as *Halophila*, occasionally occur at amazing depths towards the limits of photosynthetic life in the sea. In many areas of the world, these plants continue to occupy vast tracts of shallow waters, but in some areas they are in great peril from human activities, not the least of which is pollution from drainage, runoff, sedimentation and watercraft. Environmentally, we really need to "clean up our act" now, or not only will some of the remaining seagrass meadows in existence be lost, but the thousands of species that need seagrasses, ranging from some of the smallest fish and invertebrates up to the gargantuan mammalian Sea Cows, the amazing sirenians (Manatees and Dugongs), will also be lost. Some of the strange and recently prevalent diseases of seagrasses have been ascribed to severe population reduction of grazing sea turtles and like herbivores that kept populations in a healthy balance by their pruning activities. Aquarists should keep this in mind when they successfully cultivate grasses in the aquarium; pruning seagrasses is healthy and necessary.

Macrocystis pyrifera - Giant Kelp

Looking for good grass?

Which to choose... a quick Seagrass ID:

Thalassia: **Turtlegrass**

- Thick, wide, flat-bladed dark green leaves with round tips. Usually imported at 4-6" of length (10-15 cm)
- Slow to establish, but highly utilitarian for its wide leaves that attract and shed great nutritive matter
- The most commonly available seagrass at present in the aquarium hobby

Syringodium: **Manatee Grass**

- Thin tubular leaves that taper to a point. Usually medium green in color with yellow commonly. Imported at lengths of 6" (15 cm) or greater, and grows to more than 24" (>60 cm) tall.
- The fastest seagrass of the three mentioned here to adapt to aquarium life. Can reach full size in months and requires tall housing and regular shearing. Very hardy once established
- Delicate to transplant: take great care handling lead-tip of runner

Zostera: **Eel grass**

- Tall, narrow, flat-bladed rich green leaves (ribbon-like) with rounded points. Usually imported at lengths of 6" (15 cm) or greater
- Arguably the least adaptable seagrass of the three mentioned here to aquarium life. Favors cooler waters and strong surge activity
- The tallest seagrass and least appropriate for home aquariums, this plant can easily exceed 24" in height (>60 cm)

Halophila: **Paddle Weed** or **Midrib Seagrass**

- Forms elongate oval, flat green leaves of variable green color with a distinctive midrib pattern
- Still rare in the hobby until enough of us demand it from our local merchants, this is one of the most promising true vascular plants for the aquarium with a delightfully manageable height of 2" (5 cm)
- Tolerates lower light, weaker flow and higher sedimentation than any other common seagrass making it ideal for aquarium life

Overview of seagrasses in the aquarium: There are a few obstacles to enjoying seagrasses in a home aquarium. The first is that they are still fairly uncommon in the trade. There is a very good reason for this too and it is known as the "founder effect": most collectors have no idea that aquarists actually want to pay good money for the weeds that clog their boat propellers! There are several other common seagrass species beyond the four mentioned here. Unfortunately, very few skilled collectors currently collect and ship healthy seagrasses of even the mentioned varieties. They are rather delicate and difficult to collect properly and must essentially be harvested in a "plug" fashion. Targeted specimens are extracted by digging a wide berth in circumference around and deep below the root system. This large plug is heavy and cumbersome but necessary to protect the delicate tips of the submerged rhizomes of the plants. The runners of the root system occur deep and well-anchored in the substrate, typically below 3" (75 mm), and are quite a chore to extract. Damage to the rhizome occurs easily and essentially dooms the entire frond. For easier shipping, the soft sand and mud can be gently rinsed away, but the aquarist is then faced with a very delicate transplant at home. Never plant seagrasses by pushing them into the sand, but rather dig a deep hole, set them in place,

Another beautiful blue-hued *Dictyota* variety.

and carefully fill in around them. Root depth is a critical issue with seagrasses and is one of the most common causes for their failure to acclimate on transplant. These plants need very deep and mature substrates with a strong nutrient base. Planting seagrasses in anything less than 3" of sand (75 mm) is likely to fail. However, it is strongly suggested that the overall depth of the substrate should be or exceed 6" (15 cm) for a true deep sand bed. It is best to allow a substrate of soft sand or muddy sand to mature for 6 months or more before adding seagrasses to allow for the development of a strong nutrient base and some semblance of balanced microbial life. Supplemental fertilizer may be helpful, however it is no substitute for a mature deep sand bed. Fertilizers must be used with discretion, especially in an immature system, with concern for fueling a suffocating a nuisance algae bloom. The requirements for establishing grasses in the aquarium are simple and straightforward, but quite strict. Once these needs have been met, prepare for the likely inevitability that the old growth starter colonies will die back within two to four months anyhow, giving rise to new leaves that will grow and adapt to the physical parameters of their new captive environment. This is a very common adaptation that freshwater pond keepers will recognize as with lilies purchased from a pool where the leaves and stems are grown at a different depth than the destination pond. In this scenario, a lily with 36" leaves in a new 24" pool will not waste energy on the old growth but rather abort them and grow new leaves better suited to the new conditions. Incidentally, this strategy applies to many of the interesting calcareous "rooted" algae as well, such as *Halimeda*, *Udotea*, etc.; have faith that the dead old growth will give rise to new colonies. Once established, be sure to cultivate these grasses under at least occasionally strong and dynamic surge or random turbulent water flow. Good water flow will purge settled detritus and old growth away while imparting nutritious metabolites and epiphytic matter for filter-feeding invertebrates in the display. Rasping snails and grazing fishes such as the wonderful *Ctenochaetus* surgeonfish have a similar and stimulating effect on seagrasses. Otherwise, prune old growth as necessary by hand, pinching off dead, dying leaves at their base, assuming that you have no manatees or sea turtles wandering around the house with hunger pains.

Quick tips for growing seagrasses in the aquarium:

- Only buy plants with undamaged tips of runners (lead rhizome "roots")
- Purchase or collect seagrasses in undisturbed plugs whenever possible
- Plant pods, plugs or bare fronds with great care, never push roots into the substrate but rather place into a waiting hole and gently fill in around with soft sand
- Roots must be buried at 3" (75 mm) depth or greater
- Total substrate should be 6" (15 cm) or greater
- Deep fine sand or muddy sand should be matured 6 months or longer before adding grasses
- Realize that newly planted old growth will likely die back and give rise to new growth in months.

• *Stylopodium zonale*- **Leafy Flat-Blade:** As its name suggests, this is a brown alga has flat blade "leaves" that are irregularly split and branched. It resembles and is related to "Y-branch algae" (see *Dictyota* for more information on general care and suitability in the aquarium). Most are found attached to hard substrates and can occur at the deepest depths for photosynthetic organisms.

• *Syringodium*… see **Seagrasses**

• *Thalassia*… see **Seagrasses**

• *Turbinaria*- **Saucer Leaf Algae:** This sturdy little algae is often confused with, or likened to, another brown alga, *Sargassum*. Deliberate collections of *Turbinaria* for the aquarium trade have been uncommon, but it's unique appearance makes *Turbinaria* likely to be desirable to some in the future. Leaves are spiny, flat and triangular and may be speckled or have hues of yellow, green, light or dark brown. Some species in this genus have discreet air bladders to support their structure. They are decidedly shallow in occurrence (found in less than a full fathom, with most within 20 feet). As such, *Turbinaria* favors a strong, warm daylight spectrum of light for cultivation in aquariums. Although not many aquarists have cultured, kept, or traded this genus, it does not seem to be especially problematic under average aquarium conditions. It is not forwardly invasive, noxious or particularly useful for nutrient export unless growth is stimulated in a high-nutrient, halide-lit environment. Beyond aesthetic value to some, *Turbinaria* is unlikely to draw much attention from aquarists. This alga is fit for human consumption and also used as fertilizer on land. One of the reasons for its relatively unimpressive growth in the aquarium, yet significant presence in reef environments, is its dependence on extraordinary turbulent water flow.

• **Turtle weed**… see *Chlorodesmis*

• **Twig Algae** (the articulated corallines- *Jania, Amphiroa, Bossiella* and similar calcifying Rhodophytes): The various members and species of this group of red algae are known by many names: Flat Twig, Y-Twig, Segmented and Bushy Corallines just to name a few. Most are found in shallow, shaded and protected waters, oftentimes near or with seagrasses. In the aquarium, they will require the same

Turtle weed in the Andaman Sea... see *Chlorodesmis* for more information.

stable and mineral-rich water quality as other calcareous algae. Most forms are generally sparse and somewhat random assemblages of thin, brittle, twig-like calcified segments. Some related varieties, though, such as *Bossiella orbigniana* (formerly known as *Amphiroa orbigniana*) form rather deliberate and attractive structures that resemble sea fans or palm fronds. Twig algae are quintessentially innocuous and present little help, threat or functional use in the aquarium. They may simply be enjoyed for their natural beauty.

• *Tydemania* is an interesting but uncommon calcareous green algae in the hobby. It forms "globuliferous" twisted branch segments that remind one of a "dread-locks" hair-style. Little is published about this genus in popular aquarium literature. It is almost certainly more ornamental than utilitarian and very unlikely to become a nuisance due to its inherent needs and dependency as a calcifying variety.

• *Udotea*- **Mermaids Tea cup** or **Fan Algae:** These are fan-shaped green algae that approach 8" in height (20 cm) and emanate from a single stalk. They have similar requirements in husbandry to other calcareous algae including *Halimeda* and *Penicillus*. Stable water quality with adequate calcium and alkalinity are necessary for success. Fan algae are more adaptable to a wider range of light than most other calcareous species in the trade, tolerating lower light conditions. *Udotea* grows too slow to be of use as a vehicle for nutrient export, but its consistent growth in the aquarium is a good indicator of water quality and levels of minerals. It also provides a fine habitat for refugium micro-organisms and is attractive as ornamentation. It is rather edible by active herbivores and with consideration for slow natural growth in the aquarium, makes this alga an unlikely candidate to ever become a nuisance. *Udotea* may fare well in aquariums with even coarse substrates and coral rubble.

• *Ulva*- **Sea Lettuce** is an attractive species that forms thin, gelatinous sheets that remind one of salad lettuce. There is little variation in this alga, which grows conspicuously in flat, dark green sheets, two cell layers thick from a single holdfast. It may also occur somewhat ruffled in form. The texture of sea lettuce is delicate and gelatinous; its tissue tears easily. Most sea lettuce is found in shallow, quiet, protected marine environments in especially high nutrient or disturbed areas where natural herbivory is low. Indeed, *Ulva* is an early player in the dynamic of algal succession. In aquariums, it frequently blooms with nutrient spikes and can be fast growing. It is also quickly and readily grazed by typical herbivores. *Ulva* may reproduce sexually, however it more often spreads by fragmentation or an alteration of generations (sporophytes forming zoospores in sporangia-**parthenogenesis**). By any definition, this green alga is a candidate for use in vegetable filters for aquaria to harvest nutrients. It can also used ornamentally in small to moderate quantities with little concern for noxious exudations. Its presence on a reef proper, however, is uncommon due to its preferences for high nutrients and low water movement. In turn, it is best suited to slow-flow refugiums that are fed with raw water. Clumps of sea lettuce will grow to approximately 4" (10 cm) before degrading or fragmenting. *Ulva* is a useful food for herbivorous fishes and is also consumed extensively by people in many countries; it is used at times as a substitute for *sushi nori* (*Porphyra*) or simply as a salad or soup component.

- *Ulvaria* is superficially, if not fundamentally, treated the same as *Ulva* in aquariums except that it is structurally composed of a single cell layer. See the latter genus description above.

- *Valonia* (AKA Sailor's Eyeball)… see **Bubble Algae**

- *Ventricaria*… see **Bubble Algae**

- *Zostera*… **see Seagrasses**

Summary of Marine Plants and Algae

Many types of marine algae and true plants are readily available and can be shipped in good condition from around the world. American and European aquarists often enjoy a diverse selection from Caribbean waters in particular. Progress in aquatic marine botany can be measured by the increase in species and numbers of specimens kept by aquarists everyday. Algae use is one of the fundamental means of concentrating inorganic and organic material in captive reefs into less noxious and more useful matter. In captive systems, macroalgae serves as important habitat, food, and ornamentation. It is also instrumental in mediating biological mechanisms that have a direct impact on water quality. Given simple consideration for lighting, water quality, controlled predation, and the elusive healthy founding specimens, all marine systems can benefit from the inclusion of plants and algae. For biotope and microcosm displays, they are indispensable. Success with marine macrophytes is simply a matter of selecting the proper types from the proper places and maintaining them under adequate conditions. Ready for the challenge? Go for it! For further study and identification of marine plants and algae, the various works of the Littlers for the last fifteen years would be an excellent starting point.

"For in the end, we will only conserve what we love, we will love only what we understand, and we will understand only what we are taught." *Baba Dioum*

Understand your world and hobby and enjoy a natural marine aquarium that is complete and diverse... including algae.

An interesting assemblage of encrusting coralline and non-coralline species like the brown *Lobophora* in the center. Bahamas image.

Selecting Reef Invertebrates

Pterapogon kauderni (AKA the Banggai or Banner Cardinalfish) This darling of the aquarium hobby is but one of the many fishes that most truly can be regarded as *reef-safe* and hardy. They are gentle and compatible with just about anything that doesn't want to eat them! Quiet tank mates that allow them opportunities to feed are necessary. In good health, they readily reproduce in captivity and the event, as a mouth-brooding species, is both a wonder and joy to behold through gestation and in the ensuing nursery period for the babies. When threatened, young Banner cardinals cluster about the spines of the urchin *Diadema setosum* as pictured.

"Reef-Safe" is a slippery term. Hobbyists and merchants alike bandy it freely about when discussing selections of compatible specimens from both the vertebrate and invertebrate orders. Strictly speaking, what we mean by reef-safe is "relatively, not-too-likely to eat tankmates." But, *relative* to what? Everything that lives on a reef must surely eat something *else* that lives on the reef. We, the authors, would eat sushi... or rather, *sashimi* if we were reef denizens! So in the purest definition of the concept, nothing that hails from a reef is wholly safe to keep with any or all other reef organisms.

The practical summation, then, of a given creature's degree of *reef-safeness* is dependent largely on which tankmates you regard as desirable and wish to keep safe, and how much of a risk you are willing to take with their safety. Surgeonfishes and tangs are a good group to use to illustrate this example. Most hobby literature in the American market plainly regards Tangs as safe fishes for reef invertebrate aquariums. Hobby literature in other markets overseas, however, regards this group with caution or plainly states that they are not "reef-safe." It really is a wonder at first that there could be such a discrepancy between perceptions of the same fishes on a topic that you would expect to be clearly without controversy- either they eat/harm invertebrates or they don't, right? Alas, it is not so plainly true. If we could fairly say that 70% of all tangs kept in the aquarium present no threat to desirable invertebrates, would you regard that as reef-safe? Or would you immediately think of the 30% of all tangs that are likely to nip or harass invertebrates in time... some perhaps to the point of death? We'll tell you that tangs are essentially reef-safe fishes that serve aquarists with large aquariums (preferably over 100 US gallons) very well. It is true, however, that *Zebrasoma* species do occasionally nip or graze soft corals. *Naso* tangs have been regularly observed to nip so-called "large-polyped stony" corals (LPS). And the Clown or Pajama tang is a truly inappropriate fish for all but the largest and toughest tanks and is regarded as a high-risk species with some invertebrates.

Variation Within Species

At length, we cannot expect any individual of a given genus or species to behave exactly like a stereotype. Variations within groups with individuals are further amplified by the stresses and unnatural temptations created by the artifacts of captivity. Rather than fret over whether or not a new creature will be safe or not, simply research your targeted specimens as best as possible and have faith in the application of a proper quarantine/isolation tank to use for contingency. Presume that all life in the seas can or will sample all other life. If possible or necessary, a suspect fish or invertebrate can be tested with samples while it is held in a quarantine period. Any complications with compatibility on reef-safe issues later can be tempered by isolation of the offender and victim (separately of course!) in isolation tanks just the same. By following sensible and responsible procedures of husbandry, you'll be well on your way to making good decisions on how to successfully stock your

Acanthurus lineatus (AKA- the Striped, Pajama/Pyjama, or Clown Surgeonfish). Commonly from the Indo-Pacific. Growing to a remarkable fifteen inches in length (over 37 cm), this tang is aggressive and can wreak havoc on many tank mates including invertebrates (sessile and motile). It is not "reef-safe." Beyond concerns of compatibility, they live along reef edges in very high water movement and super-saturated oxygen concentrations and do not ship well. Husbandry for long term success requires very large and long tanks (over 100 US gallons minimum) with extraordinary water flow and aeration. Few aquarists can or should keep this species.

Natural Marine Aquarium Volume I - Reef Invertebrates

aquarium. What people really mean by deeming a mix of organisms "reef-safe" is that there are essentially no known or intended predator-prey relations amongst them.

Species Selection Strategies

At First There Was... A Plan!

Your very best chance for success with any aquarium is to have a plan before you buy a single living creature. There are a few ways to go about putting together a system and stocklist for a reef invertebrate aquarium. Too often, hobbyists are tempted to take a random approach and either use what is at hand, or what they can find, or simply what they can afford to put into the system largely by whim. We discourage this method, of course. It's much too easy to get into trouble without formulating a plan and researching your options for life support, compatibility, feeding habits, behavioral needs, adult size, etc.

System or Livestock Approach?

We understand that you might only have so much space available for an aquarium. Or maybe you have a "key species" you've just got to have, and you are willing to work the rest of your species selection, system hardware and husbandry around this favorite as well. Either of these avenues will lead you to making informed choices about the other.

While planning for an installation, you have many things to plot and dream about. Start gathering information beginning with a journal or log of considerations (livestock sequence, ultimate size, foods/feeding...) that you write and revisit. Investigate what size & shape aquariums are available and appealing to you. Consider the possibility of

The Blue Line Triggerfish (AKA Yellow-Spotted Triggerfish to science), *Pseudobalistes fuscus*, is a hardy and handsome fish from the Indo-Pacific, Red Sea, East African coast to South Africa that *fails* most any definition of "reef-safe" as an aquarium specimen. Growing to twenty two inches in length (55 cm), this beauty is pictured as a juvenile (left) of four inches (10 cm) in captivity, and a full size adult (right) in the Red Sea.

custom aquaria for animals with special needs like small fishes that need a tall column of water to spawn in or corals that can reach several feet tall within years. Glass and acrylic tanks can be ordered to your specifications and shipped to virtually any place on the planet. Barring any special consideration, the same criteria that are important for other types of captive aquatic environments apply to reef-invertebrate environments: they should be as large as possible, and more "squat" in overall shape than high as with show-type configurations.

You will need to determine if the system will feature fishes or invertebrates. The aquarium can house both of course, but a heavy and balanced mix of the two is somewhat more challenging to accomplish. Yet, we wonder... is it the Western mindset and ethic that causes us to pigeon-hole classify systems as "Fish Only," "Reef" or "Fish Only With Live rock?" There certainly can be any range of mixing of livestock groups here. We think that you will find these labels will fade in popularity in time as all will

Pseudanthias ventralis (the Longfin Anthias) has two recognized subspecies: *P. v. ventralis* in the western Pacific, and *P. v. hawaiiensis* around and about Hawai'i. These magnificently colored deepwater species (very common between 300-400 feet!) can be hardy when kept in specialized aquaria. They are not to be kept casually, but instead require very considerate species selection. This Anthias epitomizes one of the potentially hardy yet beautiful rarities that can be kept well in reef invertebrate aquariums if given due consideration of their special needs. They will be kept best in dimly lit species-specific tanks with plankton-generating, fishless refugiums to supplement the 3 or more feedings needed daily like most Anthias. An aquarium kept pair is shown here (the male on left, and female shown at right).

Selecting Reef Invertebrates

There are very few fishes that we can truly say are "reef-safe" or as nearly true as possible. If any groups fit that description, it is surely the Gobies. Most are generally small, well-behaved, neutral and utilitarian for reef systems. Be sure not to underestimate the gentle nature of many. Exceedingly peaceful fishes like gobies, firefish, and seahorses are more often outcompeted for adequate food in the aquarium than suffer direct aggression by highly active and inappropriate tankmates like damsels, anemonefish and tangs. *Elacatinus puncticulatus* (L. Gonzalez).

fall into the category of "modern marine aquarium" which inevitably will have some type of living substrate, various invertebrates and fishes all together.

About Fishes for the Reef Invertebrate Aquarium

The best fish choices are certainly species that are known almost carte blanche to leave macro-invertebrates alone. These include tube-mouthed fishes like pipes and seahorses (Syngnathidae), many blennies, gobies, some of the wrasses (like the flashers and fairy wrasses), and many damsels (Pomacentridae). We will cover these choices in proper detail in a later volume dedicated to Reef Fishes in this series of books.

As a rule, it is best to wait to add all fish livestock until after most of your desired invertebrates have become established. There are many good reasons for this that eager and impatient aquarists cannot seem to appreciate. Waiting will provide your non-fish livestock (invertebrates, plants and algae) a chance to settle in and develop their own strong defense mechanisms (chemical, physical, etc.) without the stress of curious and active fishes. Beyond any concern for literal interaction between a fish added too soon and the targeted non-fish livestock, there is also the disadvantage of fish added too early not letting desirable microfauna develop. It can take months for many plankters (natural phyto- and zooplankton) to establish colonies that can sustain predation by fishes- most of which are just cuckoo for copepods. Numerous fascinating larval organisms are also plucked from rocks too without an aquarists ever

knowing they were there or that there was a chance to watch something rare and interesting grow from seemingly nothing. Admittedly, there are a few utilitarian fishes that may be added early to serve a specific function (like controlling diatom growth) without putting much of a burden on the tanks potential development. Few of these fishes are truly necessary, however, and a good wait of at least several months for all fishes is best. For aquarists that like to see unique or strong plant and decorative algae growth, the waiting period should be longer. After nuisance species have waned in the order of algal succession (say 4 months), some incredible plant and macroalgae growth can usually be observed sprouting from live rock! If you have but only one quarantine tank and you use it *properly* (one fish per QT and 2-4 weeks per fish), you should have no trouble following our recommendations to stock slowly and you will enjoy far greater success with your reef invertebrate aquarium for doing so.

About Filter-Feeding Invertebrates

The number of filter-feeding invertebrates in the trade includes many stinging-celled animals (cnidarians like those anemones, corals and hydroids that we'll cover in a later volume in this series), most sponges, tunicates, bivalves, barnacles and tubeworms just to name a few. It encompasses thousands of organisms that we would recognize as aquarists. Their exact feeding modes and matters are covered in the dedicated chapter on this topic that follows. They are mentioned here with regard to

Natural Marine Aquarium Volume I - Reef Invertebrates

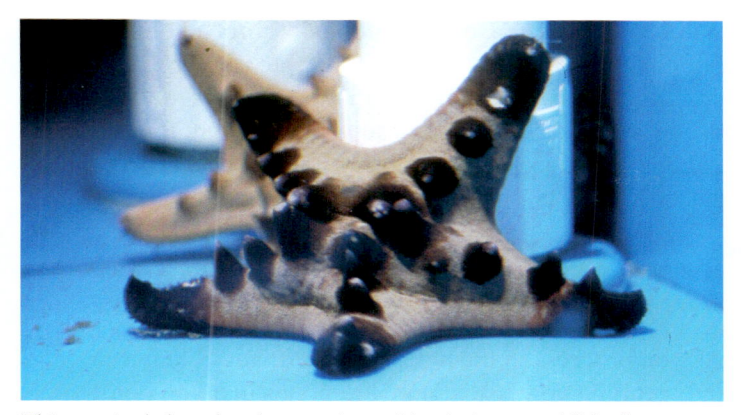

This seemingly harmless, inexpensive and hardy (once established) Chocolate Chip sea star might seem like a safe bet if you gave it any thought at all on a whimsical purchase. A few individuals are even well behaved for some months. But most of these sea stars are in fact significant risks to sessile invertebrate life forms and are never to be recommended for mixed reef aquaria.

Most imported incidental organisms are of little consequence or can even be potentially useful, but some can become a nuisance or a pest if allowed to grow too big or too numerous. In aquariums with poor water flow or nutrient control issues (lack of daily skimmate or excessive feeding/organic levels), potentially innocuous creatures like *Aiptasia* (shown) and/or *Anemonia* can suddenly spread to plague proportions and require control through limiting foods, natural predation or manual extraction.

selection, as being far more common than most hobbyists would imagine. The election to include filter-feeding invertebrates in your aquarium demands that you have a very clear concept of what they eat and how you can provide such fare. For too many, their needs are not clearly understood or readily provided for (nanoplankton feeders, for example) and their life in captivity is doomed to slow starvation that might in fact take more than a year to claim the creature's life. This reality is easily overlooked with corals, for example, believed to be wholly supported by the products of photosynthesis. The truth of the matter is that very few if any corals are purely autotrophic; there is a sliding scale of ones that eat large food items, to all sizes of plankton, to being more/less photosynthetic. Even those that will not or cannot feed organismally will still filter-feed by absorption on dissolved elements in the water. Thus, an aquarist with the "perfect" lighting system over a coral that is 98% autotrophic must still insure that food is available daily to satisfy the remaining 2% of the coral's needs. Otherwise unfed, such coral can appear to fare well, behave normally and conduct normal polyp cycles without any plain sign of duress from the daily 2% deficit. Do consider if it would even be possible to see the attrition that a 2% daily deficit would look like on a specimen after even a year! "Starvation" is not always symptomatically apparent in invertebrates that are mostly comprised of water. It is not even attrition that will necessarily take their life from the neglect but rather nutritional deficiency or suffrage by a pathogen for their weakened immune system. Ultimately, a significant portion of the life in your system will be of this **trophic** (feeding) type. Depending on how you design and stock your system, you may have to augment husbandry with "planktonic" feedings. Many things will influence this reality: the use, type, and depth of sand... the use and quality of your live rock... and especially if you use a refugium or not. We sincerely recommend the complimentary use of fishless refugiums when filter-feeding invertebrates are selected for the aquarium.

About Cold Water Invertebrates

In the course of reading the livestock sections of this book you'll notice that there are more than a few non-tropical

Although it is true that Tridacnid clams are highly successful zooxanthellate "sun-soakers," they too are hungry filter-feeders. Some specimens are in fact very dependant on filter-feeding upon nano-plankton and by absorption. Let there be no doubt, in nutrient poor systems these mollusks will starve to death. Some European aquarists have used and suggested feedings with nitrate solutions. Clam farmers feed controlled and dilute solutions of ammonium chloride in their culturing troughs not to grow their baby clams, but just to maintain them! Modern reef aquarists have taken to experimenting with clams in refugiums as efficient living animal filters. This is *Tridacna squamosa*. (B. Neigut)

Selecting Reef Invertebrates

species (cool to cold-water shrimp, snails, algae, etc) offered to aquarists by dealers in the trade. Yes, let there be no doubt that these organisms don't fare well at higher temperatures... and who knows if your supplier knows this or not. You should avoid these specimens unless it is your intent to have a coldwater system. Ultimately though, such a display will be challenging to even build as most of these creatures are regulated at some point in American waters, at least. As an aquarist, it's somewhat disappointing not to see more of these displays established. Anyone who has gone diving or snorkeling in less than tepid regions of the world knows there is plenty there to see and appreciate other than fishes. You can have an amazing, educational display of coldwater invertebrates with appropriate filtration and climate control (yes... those pesky, expensive chillers will be necessary). But evolution over time untold will not conquer the needs of temperate species for temperate conditions no matter how cheap the animal is in the dealer's tank or how badly you want to buy it. Please keep coldwater and tropical species in separate and appropriate systems.

If you don't know what it is... Then you don't know what it needs!

As you progress in the hobby, you will very likely run into exciting new fish and invertebrates that you don't recognize. Impulse purchases are inevitable for some aquarists that are new in the hobby. It is remarkably disrespectful, however, to the animals in your charge to buy them and bring them home without even knowing their adult size, needs for life and if you can provide them! How can anybody forget that these are living creatures and not furniture? They are a vital and finite resource that needs to be responsibly managed. Never forget that if we do not conduct ourselves responsibly as aquarists and police our own activities, we risk the possibility of somebody doing it for us, or worse... simply having our privilege to keep marine life privately legislated away. With the age of the Internet and tremendous possibilities for information exchange, surely you can resist the temptation to buy a strange new animal long enough to drive home to your computer or pass a library with free Internet access to research your find. If you want assurances that your living treasure will still be there on return, offer to pay for the creature with a deposit or in full (pending a refund for credit, at least, to waive the specimen if necessary). If the dealer will not take your money on deposit for a hold so that you can make an informed decision... you do *not* want to shop that dealer. You will instantly learn everything that you need to know about that merchant: they cannot educate you on the animal, and they will not support you in an attempt to educate yourself to insure the best possible life for the "commodity," as it were. As an empathetic aquarist, never buy livestock on impulse.

An amendment to the above admonition is the many fascinating incidentals that occur with live substrates like rock and sand. With more than 25 phyla of possibilities, the more you look, the more you will indeed find on close examination of your tank and most will not demand your special care or attention like the larger macro-organisms. Most of the creatures that are imported with such product can survive and fare well in captivity, and others inevitably will not. Identify all as best you can and try to encourage the continuation of maximum bio-diversity in your system. As mentioned numerous times in this text, one of the very best ways to cultivate especially small organisms like incidental plankters, micro-crustaceans and the like, is a fishless refugium.

The delicate and transient life of many sea slugs should underscore the need for careful research in species selection of reef invertebrates for the home aquarium. Most of these beauties ship poorly, have difficult if not impossible dietary demands and are naturally short-lived (1-2 years). Shown here, *Chromodoris willani* in Sulawesi.

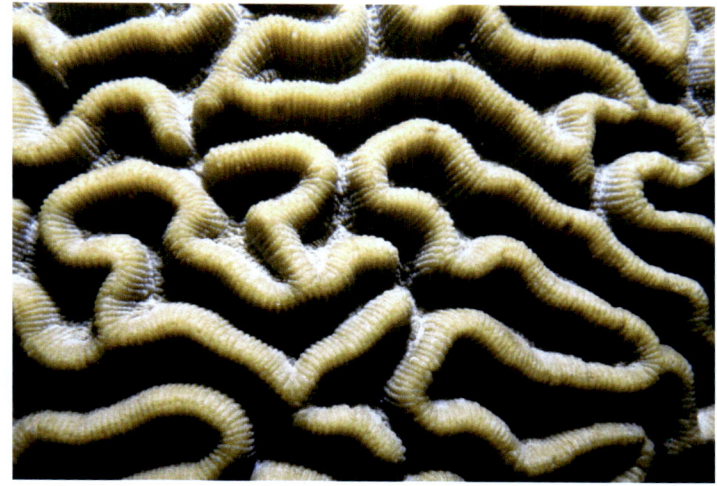

Colpophyllia natans (AKA Symmetrical Brain Coral): This beauty grows large, round to hemispherical colonies or can be encrusting. Its polyp structure is arranged in long snake-like valleys. The septa are short, equal and pointing outward. A fine ambulacral groove runs along the top between corallite walls. This image here shows the world's largest Brain Coral off of Flying Reef, Tobago, perhaps thousands of years old!

It's normal to fear some of these unknown critters when you first observe them. Rest assured though that most anything that arises would prove innocuous. The use of a proper quarantine tank, if you are diligent, will likely afford ample opportunity to screen parasites, pests and diseases from the display.

For most all of these fascinating life forms appearing unannounced with live rock and live sand, "live and let live."

About Captive Lifespans

Most marine fish and invertebrates have comparably long life spans in contrast to freshwater organisms. There are, of course, some naturally short-lived species within familiar animal groups described here. Sea hares and nudibranchs may live less than a year and octopuses not much more than two for many species.

Of the longer-lived invertebrates, lobsters, cnidarians and some larger mollusks rank the highest, living several to tens of years. The potential for many corals and anemones may very well be more than 100 years! The oldest living brain coral in the sea has been conservatively estimated at over 800 years old on the low end with many believing that it's 1200 years old or more.

Quite a few fishes can be kept for 10, 20 or more than 30 years captive too; there's a Tarpon over 60 years captive on record. Beyond the large predatory fishes in giant public aquariums that you might rightly expect to be holding some of these records, there is a fantastic list of popular aquarium specimens that have or had been kept in excess of 20 years in aquaria. That list includes the following: False-skunk anemonefish (*Amphiprion perideraion*), Harlequin Tuskfish (*Choerodon fasciatus*), The Purple/Yellow-tail tang (*Zebrasoma xanthopterus*), The Undulate triggerfish (*Balistapus undulatus*), the Saddle Butterfly (*Chaetodon ephippium*), and numerous large *Pomacanthus* and *Euxiphipops* Angelfishes including Navarchus, Semicirculatus, Six-banded, Blue-girdled and Blue-faced. Despite the youth of invertebrate keeping, largely born in the early to mid 1980's, we are now starting to finally hear reports of longevity in captive invertebrates too (specimens over 10 and even 15 years old). Statistics like that can really put your claims of "success" in perspective with captive animals. Even more so, it is inspiring to examine and improve your own habits of aquarium husbandry by it. Most importantly, it should remind you to plan for the long term care of your aquarium guests just as you would for any other pet or companion animal that lives so many years. We must expect success and growth and plan for it in the hardware and stocking selections that we make for marine aquariums at large. Although captive conditions may prove more favorable due to ready food availability and a dearth of predators, most live no longer than they would in the wild. The completion of a natural lifespan for marine organisms in your care is rewarding testimony to your husbandry and the hobby.

Given a spacious tank, a considerate diet, and due diligence in husbandry... many marine organisms like this magnificent Blue-girdled angelfish, *Euxiphipops navarchus*, can be kept in captivity over 20 years! To be clear, this is not encouragement for beginners to keep delicate animals, but rather inspiration to dedicated aquarists that it can be done. (*A. Calfo*)

Selecting Reef Invertebrates

Reef Invertebrate Husbandry

Aquarium husbandry involves the totality of what we do to set-up and maintains our captive systems. This is a BIG topic. Volumes upon volumes of work have been dedicated to the subject. Most of it is quite fundamental and readily applied to endeavors with reef invertebrates but some of it needs to be finessed for our specific needs. What kinds of reef invertebrates do you intend to keep? What sort of system would you like? What choice of components do you need to accomplish this? Will there be many or even any fish kept with your invertebrates, plants and algae? Are you including any obligate **heterotrophs** (feeders that do not manufacture any of their own food)? Of the at least partially **autotrophic** organisms (Tridacnid clams, mostly hermatypic corals, anemones and the like) how much light will you need to support them? These are just some of the many questions that you will need to address while planning your system. Also, please be sure to do this "planning" (research & preparation) well in advance- not at the merchant's register with your wallet open and your adrenaline flowing! Although there are certainly familiar types and styles of systems like the SPS tank, the LPS tank, filter-feeders, etc, there are truly no two systems that are identical in make-up, maintenance or execution. The challenges to succeeding with a reef invertebrate aquarium are wholly and uniquely founded on your ability to finesse the ever-changing living dynamic in your aquarium on a daily basis.

In the body of this text numerous details are offered concerning the needs of specific reef invertebrates and how you can provide them. In this chapter we will offer advice on commonalities in fundamental husbandry for "typical" reef invertebrate aquariums. These guidelines will serve most any reef aquarist well, although there will be exceptions of course. Yet, when we recommend strong water flow for your aquarium, it is relative to the needs of popular and common invertebrates seen and kept in our trade. *Tubastrea micrantha*, the Black Sun Coral, requires *very* strong water motion. And so, with it's mention here and elsewhere in the text, the scale is different from other reef invertebrates requiring strong water movement with a magnitude that most aquarists cannot even begin to comprehend. The corallum of this coral is so very dense that it cannot easily be broken by a hammer and chisel. When nuclear testing was done underwater in the South Pacific, *Tubastrea micrantha* skeletons remained intact and alone on the reefscape! Diving at depth in massive currents to observe these corals is challenging to say the least. This exemption is noted here when defining the unique limitations of an admonition like "strong water flow" and, again, such exceptions are noted throughout this book for animals with special needs. Most other reef invertebrates will enjoy or adapt to a more homogenized set of physical parameters that we can define as follows.

Setting Up...

Tank Size & Shape

Its a matter of physics... bigger is better- for stability, maintenance, decoration, reduction of aggression, the ill-effects of overcrowding and all ways around. Yes, you can keep reef invertebrates in small volumes of water. They are challenging and enjoy a cyclic popularity in the hobby (so-called nano-, pico-, mini- and micro-reefs). But please believe us, these systems are much more subject to environmental change and suffrage from power or gear failures, overfeeding, unexpected mortality, etc. and as such are more prone to "crashing" (complete wipe-outs). If you enjoy the challenge, by all means throw down the gauntlet. But small aquaria are not recommended for beginners or aquarists that want a lower-maintenance (low-headache!) display.

Like computers, displacement of motorcycle engines and banking accounts, we recommend that you invest in the largest system you think you could possibly want with a long-term plan of years in mind. You already do this for furry companions... or at you least should; you don't buy a greyhound or a bullmastiff for apartment-life, especially when there is little chance of owning acreage anytime in the near future. So why then are people compelled to put 50, 100 or more small invertebrates and fragments in a tank that they will outgrow in just a couple of years if not months. This, of course, assumes that they don't poison each other first or suffer from water quality otherwise. Stony coral enthusiasts often get caught up in such excitement to make this mistake when they place or even glue fragments of coral within mere inches of each other. Such tanks are doomed to struggle or fail in time by virtue of their own success- growth. Fish keepers perpetrate the same crime when they buy big fishes as babies like Unicorn *Naso* and Sohal *Acanthurus* tangs and put them in tanks of a mere couple hundred gallons or less. Please don't let anybody tell you that such fish with inappropriately large adult sizes "grow to their tanks size." It's a lie... these fishes simply stunt and die prematurely. They might live 5 years of a 15 or 20-year lifespan, but its unfortunate by any definition. For every aquarium creature that we intend to keep, it is a practical and responsible, if not moral, obligation to research and plan for a full and healthy lifespan just as you would do for a cat, dog or any other living creature. This kind of empathetic aquarium husbandry may entail waiting on the completion of the tank... buying the rest of the gear, seeking bigger and better equipment and deferring the purchase of livestock to properly "put it all together." Too many aquarists make then wrong decision when debating, for example, between buying a good book to help them understand water chemistry better or upgrading their sadly performing protein skimmer with the temptation of buying yet another new animal to suffer in the system. Yet be assured, that amongst the many valuable lessons reef keeping teaches us, patience is indeed a virtue.

The ways to optimize your tank selection are very clear. Variations on these recommendations are compromises that may be able to be negated (improved aeration/filtration, lighter bio-load, etc). Simply recognize your limitations and plan accordingly. As most reef invertebrates are benthic (found on the bottom), you will want to have a tank that is more wide than tall with the largest possible "footprint." In fact... low, long and shallow tanks are best for most aquarium organisms. These "flat" or squat-shaped tanks afford greater surface area for a rockscaping, territories for feeding, sleeping and breeding, and for maximum surface area and gas exchange (water quality... off-gassing). Many popular reef invertebrates attain adult sizes in mere months. Many corals can grow to several feet tall or wide in as little as five years. Your ultimate tank selection will depend entirely on the number and species elected for display. Nonetheless, an aquarium of at least 55 gallons (250 liters) in capacity might be fairly regarded as a safe minimum. Realistically, though, for hobbyists keeping any coral that really don't have specific plans to farm and fragment aggressively, aquariums larger than 200 gallons (900 liters) will be ideal for a 5-10 year plan or longer. Trends in the fabrication of stock commercial aquaria for the pet trade have reflected this reality with 240 gallon displays (1100 liter) found in the standard inventory of most any professional aquarium shop. Whatever size aquarium you choose, be mindful of the all-important "footprint" of your aquarium. Common 55-gallon aquariums in the US are often a scant 13" (~ 32 cm) wide and rather cumbersome if not useless for building a functional rockscape for reef invertebrates. A better choice can be found in a stock 50 gallon "breeder" tank at a reasonable 18"/45 cm wide. Although tall aquariums make a dramatic aesthetic impression, their increased height and water volume is in fact a liability for the challenges it imposes on maintaining water quality for systems at depth (aeration, circulation, etc). Shallow tanks are also easier and less expensive to light for photosynthetic reef organisms. There are many more benefits to thoughtful selection of aquarium size and shape. Choose the largest low and long tank that you can afford to keep reef invertebrates with success and ease.

Near Sea Water (NSW) Conditions

Many hobbyists and almost all dealers in marine fish livestock keep their water's specific gravity artificially low. This is assuredly a good idea for keeping *fishes* to save a little money on salt mix, for increasing gas solubility (and hence potential stocking density), and for depressing parasite loads (many parasitic, single-celled marine organisms are sensitive to low SG). But this is not a good idea at all for reef invertebrates! If you look closely you'll see that almost all non-fishes in commercial facilities are housed in a separate recirculating, batch-processing filtration system, with **specific gravity kept at or near 1.025**. You should do your best to avoid fluctuations away from this value for your seawater condition. Carefully pre-mixing and storing synthetic water ahead of use, checking and adjusting for evaporative losses daily, indeed doing what you can to maintain a stable environment overall is crucial for the osmotically sensitive reef invertebrates. Beyond that, however, do consider that most of these creatures depend on the strength of seawater (and frequent water changes in captivity for replenishment) for dissolved elements to live! Soluble inorganic compounds and bio-minerals are extracted from seawater for survival by reef invertebrates. Any dilute compromise of captive seawater is a dilution of opportunities for your invertebrates to feed and grow... and in some cases a threat to survive. Make no mistake about it... reef invertebrates demand full-strength seawater.

This Tridacnid clam went from gorgeous to grotesque in just a few hours. Unstable water conditions and a sudden change in lighting regime are to blame for weakening the animal - letting the disease take hold. (*L. Gonzalez*)

If you're going to "be serious" about keeping reef invertebrates, learn how to prepare, mix and store seawater properly in reserve. Raw tap water (for constitution) and natural seawater by contrast are probably unreliable for their seasonal variations. There are of course other concerns too about random imparted contaminants or possible noisome organisms (in NSW) from these sources. Most aquarists will benefit from using a quality brand of synthetic sea salt mixed with aerated and buffered purified water. The combination offers the kind of simplicity, consistency and reliability (quality control) that sensitive reef invertebrates need to thrive in captivity. In doing so, you are insured of a constant supply of seawater of known composition, which has a subtle but significant value in aquarium husbandry.

Any protein skimming is better than none, but choose your skimmer according to the needs of your aquarium. This huge (6 ft/1.5m) downdraft model handles a 450 gallon aquarium with ease. (*L. Gonzalez*)

Purified water (R/O, DI) must be vigorously aerated for hours before being used for evaporation make up or constitution with salt. Buffering in advance may be helpful (monitor with quality test kits). Salted water should be mixed for hours before being used with livestock; full dissolution and completion of chemical "reactions" in the mix take time and can be irritating to fish and invertebrates if used fresh (mixed and poured right into the tank). Seawater that is held for some days or more should be kept sealed and dark. A dedicated, sturdy trash can (a favorite with aquarists- Rubbermaid's "Brute" line) will work well and inexpensively for this purpose. Heat water slowly (not too fast for fear of reducing oxygen) and aerate it always before use in the aquarium. Lastly, be sure to use a professional hydrometer to verify salt concentrations. Plastic hobbyist hydrometers can be variable and easily corrupted (air bubbles, deposits, disturbance of the metered arm, etc). Always keep a good glass hydrometer on hand at least for reference or calibration and comparison. Refractometers are fine instruments for a more exact measurement of salinity. All of these considerations are very important for consistency in husbandry with reef invertebrates. Most marine species of invertebrates are much more sensitive to changes, extremes in water chemistry at large than fishes. Specific gravity needs to be matched closely on acclimation and, if need be, changed very slowly with any kind of transition.

Other aspects of water chemistry are readily defined. A reasonable range for pH in the aquarium is 8.0-8.6, but ideally keep a system between **8.2-8.4 pH** and steady. The pH scale is logarithmic and as such changes are very stressful (10-fold orders of magnitude with a single point change). Sufficient buffering (alkalinity) is as important as pH and will contain pH drift due to overfeeding, changes in stocking density, detritus accumulation, natural acid production, and extraction for growth by organisms in the water.

Tropical reef invertebrate systems can tolerate or adapt to a wide range of temperature. As with all aspects of water quality, though, stability is crucial. We recommend that you keep your aquarium between the mid to upper seventies Fahrenheit in temperature (**76-78F/24-26C**). There are others who advocate much higher average kinetic energies (like the 84 F/29 C or higher), and indeed there are some animals that are found and collected in such warm water. However, the vast majority are not, and what little benefit one "gets" for such higher temperatures (perhaps more activity or growth) must be weighed against reduced lifespans, lower dissolved oxygen, more precarious existence of the bio-load, and a very real possibility of accelerating or amplifying potential disasters. Metabolic rates of these cold-blooded (**poikilothermic**) animals are increased at elevated temperatures, but oxygen et al. gas solubilities are diminished as well. Don't overheat your tanks.

Bio-mineral concentrations will need to be monitored and supplemented by dosing or simple water exchanges. Alkalinity should be 8-12 dKH, calcium to be held somewhere near 350-425 ppm and magnesium about triple the calcium level of the system. Aquarists eager to maximize growth often try to push these levels higher but that does not necessarily make an invertebrate grow any faster! It's not like the system is drained and replenished every day of its sum total of minerals! No! The system simply has a net daily demand of what usually amounts to mere parts per million daily. But this is never more than a fraction of the available pool. These recommended parameters above are for stability and a strong pool of reserve for safe aquarium keeping over a period of days or weeks.

Water Circulation: Most places where reef invertebrates are collected have the equivalent of hundreds of "water changes" per hour in their given space. Aquarists that are divers are soon impressed by the extraordinary amount of water movement on even the calmest reefs. Your aquarium should have vigorous moving water of at least ten to twenty times the tank's volume turned over each hour through and about. Ideally, there should be no "dead spaces" in your system, when there is insufficient water flow. Accomplishing this is a bit of an art form and really is specific to each aquarium as pump and plumbing positioning as well as the various rockscapes and artifacts of the display proper dramatically change the dynamic of flow in the aquarium. You will need to be creative and flexible in the delivery of water flow to the aquarium and be mindful as well of changes in time as invertebrates grow, move or are removed. Some aquarists let themselves believe that they have too much water flow, but we can assure you that such a thing is very rare in typical home aquaria. What occurs more often is in fact *in*adequate water flow that is applied improperly (laminar, *uni-directional* and continuous) which appears to be excessive. Most water moving devices available in the hobby produce a unidirectional, laminar flow that can literally blast the tissue off of some invertebrates and irritate most at any rate. This is commonly the case with power heads pushing water directly into a reefscape. A well-intended aquarist might instead direct such flow over the top or across the reef but this is no more helpful when such useful energy is diffused as it travels or wholly stifled when directed into the glass. The best flow for most reef invertebrates is surging but it is challenging and generally inconvenient to produce in the home aquarium. Devices to produce such flow are often unsightly or noisy to operate. Wave-timers are also popular but even these are a rather dreadful waste of energy as they stagger the operation (through on-off cycles) of potentially full-time pumps. If you are an aquarist that simply wants to produce very good water flow without spending a lot of money (as with oscillating devices and wave-timers) or hiring a rocket scientist to engineer a wave maker, there is a very simply solution. Take your standard water moving devices (water pumps, returns from the sump, powerheads, etc) and simply direct their outputs to converge upon each other to create a random turbulent pattern of water flow. This will satisfy most invertebrates admirably and simply an aquarists application and obligations.

The authors vary on their opinions regarding the use of powerheads versus dedicated water pumps, but all agree on the need for sufficient flow at large. Powerheads offer convenience and initial affordability but sacrifice efficiency (heat imparted and longevity principally) and aesthetics (unsightly to some in the aquarium). Dedicated water pumps offer efficiency and aesthetic discretion but are inconvenient to install and initially expensive to purchase. Either or both strategies can serve you well. Consider which benefits suit your application and rest assured in the decision. Please be conscious too about the need for protection of invertebrates and fishes from the powerful intakes of modern pumps. Too often an aquarist overlooks this simple trap. A huge source of captive loss of these animals comes from their being "sucked up" against such intakes. Guards for pumps are readily purchased or easily made, never run a pump without intake screening. Beyond that, issues of water flow are a relatively simple matter... more is usually better and keep it chaotic (random and turbulent)!

Aeration is intimately related to water flow although it can be addressed separately. Issues of gas exchange are crucial for aquatic life. This is especially true with marine livestock where oxygen is harder to keep in saturation. Although invertebrates don't "breathe" like vertebrates, they do respire and the dangers of anoxia (as evidenced in the challenges of shipping and transporting these animals) are very real and significant. Some animals use simple diffusion over their tissues to supply oxygen, while others have more organized ventilating structures (pumping, internal circulation and respiratory apparatus, and the like). Nonetheless, all the reef invertebrates that we're interested in are aerobes. It is our duty to insure the highest possible degree of oxygen saturation for their optimal health.

We do not specifically need air bubbles to oxygenate water. Aeration is all about gas exchange and surface area, which occurs mostly, and best at the surface of the aquarium. Indeed, as mentioned above... gas exchange is a limiting factor in stocking potential for livestock and one of the biggest reasons why a low long aquarium is better than a tall one of the same volume. Vigorous air bubbles in protein skimmers (injected or aspirated) are very helpful for aeration but most saturation occurs simply from the turnover of oxygen poor water near the bottom of the aquarium to the top. This is really how an airstone succeeds in aerating a tank... by lifting oxygen poor water (with bubbles) towards the surface of the aquarium and oxygen rich from the surface downwards. Better for this purpose are water pumps by virtue of the sheer volume of water they can move. Pumps and outlets can be directed to move water upward and disrupt the surface area. An overflow that constantly draws a thin volume of water over a dam to the sump for return is also a great way to accomplish aeration. In turn, any obstruction or stifling of the surface of the water can seriously and suddenly stress or kill your livestock. Do what you can (surface skimming, over-flow boxes, stand-pipes, wicking off the surface with a towel, etc) to remove surface films. Aerosols in the rooms air can collect and lay upon the surface of the aquarium, oily food cooking in a home or being fed or pump leaks in freak events can do the same. Know that the surface film issue is common and urgent any time that it occurs. A great old-fashioned trick during power outages with aquariums (when batter operated airpumps were not available) was to take a coffee-can or plastic bowl and poke a single hole in the bottom of it; you could then scoop up a can of water and position the full vessel in a straddle over the surface of the tank to agitate the water's surface as the bucket drained. Modern engineering has helped us improve on the means to provide aeration, but the matter stays the same. Like water flow, there is almost no such thing as too much aeration.

*Note: with modern high-pressure water pumps, problems with excessive micro-bubbles have occurred that can cause irritation with many corals, for example, and risk a condition of super-saturation. Such events are rather like "the bends" with divers from nitrogen poisoning. These pesky air bubbles and the supersaturation of oxygen will be covered in our future volume on "Reef Corals" in the series.

Lighting is a fundamental parameter of life support in aquariums with photosynthetic reef invertebrates, plants and algae. Technologies to provide light for aquariums is one of the fastest growing and evolving sectors of the industry. An aquarist researching the subject can easily get overwhelmed or confused by the choices. There have been some wonderful and concise "consumer" reports and studies of reef aquarium lighting by Dana Riddle of Riddle Laboratories and Sanjay Yoshi (excellent and objective work from these fellows). We, the authors also have more detailed and current articles and information on our WetWebMedia.com website. If you are the type of hobbyist, however, that does not want to learn the nuances and science of it all... we can simply give you the facts on very reliable and foolproof lighting for photosynthetic reef organisms.

First of all, there is no one "best" lamp (fluorescent versus halide) for the aquarium. **Lighting hardware is to be determined solely by the needs of the invertebrates and the depth at which you keep them**. For that reason, you will not hear a "watts-per-gallon" rule of thumb touted loudly here. Fluorescent lamps can grow "high light" corals every bit as well as metal halide lamps, but they simply cannot do it through the same depth of water. Conversely, more light is *not* better than less, as has been the trend with the dreadful popularity of 400 and 1000 watt metal halides over shallow aquaria (less than 30"/75 cm). The opposite is true in fact! Corals that do not receive enough light to reach their compensation point can be easily and successfully kept healthy indefinitely if their need for carbon (that which is not sufficiently translocated by the products of adequate photosynthesis) is met by feeding. Yes... to be clear, under-illuminated can be fed to compensate for a lack of light within reason. However, the opposite is not true. Over-illumination cannot compensate for a lack of adequate feeding and some photosynthetic corals can ironically starve to death under a bright lighting scheme! You must research the needs of your specific photosynthetic denizens and not put much or any faith in a standardized lighting package or scheme. Zooxanthellate invertebrates come from all over the reef and had widely variable needs. It is far better to have a moderate lighting scheme with well-fed corals than a precariously bright scheme in a nutrient poor reef aquarium.

Reef invertebrates are accustomed to quite regular light/dark cycles. We recommend you buy and use timers for the lighting of your system. The photoperiod is a matter of finesse and yet another case of more *not* being better. Inadequate lights left on longer will not compensate most needy animals. In some cases, the above-mentioned extra feeding will not help (be "enough" to save the underlit) or be practical either... many of these organisms are fundamentally photosynthetic. Again, this comes back to doing proper research on the lighting needs of your specific guests. Many corals and anemones have a minimum amount of illumination required to stimulate photosynthetic activity by their hosted colonies of zooxanthellae. It's true that most cnidarians are quite adaptable, but lets not abuse their hardiness. Halide driven systems might have an average range of 7-10 hours while fluorescent driven systems are inclined to run somewhat longer at 10-14 hours. The color of the lamps used is also a matter of controversy driven by legend and marketing. Fortunately, the truth is a simple matter. Most aquarists like a noticeable to significant "blue" factor to their color scheme. Yet, every good metal halide on the market (including popular 6500K lamps) has more than enough blue components to the spectrum. Actinic supplementation is not necessary with most any halide bulb you buy. The extra blue light just looks cool and there's nothing wrong with that. Most photosynthetic reef organisms you buy are collected in relatively shallow water and as such favor a "warmer" color scheme. Lamp combinations in the 6500K to 10,000K ranges will let you keep most cnidarians, plants and algae in the trade. Aquarists favoring blue light with a desire to experiment can apply 20,000K halides with warmer colored lamps. Some coral pigmentation will improve aesthetically with 20,000K light while others are corrupted. It is just further evidence of the variable needs and nature of corals and other reef invertebrates collected over a wide range and depth. Metal halide bulbs are more economical and accurate (color rendition and resistance to straying) with lamps having a useful lifespan of perhaps as much as two to three years. Halides also generally deliver more usable per light per watt than fluorescents. Most halides should be mounted 6-9 or more inches off the water's surface (~15-22 cm minimally) with a lens protecting the bulb. Fluorescent lamps instead have a useful lifespan of 6-10 months for most. They should be mounted not more than 3" off the surface of the water (75 mm). Regarding lamp type and heat generation, all lamps produce heat and all light canopies should be sufficiently ventilated enough to make the point of amassed heat moot for whatever type you employ. The main advantage of fluorescent types of lighting beyond any aesthetic preference is that they allow an aquarist with shallow water (24" or less) to "buy into" a less expensive package that will work well. Larger and deeper tanks will benefit from an investment in metal halides. Most marine aquariums are in the 75 to 200 gallon range. For these tanks, 150-250 watt metal halides in the 6500K-10,000K range will provide excellent light for most reef invertebrates in the trade (double-ended HQI lamps have performed admirable here across the board).

Monitoring Water Quality

Daily observation of fundamental water quality and system performance is a necessity for literally any type of aquarium. It is not only an absolute "must" for the well being of your livestock and system, but for your enjoyment as well. Best of all... it only takes a few minutes each day to accomplish. A simple checklist that is clarified or expanded with extra weekly or monthly chores is a good idea. The best time to take stock of all is during feeding time. Is everybody present and accounted for? Are they feeding normally and with vigor? Spot-check of your mechanicals and controllers. Are all pumps working as expected? Is the specific gravity and temperature about right- they can both be read from the same floating glass instrument with some models? Take the time to verify that any automated gear likes float switches, dosers, evaporation make-up devices or timers are clean and tuned.

On a **weekly** basis most aquarists do their "regular maintenance." Aspects like water changes, a more-thorough check of water quality (Calcium, Alkalinity, pH, etc), and

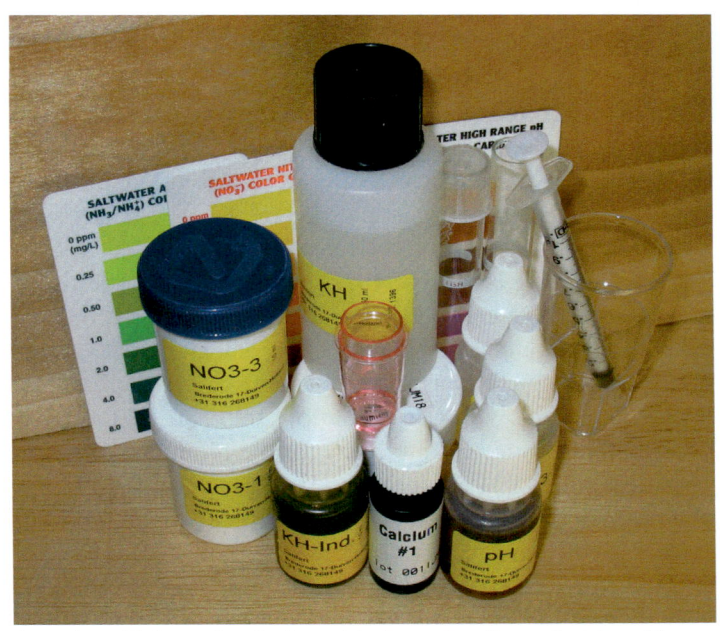

A good test kit is worth twice its weight in precious livestock. A broad battery of tests is an essential part of every aquarist's toolset. (*L. Gonzalez*)

filter maintenance are conducted then. Finessing husbandry includes attention to details and always finding a better way like exchanging a small weekly amount of carbon or other chemical media rather than one large portion monthly. It is also recommended that you wipe clean all light bulbs and lenses (when lights are off and cool) to insure the uninhibited delivery of light to photosynthetic organisms in the display. Charting and tracking these duties and water quality trends (test kit readings) can be very helpful in the future for replicating successful runs or diagnosing difficult or lackluster periods in the tank.

Monthly routines incorporate checking on supportive supplies like reagents, standards for water quality testing, and hardware upkeep like cleaning heaters, probes and protein skimmer venturis. We encourage you to make an actual list of overall "things to check" and include going over all hoses, connectors, valves and electrical connections on this monthly basis. Having extra equipment (like power heads, water pumps and heaters) to exchange during cleaning intervals can be quite helpful (beyond their value as back-up/emergency hardware). Tough deposits (especially internally) can usually be etched or scoured away by running the device in a bucket of strong vinegar water for a day or so. Another handy reminder is to date all light bulbs used and keep them fresh for your photosynthetic creatures (heeding the practical lifespan for each lamp). The exchange of aged or exhausted calcareous media periodically is also quite stimulating for the system in time. If you follow the fundamental suggestions and guidelines in this chapter, you can be assured of having a good foundation for a program of husbandry in keeping reef invertebrates.

Husbandry

Feeding Reef Invertebrates

Due to considerable variability of modes and manners of feeding by the enormous group or organisms that we collectively call "reef invertebrates," we splay the address of pertinent topics on this subject between an overview of fundamentals here and specifics in dedicated sections elsewhere in the text (per interest or species). For the casual hobbyist that simply wants to know how to accomplish adequate feeding with minimal effort, there are methods described herein to put into place for the production of natural foods in a complimentary regime of nutrient cycling. For the dedicated aquarist, these techniques, which largely revolve around modified refugium technologies and specialized target feedings, can be refined to support very specific organisms in a collection. Like so many other aspects of good aquarium husbandry, successful feeding of reef invertebrates is not about force, but rather finesse.

As with nutrifying captive fishes, "less is more" is the route to go with feeding inverts. Skilled feeding techniques are not about getting maximum food into your animals, but getting maximum nutrition into the food that gets into your animals. We are looking for dense and nutritious matter here. There are not so many "cows" on a reef seeking massive volumes of low-grade food; remember that the living reef is not nutrient poor... but rather nutrient concentrated. Even the algae grazers are getting significant protein from the micro-crustaceans in the turf. Certain fishes like some triggers have evolved to exploit this relationship when grazing algae in search of crustaceans. The importance of the food-value dynamic is realized sadly too often by aquarists that allow themselves to get into a habit of feeding significant portions of adult brine shrimp, for example, to their marine fish and invertebrates. Corals and fishes can literally starve to death feeding on it if it is used as a staple. Although fresh baby brine shrimp (hours old) is rather nutritious and a crucial foodstuff in aquaculture, the popularized frozen adult brine shrimp packaged for aquarists is often essentially useless food in contrast to the many excellent alternatives available like mysis shrimp, gammurus, pacifica plankton and minced krill. Without getting into a long rant or extended dissertation about nutrition, let us ask you to simply read the nutritional analysis of the foods that you use... you will likely be shocked. Adult brine shrimp, in this example, is a hollow food comprised of mostly water (see moisture content) with a percent of protein generally in the single digits, in stark contrast to alternatives. If it serves your purpose to stimulate a hesitant feeder to begin accepting other foods, so be it. But do seek more nutritious fare soon enough after beginning to use adult frozen brine shrimp. Protein need not be your limiting factor on food choices, although it is quite necessary for many organisms. Rather, seek a balance of nutrition with an emphasis on the essentials. HUFA's (highly unsaturated fatty acids), vitamins, iodine, proteins and so much more are necessary for good nutrition. As an example, lipid levels in corals are an indicator of readiness to reproduce. For photosynthetic invertebrates, the feeding dynamic necessarily embraces other factors too, like carbon dioxide levels and bio-mineral content to ensure growth and vitality. Indeed, keeping the whole invertebrate system healthy and manageable has more to do than simply making sure all organisms are simply fed.

Good reef aquarium husbandry is intimately attuned to the import and due export of nutrients in the system. We must strike a balance between feeding desirable organisms enough without leaving excess nutrients for nuisance organisms to thrive. There is no excuse for uneaten food stuffs (rotting algae, prepared foods, etc) to lie about and cause pollution. Throughout this reference and beyond, you will learn about the critical importance in reef-keeping of aggressive nutrient control that depends on good and proper water flow to keep organics in suspension for cycling (use) by desirable organisms or simply to export via artifacts like animal or vegetable filters (harvest in refugiums) or protein skimming. With careful planning as to the types of life forms kept, the system design, and set-up and maintenance, this is not difficult to achieve. It bears repeating that the primary cause of death of captive aquatic life is poor water quality mitigated by metabolite anomalies... basically the ill-effects of too much nutrient concentration! And what do you suppose the origin of these high levels of dissolved organics is? Mostly overfeeding and mis-feeding. It is exceedingly rare that reef invertebrates die from starvation in an otherwise well-set up and maintained system. This is especially true of specimens that feed both organismally and by absorption. Alternatively, offering too much food is a major cause of mortality.

An Overview of Feeders and Feeds

Let us detail some of the types of foods and feeders that you will see mentioned throughout this book and popular aquarium literature. Reef creatures can be categorized as either autotrophic (they manufacture their own food through symbiosis), heterotrophic (utilize external sources of food and cannot manufacture any of their own) or mixotrophic (yes, you guessed it! They do a little of both... self-feeding and seek food). Most of the reef invertebrates described in this reference are dedicated heterotrophs with a few mixotrophs mixed in for good measure. Even among "autotrophic" corals and anemones, very few are entirely self-supported in symbiosis as zooxanthellate organisms. Most of our captive cnidarians need some form of nutrition delivered organismally or through absorption. This brings us to the two primary modes of feeding by heterotrophic invertebrates. Organismal feeders eat particles, chunks, pieces or some kind of solid matter. Absorptive feeders draw sustenance from dissolved matter in the water. Many reef invertebrates utilize both strategies.

Addressing autotrophic needs is rather simple and straightforward. There is considerable data on the matter for those interested in looking for it. In practical applications, you must simply take heed of the fundamentals of good aquaristics on the matter. Discover the lighting needs of the photosynthetic organism that you wish to keep (daylight shallow water needs, deepwater blue dominated radiance, heavy UV versus filtered UV, etc). Once you have secured proper lighting hardware, maintain it! Keep bulbs fresh and clean. Maintain optimal water clarity for the penetration of light in to water at depth, and so on. Beyond address with specific photosynthetic organisms in this text, we will explore the general dynamics and technologies of light for cnidarian reef invertebrates in the dedicated volume, "Reef Corals and Anemones" in this Natural Marine Aquarium series.

Addressing the needs of organisms feeding by absorption is also rather brief, although not nearly so well understood as dynamics of lighting for autotrophic species. We understand quite clearly the compositional make-up of natural seawater. And we recognize the essential components that are necessary for reef invertebrates in it. What we do not clearly know yet is the recipe of nutrition for satisfying each and every species of reef invertebrate kept that feeds on dissolved matter (organic and inorganic). This last comment is a gross understatement by any definition. Fortunately, the impact on successful reef-keeping is not so severe and it will be clarified and improved in time. For now, we can simply provide the inorganics, bio-minerals and the like via a regular water change schedule and controlled dosing. Trace elements and key elements are readily replenished in this manner and spares most aquarists from having to test and dose for many things. Weekly water changes are better than monthly in this regard... and daily is better than weekly in due course. Automated water change schedules (solenoids, overflows, timers, float switches, etc. for the "techies") can endeavor to conduct daily water exchanges. For all others doing manual exchanges, weekly water changes can be conducted in small and manageable portions to support mineral replacement. Organic components on the other hand will require some experimentation by the aquarists

This phytoplankton "green water" reactor designed and built by hobbyist Ken Gosinski fortifies his 450 gallon reef system (750 total system gallons). Complete plans and hobbyist tech-support for such do-it-yourself projects are readily available on the internet and at many local aquarium club meetings. (*L. Gonzalez*)

for invertebrates that feed on such dissolved matter. Even here, though, the matter is easily resolved for many with the common reality that most aquariums are overstocked, overfed or inappropriately fed. Fish and invertebrate fecal pellets and excretion are a tremendous source of nutrition for many other reef organisms. For light bio-loads, the common circumstance of inferior or non-existent protein skimming also quickly contributes to the accumulation of dissolved organics. At any rate, most aquariums have precarious if not problematically high levels of dissolved organics and most invertebrates that feed by absorption have little trouble finding sustenance in this category.

Organismal feeders comprise the majority of invertebrates and all fishes in the marine aquarium. There are many different types of organismal feeders that have evolved to exploit different niches and prey possibilities. We describe the nuances of each method with the topics of interest in the species surveys and applicable chapters (live sand, refugiums, etc) throughout the text. To familiarize you with just some of the strategies employed by reef invertebrates we list the following:

- **Filter-feeders** (AKA Suspension feeders) rely on currents to bring specific prey of specific sizes that may be live or dead matter (collectively **seston**) and includes phytoplankton and zooplankton for most.
- **Deposit feeders** - derive sustenance simply from deposited organic matter. Grazing snails and urchins are prime examples. Prey may include lower forms like algae or higher forms like other benthic invertebrates.
- **Detritivores** are a specific group of feeders often including those known as bioturbators that essentially forage in soft substrates for various organic matter that may include diatoms, polychaetes, micro-crustaceans and more. Some are sediment de-stabilizers preying on sediment stabilizers (tube-building worms and various algae binding the substrate together, and the like).
- **Herbivores** like *Diadema* urchin species are crucial for reef ecology, performing as a key participant in nutrient cycling.
- **Carnivores** like crabs, predatory snails, cnidarians (on zooplankton) sit atop the invertebrate food chain.

This list is indeed brief and necessarily only grossly depicts common feeding modes of reef invertebrates. Many species combine strategies of course. But you can begin to understand the possible variations of how an organismally feeding creature might get its food and how very complex the mode of delivery or acquisition can be.

What About Balanced (No Purposeful Feeding) Systems?:

There are indeed aquarium systems that can be designed to require little or no food delivered (target feeding) by the aquarist. Such communities are made up of a mix of photosynthetic and non-photosynthetic life that are furnished with small to microscopic foods via live sand, live rock, deep sand beds, mud filtration, algae growth, and especially refugium culture of live foodstuffs. Such a balanced microcosm is an ideal situation, but is rarely realized. We cannot encourage you strongly enough to exploit the many different possibilities with refugium methodologies, and we have gone to great lengths in this reference to provide fundamental and comprehensive coverage of it and related topics to get you started. Nonetheless, please do count on supplying at least some foods (target feeding) to most of your invertebrate livestock.

Types of Foods

The variety of foods available to hobbyists for feeding their aquarium guests is remarkable. Fresh, live, thawed-frozen (whole-prey and formulated mixes), and dried-prepared foods of all types (pellets, sticks, flakes, baked, extruded, freeze-dried, etc), abound... who could ask for anything more? Yet, from the likely perspective of residents of your aquarium, the smorgasbord is unfamiliar and takes some getting used to. Beyond alien prey, there are huge issues of texture, color, "smell" and movement (or lack thereof) that can give a newly imported and hungry creature cause for pause. Imagine going on a strange, unscheduled and unprecedented trip perhaps like an alien U.F.O. abduction. Then consider your reticence in light of the turmoil when offered "food" by your captors that was totally unfamiliar. Too large, hard, stinky, and proffered in the middle of the night (off your feeding schedule) and with other bizarre creatures you've never seen before darting about and possibly eating it all before you could while pushing you about! Not a pleasant proposition, eh? This may be roughly analogous to what our aquatic livestock face, coming from the wild and being placed in the small confines of a clear box (aquariums) with a foreign mix of other life forms. As empathetic aquarists we should be mindful of the stress of acclimation and the importance of patience to get the best and most appropriate foods into our acquisitions... and to do so quickly.

There are a few things that aquarists can and should do to accommodate their aquatic charges regarding food and feeding opportunities. To begin with- study! Discover what sorts of things they eat in the wild and assess how well you can reasonably provide a substitute. The lives of many reef invertebrates could be saved if aquarists had patience to resist stocking the tank with carnivores while their living substrates and refugiums matured. Also consider how your creatures feed and from where: on the bottom, in the bottom, in the water column... and at what times? Be mindful of how these needs can and will blend, if at all, with the needs of the other tank mates. Case in point, there are many heterotrophic filter feeders that are quite hardy and easy to keep in captivity but suffer high rates of mortality for being kept inappropriately in reef tanks dominated (and operated for) autotrophs! Expect that new acquisitions are going to be likely to be shy at first... have patience for polyps and feeding structures to be displayed. These behaviors do not occur on command or with the snap of a finger just because you think you have "perfect" water quality. For these and other reasons, please be sure to establish a diet and menu for newly acquired organisms in the peaceful and controlled confines of a proper quarantine period on arrival.

A Brief Overview of the Merits and Demerits of Various Foods for Marine Invertebrates

Live or Fresh (*in situ* grown micro-crustaceans, algae, plants, larvae, etc)

- \+ Ideal quality and nature for invertebrates. Most staples can be cultured in an appropriately designed refugium
- \- May be labor intensive to culture, expensive and risky (disease & availability) to import

Thawed fresh-frozen (whole-prey and formulated mixes)

- \+ Excellent balance of quality overall, fresh vitamins not readily available in dried foods
- \- Expensive to use as a staple, easily mis-fed (pack juice fuels nuisance algae if not decanted or limited)

Dried-prepared foods - *baked*: traditional pellets, sticks, flakes (high processing temperatures)

- \+ Dense, convenient and nutritious matter
- \- Texture is not always well-received at first by consumers, high processing temperature deteriorates some quality (vitamins)

Dried-prepared foods - *extruded* (low processing temperatures)

- \+ Dense, convenient and nutritious matter... arguably better in some ways than high-temp baked foods
- \- Texture is not always well-received at first by consumers

Freeze-dried foods

- \+ Very dense nutrition and some of the highest protein (for growth) of any aquatic foodstuff
- \- Texture is not always well-received at first by consumers, floating nature is quite unnatural as prey

Bottled suspensions - phytoplankton substitutes

- \+ Convenient for all, useful and very nutritious if fresh (weeks old)
- \- Highly perishable, most demand close attention to shelf-life and holding temperatures, may require additional processing to reduce particle size, easily abused

Bottled slurries - zooplankton substitutes

- \+ Convenient for some dedicated heterotrophs
- \- A messy and easily abused category of products. Many in this group have earned the title "Pollution in a bottle!"

Delivering Foods - "Getting it Right"

One of the most commonly underestimated dynamics of feeding aquatics is the actual delivery of foodstuffs. Numerous aspects of delivery are crucial for optimizing feeding success and efficiency. Frequency, amounts, place, time and particle/prey size are of enormous consequence. Even given palatable, nutritious foods, in agreeable formats... what good will having foods "that work" if you can't deliver them in timely ways to the intended stock? You should be able to see your motile invertebrates foraging and take note of their habits (where and when) and meet them there with a feeding stick (store-bought or home-made from a wood or plastic dowel) or simply with the use of plastic tongs. Most feeding behavior is species-specific, and easily determined by careful observation. Its the old adage... form follows function! If a sessile invertebrate puts out feeding structures that are large, formidable and only displayed at night... it is likely targeting zooplankton. On the contrary, if an invertebrate's feeding structures are very small and displayed throughout the day or at random times... it is likely targeting phyto and/or nanoplankton (and may include non-living matter). We assure you that it will not be difficult to make a reasonably accurate estimation of such preferences with thoughtful observation and a little bit of sincere research.

Filter-feeding strategies certainly do abound amongst reef invertebrates and this aspect of feeding will need to be addressed by most modern marine aquarists keeping live substrates (rock and sand) replete with such organisms, even when corals and other invertebrates are not kept.

Refugia can be designed to culture all manners of natural and nutritious live foods including microcrustaceans, epiphytic matter, gametes and other types of plankton large and small. This 120 gallon mud and macroalgae refugium supports a beautiful, thriving 240 gallon reef aquarium. (*L. Gonzalez*)

Most every observed phylum of reef invertebrates has at least some species that derive their food intake by passively sieving (like featherduster worms) or actively pumping the water column (like bivalves and sponges), or producing mucus nets that they periodically engulf (like *Fungia* corals). Know that a good deal of the life in your system (more than half by mass and number) are truly filter-feeders. Plan accordingly to either provide such foods yourself, have ample foodstuffs generated (refugiums), or do not overcrowd your system to the extent that plankters produced *in situ* (from live rock and sand) are enough to sustain them.

Aquarists have been experimenting with commercial and home-made suspensions in an attempt to produce plankton substitutes with varying degrees of success. Using prepared food is a favored technique for systems with a modest or thin live sand bed, or without adequate amounts of live rock or refugium product. You can make a DIY mash of meaty and algal foods via an electric "blender" then ameliorate the mix (add water to dilute) to be dosed in the tank with a baster, pipette or like vehicle of delivery. It may be helpful to temporarily suspend some filters or pumps during brief feeding events too (consider using a delayed timer to resume pump functions). Some animals, however, will only feed with certain types of water flow (more is usually better). Thus, some experimentation will be necessary to produce flow during feeding times that does not waste food while sufficiently stimulating feeding activity. The biggest obstacle to commercial and home-made preparations is particle size. Most filter-feeders need very small prey sizes measured in microns which cannot easily be produced. Even when foodstuffs can be reduced adequately in size, aging of mere days and certainly a matter of weeks causes clumping and inevitably increases in particle size which can make the food-matter wholly unusable for some filter-feeders.

Herbivorous and carnivorous macro-invertebrates and fishes are generally easier to feed. Challenges with these groups are more a matter of providing the right quality of nutrition than particle or prey sizing. Herbivorous invertebrates will feed on algae and perhaps to some extent, true vascular plants like seagrasses. Such animals are best served by stocking in low densities, promoting either micro- or macroalgae per their preferences in the system, or importing such matter grown elsewhere. A refugium or other remote vessel can serve as a vegetable filter to not only grow suitable food for herbivores, but to also improve water quality and cycle nutrients for their participation (see more on this dynamic in the Refugiums and Plants & Algae chapters). New companies in the trade are developing to serve aquarists that seek living products for natural marine aquariums with refugia. Fantastic species for culture like *Gracilaria* and *Chaetomorpha* algae have become deservedly popular and can be found from mariculture specialists and even many local aquarium stores. Please be sure to resist the use of terrestrial greens like lettuce and spinach to feed aquatic herbivores. There are few benefits to the use of farmed vegetables and many potential dangers. The first problem is that most cultivated vegetables are grown with fertilizers (nitrogen and phosphorous based) that pollute the matter with phosphate and nitrate in concentrations that measure parts per thousand (!) which is a tremendous fuel for nuisance algae growth. The use of these land plants as food for aquariums is indeed a source of various and concentrated contaminants. Even if that were not an issue (organically grown produce, etc), terrestrial greens have very little food value for marine life. Their cellulose component is an impediment to digestion of the very little nutrition that they have and proffering such matter without breaking it down (boiling, freezing, etc) will cause the aquatic consumer to pass the matter almost wholly undigested. Like other hollow foods mentioned, terrestrial greens fed to marine organisms can starve the consumer and truly have little place as aquarium foods.

It bears mentioning that there is some concern about allowing the pack juice of nutritious thawed frozen foods (primarily whole prey items with a lot of extra water) to enter the aquarium. In time, the admittance of this juice is a serious burden on water quality and catalyst for nuisance algae. To begin with, please be sure to thaw all frozen foods in cold water to preserve nutritional quality (never thaw at room temperature or warm water). Know that the nutrient rich pack water is not readily utilized (or needed in such concentrations) by reef invertebrates and certainly not by fishes. Meaty fare should be decanted of all thawed juice (strained with a nylon net, for example) and this liquid discarded. The argument that such protein-laden water is useful to some filter-feeders is true but weak. By the same line of logic, a dissolved hamburger in the tank will also provide useful nutrition! Incidentally, this is a similar basis to the appropriate criticisms of feeding abuses with bottled supplements (such products are often just "Pollution in a bottle"). If you choose to use a bottled phytoplankton or zooplankton substitute, be sure that the creatures you feed it to actually eat it! Feeding meaty slurries to gorgonians and Neptheids is about as useful as feeding phytoplankton to many popular corals (that are predominantly dependant on zooplankton and other fare). Explanations that the uneaten portions (which can be most or all of the matter) will degrade into still useful dissolved organics or fodder for micro-fauna is a sloppy argument. "Getting it right" with feeding invertebrates means the right food gets to the right organism.

At length, whatever food you feed... and however often you feed it, please remember one thing - "what goes in must come out." Overfeeding does not only mean uneaten food; it also includes overeaten food- excessively fed and consumed. Relative to water quality issues, and this is a bit of an exaggeration, it makes little difference if the food you feed falls into the aquarium uneaten or passes wholly though an invertebrate or any combination thereof: the matter still was imported into the system just the same and needs to be reduced, converted and/or exported. Too often, much of it simply accumulates and degrades water quality as evidenced by nuisance organism growth and ever-increasing levels of dissolved organics. There are so many subtle aspects of water quality that we cannot conveniently test for, if at all, that can compromise aquatic life. "Perfect" water quality cannot be measured by reports on the traditional dozen or so parameters (pH, ammonia, nitrate, etc). Without large and frequent water changes to dilute the unknown or unmeasured burdens of water quality, acute attention to nutrient export mechanisms like skimmers and vegetable/animal filters will be necessary in the confines of a closed aquarium system.

Develop A Feeding Routine

Just as the sun rises and sets, the world's reefs have a rhythm. Establishing and maintaining feeding routines regularly (same time, place, types of foods, amounts, etc) pays dividends. It's helpful to feed more active and greedy animals first (even a different food if necessary) else you may find that shrimp stealing the anemone's food! Beyond good husbandry, reef invertebrates really do develop a rhythm and can be "trained." Animals starved and not inclined to feed soon after import will often come around nicely after several days or even a week or more of feeding the tank on a rhythm. Be sure in such cases to siphon out the uneaten food later. With some finicky reef creatures, there are cases where a simple change in one aspect of feeding has led to starvation. After this chapter overview *en toto*, we have hopefully convinced you to develop and maintain a regular routine in your aquarium husbandry for feeding reef invertebrates.

Despite our passionate appeals to aquarists to culture natural foods at home, many useful and nutritious prepared foods are available for your convenience. Examine the quality, texture and shape of the matter and consider which ones will serve your animals best. (*L. Gonzalez*)

Reproduction in and of Reef Invertebrates

A series of pictures showing a *Diadema* long-spine urchin spawning in the aquarium. (*K. Gosinski*)

How do you measure success? By how much money you have or earn? By the speed or stylishness of your car? Perhaps by the sophistication of your ideas, or depth of your close friendships? For many aquarists and biologists, success is measured by the reproduction of their livestock in captivity, whether directly by their influence and direction or not. The successful replication of marine invertebrates is a hallmark of good reef husbandry, perhaps its very zenith. Other than one's very survival, the drive to reproduce is the strongest instinct of life on this planet: Continuation of the Species. Thus, and not surprisingly, given reasonably good care and conducive circumstances, or in some cases deleterious triggers (stress-induced spawns), captive invertebrates will propagate in your systems.

Exactly how various invertebrates reproduce in the wild and in aquaria is an expansive topic with interesting variations. It may be conducted **unassisted** in natural settings or **assisted** by the aquarist through **passive** techniques (manipulation of photoperiod, water temperature, and other physical cues, e.g.) or by **imposed** techniques (the introduction of hormones or actual fragmentation). If the act is sexual the participant(s) may conduct themselves as separate sexes or hermaphrodites (acting in a dedicated gender, or in rare cases self-fertilizing). In other circumstances the participant performs simply as a sole asexual individual.

The evolution of marine aquaria, to include a more complete and diverse representation of the reef environment, has facilitated reproduction in invertebrates faster than we can document. Alas, too many uncommon events of reproduction occur in a hobbyist's tanks with only a passing mention to a friend, at club meetings or on the Internet message boards. We beseech you to please (!) photograph and document any stage of reproduction that you observe in your aquarium and share that information. It is interesting to fellow aquarists and to science at large. Thanks to the use of live sand, live rock, refugiums, and a diverse compilation of plants, algae and animals in our aquariums, we can witness every mode of reproduction described here.

To date, most successful acts of reproduction in aquaria are asexual in nature. Successful *sexual* production has been the most challenging mode to harness, and there are many reasons for this. Mind you, stimulating reef creatures to release gametes is not difficult for many aquarists to accomplish. The biggest challenge to aquaculturists is the navigation of larvae through their developmental stages to ultimately reach adulthood. Providing the circumstances and principal foods in exact sizes at the exact times necessary in the right concentrations and types has proven to be an extraordinary challenge for some species. The organisms that we have had the most success rearing by events of sexual reproduction have had direct development or very short larval cycles (hours in many cases, not weeks). To be specific, the challenges that an aquarist will face in these endeavors include:

- **conditioning brood stock** (feeding, water quality, maturation, critical mass, etc.)

- **identifying and providing necessary cues to trigger a spawn** (parameters including: water temperature, photoperiod, lunar cycles, hormones)

- **navigating larvae through developmental stages** (providing specific substrates necessary for settlement, reckoning the challenges of pumps, filtration and other artifacts of system hardware)

- **identifying and providing larval foods** (incremental and changing with growth; can be very specific by size, type and concentrations to even stimulate feeding behavior in larvae)

- **Rearing** (having the housing necessary to grow out potentially large spawns and judiciously culling the rest - ultimately finding an outlet for mature offspring)

Endeavors in rearing the products of sexual reproduction may play a vital part in the survival of our hobby and industry. Regardless of how great or small our current impact is on limited reef resources, we simply should be pro-active in establishing a self-sustaining industry. Many people are hard at work in this regard and we hope that you will be, too. There is perhaps no other hobby so large or with so much potential as that of aquaristics to pay dividends back to the environment and to science for our efforts.

For now, we shall continue to refine techniques of asexual reproduction at large. Fissionary events and imposed fragmentation of reef invertebrates have become delightfully commonplace as evidenced by the proliferation in the trade of handsomely colored Cucumarids (fissionary split sea cucumbers), "unstoppable" Bubble-Tip Anemone (BTA) splitting (clones are becoming quite common in aquarium clubs), and of course the strong trade in home-propagated soft and stony corals. In this and so many other ways we are reminded of a favorite saying, "information is the oxygen of understanding." As we are all pioneers in the care and culture of marine invertebrates, let us pursue and share our discoveries freely.

Reproductive Strategies:

<u>Sexual</u>

Egg-laying:

direct release (deposit or broadcast) or brooding

- external fertilization (usually sessile organisms)

- internal fertilization (usually motile organisms)

Live- bearing:

<u>Asexual</u>

- **vegetative** (various modes of fission - quite common in many reef invertebrates)

- **fragmentation** (common in algae and reef coral especially)

- **parthenogenesis** (from the Greek *parthenos*-virgin, *genesis*-origin: "virgin birth") a form of reproduction in which an egg develops without fertilization. It's rare in nature but not uncommon in the sea as with some plants and algae (parthenocarpy: from the Greek *karpos*-fruit), micro-crustaceans (rotifers) and even coral (as with the common *Pocillopora damicornis*)

Know Your Livestock

The more you know about how the life you keep creates its own progeny, the more you will appreciate it, cherish it, and are able to understand the organism at large. Perhaps read the previous sentence over again until you can repeat it. It is remarkably important to have a complete understanding of how your charges live, and to know how to care for them in turn. Ask yourself if you can even grossly identify which of the creatures in your care are egg-laying or live-bearing.

Over the many years of trying to help others to be successful in keeping aquatic life, we've heard repeatedly from folks who want to know the appropriate foods, feeding, compatibility, lighting, water movement, and other requirements of stock they've *already* acquired. Such questions are often followed up with requests for advice on how to propagate or breed them. The questions indeed are fine and fair - it's the *timing* of such questions that's absolutely dreadful. It is the consensus of the authors that it is irresponsible for those who endeavor to take into their care creatures much plighted for a variety of reasons.

Among other things, this shows a disregard for life as the answers to these questions upon which a living creature's very life hangs must, by their very nature, rely on the timeliness and accuracy of a reply from random consultations for a successful outcome. Simply put, impulse purchases ought to have no place in the livestock trade and the authors strongly urge all aspiring and current causal aquarists to research and carefully consider aspects of husbandry before making any purchases.

Having worked in this trade together for decades, collectively, we can honestly tell you that *nothing* you will ever come across is so rare that you should buy it without taking the time to research its needs and determine if you can meet them first. There really are "plenty of other fish (and invertebrates) in the sea." In fact, almost every organism that is rare in the trade is still extremely common in the wild; it is merely the workings of the trade (supply and demand, chain of custody, shipping realities) that limit the availability of such specimens. If you have ever been diving on a reef in the midst of a school of hundreds to thousands of "rare" tangs over square kilometers of "rare" coral, you have a much better appreciation of how easy it should be to be patient for getting aquarium specimens.

Reef Invertebrate Reproduction by Group:

In support of efforts with invertebrate reproduction, we offer the following summaries of some key invertebrate groups. Here we have detailed the fundamentals of their nature and habits regarding reproductive activity. Overall success with propagating any species of invertebrate is a matter of providing proper foodstuffs within an environment conducive to meeting their needs, as previously stated. Specifically though, some of the first information that you will need to gather is data on sexing brood stock, as well as the required inclinations or triggers for specific interests. Practical considerations will need to be addressed, too, in the habitat that you design with regard for predators on gametes, larvae, parents, etc., as well as the dangers of intake screens, pump impellers, and more. Please review the husbandry surveys of these groups later in the text for further insight.

Sponges - Phylum Porifera

This most primitive group of multi-cellular animals, like all living organisms, aspires to "make copies" of itself in space and time. They do so both sexually and asexually. Asexual reproduction in Porifera is accomplished in at least two ways. Buds can be produced and liberated from parents, and packets of "essential cells" (**gemmules**) can be isolated or cut away (as by an aquarist) to form a new individual.

This last method has been employed in commercial sponge harvesting with natural source-sponges for household use off of the coast of Florida. Sponge farmers used to attach cuttings to cement blocks and dump these seed colonies into the ocean for regeneration, repopulation, and replacement in over-collected areas. In scientific literature there is a classical description that illustrates the remarkable powers of sponges to re-associate, or rebuild themselves. It's said that a colony of sponge "strained" or passed through a sieve can rejoin itself from among the pieces on the other side and grow back together - re-aggregation. Thus, let us reassure you that, if you can get beyond the fundamental challenges for keeping given species in captivity, propagation, if not reproduction, can be *very* easy.

Sexual reproduction in sponges takes place by using both hermaphroditic (both functional sexes in one individual/colony) and **dioecious** ("two houses," separate sexes) modes. Most sponges are hermaphrodites, but produce eggs and sperm at different times, which facilitates out-breeding between colonies (self-fertilization is rare). The formation and release of sex cells is triggered by photoperiod, water temperature and cell regression particular to each species. Sperm are transported out of the animal by the same mechanism that food and oxygen are moved in and wastes and carbon dioxide move out. This is accomplished by way of the collective spiral motion of the flagella of their choanocyte cells. Fertilization occurs *in situ*. Sperm are carried by water currents, if they are fortunate, into another member of the same species where they fuse with an egg. With most species, development goes on inside the mother colony (**viviparity** - from viviparous: live-bearing), though some of the Demospongiae release fertilized eggs (**oviparity**) with further development taking place in the sea. Larval development is short, at about two days for many, with the free-swimming individuals settling on a prospective substrate.

For aquarists: the growth rates of encrusting sponges, with a few exceptions, can be excruciatingly slow. In the wild some have been shown to grow less than two inches in ten years. Yes, large specimens may be many decades old. Although modern methodologies with marine aquaria as diverse microcosms has made the keeping of sponges much more successful, basic husbandry for the group overall is quite challenging. Success is further hampered by significant import and shipping obstacles (such as time of transit, lack of water flow, noxious exudations, and air exposure). As you will hear time and again, in this reference and in popular literature at least in the near future, we recommend that the culture of sponges be left to experienced aquarists and those with a desire to establish species-specific displays for study.

Worms - Let's talk Polychaetes

Although there are several other groups of worms that occur in marine systems, the most prominently desirable category are the segmented "bristle" worms and their tube-dwelling brethren, the featherdusters, of the phylum Annelida, class Polychaeta ("many bristles"). Another familiar group includes the flatworms, some called "planaria," that hobbyists generally wish to avoid due to their specialized diets, difficulty of care, or outright predatory or nuisance tendencies. At first glance, the errant (free-moving, surface-scouring, burrowers, planktonic) and tube-dwelling (sedentary) polychaete worm species may appear quite different, but they are in fact closely related and essentially share the same modes of reproduction.

Asexual reproduction is known amongst some of the families of polychaetes (some Sabellid fan worms, and errantiate Cirratulids, Syllids, Spionids). This takes place either as a budding or a division of the body into two or more fragments. Sexual reproduction is the popular mode of reproduction for polychaetes at large. Most species are separate sexes (dioecious *versus* monoecious), although there are a few hermaphroditic species including some of the featherdusters. Eggs and sperm may exit the body through specialized pores - ones shared with the excretory system (nephridiopores) - or by rupturing through the body wall leading to death of the adults. Aquarists have reported extraordinary events of sexual reproduction in worms by cues from the first ripe individual to release gametes (a common trigger among reef invertebrates). We have heard fantastic stories of hobbyists tugging and pulling at a gravid bristleworm (unbeknownst as gravid to them) with tweezers in an attempt to rid them as vermin from the tank.

Much to their chagrin, however, the worm tears apart and gametes are shed into the water which promptly stimulates tens or hundreds of other worms to crawl out of the rocks and sand to spawn *en masse*. The whole pride and purpose of the aquarist's "attack" could make you laugh yourself to incontinence in hindsight (well, do resist).

Epitoky is a term applied to another reproductive phenomenon of polychaetes. In this manner the worms form worm-like epitokes from transformed segments that break off as a metamorphosed units (or bottoms of tube-dwelling species) that go swimming about in swarms at the water's surface, shedding eggs and sperm. Epitokes are cued by environment (mainly lunar periods and light intensity) and collected by native peoples as food in various places in the tropical South Pacific. Ciliated larvae are the result of such swarming events, and said larvae settle to the bottom of the sea within a few days.

Some polychaetes are egg depositors, rather than broadcast spawners. They retain their spawn within their tubes, burrows, or mucus masses. Embryos typically develop into ciliated trochophore larvae, which develop into multi-segmented young during their short planktonic lives, and ultimately into a settling phase. Some species have no planktonic larval phase and emerge instead as miniature adults upon hatching from their egg cases. As is common on this planet, short-lived species (say, a lifespan of a year or two) tend to produce large numbers of small eggs, which have a planktonic feeding phase of a week or more. The more perennial species produce fewer, but more "yolky," eggs; the larvae are non-feeding or weakly so, and they tend to be benthic.

For interested aquarists, estimates of worldwide infauna (the life found in and amongst substrates) put polychaetes somewhere between 40 and 80 percent of the total. One study found an average density of 13,425 individuals representing 37 families per square meter of substrate in Tampa Bay, Florida. Additionally, there are numerous additional species inhabiting hard substrates like our live rock. To put it bluntly, if you're going to have a marine system, you will have some of these worms. Most are harmless or even beneficial (including most bristleworms). Hopefully you can avoid large specimens (true fireworms from the Atlantic) that might bother your other livestock. Fortunately, many polychaetes are almost microscopic.

Most polychaetes are quite useful for both binding and/or stirring up the sand, eating wastes, excess food, and becoming foods themselves. If you see them, don't panic. The infamous and dreaded bristleworm is actually a remarkably useful member of the infauna; they only become problematic when an aquarist neglects the system or when the system has a flaw which allows their proliferation to plague proportions. Consider carefully: if you have a complaint that they are reproducing at an alarming rate and becoming a plague or nuisance, what do you think they are eating? They don't grow from thin air any more than they can grow from barren water. These useful scavengers flourish in the presence of excess organics, and their presence in large numbers indicates an all too common problem with nutrient export such as over-feeding, poor skimming, poor water flow, accumulation of excess detritus. As it relates to propagation, however, be sure to feed them well to enjoy their utility or beauty (as with fanworms). Unless you see your worms actually attacking your featured livestock, leave them be.

* A passing note about featherdusters and fanworms "losing their heads": this is *not* a reproductive event, but rather a sign of stress or duress. It is not necessarily the end of these specimens either, though. Featherdusters do shed their feeding crowns and can easily regenerate the appendage in time. If their tube body feels firm (indicating that the worm is "at home" and living) and there are no signs of mucus or decay from within the tube indicating death, be patient. It may take weeks or months, but a new crown of "feathers" will be forthcoming.

Mollusks "a–plenty"

Gastropods: Snails, Slugs, Nudibranchs

With some 35,000 described species, the gastropods are the most successful (largest, most diverse, and best distributed) group amongst mollusks. Most of the Prosobranch shelled gastropods are dioecious (either male or female), placing eggs possibly in a gelatinous mass or fertilizing them individually with the sperm of a single male or males in the environment, although not through copulation (internal fertilization). Opisthobranchs (shell-less or reduced shell - bilaterally symmetrical gastropods) do engage in copulation, often through cross-fertilizing of each other as hermaphrodites behaving either simultaneously or in protandric mode (male first).

Embryogeny - The most primitive gastropods, the Archaeogastropoda (abalones, limpets, trochids), disperse their eggs and sperm into the water column where the ciliated trochophore larvae develop. All higher gastropods suppress this larval stage, passing through it before hatching instead. The majority of gastropods have larvae called **veligers**, named for the swimming organ the *velum*, which are two lobe-like semicircular body parts whose cilia move the individual about.

There are two, basic larval nutritional strategies of veligers. **Planktotrophic** forms feed on plankton and may stay in the open sea for up to three months. **Lecithotrophic** types are endowed with yolk reserves, do not feed while pelagic, and generally settle within a few days. On settlement, the velum and foot develop with a twisting of the body (torsion) occurring, and the individual resembling a miniature adult of the species is formed. Chemical clues about likely provident settling spots are recognized by relayed secretions of hydroids, bryozoans, and ascidians, which are the foodstuffs of the species. This complex mode of reproduction is quite fascinating, particularly how the larvae choose a specific place to settle based on chemical cues. It is also a serious impediment to the successful rearing of gastropods in aquaria by casual hobbyists.

Polyplacophorans: Chitons

Most polyplacophorans are separate sexes (dioecious). They shed their sperm into the sea, fertilizing the eggs of females either there or within their mantle cavities, from where they are afterward shed into the open sea. In light of this and the small size of the animal, one can understand the natural clustering of individuals of these species in light of their need for proximity in reproducing. Chiton young develop as trochophore larvae (no veliger stage), and settle down for extended periods of metamorphosis on the bottom. There are some species in this group, however, that brood. Many aquarists have had experiences with the successful reproduction of chitons in aquaria, although it is more a matter of conspicuous occurrence and popular anecdote. More documentation of the activity would be nice to see in time from aquarists.

Bivalvia: Clams, Oysters, Mussels

Most all bivalve mollusks are dioecious and reproduce without copulation. The few hermaphroditic species in this group do, however, include aquarium specimens like some of the oysters, scallops, and cockles in the trade. There are brooding (inside the shell) species of bivalves, although the majority shed their gametes into the environment where they become trochophore larvae (if they fuse successfully). In time they will develop the two recognizable shell halves of the group's namesake, and then settle in for a benthic lifestyle. As with the gastropods, both feeding and non-feeding veligers are possible here. For the long list of reasons that revolve predominantly around an extended or complex larval development, few bivalves can practically be cultured in aquaria. This is certainly true of the most popular members of this group, the Tridacnid clams. The actual event of bivalves spawning in captivity is rather common, but a commercial facility with adequate space and hardware for support will be necessary to successfully rear the products of sexual reproduction.

Cephalopods: Octopuses, Squids, Nautiluses, Cuttlefish

These species are all dioecious with a single gonad positioned in the rear of the body. Males have a storage vesicle (Needham's sac) where they store their sperm in a packet (the spermatophore), which opens on the left side of their mantle cavity. Fertilization occurs either inside the mantle cavity or outside following displays of courtship, interaction, and copulation. The act involves the placement of the spermatophore from the male by way of a specialized arm, the hectocotylus. This arm is modified as an intromittent organ for penetrating the mantle cavity of the female. Fertilized eggs and their external membranes are attached in clusters onto hard substrates.

Larval development is direct (no trochophore or veliger larval stage) but young may exist as plankton for some time until they reach a specific size (as with octopuses). Species in this group are usually short-lived with a lifespan of one or two years as the norm. Three to five years might be the maximum for the group, with rare exceptions being unsuitable species for aquaria. Most cephalopods die after reproducing. Successful reproduction in captivity is not wholly unrealistic and has been accomplished with more than a few species by private and public aquarists. The main obstacles to successful reproduction with these magnificent and intelligent animals are to be found within the context of husbandry. Water quality and behavioral issues (such as conspecific aggression) are quite challenging hurdles to overcome with cephalopods to realize success in captive reproduction.

Crustaceans

Most, but not all, crustaceans have separate sexes. The gonads are paired in both, and empty out at the base of paired thoracic appendages; which pairs of appendages have this vary by genera. Copulation is the rule for the group, with males grasping females via specialized leg aspects. Sperm may be flagellated or not, and transported as spermatophores or not, or the intromittent organ may be modified as a penis for genetic intromission. Some females have a seminal receptacle, but most have prominent, pouch-like, ectodermal invagination for copulation.

The majority of crustaceans are egg-brooders. They hold onto their fertilized eggs in a brood chamber or even just clutched on their abdominal swimmerets (**pleopods**). Free-swimming planktonic larvae are most common for marine species (and some freshwater ones, too). Folks who've raised tropical fish with freshly hatched baby brine shrimp will be familiar with the most basic larval form of crustaceans: the nauplius. Nauplii are characterized by having three sets of appendages (the first and second set of antennae and mandibles) that are unsegmented. The second pairs bear setae for swimming. These young also have a single median (naupliar) eye. In more advanced species, such as crabs, the nauplius larval stage metamorphoses through molting into larger forms (zoeae) with more

This series of photos shows gamete release by a newly purchased *Astraea* snail immediately after acclimation and release into the aquarium. (*L. Gonzalez*)

segments and appendages resulting to form a unit that bears the same number of legs as the adult, and is then called a post-larva. Depending on the group, any or all of these stages of development may be suppressed, and one can become overwhelmed by the numerous names applied to instars (molt stages) of particular groups.

The successful propagation of crustaceans in aquaria at large is highly variable. At length, there are many desirable species to work with that already have established protocols for the earnest aquaculturist to follow. For the more ambitious aquaculturist, the foundation has been laid with similar species, no doubt, to support pioneering work disseminating procedure for reproduction in undescribed species.

Stomatopods: Mantis "Shrimps"

Some readers of this brief passage here will surely look upon it with chagrin, for they have had their own unfortunate experiences with predatory Stomatopods. Nonetheless, there are members of this group that are fascinating, useful (for display and study) and, yes, even beautiful. When one considers their reputation of real and perceived aggression ("thumb-splitters" to unwary SCUBA divers), it's no surprise that reproduction in this group is not well documented in popular literature. Still, there are enthusiasts who appreciate these animals. We do know a bit about their reproductive activities, too. The eggs of mantis shrimp are agglutinated into a ball that the female broods, foregoing feeding, and are held either under her abdomen or upon her back. This ball may also be kept in a burrow where it is turned and cleaned continuously. The zoeal larval stage is that which hatches out, and these young spend as long as *three* months as plankton before settling down. Hobbyists with an unhealthy fear of mantis shrimp breeding in their tank, taking over their system, and threatening dogs, cats and small children can relax; Stomatopods are not going to readily propagate anytime soon in the displays of casual aquarists.

Decapods: Shrimps, Crabs, Lobsters

Most Decapod reproduction involves intimate placement of eggs, sperm and offspring. They have non-flagellated (immotile) sperm, and as such must engage carefully in genetic intromission/fertilization. The need for identifying and interacting with another of its kind (not always the opposite sex; as with shrimp, most notably) is key to their survival in space and time. Most use the handy encapsulated spermatophore transfer, mentioned above, to ensure successful transmission of the acting male counterpart. Sexing, when appropriate, of these species is often done through the identification of the anterior two pair of pleopods (swimmerets) used by males during copulation. Females have differently shaped, second-paired appendages, and sometimes seminal receptacles, for receiving and storing spermatophores.

Mating occurs in most decapods immediately following molting. Typically, there is some degree of courting behavior preceding the event. Hermit crabs vulnerably leave their shells to some degree to mate and press their ventral surfaces together, releasing eggs and spermatophores simultaneously. Species with seminal receptacles may store and use the sperm contained therein to fertilize eggs weeks later. Species without receptacles generally lay their eggs at once. Penaeids, or true shrimps (human consumption species), break ranks and release their eggs directly into the water. All other decapods attach their eggs to their pleopods under their body and larvae do not become pelagic until later. Hatching stages are highly variable in this group. Lower marine forms (in contrast to freshwater species) of Decapod have naupliar or metanaupliar larvae, while more advanced forms have protozoea and zoeal stages. These planktonic stages and their metamorphosing instar phases take days to weeks before changing to a settling stage. This makes the successful harvest of sexually produced offspring challenging for aquarists to realize with most species.

Cleaner Shrimps

Lysmata species are popular marine aquarium specimens, and are actually quite practical to culture. They are hermaphroditic - both sexes in one individual. Cross-fertilization is the rule, with partners or groups taking turns laying eggs and fertilizing each other's. For captive propagation attempts, any two individuals that get along will do, so pick a pair of healthy specimens (preferably in the same tank and apparently tolerant of each other) and you are ready to begin.

Eggs are fertilized as they are extruded by the stored spermatophore, and then placed on the rear parts of the four distal pairs of pleopods. If the sperm supply is insufficient, one more batch of eggs will be produced (infertile) and dissipate in a day or two with egg-extrusion ceasing until a new spermatophore is inserted. Incubation generally takes between 14 and 20 days with the initially bright green eggs changing to light amber, and then to light silver right before hatching. In actual aquaculture practice, hatch-ready individuals are moved to a separate system with an airstone and not fed, as this will only pollute the water. Developed eggs hatch into planktonic larvae toward the evening during first few hours of night, and the "mother" promptly molts. The spawner is then removed to another tank and treated with an iodine supplement. Day old zoeae larvae are easy to feed and have large mouths capable of taking newly hatched brine shrimp. They pass through ten or more

Upper: A heavily laden Cleaner Shrimp, *Lysmata*, clutching silvered eggs within its distal pleopods. **Lower:** the freshly hatched zoeae. (*L. Gonzalez*)

molts in 39 days before settling. *Lysmata amboinensis, L. grabhami, L. debelius, L. wurdemanni,* and *L. californica* have all been spawned and reared in captivity. All are still wild-collected for the trade as well.

Boxing Shrimps: Stenopodidae

The **Coral Banded Shrimp** (CBS), *Stenopus hispidus*, is one of the most prominent aquarium and naturally occurring species of shrimp. It has been spawned, but the young have proven rather challenging to rear. The species is collected in the tropical West Atlantic and Indo-Pacific for aquarium use. Of highest concern is the collecting and continuous bonding of breeding pairs. If *Stenopus* pairs are separated for even a day or longer, it is too often the case that males will fight with females. Females can be distinguished from males at times of breeding by the unmistakable green-blue color of their ovaries. At any other time, look for characteristics of her pleopods: she has "hairier" setae, and a longer third pair pleopods (swimmerets). Breeding pairs of this characteristically aggressive (to conspecifics) shrimp can be forced in captivity by introducing a just-molted female to a male *with* supervision. Eggs are fertilized as they are extruded, attached to the pleopods, and hatch out in about ten days (80 F.).

Echinoderms - The Spiny-Skinned Animals

Asteroids: Sea stars

Sea stars (starfish) are renowned for their powers of regeneration. Arm sections can be replaced or whole animals re-grown from a segment containing an arm and a sufficient portion of the central disc. Regeneration isn't always assured, however, and it can take upwards of a year or more to complete. However, there are seastars, including *Linckia spp.*, that employ division into two or more parts as a natural and deliberate reproductive strategy in the wild. Sexual reproduction in seastars involves a "traditional" shedding of gametes into the environment, which typically occurs only once per year. The amount of gametes can be prodigious. A single female seastar may produce more than 2.5 million eggs. Most Asteroids have planktonic larval phases, but some species engage in brooding, with their young proceeding through direct development. And so, as strange as it sounds, most seastars go through *swimming* phases in the upper water column at one point in their life. Their ciliary bands function for both locomotion and food gathering. Weeks to months later, going through metamorphoses, young seastars settle down to the bottom, with most being less than a millimeter in diameter.

Ophiuroids: Brittlestars and Basketstars

As with seastars, many Ophiuroids can regenerate cast off arms and undergo asexual reproduction by division. Most are dioecious, but protandric (male first) hermaphroditic species are not uncommon. Both brooding and spawning strategies are employed.

Echinoids: Sea Urchins and Sand Dollars

All echinoids are separate sexes, either male or female. Eggs and sperm are generally released into the environment where fertilization occurs, although there are some coldwater brooding species. Planktonic larvae (echinopluteus) swim and feed like non-brooded Asteroid young. Progeny settle literally on decades old beds of adults, cued chemically by them. It is but one of the many marvels of nature.

Holothuroids: Sea Cucumbers

Holothuroids vary from all other echinoderms in possessing a single gonad. Most are dioecious, a few are brooders, but most shed their gametes into the environment. Vegetative

Numerous micro-organisms are inclinded to develop in a natural marine aquarium with successful live sand, live rock and refugium methodologies. Many larvae can be identified in time as they develop, some will simply wax and wane as fleeting characters in the marvel and mystery of captive reef life. (*L. Gonzalez*)

fission (asexual) is also commonly employed in some species. Non-brooding Holothuroids have planktonic larvae that may or may not include feeding types.

Crinoids: Featherstars

The regenerative properties of crinoids match those of the sea- and brittlestars, and thank goodness for this. These animals are very easily damaged by handling. Featherstars are all dioecious and lack discernible gonads with sex cells arising instead from germinal epithelial cells. Again, spawning is the rule, but there are some brooders here as well. Free-living, planktonic featherstar young may take months to settle and they remain attached for up to several months more before "breaking off" and becoming mobile.

Ascidians: Sea Squirts "Do It" Too

There is tremendous variability in the reproductive modes of these "almost-vertebrate" animals. Asexual modes are known amongst many, but not all. Budding (blasotozooids) originates in different parts of the body depending on taxa, and they separate from the parent to become a new individual in time. Ascidians are almost all hermaphroditic, with most species having a single ovary and testis. Sexual reproduction in solitary species may involve brooding. If shed into the environment, their eggs tend to be "yolky." Dispersed larvae develop into telling "tadpoles" that possess the characteristics that lead to their inclusion in our own phylum, the Chordates: a dorsal stiffening structure (the notochord), a dorsal hollow nervous cord, and primitive gill-like pharyngeal clefts. These larvae remain planktonic for minutes to a day and a half before settling and metamorphosing into sedentary forms. Reproduction in captivity has been observed in some "incidental" species, supported most often by systems fairly described as large, mature, and well-seeded (refugiums, deep sand beds, copious amounts of live rock and numerous filter-feeding opportunities in support).

Summary

Presently, given the use of mechanical filtration, protein skimming, and the typically impoverished nature of live substrates in aquaria (such as excess fish load, or lack of adequate refugia) there is not much chance of larval forms of captive reef invertebrates surviving to the point of adult development. The future of our hobby will see many changes here with the adaptation of less physically destructive water movement and filtration technologies. Several manufactures have developed products that reduce impeller shear, and better days with such hardware are sure to come. Already, the use of refugiums has facilitated the beginnings of this evolution. With the sophistication of individual aquarists (well-researched and networking), associations in aquarium clubs, and the tremendous resources on the Internet, there will be more, and more successful, attempts at purposeful reproduction of non-fish livestock in our care. The culture of natural foodstuffs via "plankton reactors" alone will open doors for the care and reproduction of species never dreamed possible before. We look forward to these developments and discoveries.

Not all reproduction occurs on and above the substrates. Much activity occurs *in* substrates like rock and sand by dense living infauana. (*H. Schultz*)

Nassarius snails spawning in an aquarium. (*P. Ponder*)

Tiny *Trochus*, or giant powerhead? (*L. Gonzalez*)

Juvenile *Turbo* snail in an aquarium. (*L. Gonzalez*)

Reproduction 141

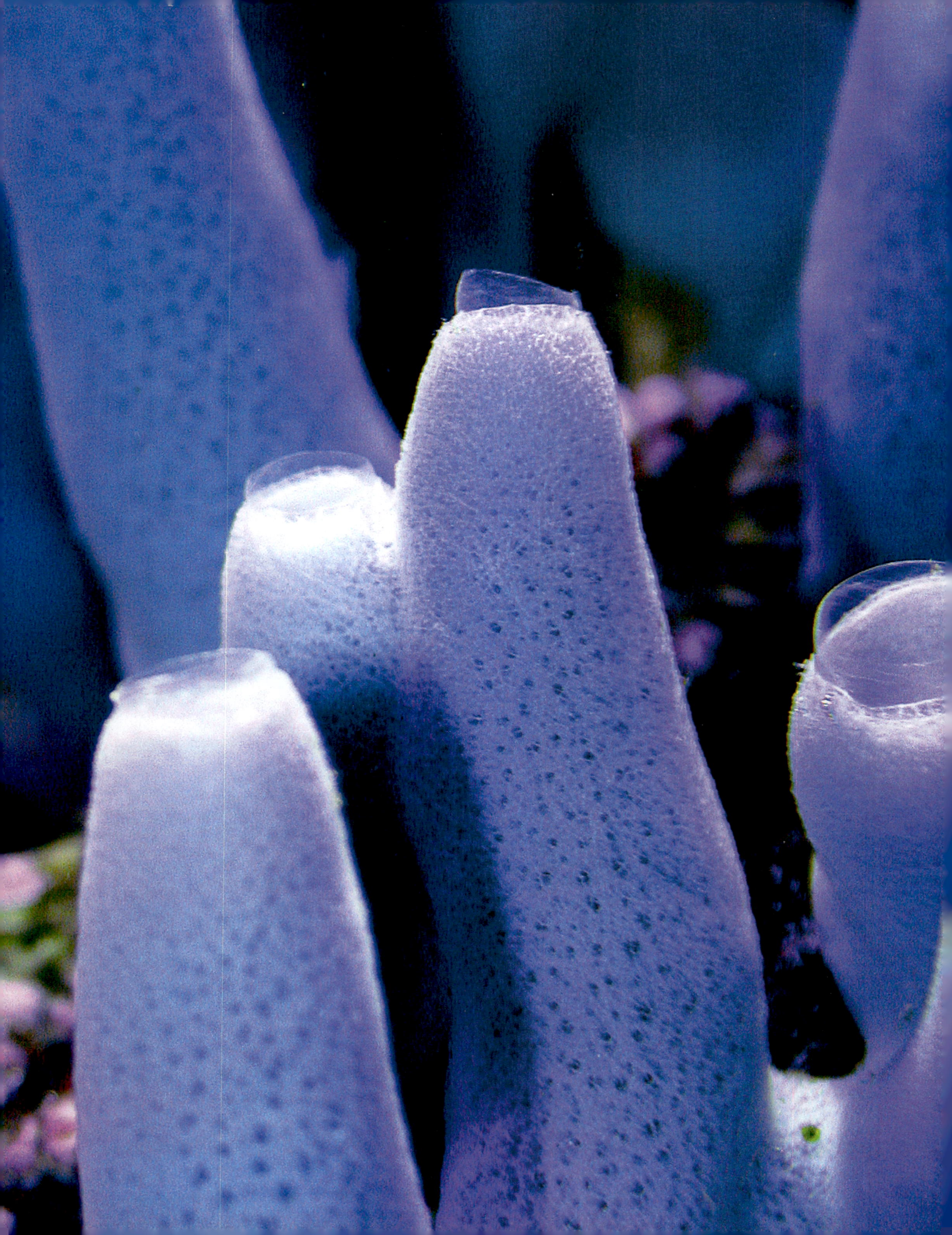

Sponges
Phylum *Porifera*

Scientists classify the sponges as the simplest of the multi-cellular animal groups, but please don't think these living filters are primitive or unimportant to us as aquarists. The nearly 10,000 known species of Poriferans play essential roles in wild and captive environments alike. Clockwise from left: ***Callyspongia pilcifera***, the "Azure Vase Sponge," in the Bahamas, ***Monanchora unguifera***, the "Fine Lumpy Sponge," also Bahamas, and ***Phyllospongia lamellosa***, a phototrophic sponge, in Bunaken, North Sulawesi.

Left: The incurrent pores of a Poriferan may not be so very clear to the naked eye, but as you can see on the top of these specimens pictured here, the excurrent osculae are large and umistakable "trumpets" crowning this specimen (***Haliclona***). (D. Lebrun)

Living sponges are not only remarkably beautiful but are also of considerable use to aquarists. They are very efficient living filters that can process such great volumes of water through their bodies that we hardly conceive of its true magnitude. There are some estimates that the amount filtered *daily* through the living sponge mass within the water of the Caribbean is equivalent to the entire volume of those waters. Other reports have stated that individuals typically cycle the equivalent of their own body volume in water several times per hour; this is a staggering amount. One may only begin to conceive what a boon that could be for water quality in the aquarium. It should also clarify one of the inherent challenges to keeping sponges: keeping them well-fed.

It's amazing to think what a difference a few years makes in our fast-evolving hobby. Looking back even just a decade ago, you'd find that very few people successfully kept live sponges alive. For those attempting what was then considered to be impossible, they either couldn't secure healthy specimens, or simply didn't know how to keep them alive after import. In most cases, they did not have the practical means either in the popular "sterile" Berlin-style reef systems of the day. Bare-bottomed reef displays without refugia and stocked heavily with hungry fishes (commonly unfed to control nitrates) and coral are a recipe for an exhibit barren of plankton. As such, these aquaria were also sponge-less. What sponge matter that did incidentally arrive was often scoured off with the well-founded

Sponges 143

Ircinia felix: The "Stinker Sponge" is one species that you do not need to be in a hurry to acquire. As its common name implies, this sponge smells very bad upon removal from the water. Stinkers occur in light gray or brown encrusting globes with conspicuous hexagonal markings on the surface. Cozumel image. (*D. Fenner*)

Opposite page, **Left:** *Pseudoceratina crassa*. **Center:** *Niphates digitalis*, the "Pink Vase Sponge" occurs in pink, blue, and gray, and grows to twelve inches in height (30 cm). This sponge often is acts as a host for the zoanthid *Parazoanthus parasiticus*. **Right:** *Dysidea sp.* **Bottom:** A veritable garden of sponges including *Aplysina*.

expectation that it would die and foul the system.

But we are now in a very different situation with state-of-the-art, modern aquarium keeping. Natural methodologies utilizing deep beds of live sand along with copious amounts of live rock and refugiums have opened the door to keeping a broad range of fascinating filter-feeding invertebrates. Today's reef aquarists are kinder and gentler, more empathetic, experimental, and much more knowledgeable. Sponges are finally being recognized by aquarists for what they truly are: wondrous, filter-feeding adjuncts to modern reef keeping that give benefits to aquarists, both as ornamentals and as substantial components of the living filtration dynamic.

There *are* some species to be avoided; some varieties are parasitic, stinging, and noxious, and many others are too delicate or unsuitable for captivity. But by and large, the keeping, and even propagation, of sponges is very much a part of the complete reef experience. With thoughtful care in selection and husbandry, you can enjoy the beauty and benefits of this unique and efficient filter-feeding group.

Sponges comprise the phylum **Porifera**, meaning "bearing openings," a reference to their overall porosity. It also best describes their general mode of feeding (filtration), respiration, and excretion of wastes,

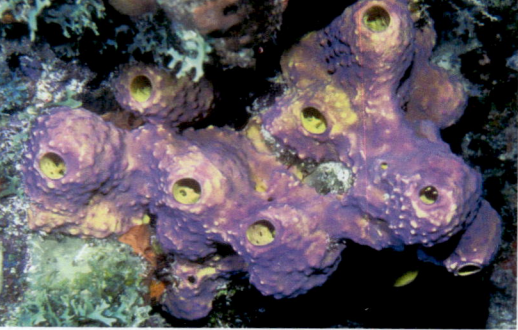

Aplysina fistularis, the "Yellow Tube Sponge," is an example of a polymorphic species. In shallow water it forms yellow to orange tubes that grow antler-like extensions, but grows longer tubes without these at increasing depths. Be warned: don't touch this messy organism. A purple color will stain your hands for days! Clockwise from left: a deep-water colony, with tube-shaped form, a young colony, and a shallow water colony with "antlers." (*R. Fenner*)

144 *Natural Marine Aquarium Volume I - Reef Invertebrates*

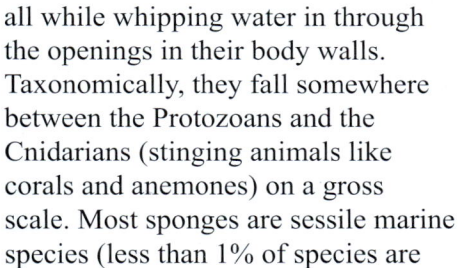

all while whipping water in through the openings in their body walls. Taxonomically, they fall somewhere between the Protozoans and the Cnidarians (stinging animals like corals and anemones) on a gross scale. Most sponges are sessile marine species (less than 1% of species are freshwater dwellers) living attached to both hard and soft substrates, although a few actually do move about. Collectively, they occur in a wide range of sizes and shapes, from tiny specks to massive barrels, tubes, spheres, encrusting plates, rubbery sheets, and even penetrating parasites. They are found circumtropically, from the shallows to the depths. A cursory glance at the structure of a sponge reveals a high percentage of non-living, mineral elements called **spicules.** These provide a semi-skeletal integrity to the animal and are often relied upon by taxonomists

Sponges 145

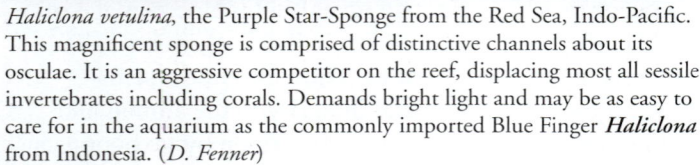

Haliclona vetulina, the Purple Star-Sponge from the Red Sea, Indo-Pacific. This magnificent sponge is comprised of distinctive channels about its osculae. It is an aggressive competitor on the reef, displacing most all sessile invertebrates including corals. Demands bright light and may be as easy to care for in the aquarium as the commonly imported Blue Finger *Haliclona* from Indonesia. (*D. Fenner*)

for species identification. For such purposes, a part of the sponge is reduced ("melted down" or separated from organics) and the spicules are examined through a microscope. The presence of various types of spicules (calcareous, siliceous, a combination thereof, plus the organic matter

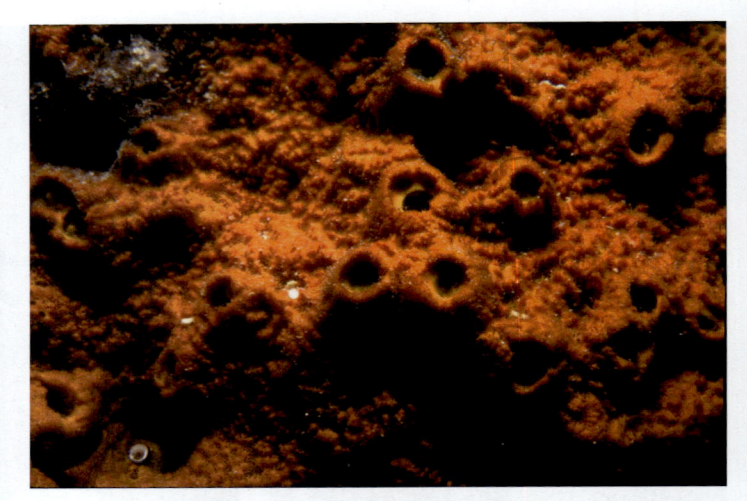

Top right: Inside the excurrent openings (osculae) of some sponges like this ***Cliona delatrix*** you will commonly find commensal organisms. Shrimp, crabs, brittle starfish and small fishes make frequent appearances. Large bored sponges may even house lobsters! **Middle right:** Rope sponges like this magnificent *Aplysina cauliformis* are regrettably more difficult to keep than the encrusting-form Demosponges. They generally have strict and narrow requirements for water flow, nutrient levels and food particle sizes that are challenging to satisfy in the confines of a small home aquarium.

Left/Below: *Atergia sp.* This sponge is a unique species distinguished by octopus sucker-like papillations. It occurs in a range of colors on dead coral and in protected regions.

spongin) is used to differentiate Porifera into its respective Orders. Along with proteins, these remnant artifacts of the non-living Porifera are the skeletons of both bathroom and faux-finishing sponge fame. Another set of terms you'll run into if you delve into sponge biology describes their physical organization. **Asconoid** sponges have just one central chamber and are small (less than 8 mm). **Syconoid** sponges have infoldings of the body wall that greatly expand their internal surface area. This may be a strategy to compensate for their typically small size (most max out at about 4 cm). Most sponges, however, are of a type termed **Leuconoid**. These forms have many incoming (incurrent) and outgoing (exhalent) channels within their bodies. They are the varieties most commonly found in the commercial trade, other than incidental species imported with live substrates like rock and coral bases.

The identification of sponges at any level can be challenging; they are **polymorphic** like many stony corals (Scleractinia). This means that a given species can occur in various forms, depending on current, light, and availability of food, looking nothing like each other at times. Members of the same species may occur in quite different colors (shades of brown, black, green, yellow, purple, pink, red, and more), different shapes (boring, encrusting patches, ropes, balls, barrels, tubes, vases, etc.) and sizes. Any or all may be influenced by factors like lighting, water depth, and especially current and nutrients. Aquarists will want to seek specimens

Callyspongia plicifera, the "Azure Vase Sponge." Bahama image.

Selection Guidelines for Sponges

- **Be sensible and realistic** about your ability to keep a given species, and always research the needs of an organism *before* you buy it. Seek the best, hardiest, and most appropriate specimens for your system and means. Know that husbandry for sponges run the gamut for water flow, light, and nutrient levels.

- **Determine if the specimen has been exposed to air**. *It should not be*. Air is easily trapped inside the many channels of this animal, killing off those parts. That said, we will state once again: **Sponges should never be lifted out of the water.** Recent exposure will not typically be expressed symptomatically (days), but after some time the old familiar white hazing on the tips and in spots indicates that there is a pending demise. While you may have to trust the dealer's word for how the specimen was delivered to the point of purchase, you can at least ensure that the sales clerk bags and moves all specimens under water for you, and that you continue the process at home during acclimations and transfers (QT to display).

- **The excurrent (exhalent) "mouth(s)" of a sponge should be open** indicating normal regulatory functions (internal water flow = life). Impugned sponges close these pores (**osculae**).

- **Do not purchase any sponge with obvious dead or dying parts**. Decay will be evidenced by gray or white spots, white fuzz, or a haze to the colony in part or whole. In advancing stages the dying portions become almost translucent. It may be possible to simply carve out necrotic spots, but the need to do so does not bode well for a successful prognosis. To avoid large losses due to noxious exudations, be sure to put any newly acquired or stressed sponge directly into isolation (QT). Barring this, you should leave it in the dealer's tank to keep watch on during its acclimation, thus ensuring a relatively safe purchase.

- **Evaluate the integrity of the specimen that you are considering**. Most healthy and desirable sponges are relatively firm, consistently colored, and do not impart a noticeable odor to the water. Offensive, soft (vacuolations - missing areas), or irregularly colored sponges are unsafe risks.

- **Quarantine all new specimens!** This mantra is repeated throughout this text, but it is especially crucial with Poriferans due to their precarious acclimation to captivity, which is further amplified by their potential to severely impact water quality negatively by noxious exudations or simple gross organic matter upon death. Although a specimen may look very well in the dealer's tank, you cannot be certain that it was not fatally exposed to air days or even minutes prior to your purchase. The risk of sudden and catastrophic death is calculated and requires a respectful address with some species. There is also, always concern for the carriage of incidental pests, predators, and disease into the display without proper quarantine.

- **Buy from a competent and reputable dealer**, who understands the special needs in handling and husbandry for sponges. A good dealer should be willing to cure them for a week or more for you to assure their vitality and acclimation for sale.

- **Never buy ripped/stripped sponges that have been improperly torn from the reef**. Demand specimens that are smartly collected on natural substrate (sand or rock). A ripped sponge is unmistakable and undesirable.

Sponges 147

Clathrina canariensis is a lovely calcareous sponge that grows well in aquaria and is usually acquired incidentally with live rock. Occurring in the recesses of a reef, this sponge prefers to settle on the undersides of rocks and caves in protected areas. Photographed in St. Thomas. (*R. Fenner*)

of species most forgiving of varied water flow and nutrient levels found in their aquariums.

On microscopic examination, one can see that the Poriferans are comprised of relatively undifferentiated cells (they are compositionally similar throughout) in a dense matrix - the aforementioned proteinaceous **spongin**. Sponges are animals without organs or tissues, but they do have specialized cells that work to help move food items, repair damage, and clean or attract fouling matter. Each cell independently provides for its own metabolic needs, making a sponge more like a commune than a colony. Living up to the meaning of its name, the holes in a sponge colony are comprised of many incurrent pores (**ostia**) formed by **porocytes,** which allow water into an open space (**atrium**) in their bodies, and then out of one or more larger openings (**oscula/osculae**). Study of the anatomy of a sponge makes clear the severe admonition heard so often by aquarists, "Never take a sponge out of the water." Although this is not true for all species, it is a reasonable warning for most, and just good-sense husbandry. The problem with exposure to the air specifically is that any air trapped inside the atria is exceedingly difficult for the flagella-equipped collar cells, or **choanocytes,** to purge.

The choanocytes are responsible for producing the water currents through the animal which carry oxygen, carbon-dioxide, sex cells, and food (largely ammonia) in and out of the colony. Sensitive Leuconoid sponges have not evolved to handle the unnatural transport of air through the body.

Selection

About the only thing that you can depend on, regarding the selection of a healthy collected sponge, is that you can't depend on anything. The best sponges for your aquarium are almost always the encrusting types that hitchhike into your aquarium *gratis* with your live rock. If they have survived the rigors of import for those many days without light, under variable temperatures, and out of water, you can be sure that they are hardy. Instead, it is the deliberately collected, decorative species, that have atrocious reputations in captivity. Here

Ectyoplasia ferox - The "Brown Encrusting Octopus Sponge," occurs as both encrusting and multiple-armed arborose morphologies. Another polymorphic species. Here in the Bahamas. (*R. Fenner*)

we are referring to the brightly colored orange, red, and yellow varieties from the Caribbean. Very sadly, these are the most commonly collected species for the American trade as they are colorful, plentiful, and inexpensive. They are also collected too often by ignorant personnel, transported poorly, held improperly, and given altogether poor husbandry. Even beyond all this, they are extremely challenging, if not impossible, to keep with the present systems and skills of casual aquarists. Alas, this is lost on many because such sponges may live for some months, even upwards of a year, leading many an aquarist to think they are having some success. In fact, the truth is that it simply took longer to starve to death than most colonies. Most unsuitable specimens die within just a few weeks of import.

Success with sponges in captivity must be measured in years, and more than just a few. They are amongst the "immortal" animals on our planet - apparently having no defined senescence. Colonies do not die of old age, *per se*, but perish only from predation, pathology, or extreme natural climactic conditions. Sponges are also extraordinary in their regenerative properties. Even after a severe insult or destruction to the colony, the tiniest living remnant can re-build all that was lost. Aquatic mammals (in the form of divers and snorkelers) are often amazed to re-visit areas that they've seen damaged, after only a few months or years, to find the sponges fully recovered, as if nothing had happened.

As previously mentioned, not all sponges will be acquired specifically or deliberately. When you purchase live rock for your aquarium, you may very well be purchasing a considerable amount of live sponge material with it. These are truly the hardiest species of Porifera, for they are the specimens that have proved themselves by surviving the chain of custody through import. We strongly encourage aquarists to grow them. Successful and *stable* sponge growth is a compliment to good aquarium husbandry.

> **Special Note:**
> Take heed that sponges are *very* sensitive to any chemicals, supplements, "tonics" and medications in the water. This is not only a reality for their gross and efficient processing of great volumes of water as filter-feeders, but also for their symbiotic residents. We hope we don't need to tell you what happens to their desirable symbiotic algaes when an antibiotic is used. Many of the dreadful products on the market that claim to kill nuisance organisms like pest anemones, snails, and algae often take a severe toll on sponge life. Always be very mindful of dissolved chemicals and solutions applied to the tank. For aquarium treatments, simply remove and isolate sponges in QT for four or more weeks to allow sufficient time (letting it lie fallow) to screen for undesirable communicable organisms.

Stylotella and *Stylissa* species are, occasionally to commonly, imported from the Pacific as "Cork Sponge." Alas, their navigation of import and survivability is rather dismal, akin to the problems aquarists experience with Caribbean tree, finger, and ball sponge species. These should presently be avoided by casual aquarists. *Stylotella aurantium* pictured here from Fiji. (*R. Fenner*)

A common import from Indonesia, *Haliclona* (the photosynthetic "Blue" or "Purple Finger Sponge") has a dismal record for shipping tolerance, but is remarkably hardy and fast growing once established in a well-maintained reef aquarium. Use high light and high water flow to compliment deep sand bed or refugium methodologies.

Care

This far along into the chapter, you should have a clearer understanding of why sponges do not enjoy a good reputation as hardy specimens in the hobby. Issues that revolve around air exposure and damage in collection (cleaved without substrate) need to be screened in your selection process.

The Spiny Ball sponge (*Leucetta sp.*) from the Caribbean/Atlantic, a common live rock hitchhiker. Aquarium photograph. (*L. Gonzalez*)

Confirmation of successful selection will be revealed in a proper quarantine period. Let there be no doubt, most disasters with new aquatic specimens can easily be avoided by patiently committing to a two to four-week isolation period. Moving specimens into and out of quarantine can be accomplished easily with a jar or bag underwater. Quarantine isolation will help acclimatize them to your captive water conditions. Also, be warned that there are many common hitchhikers with Porifera: commensal and symbiotic species from bacteria to worms, crustaceans to brittle-stars, and even fishes that naturally use sponges for habitat. Most are harmless or may even be beneficial, and as such should be preserved.

Most sponges collected for aquarium use have broad environmental tolerances. Nevertheless, a few concerns must be addressed. Your chances of keeping sponges are greatly enhanced by placing them *only* in well-established reef systems (preferably a few years old, or one year as a bare minimum). Among other reasons, this qualification ensures that at least some organic material and mature bacterial populations exist to nourish them. Fishless refugiums (and those without any other planktivores) are highly recommended as support, to improve chances of sustaining Porifera in captivity. Detritus and sediment (siltation) is a scourge for most species (although there are prominent exceptions). A slow-flow refugium used to house sponges will limit the possible species that can be kept. Traditional reef water quality (NSW) should be maintained. The very best captive environment for these remarkable filter-feeders is a cryptic, species-specific tank.

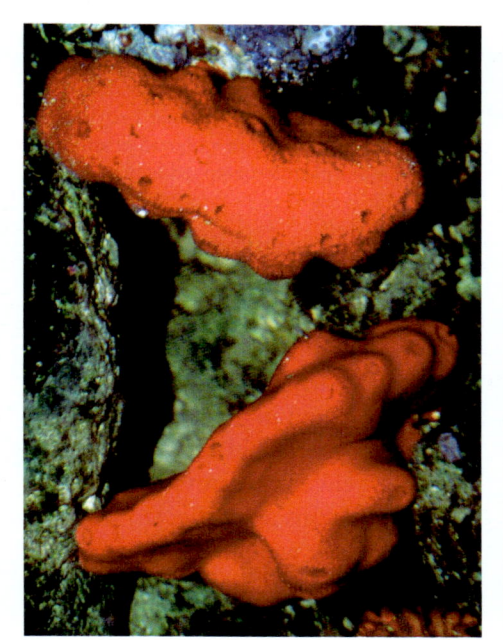

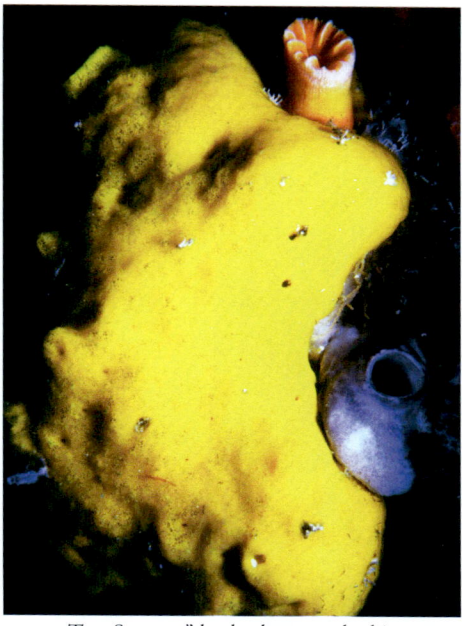

Left: *Pseudaxinella sp.,* the commonly encountered "Red Ball Sponge." **Middle:** *Ptilocaulis sp.,* the "Red Finger or Tree Sponge," both photographed in aquariums. (*A. Calfo*) **Right:** *Leucetta sp.,* this calcareous species occurs in stout, boulderesque shapes of an opaque lemon-yellow color with sparse and conspicuous osculae (excurrent openings). North Sulawesi image. (*R. Fenner*). Several genera of finger, ball and tree-type sponges are popularly kept by beginners. Too often, however, these red, orange and yellow beauties are challenging for even the most advanced aquarists to keep for more than a few months.

Accompanying the aforementioned is the recommendation not to be over-zealous regarding absolutely "pristine" system water quality. The use of silicate removing chemical filtrants, "super-skimming," and/or ultraviolet sterilizing is counterproductive to sponge culture.

Did you say Lighting for Sponges?

It might be surprising for you to discover, but most sponges collected for aquarium use are variably photosynthetic, utilizing symbiotic cyanobacteria (Blue-Green Algae or BGA) to provide sugars and sustenance directly to the colony. That makes many Porifera **mixotrophic**, and not wholly dependent on heterotrophic feeding, although they are still generally so. Thus, let's dismiss the old notions that sponges kept in light are harmed by it. The truth of the matter is than many sponges do not *naturally* occur in high light environments, and when forced to do so in aquaria with excessive nutrients, ensuing algal overgrowth often overwhelms them. Like many low light aposymbiotic corals, they simply have not evolved sufficient means to defend against algal

Sponges, like many corals, often grow morphologically to suit physical paramters like water flow. This unique Demosponge has grown to look remarkably similar to the calcareous algae, *Codium*. This is a hardy group that grows well for some once properly acclimated. Aquarium photo. (*L. Gonzalez*)

Aplysina cauliformis, the "Row Pore Rope Sponge." Many of the magnificent rope sponges occur towards the depth limits of recreational diving. Common in waters between 1 to 3 atmospheres. They occur in a myriad of striking colors. Blue-Purple photographed here in Antigua. (*R. Fenner*)

is *not* reliable as an indicator of any particular light or water-flow preferences to guide you in proper placement; sponges can be found randomly distributed, in all colors of the rainbow, and in most any niche - from full sun in shallow water to deepwater caves and crevices. Most encrusting forms are collected from areas of strong water movement, whereas tube and vase-like ones generally hail from calmer (and deeper) waters.

Few aquarium sponges will tolerate slow water flow, and almost none commonly found in the trade can sustain siltation (with the notable exception of *Cinachrya* Moon sponges mostly imported from the Atlantic). This is not to say that there aren't many Porifera found in high silt environments - quite the contrary. But only a few of the species found occurring in ports, bays, and other high-sediment niches are desirable, if able to be kept at all, in a pristine reef aquarium. We need to provide good circulation in order to deliver food, remove wastes, and purge detritus and other stifling matter from these filter-feeders. If you notice nuisance algae or sediments accumulating on your sponge, do not hesitate to baste or blast away with water the offending settled matter. In much the same way that air can become trapped and fatal for a sponge, particles can clog vital pathways irreparably, subsequently leaving them without adequate water flow.

Foods, Feeding and Nutrition

By now you know that many sponges are partly photosynthetic, and that all sponges filter-feed heavily. Despite the presence of symbiotic algae in many sponges, most are *limited in growth*

succession like other naturally high-light species have. But that does not mean they should be deprived of light. Most any illumination will be tolerated if encroachment from nuisance organisms like microalgae is not an issue. Deliberate lighting schemes for sponges should be moderate: the same that one would apply, for example, to corallimorphs, zoanthids, and many soft corals. For very brightly lit aquariums, there are some species that are quite well-adapted and well-suited like the lovely and popular blue symbiotic sponges (sometimes purple) of the genus *Haliclona* for such dynamic environments. *Haliclona* is indeed one of the more photosynthesis-dependent sponges in the trade. They are found in shallow water and bright light often attached to (and killing) digitate Poritid species of stony coral.

About Water Flow

Sponges often need strong water flow. Although they have mechanisms to create their own flow throughout the colony, many species depend on the action of water currents sweeping across them to affect pressure and enhance this dynamic as a whole. We can usually look to their morphology for indications of their preferences. Unlike some corals, though, color

Natural Marine Aquarium Volume I - Reef Invertebrates

Right top: Avoid buying sponges that have been cut or cleaved from the reef. A properly collected specimen will at least have a small amount of natural substrate intact. For some species, the media may be sand, but for most it will be carbonate rock. The image here illustrates a good specimen in this regard. **Center:** *Callyspongia sp,* the Blue Callyspongia is a unique species of Hawaiian sponge found living exposed on the open reef, in contrast to cryptically hidden sisters. **Bottom:** The potent chemical punch of Poriferan life forms is well demonstrated by the common and shallow water Atlantic-Caribbean Fire Sponge, ***Tedania ignis***, which packs a painful sting that divers do not soon forget. Chemical potency in sponges at large ranges from stinging to fatally poisonous and foul smelling to dye-exuding: anything necessary to protect this otherwise defenseless sessile invertebrate group. Their chemical warfare is legendary and a force to be reckoned with in the sea and aquaria alike. (*Photos: R. Fenner*)

by available food. When making attempts at cultivation, providing the food necessary for healthy sponge life in captive aquariums, we must address some very specific and somewhat challenging needs. Naturally-occurring prey necessarily includes some of the finest plankton (bacteria and nano-phytoplankton, for example) and dissolved organics. Such prey items are not readily delivered, if they can be at all, through the utilization of prepared foods like bottled supplements. Ideal sources in aquaria include plankton reactors (like sterilized perpetual drips from live greenwater cultures) and mature refugiums. The natural or deliberate stirrings of deep sand beds may also be quite helpful for feeding Poriferans. As previously mentioned, the literal mechanism by which sponges are fed is a constant whipping of their choanocyte flagella. This action draws in particulate food and dissolved organics, oxygen, and more, while creating an excurrent flow of water through the colony that serves to flush out carbon dioxide, other wastes, and occasionally gametes.

Sponges are rather ideal bio-indicators of natural plankton levels in the aquarium (micro-, nano-, phyto- and zoo-) as evidenced by their good health, or reciprocally, by their

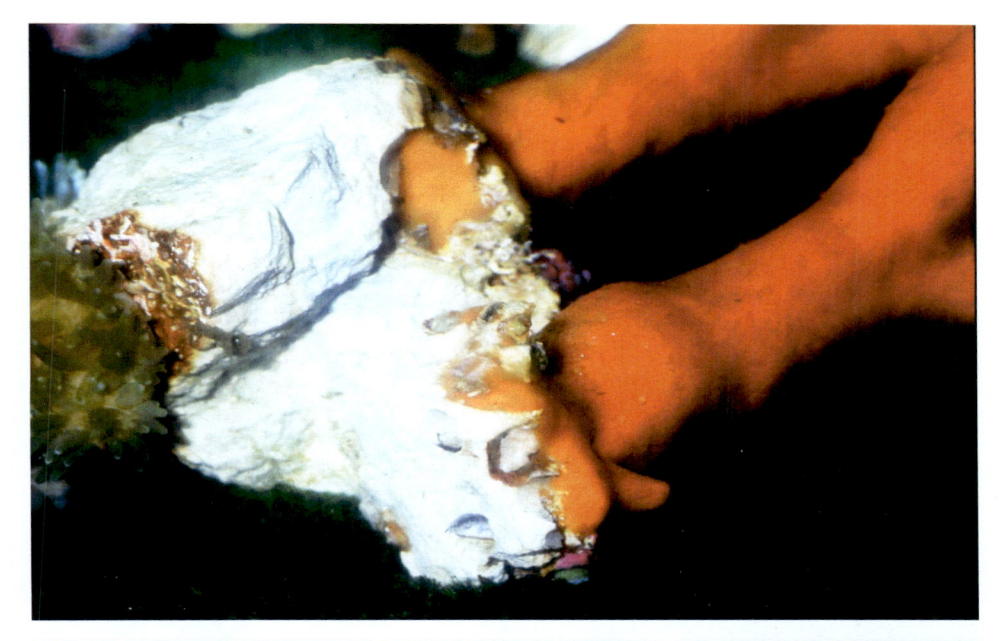

Sponges

Cribrochalina vasculum, the "Brown Bowl Sponge." Bahamas image.

waning mass. While the more heavily photosynthetic species fare better, the decidedly heterotrophic specimens will simply waste away and shrink in size over time. Some aquarists have tried to target feed their sponges directly (or indirectly, as with a suspension in the water of slurried meaty foods). Unfortunately, such techniques tend to either clog the sponge or, at best, do not help while also more likely being a burden on water quality. This is not to say that there isn't a prepared food, now or ever to be, that we can proffer when natural methodologies fail (refugia, DSB, etc.), but rather that casual aquarists are not usually going to want to or be able to target feed sponges successfully.

Aquarists have experimented with applications of thawed pack juice from frozen meaty foods (normally to be discarded for fear of fueling nuisance algae growth) with variable degrees of success. It is likely that some sponges will even eat fine zooplankton like freshly hatched brine shrimp and especially cultured rotifers. You might recall dramatic news reports of "killer" sponges that capture larger prey like shrimp. Feeding on such large food particles is rare among sponges, in general. At any rate, poor nutrition is possibly one of the most significant factors in captive sponge mortality. There are even some concerns that excessive filtration or aggressive protein skimming actually could be an impediment to achieving success here. There is some merit to the admonition, "Cleanliness is not sterility." It certainly does take finesse, as well as a concerted effort, with specialized techniques and hardware aspects to properly feed Poriferans.

Placement

Placement of a sponge in the aquarium is a matter of great importance,

Theonella swinhoei, Red Sea image. (*D. Fenner*)

154 *Natural Marine Aquarium Volume I - Reef Invertebrates*

Latrunculia sp. Red Sponge from the Red Sea is reported to be very poisonous (Baensch Marine Atlas). Quite a few sponges have been studied for their noxious chemical composition and potential in medical research for developing anti-viral and anti-cancer treatments among other things.

although not literally analogous to their occurrence on a reef, when considering our necessarily homogenized conditions (by virtue of aquarium sizes and depths). Popular species in the hobby hail from just a few meters depth under full tropical sun to the chilly depth limits of recreational SCUBA diving. As such, the deep water rope sponge cannot realistically be maintained in a "dark" cave in a 30" deep (75 cm) aquarium under a bank of blazing 400 watt metal halides. Due diligence in research for the specific needs of sponges is critical. As much as any species, these

Siphonodictyon coralliphagum, the "Variable Boring Sponge." This species is not welcome in most modern marine aquariums as it bores into living coral and carbonate surfaces. It occurs in various physical forms, from fingers to bowls, and as encrusting or tubular forms. Cozumel image. (D. Fenner)

organisms will likely suffer and die if purchased with impulse and ignorance.

Our goal is to help determine what the best placement is for your new acquisition in order to ensure that you meet its most critical demand: nutrition. Depending on species, you may find that your sponge will fare better in the high flow, low sediment final chamber of a sump or perhaps the bacteria-laden mud refugium. In

Sponges

156   *Natural Marine Aquarium Volume I - Reef Invertebrates*

Top: *Grayella cyathophora*, and *Hymedesmia*, both Red Sea. (*D. Fenner*) **Bottom:** *Ircinia strobilina*, and *Siphonodictyon coralliphagum*.

other situations a proper zooplankton reactor, refugium, or the display itself may work best. Consider where your species has grown naturally on the reef, and then how to best emulate the physical parameters of that environment. If your system or display does not allow for the modification of hardware and filtration dynamics to accomplish this (as with a garden reef aquarium), then you should resist buying an organism with such special needs.

Another concern with the placement of sponges in the aquarium is their participation in celebrated chemical warfare, like Cnidarians. While any and all species impart some noxious exudations naturally, the unnatural stimulation of a crowded tank incites ever more copious or potent production of toxic compounds used to deter the encroachment, settlement, and very ability of competitive species to live without direct contact. Such warfare is not limited to other

Opposite - left column, from top to bottom: ***Monanchora barbadensis*** - the "Red Encrusting Sponge." This species forms bright red sheets with canals that radiate from the excurrent pores. Image here from Cozumel. ***Ianthella basta*** - from North Sulawesi and Whitsundays, Australia. ***Leuconia palaoensis*** - a discreet exterior of lovely pale pink to light blue tubules with a soft, thin texture. Austro-Malay distribution. North Sulawesi image. ***Gelloides fibulata*** - the "Thorny Horny Sponge" (really). It hails from the Indo-Pacific and occurs in two forms: encrusting and tubular. Grows to fourteen inches in height (35 cm). Pictured here off of Pulau Redang, Malaysia. **Right column, top:** ***Agelas conifera*** - the "Brown Tube Sponge." This bulbous species is typically smooth-walled and brown to tan in color. Bahamas image. **Bottom:** *Spirastrella coccinea* is a relatively common encrusting sponge photographed here in Cozumel, by Diana Fenner.

Clathria mima. Indonesia. This one photographed to the north of Lombok.

Sponges 157

Genus *Phyllospongia* is an exception to the rule of sponges being sensitive to captive life and heavily dependent on filter-feeding. This genus is phototrophic, essentially deriving all its nutrition from endosymbiotic blue-green algae. Although they commonly occur throughout the Indo-Pacific, specimens rarely find their way into the aquarium trade. They are amongst the hardiest sponges for aquarium use. These sponges thrive in typical and traditional reef aquarium conditions, and are easily propagated asexually by fragmentation. This is *Phyllospongia lamellosa* in North Sulawesi, Indonesia. (*R. Fenner*)

sponges and can apply to any other reef invertebrate, including corals. Be especially mindful of placing encrusting species too close to sessile invertebrates, and use a "break" (like a rock or rubble firewall) on the substrate to stop them from spreading excessively.

It is not surprising that Poriferans are some of the most toxic creatures in the sea. Sitting soft, sessile, and defenseless on the ocean floor, it only stands to reason that they would have evolved toxins to inhibit predation and noxious exudations to slow the growth of neighboring competitive organisms.

A number of sponges, in fact, have been targeted by science for the study and development of anti-viral and anti-cancer treatments, among others, addressing specific conditions as well, including malaria, tuberculosis, herpes, and non-Hodgkin's lymphoma. Chemical complexities in sponges are yet another part of their natural wonder and a value to the world beyond hobbyist enjoyment.

Overall, aquarists with large or significant sponge growth in the aquarium must keep the potential for

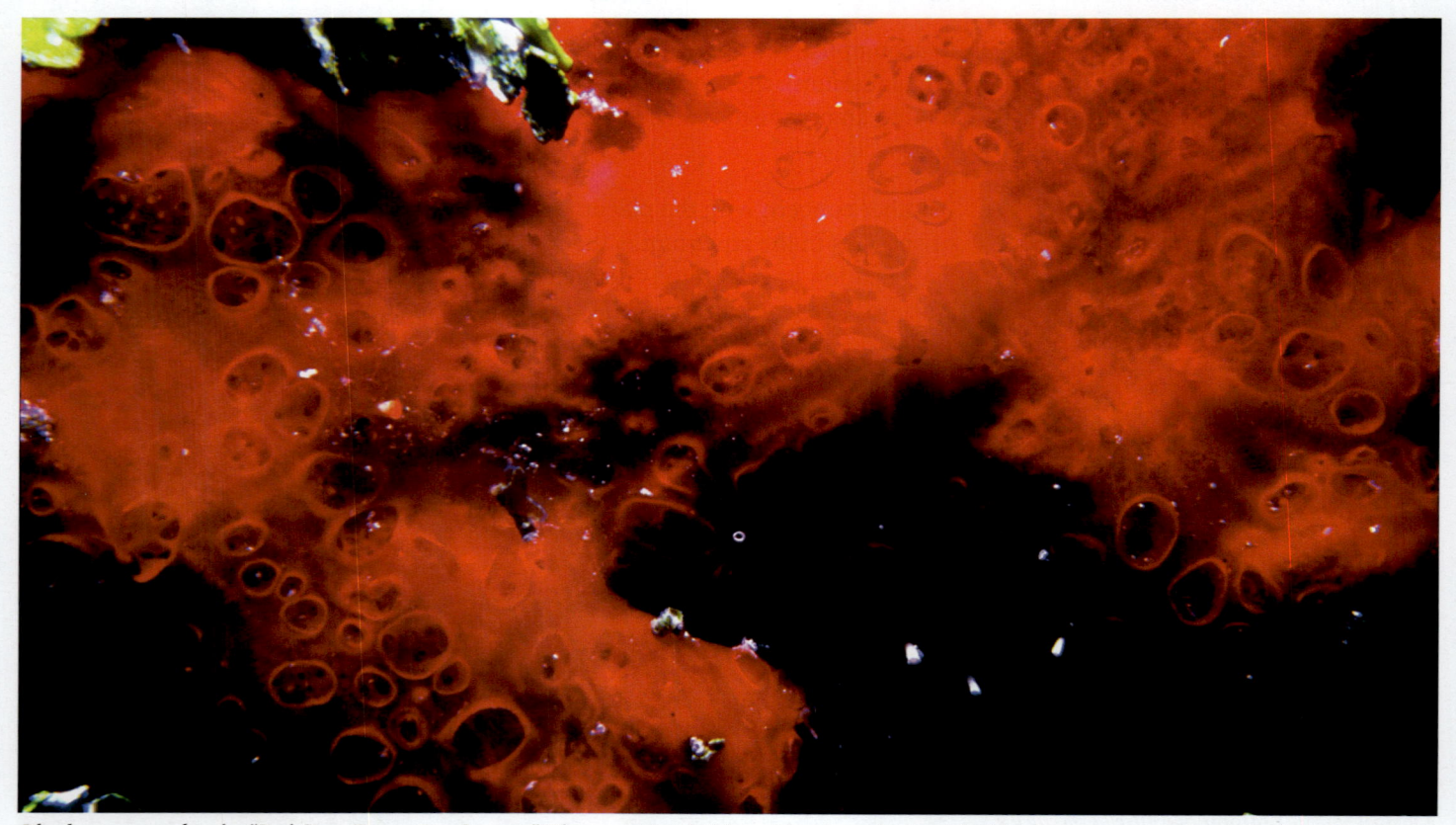

Phorbas amaranthus the "Red Sieve Encrusting Sponge." This attractive, bright red species is distinguished by the numerous craters (raised circular sieve-like areas) on its surface, and small incurrent pores surrounding the extended excurrent openings. This sponge reacts to contact by closing its pores. (*D. Fenner*)

Parasitic Sponges: There are pest sponges that attack bivalves, corals and more. Although uncommon in aquaria, a suspected organism should be removed (unless it's your desire to observe and study this relationship). **Left:** *Mycale laevis*, the "Orange King Sponge," boring through a digitate *Porites* in the Bahamas. **Right:** Another boring sponge encroaching upon a stony boulder coral in the Red Sea.

what can effectively be considered a "toxic soup" in mind at all times. This means that close attention should be paid with regard for water quality and the potential for a catastrophic event in the aquarium should one, some, or all of the sponges become stressed or die. Conduct regular partial water changes and make use of chemical filtration (carbon, ozone, etc.) to temper the effects of degradation of water quality. Also exercise preparedness by having standing seawater on hand at all times in case of the need to perform an emergency water change.

General Health and Maintenance

Getting to the root of the matter, the single greatest danger to your aquarium sponges is you. Challenges with keeping sponges are less a matter of the unknown needs of Porifera, but rather lie with aquarists *unwilling* to provide for their known needs. If you think that you can simply throw a sponge in a sump or refugium or in the display proper, but cannot explain why it will work, you are summarily ruled out as a "viable parent" for Porifera.

One of the most common causes of mortality in acclimated specimens is algal overgrowth. Nuisance overgrowth is always a matter of poor planning or execution of husbandry. There are simply no mystery plagues afflicting so-called "perfect" reef aquarium systems. There is a fine line to walk when caring for sponges. On one hand you must provide enough nutrients for them, without accumulating too much and thus degrading water quality or otherwise causing harm. Low nutrient levels are still recommended - with a close eye on minimizing deleterious phosphate and nitrate concentrations. For trace elements and bio-minerals especially, regular partial water changes are possibly the best vehicle for delivery.

As with most reef invertebrates, you must be careful not to cause osmotic shock with new seawater or freshwater influx to the aquarium. As far as physical maintenance of the colony itself, little or no contact is necessary short of pruning excessive growth. With a lapse of water flow or delivery, you may need to vacuum or baste away detritus, sediments, and settled organics from the body of a sponge periodically. Beyond that, be sure to keep up with ancillary tasks of maintenance, like ensuring water clarity along with cleaning of the lens and lamp fixtures (for photosynthetic species), and regular pump cleanings to maintain optimal water flow.

Reproduction

Sponges are known to reproduce in both sexual and asexual modes. In

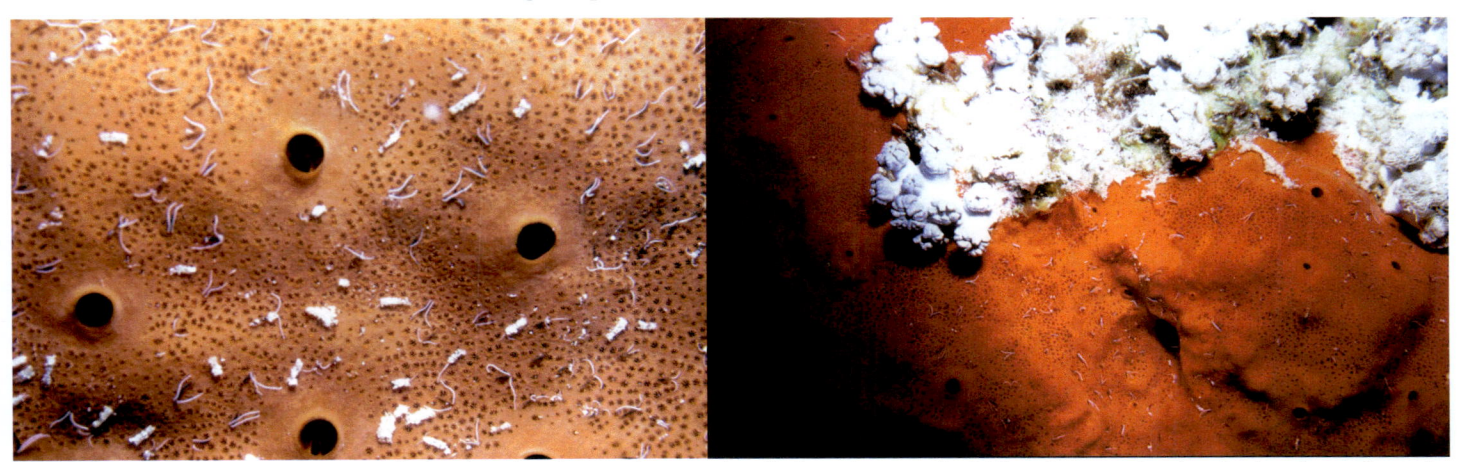

Cliona vastifica: Boring Sponges are widely distributed with more than 150 described species recognized. These have no place in mixed garden reef aquaria. Pictured here in the Red Sea. (*D. Fenner*)

sexual strategies, offspring may be produced as dispersed planktonic larvae or brooded (fertilization and development internally). Like so many other reef invertebrates, aquarists cannot easily harness the products of sexual reproduction in the limited dynamic of captivity. Issues of crowding, filtration, predation, and inadequate means of support for subsequent larvae have been challenges to the endeavor in aquariums. Nonetheless, significant progress can be made via asexual propagation strategies. Budding and fragmentation occur commonly in the aquarium, as well as on the living reef. Deliberate attempts to fragment a colony should always keep in mind standard coral propagating techniques. A scrupulously clean cut, and fast and secure attachment of the division are crucial for success. Stolen buds or cleaved fragments may either be tied or stitched to a hard rock with fishing line (monofilament), or even glued lightly with cyanoacrylate "super glue." As with coral propagation, donors should ideally have been established and growing for some period of time (preferably 6 months or more) before an aquarist begins any imposed propagation technique. In time, as aquarists shed their fear of keeping Poriferans and begin to study and culture them in the specialized systems these species deserve, we can be sure to hear of many more fascinating natural and imposed propagation strategies in sponge life. Please see the Porifera passage in the chapter on general reef invertebrate reproduction for further insight to sponge reproduction.

Compatibility

There are several ways to look at compatibility issues with sponges. Let us first break them down into two categories: first, direct interaction by contact, and second, chemical influences between sponges and other organisms. We have already briefly mentioned the toxic chemical nature of sponges and the noxious exudations used to inhibit the growth of other reef invertebrates and inhibit predation. Some of these compounds have been

Xestospongia muta, the "Giant Barrel Sponge": This Poriferan megalith grows to six feet (180 cm) in height. Photographed in the Bahamas.

partake of Porifera we can include: Angelfishes, Triggerfishes, Filefishes, Asteroids, Echinoids, Crustaceans, and a very long list of Mollusks (snails, cowries, sea slugs, etc.). This list is far from all-inclusive; you will have to research mostly, experiment conservatively, and above all be observant.

Symbioses: Partners and Parasites

All manner of other sea life alive are represented in, on, and with sponges or their remains. So much could be written at length on the relationships and participants interacting with Porifera that it would be an entire tome in itself. Many are familiar to aquarists, like the crabs that use sponge for camouflage (numerous genera). Other partners will be more familiar to divers, like the gobies and lobsters they observe inside great tube sponges. In residence they seek hiding, or a convenient perch to capture food on the currents (including matter which is drawn in specifically by the sponge). Some very specific

Acervochalina sp. (left) and *Siphonochalina sp.* (below), both Red Sea images.

identified, many more are suspected, but none have been quantified in aquaristic terms. Like allelopathy in plants and corals, we must simply be mindful at present that this intangible dynamic is very real, and that it can be tempered primarily by dilution - good water quality, chemical media, and weekly water changes, along with other necessary maintenance of such a system.

Most important to aquarists is an understanding of direct interactions between sponges and other invertebrates and fishes. Despite being one of the most noxious groups of animals in the sea, many organisms have evolved to consume sponges. As a primary group of reef organisms, sponges are a food staple for numerous fishes and motile invertebrates. Among the many creatures that are known to

Sponges 161

feeding relationships have been formed in some cases, such as the remarkable Synaptids (Holothuroid "worms" - part of the sea cucumber group), which collect in large masses to graze on the surface of the sponge seeking metabolites, and are perhaps incidentally attracted to micro-organisms and organic matter. An even closer examination of life on living sponges will often reveal a plethora of microscopic forms including tiny starfish, micro-crustaceans, worms, and the tiniest of fishes. Most of these relationships are commensal, but all told they vary widely, from mutualistic to predatory or even parasitic (both by the sponge, and upon it).

Summary

The benefits of sponge culture in the natural marine aquarium cannot be overstated. They are major constituents of the world's seas, performing as extraordinary living bio-filters, providing habitat for a wide variety of micro- and macro-organisms, and producing biologically important molecules often used by other life forms. And although they pose potential dangers in culture, including noxious exudations and sudden necrosis, they are no greater a challenge than commonly overstocked reef tanks with overly large numbers of coral fragments (crowded and left to fight it out, or worse, grow into a bigger, destructive bio-mass). Aquarists are beginning to see their potential and flaws with a clearer understanding. Sponges are no longer summarily dismissed as delicate, short-lived, and dangerous. Aquarists can embrace them as an important factor in a complete reef display, and encourage their growth, even if only as bioassay organisms (water quality indicators), incidental food, and efficient, living filter adjuncts.

Spirastrella (*Sphenciospongia*) *vagabunda*, the "Vagabond Boring Sponge." A boring species that anchors itself in carbonate places (rock, corals, etc.) by acidic secretion. They are an important group of organisms for recycling limestone on the reef. Hawai'i photo.

Ircinia strobilina, hosting *Parazoanthus*. Caribbean image. **Right:** *Callyspongia vaginalis*.

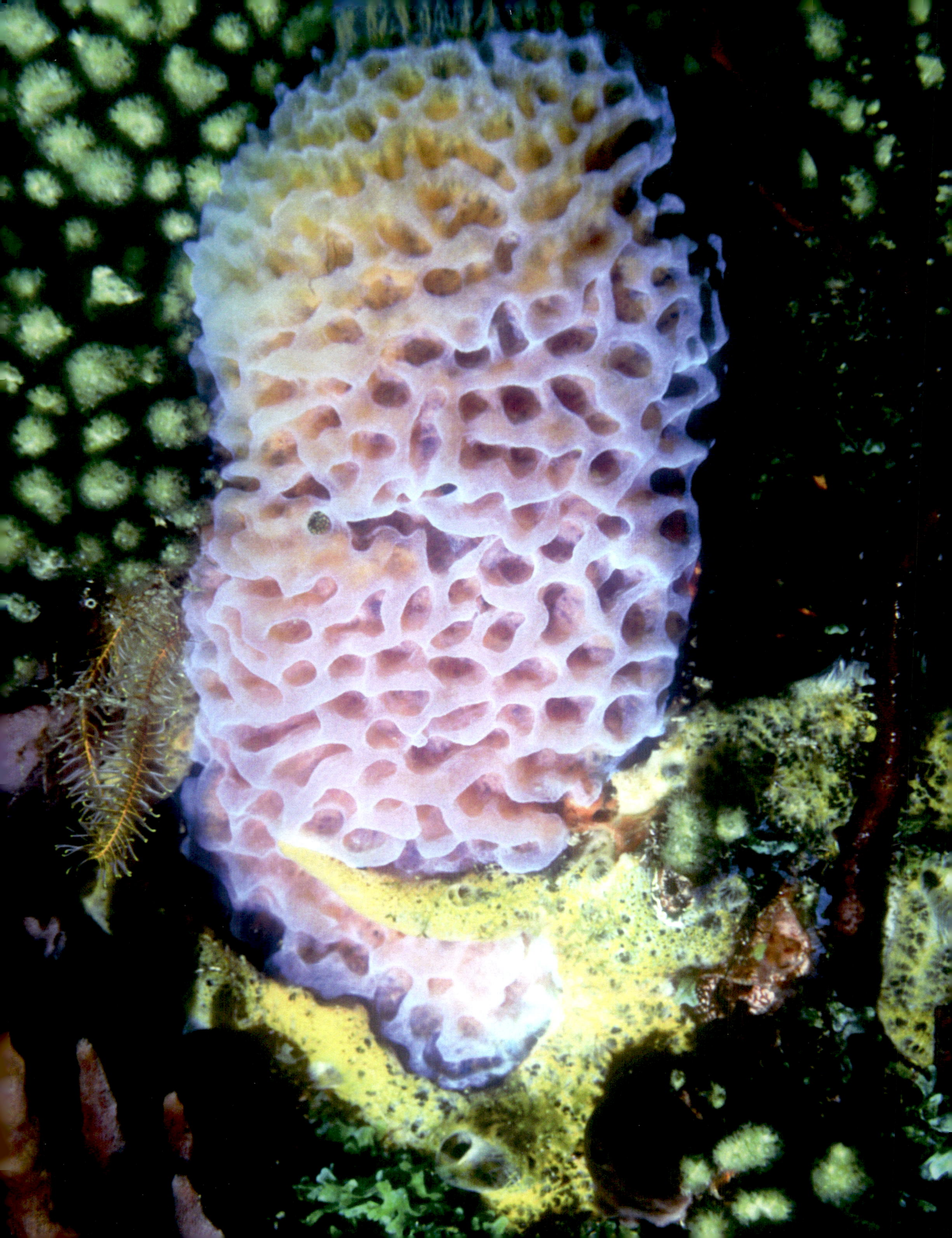

Marine Worms
Feathers and Fans, Bristleworms, Flatworms and more!

Marine worms are one of the largest and least phylogenetically understood organisms in the sea. In this text, we have hopes of proffering a practical guide full of useful information to simply improve their care in aquaria. As such, if the "worm" taxonomists are still unclear about the classification of worms worldwide, then surely we have no place or interest in trying to delineate them here. More to the point, aquarists generally have less interest in issues of taxonomy and more interest in husbandry and simply identification ("What is it, and is it good or safe for my tank?"). With regard for this reality, we address here the creatures called "worms" in their many guises and taxa of interest in the aquarium. Some are purely ornamental, some are practical, and many can be predatory or nuisance organisms. Understanding their individual roles and values is our first step.

Like them or not, marine worms are abundant and necessary in the seas and modern aquariums alike. They are a critical component of the living reef ecosystem rather like earthworms in the garden. When we take a step back and admire a terrestrial garden we see the flowers and leaves and landscape, but give little thought to the insects and worm species that worked to help or hinder the culture of the garden. In kind, marine worms are ubiquitous and pervasive in aquatic ecosystems... playing important roles of predation, decomposition, scavenging (detritivores), filter-feeders, substrate agitators, reef "binders" (through hard and soft tube production) and reef recyclers (dissolving, destroying and/or consuming inorganic and organic matter). Their enormous biomass and activities at large process a significant measure of nutrients on the reef.

Much of the worm diversity on a reef gets translated to the home aquarium with imports of collected live substrates in rock, coral and sand. For perspective, a single piece of coral studied on the Great Barrier Reef (Allen 1994) was found to harbor some 1400 worms of 103 different species. Can you image how many worms you would have if every coral in your reef tank came with even 10% of this abundance and diversity? Be assured, from live rock and sand alone, you likely have many thousands of beneficial worm species in residence.

Tiny Sabellids like these white fan worms with orange zoanthids are common and prolific in some systems. They spread easily between aquaria by sharing live rock and sand, and may fare well even without target feeding if the tank is mature and has a hearty bio-load. (A. Calfo)

On the most fundamental level, worms consume excess food in the aquarium and they *are* an excellent food for many fishes and other invertebrates. In many modern aquariums with deep sand beds and refugiums, they serve in crucial roles for "aerating" sand and mud substrates (agitating, liberating and consuming organic matter) and they prevent harmful impaction or stifling of the bed. Beyond any utilitarian purpose though, some worms are simply attractive and ornamental and sought simply in admiration.

There are many "worm" groups familiar to aquarists and most of them are remarkably distant in relation. Some things we call worms (like Synaptid sea cucumbers, AKA Medusa Worms) are not even worms at all. Nonetheless... if it's squishy or roundish or both (vermiform), we just may fancy to call it a worm in the hobby. The two most prominent groups of real worms in the hobby are the Polychaete Annelids and the Platyhelminthes ... or in layman's terms: the segmented worms and the flatworms. For the sake of simplicity, we'll break our address here of familiar marine worms for the aquarium down into four basic categories: 1) the sedentary polychaetes, 2) the errantiate polychaetes, 3) the flatworms, and 4) all others. Most will enter the aquarium incidentally and, again, rest assured they are likely harmless if not beneficial. We categorize the few undesirable groups into one of three categories: 1) parasites and predators, 2) nuisance growths, and 3) toxic risks.

When you discover worms in your aquarium, please don't be alarmed or reactionary. Case in point, the sight of a common segmented worm with just a few bristles is enough to send many aquarists with just a little bit of knowledge and a whole lot of fear in search of a flame-thrower and napalm dispenser. We ask you to pause in such cases and put down the military hardware. Noxious bristleworms, if indeed that's what you have and fear, are not only overstated as "bad guys"... but they are tremendously beneficial for the living substrate! They have *un*-deservedly been shouldered with the deservedly bad reputation mainly from the True Bearded Fireworm (*Hermodice carunculata*) from days long ago when Fireworms were common in the trade from wild live rock out of Florida and the Caribbean. But that wild rock has been illegal to collect since 1997. Other than immature (1-3 year aged) aquaculture rock from the region... little wild substrate and very few of the real "bad worms" even make it into the hobby anymore. Arguments too that bristleworms can grow to plague proportions are skewed by the omission that such plagues need fuel to grow- only in neglected or overfed tanks will these worms flourish to nuisance levels. Like many reef organisms in the aquarium, due diligence in husbandry is necessary to keep these and most other organisms in a balance that benefits the system. It's another friendly reminder of the importance of identifying and understanding the animals in our charge always before reacting. Expelling unknown marine worms from the aquarium is not a matter of "better safe than sorry" when the group is overwhelmingly helpful and harmless.

Polychaetes

Polychaetes belong to the phylum Annelida, which is composed of 3 main classes: Polychaeta (marine worms), Oligochaeta (earthworms) and Hirudinea (leeches). Aquarists will recognize many species here and favor most all of them, in

---

The Who's Who of Polychaetes: Some notable families

Errant Polychaetes

- Nereid "Bristleworms" and Amphinomid "Fireworms"
- Polynoids- Scale Worms (*Halosydna,* for example)

Sedentary Polychaetes

- Sabellid fan worms (soft-tube "Featherdusters"- fan worms)
- Serpulid fan worms (fairly straight hard-tube "Featherdusters"- fan worms)
- Spirorbid worms (spiral hard-tube "Featherdusters"- fan worms like the small white species that appear on aquarium glass, etc)
- Spionid worms (two-tentacled worms that make a silt or sand tube and sweep the substrate looking for food)
- Terebellid worms (AKA "Spaghetti" and "Medusa" worms... burrow in sand and sweep for food with numerous tentacles)

Characteristics of the most Polychaetous Annelids include:

- Their bodies are segmented (**metameric**) and bilaterally symmetrical
- Most have fleshy "legs" (**parapodia**) attached to each segment with "bristles" (chitinous **setae**)
- They have a nervous, digestive and circulatory systems that are redundant per segment, thus fragmentation is not fatal (quite the contrary)
- They are firm or turgid by virtue of a hydrostatic "skeleton" (**coelomic fluid**)
- Feeding strategies are highly varied ranging from predatory through sediment-sifting to filter-feeding
- They may be hermaphroditic or dioecious, and can reproduce by budding or by gametes

Seafrost (*Filograna* and *Filogranella* species) is a common, calcareous worm colony collected in both Atlantic and Pacific waters, from temperate to tropical regions. It forms delicate colonies in quiet, protected areas of the reef. Although not regarded as especially hardy in the aquarium, they prosper if given adequate nutrition, bio-minerals and protection from suffocating algae growth.

stark contrast to the flatworms (Platyhelminthes) addressed later in this chapter. A few polychaetes are *bona fide* nuisance creatures that eat desirable invertebrates like corals and even catch small fishes (rare). Most however are ornamental, utilitarian or both.

As aquarists, we first break this category down into two basic categories: errantiate, and sedentariate. In other words, we classify them as motile (moving across and through substrates) or sedentary species (living *in* substrates).

Errantiate Polychaetes: Most are creepy, crawly, harmless little worms that simply plow through substrates as detritivores. A few eat coral and more than a few savor bivalves (Tridacnid clam keepers beware). Most though pose little threat to desirable reef invertebrates in the aquarium. All can be observed with some fashion of bristles (setae); other characteristics include many conspicuous segments to the body and a distinct head with a pharynx (jaws or teeth).

Sedentariate Polychaetes: The sedentary polychaetes include some of the most desirable worms in the hobby. Sabellid (soft-tube) and Serpulid (hard-tube) fan worms, like the *Spirobranchus* "Christmas-tree" species, are highly sought after for their beauty and ornamental value. Most *cannot* be kept successfully for a full captive lifespan by casual aquarists though. Spirorbids are the common white-tubed spirals that appear on the aquarium glass, pumps, in sumps, etc.; they are prolific and hardy without target feeding in most tanks. Terebellid worms (AKA "Spaghetti" or "Medusa") and Spionid worms (AKA "Hair") are likewise prized by aquarists, but instead they have become valued for their exceptional utility as detritavores in modern aquaria with refugiums and deep sand bed methodologies.

Unlike the errantiate polychaetes... the sedentary worms have no teeth or jaws, but feed mostly instead on fine particulates- often by mucus or filter-feeding strategies.

Fan Worms and Feather Dusters

Selection

Fan worms have a wide range of hardiness and suitability for the marine aquarium. Unfortunately, the larger and more ornate "Featherduster" species (notably the sabellids and the beefy *Protula* serpulids) tend to be somewhat to very difficult to keep without large and mature aquariums. For all the known benefits of natural filtration methodologies and our efforts in this reference to encourage you to use live rock, live sand and refugiums, be assured that success with keeping fan worms will be supported by these aspects. Beyond the high profile species, though, there are many smaller hard and soft tubeworms that live and reproduce easily in captivity. The tendency for many aquarists to overstock and overfeed their tanks to some extent is a boon to many fan worms and is also indicated by plague growths of some.

Fan worms all build tubes to support their sessile lifestyles. Sabellids form a thin membrane-like tube with the help of some mud, silt or sediments in the wild. In home aquariums, some stressed individuals that survive the

The Indo-Pacific Fireworm, *Eurythoe complanata*, feeds indiscriminately. It is unsuitable for the coral reef aquarium and like *Hermodice sp.* is venomous. There are a variety of methods available to trap these undesireable aquarium additions and wearing gloves to handle them is a necessity. (*H. Schultz*)

Sabellid fan worms (AKA Featherdusters) form non-calcareous tubes of silt, sand, sediments or mud. Pictured here (right), *Sabellastarte indica*, the Banded Fan worm. They are the most common "Feather Duster" sold in the aquarium industry. They are found in all tropical seas and grow rather large to 4"/10 cm in diameter of the coronal head. Photographed here in Indonesia.

Some **Serpulid** worms live embedded in live coral or rock like *Spirobranchus giganteus* (AKA Christmas-tree worms) pictured above. Also known as "Coco," or "Jewel Stone" worms, there is an unsubstantiated legend that colonies will die if their "host" coral (typically *Porites* sp) dies. There is likely no merit to this claim. The source of the lore is more so because any aquarium conditions poor enough to kill a hardy *Porites* species are surely bad enough to kill these sensitive filter-feeding worms. And if the water conditions don't kill the worms, then the proliferation of bacteria consuming the dead coral tissue simply becomes pathogenic on the worms instead.

Spirorbids (below) build conspicuous, tiny, bright-white calcareous tubes (hard) that are commonly seen on aquarium glass walls and throughout the tank on pumps, hardware and anything else that wasn't moving fast enough when they settled out! They are indeed prolific in aquaria (more so in systems with enough- or excess nutrients to support them). They are harmless filter-feeders that are somewhat difficult to control if allowed to bloom in the aquarium. Other than limiting nutrients to reduce future spawns, there are no practical "reef-safe" predators on these worms that won't harm other desirable invertebrates in the aquarium.

Terebellid worms are especially popular in modern marine aquariums for their prolific nature and their active detritavore lifestyle. A single worm has many tentacles that it uses to actively sweep the substrate in search of food.

Commonly seen in the trade are "Spaghetti" worms (*Eupolymnia crassicornis*) and "Medusa" (*Loimia medusa*). They are harmless and ideal for deep sand beds and refugiums. Some are cultured and sold in the trade but most starter cultures are simply acquired with live sand. Above image: *Loimia medusa* in Hawai'i.

Feeding strategies vary among polychaete worms. Some **Serpulids** species trap and consume microscopic food particles by **ciliary action** with the use of their "feathers." Other Serpulids employ a **mucus net strategy** to trap their food much like the Vermetid snails with which they are often mistaken.

Worms 167

Bispira brunnea, Social Feather Duster Worms (AKA "Cluster dusters") are small Sabellid soft-tube colonial species found in areas of moderate to brisk water flow. Colors are highly variable but orange and purple varieties are popular in the aquarium trade. This species has great potential as an aquarium specimens in part for its common asexual reproductive strategy. Pictured here are lovely pale colored specimens in the Bahamas.

evacuation of their tube will even exert great effort to build a new tube with any available materials, which may include sand or fine gravel (not ideal). Serpulids secrete a hard calcareous tube and of course need adequate bio-minerals for such calcification. Maintain very stable and appreciable levels of calcium and alkalinty to grow hearty populations of serpulids. Even the boring Serpulid species, like *Spirobranchus*, build an obscured tube that is visible in broken coral or rock.

The fragmented home of this Serpulid fan worm is a living scleractinian coral. The imposition of such worms on living corals is common and harmless. (*G. Rothschild*)

Whether you seek a hard or soft tube species, inspection of the body is important to screen for tears or injury to the tube-shell that would likely have hurt or compromised the worm inside. Fan worms can repair such injuries, Sabellids more easily than Serpulids, but the demands on energy and resources of an already stressed animal in captivity do not bode well for its survivability. Fan worms have a pair of large pads that supply mucus to mix with inorganic materials for building their tubes from the inside out. The process is often likened in application to pottery making as they continue to lay a coating on the "growing" (building) edge of the tube's lip. As a step in the selection process, however, it is best to simply avoid these animals under less than optimum conditions. Never tear a worm away from a rock, glass or another worm on purchase! There is a high risk of fatal injury to the worm. Its best to acquire all fan worms undisturbed and attached to their natural substrate.

Feather "heads" popping off

They can't help it. When they get really upset, they just lose their "heads." It might surprise you that this extreme expression of duress is not fatal. At least, it is generally not fatal the first time. Fan worms that have been severely stressed can abort their feathery feeding structure (autotomization... like a lizard losing a tail) and simply re-grow a new crown within weeks. This is usually induced by poor water quality and/or handling but it is also evidence of attrition from starving animals. In fact, it is said that the head can re-grow a new "body" under ideal conditions although this is very unlikely, if for no other reason than the fact the feathery crown is too tempting of a morsel to eat in the confines of an aquarium or upon a crowded reef. Please do not throw your worm away if it loses its crown... it is not likely dead. If the stress that causes autotomization persists and the worm is stimulated to lose and try to re-grow a new crown repeatedly, the animal may very well just die. Assuming the candidate you are considering in a dealer's tank does indeed have an undamaged tube and its crown is intact, the next thing to look for is a simple and prompt fright response. To be clear, fan worms have no formidable means of defense on a reef; they possess no thorns, spines, claws, teeth, venom or poison. They simply stick their vulnerable feathery appendages out into the water for all to see. Needless to say, for all their vulnerability, they have evolved a "lightning-fast" fright response. Healthy animals will draw their feathery crown into their tube

Sabellastarte sanctijosephi: one of the species commonly shipped as a "Giant" or "Hawaiian" featherduster, distinguished by their dual crown. They have a poor record of survivability in captivity and most specimens starve slowly over a period of months without adequate nutrition.

immediately upon sensing a change in light intensity as with a hand passing over them to make a shadow. This of course might simulate a large predatory fish bearing down to prey upon the worm. Of course, the autonomy is a second line of defense if a predator should get a grip on the worm. Fragmentation is much better than being eaten alive! Nonetheless, unnatural water movements, vibrations and changes in light should all stimulate a fan worm to quickly retreat into its tube. Make no mistake that a newly imported worm, if healthy, will respond to such stimulation.

The last fundamental criterion of selection is a practical matter of research or inquisition. How long has the worm been held captive? It's important to discover how long it has been since import and under what conditions as best you can tell. Fanworms are decidedly heterotrophic and cannot manufacture any of their own food. They also are not like to get much or any useful food in the tanks of dealers and wholesalers. Thus, the further removed from the day of import that your specimen is, the more likely it has been at least somewhat deprived of nutrition. We strongly recommend that you buy the "freshest" imported worms possible. When you do finally decide on a good specimen, be sure that it is bagged, acclimated and released underwater. Do not remove fan worms from the water as this puts great stress on their hydrostatic structure (a "skeleton" by virtue of **coelomic** fluid).

Summary of selection tips to consider:

- Injury to the tube? (hard or soft)
- Condition of crown ("feathers" damaged or aborted in autotomy)?
- Very responsive to movement and light (fast retreat response)?
- Adequate conditioning (duration held since import)?
- Do not remove worms from the water (prevent air-exposure).

Care of Fan worms

Fan worms present similar challenges in husbandry as many other filter-feeding animals like aposymbiotic corals. They need very high water quality... but they also need a large and constant supply of food. Those two realities are a challenge if not a nightmare to manage in closed aquarium systems. Heavy feedings are of course a great burden to water quality (DOC levels, Redox, O2, nitrates... even sanitation!). Although these worms can sometimes be found in very high nutrient or even "polluted" areas, the seawater in their immediate area is exchanged (turnover or dilution) by an inconceivable magnitude. Literally hundreds of thousands of gallons of water are moved in an out of a given area in one day... even a matter of mere hours. Thus, aquarists have a dilemma in supporting water quality while providing adequate food to featherdusters and the like.

Ultimately, it is recommended to

Protula magnifica, the Magnificent Tubeworm, is a Serpulid species from the tropical Indo-Pacific that has not yet proven to be easy to keep in aquaria. They are large, hungry and poor shippers best remitted to advanced and specialized aquarists. Experimental target feeding will be necessary to better understand and improve success with *Protula* in captivity.

maintain near reef conditions for fan worms. Full strength seawater (no low salinities for most) and a low to medium organic load. As with most sessile invertebrates, they depend on good strong water flow to bring food and oxygen to them and to carry waste away. For the Sabellid soft tube species, be mindful to only use very fine sand or mud as a substrate. Fan worms are often more secure and protected from curious arthropods (shrimp and crabs), fishes, or other

It is natural or healthy to bury most Sabellid soft-tube fan worms in a soft substrate. Use very fine sand or mud and always avoid course sand or gravel. They are protected in this manner from known predators and curious opportunists. They also derive nutrition (micronutrients) from the local environment in deep sand beds. Many fan worms are indeed naturally found in the wild nestled in mud and sediments like the *Bispira* species pictured here in Cozumel.

*This **Timarete** medusa worm has ventured up and onto the aquarium glass. (R. Hilgers)*

predators if you slightly bury their tube in a soft substrate- only a few inches deep.

- maintain near reef quality water (full-strength seawater, low to medium organics)
- provide strong water flow to bring food and oxygen to these sessile creatures, and to carry wastes away
- bury Sabellid soft-tube species just a few inches deep for their protection from arthropods, fishes and other predators

Feeding

What we know about how fan worms feed in the wild and what we have to offer them in aquariums unfortunately has not evolved into synchronicity just yet. As mentioned above, fan worms are sessile heterotrophic invertebrates wholly dependent on the currents bringing the right food of the right size and enough of it to them every day. Solutions to feeding them are not anywhere near as easy as feeding fishes or even many corals. Fan worms are categorized specifically as **ciliary-mucous suspension feeders**. If that phrase were in Latin it would probably translate back into English as "not recommended for beginners." Ha! Truth be told, the needs of fan worms and the ability to access how those needs are being met are not easy or apparent to new aquarists. In many cases they are not even possible to satisfy in small or young tanks (under 1 year old or less than 100 gallons in size). Without a significant sized refugium or very deep sand bed, it may be wholly inappropriate to keep large fan worms in tanks under 50 gallons in size (as an estimate).

The feeding tentacles (feathers) of these worms are complex structures that have evolved to act as filter-feeding support. This crown is an assemblage of branches that have side branches (**pinnules**) in turn. This collection altogether resembles a feather but performs more like an intricate unfolded "sieve." Each tentacle is lined with a conveyor-like assembly of mucus-covered hairs (**cilia**) that beat to move water across the tentacles. In this manner, a fan worm derives oxygen, sheds metabolites and, of course, captures food. Some of the prey known or suspected to get conveyed down the ciliary food groove include: bacteria, nanoplankton and organic particulates. Particles are even sorted by size in the food groove of the tentacles as they make their way to the mouth with inappropriate fare rejected. The largest particles are channeled to the base of pinnules and simply discarded. Thus, aquarists must be mindful, particularly with bottled food supplements, which are largely ineffective because of clotting or particle size, that seeing food stick to a feather does not mean the food is going to get eaten. Most prepared foods (bottled slurries, thawed frozen purees, etc) are in fact too large for fan worms to eat. This is the biggest obstacle to keeping "featherdusters." Target feeding is all but impossible while we cannot provide bacteria and nanoplankton readily yet, without live cultures or naturally (large mature systems). To continue with our detail with feeding aspects of fan worms, we address tube building. Medium sized particles captured by the tentacles are kept by Sabellid species for building soft-tubes (mixing matter with mucus), but they are rejected like large particles by

*Most fan worms, like the **Spirobranchus giganteus**, are difficult to target feed and often will not survive on the liquid slurries of commercially bottled zoo- and phytoplankton substitutes. Marketing claims have countered that wasted food helps by increasing the levels of dissolved organics; others simply call this pollution. There is no definitive or practical use for prepared foods yet with this group of reef invertebrates. Live plankton drips are much better. Very large and mature aquariums with refugiums and deep sand beds are also strongly recommended to provide at least some bacteria and nanoplankton. (L. Gonzalez)*

Anamobaea orstedii is a strikingly colored fan worm seen throughout the Caribbean and pictured here in Belize. It is commonly known as the Split Crown Feather Duster and can be found in groups or solitary. They are Sabellid soft-tube species.

Pomatostegus stellatus is known as the Star Horseshoe Worm. It's commonly found throughout the tropical West Atlantic in all reef areas. They are calcareous and often found embedded in coral or overgrown with algae.

Bispira variegata, the Variegated (Pink) Feather Duster is a common import with a good reputation in captivity. As with all feather dusters, defer adding these animals to mature aquariums at least 6 months old, and preferably one year.

various Serpulid species. In all species, only the finest particles are kept for food. In essence, feather dusters need naturally occurring bacteria and microscopic plankton that only occur in large natural marine aquarium systems if they are to survive for more than months. Culturing live plankton may be helpful especially if fed to the system in a slow and continuous drip. Brine shrimp-sized prey is too large for most, rotifer-sized prey is better and phytoplankton-sized prey is rather good. Sand stirring of deep beds (manual or by natural activities in the tank) has been very helpful for fan worm species through the liberation of bacteria and numerous micronutrients.

Reproduction

Little is known about the successful and deliberate culture of fan worms in aquaria. To date, events of propagation have been a simple matter of inevitable procreation in some hardy species and no activity to speak of in most. Sabellids and Serpulids can reproduce sexually (they are dioecious- the sexes are separate) and typically the males spawn first with a release of sperm into the water. Neighboring females take the chemical cue (male gametes) and release their eggs. With successful fertilization, pelagic (swimming) larvae develop within days. Fan worm larvae are said to feed on phytoplankton (Strathmann 1987). This has given many aquarists hope for breeding, feeding and rearing more tubeworm species in the future,

as plankton-friendly methodologies are refined in aquaristics. Like so many other species with pelagic larvae, however, it is presently unlikely for fan worm spawns to successfully be reared through developmental stages in the filtered confines (living feeders and aspects of hardware) of a closed aquarium system. Still, there are at least a few species of fanworms that succeed in reproducing well in captivity. Again, the best success is achieved in large and more mature aquariums with better natural feeding opportunities.

Compatibility of Fans and Feathers

Tube-dwelling polychaete worms are not at all aggressive towards each other or towards any other creature you can imagine, for that matter. They are truly "reef-safe" by aquaristic standards of measure. Compatibility issues with sedentary polychaetes revolve entirely around discerning what eats them. The answer to that question in broad terms is: a lot of creatures eat polychaetes! Sedentary worms have no appreciable means of defense and seem to be very palatable and digestible by numerous organisms

Loimia sp. - a medusa worm on aquarium glass. (*R. Hilgers*)

Worms

including humans (!) (read more about epitokes below). In fact, we cannot practically list all of the potential predators of polychaetes, but there are more reef organisms than not will prey on them. Even standard "reef-safe" fishes that do not harm cnidarians like corals and anemones may attack fanworms (dwarf angels and some tangs, for example). And most any reef fish will feed upon Terebellid and Spionid worms in the wide-open given the chance (wrasses, damsels, clowns, pseudochromids for certain). Predation is hardly limited to the vertebrates; arthropods are also heavy predators on polychaetes. Hermit crabs can outright destroy the desirable worm fauna in living substrates.

The successful long-term care of fan worms and other sedentary polychaetes requires careful planning of tank mates. Providing suitable habitats like deep sand beds for burying, or bolt holes in rocks are also crucial consideration for their protection and good health.

Errantiate Polychaetes

The "Motile" Segmented Worms

Wherefore Art Thou Errantiate Polychaete Annelids?

The errant ("wandering") segmented worms: why do they travel? Because they can. They rage in size from microscopic wisps to formidable monsters at more than ten feet in length (3+ meters). They occur in huge numbers in the wild amassing concentrations in some areas of hundreds of thousands per cubic meter. Although there are literally thousands of different species of "wandering" polychaete worms known in dozens of very different-appearing families, the vast majority of those in reef aquariums are of the family Amphinomidae. It may be fair in some sense of the word to call these segmented worms "bristleworms," as they all have some number of "bristles" (setae) at each body segment, used in leg-like fashion to facilitate movement. However, other worms have better descriptive names like the legendary Fireworm (*Hermodice*) with its dense clusters of setae, which may be shed causing pain or discomfort to a would-be predator. The spines and bristles don't stop with even the furriest bearded fireworm, however. There are large errant polychaetes called Sea Mice that grow bigger than a man's fist (cited as large as 20 cm) and are armed to the teeth in a porcupine-like mass of spines. Aquarists need not worry about encountering these bristles, though. Sea Mice avoid light and are generally not found in anything less than a few feet of substrate (1 meter or more) and at great depths (200m). Even the fireworms are rare in the hobby now that wild live rock from the Caribbean is protected and aquacultured live rock is still fairly immature. Again, far and away, most segmented worms that aquarists encounter in the hobby are harmless or even beneficial for their predominantly scavenging behavior.

What Makes a Bristleworm so Bristly and Fiery?

Within the this class of worms there is great variation in body forms and length, but all share the aspect of numerous bristles (setae/chaetae) in two general packets or podia (above and below) of positioned clusters. It is important to note here that some of these bristles are formidable weapons whose danger extends beyond the obviously sharp or irritating physical structure. There is a very good reason why many larger species are called "Fireworms." The bristles of *Hermodice* are hollow tubes filled with venom that varies in potency. For aquarists that have come unfortunate contact with a fireworm, the symptoms can range from a mild itch to a severe burning sensation. Home remedies for extracting lodged bristles in the skin vary, but the premise is to use a product like "liquid bandages" (various brands of topical dress that are painted over a wound and dry to form a coating). Once the liquid coating has dried, it can be peeled away to pull the embedded chaetae (tiny bristle spines) out. In the absence of this medical product, you can use very sticky tape, like duct tape, to pull the spines from skin. Be sure to disinfect the afflicted area afterwards and seek a

Bristleworm or Fireworm? What's the Difference?

These common names are exchanged freely and quite inaccurately to describe two *very* different polychaete worms. It is very important for you to understand the differences between them, and there are many. Bristleworms (**center**) are relatively small (to several inches) and mostly harmless detritivores. They can, in fact, be enormously beneficial to natural marine aquarium methodologies for supporting the health of live substrates. Only in neglected aquaria (excess detritus or food, inadequate water flow, etc) do they reach nuisance populations. The real "bad guys" are true Fireworms (left and right images: notice the more densely clustered "bristles"). Fireworms can grow large to 12"/30 cm or more and eat many desirable reef creatures including corals, bivalves, and even fishes when available. (*middle & right images: H. Schultz*)

doctor if swelling, numbness or pain persists. An even better solution is prevention. We tout this almost as much as quarantining animals, to **wear gloves** when working in aquariums with reef invertebrates. There are some dangerous things that can breach the skin in a tank full of spiky, spiny and stinging animals. Even beyond risk of physical injury, there are some concerns with infections like *Vibrio* or *Mycobacterium* in all marine aquariums and environments. With or without gloves, be sure to handle all bristleworms with forceps, tweezers or a like instrument.

Feeding errant polychaetes:

You may be wondering why anybody would want to feed and encourage bristleworms to grow in the aquarium. In light of the overwhelmingly unfavorable tales about them in popular literature, it is little wonder. Yet the truth of the matter is that small bristleworms are generally harmless and peaceful detritivores. They actively burrow through sand and gravel feeding opportunistically on organics they find, and contribute overall to the stability of the system. In the process, they agitate the substrate to the benefit of many other micro-organisms. Frankly, their natural and incidental presence is to be encouraged. If you have good husbandry and neither overstock nor overfeed the aquarium, you can simply allow their growth to be limited by available nutrients, and mild predation by invertebrate or fish species. If the system is neglected, however, you can expect populations of possibly nuisance species to flourish. To

Hermodice carunculata, the legendary Bearded Fireworm, is so-named for its large, pleated structures (caruncle) on all segments. Their dense clusters of bristles on the flanks are hollow tubes filled with venom that may have variable effects on bare human skin ill-fated to come into contact. Fortunately rare in the hobby, this denizen is common in the Atlantic and does feed on corals.

avoid such plague populations, maintain vigorous water quality throughout the system. Good water flow keeps detritus and other organics in suspension for direct ingestion by more "desirable" creatures (like ornamental filter-feeders) or export by filtration gear like protein skimmers. Careful pace and portions of feeding to the fishes and other invertebrates in the system will also go a long way to avoiding excess particulates from feeding nuisance growths of bristleworms.

For their categorically opportunistic behaviors, it is easy to mistake some harmless bristleworms for being "predatory." Aggregations of *Eurythoe* for example, attracted to the dead and dying (fish, coral, etc) are often believed to be killing them. We assure you though that they almost never attack healthy animals and their presence indicates their instinctive sense of organisms with not much time left in this world. Any healthy fish can swim away from a worm, and most any healthy coral is not interesting prey for bristleworms. Again, it is the *Hermodice* bristleworms (AKA Fireworms) are decidedly predatory

Chloeia fusca (an Amphinomid): all Fireworms are bristleworms, but not all bristleworms are Fireworms. Let's not take our chances though with this well-equipped specimen from the Indo-West Pacific. Indeed, not all Fireworms are *Hermodice* species from the Caribbean, although the Atlantic forms are more familiar and common in aquariums. These polychaetes are usually found on sandy, muddy bottoms, where they actively seek prey- a positively phototropic predator.

Worms   173

> ### You did say errantiate, didn't you? **Swimming** Polychaetes
> ### (AKA- Bob eats worms and likes them!)
>
> "As a boy growing up in the Philippines I had many novel adventures - one whose remembrance struck me right between the eyes as I sat in a Marine Invertebrate Zoology class one day. The local island people in the smallish village where I lived for two years were pretty nonchalant where time-keeping was concerned. We were poor and had few watches to keep time. We pretty much ate when we were hungry and slept when we were tired... except for certain celestial and tidal events. When the tides and moon were propitious, everyone would get whipped into a high state of frenzy and collect "palolo" at night. These "things" could be scooped up in the reef shallows with our baskets by the bundle, and were they good to eat! Now skip ahead a few years and I'm at a lecture hearing about the same previously mysterious and forgotten organisms. Those "things" were **epitokes**: s*egments of errantiate polychaete worms* that on cue develop sets of eyes, break off, and swim to the surface to mass-reproduce. Yuck! I mean yum... I mean yuck! I mean yum. Oh blissful ignorance!"
>
> There are more than a few worms that can be observed in the black of night in your aquarium swimming about. Not all or likely any of them will be epitokes, although it is suspected of occurring on occasion in captivity, from *Eunice* worms like described above. It is still a fascinating story and reminder to peer into your reef invertebrate aquarium at night. Some animals will be attracted to your beam of light and others will flee from it. In general, some nocturnal species (not all) seem to be more indifferent to a red lens covered light if you care to improve your spying ability at the expense of a beam of daylight. It may be fair to say that there are far more organisms in your aquarium at night than by day. You will also get to enjoy seeing the nocturnal behaviors of your display animals, not the least of which are dramatic feeding tentacles of plankton feeding cnidarians. Hedge your bets for a good show by feeding the dark tank minutes before you spy it with light.

and will seek healthy cnidarian tissue, bivalves and some fishes if the worm is big enough or the fish small enough. They have visible and intimidating mouths and jaws to gnaw away at flesh indeed! Identification is the key to understanding what you need to know about these worms feeding preferences and inherent risks.

Reproduction in errant polychaetes

Polychaete worms have exciting sex lives. There's no other way to say it with the options and strategies they have been known to employ. They conduct group sex, cloning, fragmentation and budding, and the wildest of all - epitokes, as described above. Some of the Nereid epitokes of "clam worms," by the way, exhibit bioluminescence during acts of reproduction in a magnificent natural "light show."

In general, most polychaetes produce larvae through events of sexual reproduction triggered by a chemical cue, which are usually gametes from another. This reminds us of a story we heard from an aquarist that wrote in to us at **wetwebmedia.com** regarding an "odd" occurrence in their aquarium. This aquarist recanted a tale of a "cat and mouse" game he had been playing for several days trying to trap a large bristle worm. When this aquarist finally lured the worm far enough out from under a rock, he grabbed onto the worm with a pair of tweezers and tugged to extricate the little vermin from its lair. In the process, the worm tore apart and released what appeared to be gametes into the aquarium. That fear was confirmed for this aquarist with profound irony when hordes of bristleworms all rushed out from under their rocks and from within the substrate to spawn *en masse*! The good news for all is that reliably successful larval development in aquaria is still some time away in the future.

Yet another fascinating strategy of polychaetes is **ameiotic** reproduction. This is a form of parthenogenesis whereby females produce unfertilized eggs that develop into juvenile polychaetes. This strategy is rather uncommon and is also believed to often be triggered by unfavorable environmental conditions.

Another strategy that is perhaps more common in the aquarium than the wild is simple fission or fragmentation. As we know from science that the segments of polychaete annelids have redundant vital systems, it comes as no surprise that a worm can break in half and grow a new "head" and "tail" respectively.

Controlling bristleworms

Options for limiting unwanted species and populations of errant polychaetes can be summarized as follows:

- **Limiting food source** (the simplest and best long term solution)
- **Manual removal** (before and after quarantine)
- **Trapping** (often does not work as good as you'd hope... often only catches a portion of individuals)
- **Natural predation** (variable effectiveness)

Avoiding "Bad Worm Days"...

"Oh my gosh, I've got worms. *Help!* What are they? How do I get rid of them?" So go the all-too common cries of newer marine hobbyists utilizing live substrates in their aquariums. Are bristleworms truly a pestilence, you may ask? Will they eat your livestock, turn over the furniture, or imbalance the economy? The answer is that only one of those concerns is remotely possible... and its pretty cool to see a 20 mm worm flip over a sofa, anyway.

An answer is that there is no (safe) way to avoid getting at least *some* errant polychaetes in the aquarium. And why would you really want to? They are pervasive and necessary in the reef environment. Short of hostile measures to sterilize vehicles of import for polychaetes that will kill every other good organism with them in turn, the control of bristleworms is a simple matter of good husbandry. They simply do not survive, grow and breed without adequate nutrition. Excess populations mean that there has been excess available food, plain and simple. In turn, the best long-term solution to limiting any population of hungry "pest" organisms is to simply **limit their available food**. You may agree that your tank is overstocked or overfed, or you may simply have inadequate water flow (target 10 to 20 times tank volume turnover) or a misdirection of effluents in the tank, which allow the accumulation of detritus. Perhaps there is a messy fish that mashes and spits many particles of food. Whatever the culprit may be, identify the source of solid organic matter that is feeding the worms and reduce it. You will then be able to manage a tolerable population of polychaetes.

Manual removal and trapping

To reduce the possibility of getting excess or dangerous bristle worm species from the onset, prevention by proper screening and handling of living substrates is the best means of prevention. Live rock with or without coral or other reef invertebrates attached should simply be held in a proper quarantine tank for screening. During isolation, the rock can be suspended off a bare-glass bottomed tank with a simple platform made of PVC, "egg crate (plastic light diffuser), or any like open structure. The shelf should not be too high (less than 1"/25 mm) so that bristleworms, mantis shrimps, unwanted crabs and other predatory carnivores can climb down to the glass bottom of the tank for meaty bait. Bait the tank at night and check for lured scavengers in the dark of night. Net or siphon out the unwanted animals off the glass-bottom as necessary. This is a remarkably easy way to prevent larger predatory species from entering your main display tank.

A modification of this technique, which can work in infested display tanks too, is to take the bait (ocean meats like fragrant shrimp or krill) and wrap it tightly in a small nylon satchel (like sterilized women's hosiery or a bit of cloth net). Many bristleworms and stomatopods will get their spiny aspects snarled in the nylon and stall their escape long enough for you to siphon or net them out of the tank. An ingenious trick that takes this concept even a step further is the "packed food in a tube" method. This trap is made with a section of pipe, which is loosely filled with filter "floss" media (polyester fiber fill like Eheim's **Fein** and **Grob Flocken**). PVC plastic is OK for the pipe, but clear 1" lift tube is better and more fun. In the center of this pipe (say 6"/15 cm long) is a bit of meaty food (again... shrimp, clam, etc). With this contraption set on the bottom of the sand, numerous bristleworms will become trapped or embedded overnight. It is a very easy and reliable way to collect worms. There are many other effective DIY bristle worm traps known and waiting to be discovered across the Internet, shops and aquarium societies. There are even commercially produced traps, although most of these have been ineffective to date. Some folks have even made comfortable livings re-selling all plastic mousetraps you can pick up at the hardware store. One high-profile company in the 1990's hilariously scratched the name and patent number off the plastic mousetraps (not their company's product!) and simply resold them as bristle worm traps. We're not sure which is funnier- that they did it or that people continued to buy them despite the fact that they were mediocre for the job at best. Of course, all of this is better than removing the

Stenorhynchus seticornis, the Caribbean Arrow Crab is one of the most celebrated predators on undesirable or excess bristleworms. Unfortunately, they may also consume favorable polychaetes including fan worms (featherdusters). Large or aggressive crabs may also attack and kill small fishes too. They are delightfully hardy, long-lived and easy to feed arthropods, but you must know that they are not without their risks in a mixed invertebrate aquarium. (*D. Fenner*)

offenders out singly with tweezers, or worse.... with hook and line contraptions for big specimens!

A warning: There have also been some radical suggestions for removing bristleworms including various chemicals or stress-induced treatments. Rinsing uncured live rock with freshwater can extricate many undesirables from the rock but also threatens desirable micro-crustaceans, desirable worms and other live organisms on and in the rock. Other writers have advocated a deliberate increase of QT water temperature (with no other living organisms like fishes or corals with live rock) to approaching 90F, to drive them out from their hiding. This, of course, is likely to stress and kill other desirable organisms and is not recommended. Some other clever techniques have been advised like making high CO2 or Magnesium chloride "seawaters" to dip live rocks in for the purpose of driving larger organisms from within the rock. Unfortunately, many worms deep in the live rock still will not come out during this treatment, or worse... they will die inside and decay. Furthermore, most all of the faster motile, desirable species like starfish, shrimp and zooplankton will be expelled. Again, these are drastic measures that should be unnecessary with proper screening of all livestock and living substrates in QT for a few weeks to a month, and good husbandry in the display afterwards.

Natural predation

Natural predators on bristleworms, of which there are many, can be used but only with careful consideration. Any fish or invertebrate that will heartily prey on "bad" polychaetes will at the very least prey on "good" polychaetes as well. In many cases, other desirable reef organisms may also be at risk in time from these predators. Even under the best of circumstances, controlling a nuisance or plague population of bristly polychaetes biologically is challenging. Prevention and good husbandry is always the best method of control. Limiting nutrients and starving them into submission over a period of just a few weeks is the next best thing. Adding predators to the bio-load for control of excess worms must be the last resort. It seems counterintuitive and ironic in some ways, as it is issues with a burden on the bio-load (excess organic particulates) that allows a population of bristleworms to grow large enough to want a predator in the first place! Adding another hungry heterotrophic animal into the mix makes little sense even in light of the fact that it does not require "extra" food but will hopefully eat the worms instead. We can say this because we know that the plague population of worms will either stay the same, get worse or better. If the pest population continues to stay the same and supports the predator, that means that the predator is simply recycling the still uncorrected flaw of excess food in the aquarium that allows the population of worms to continue. If the pest population of worms increases, it suggests that the burden on the system with the new predator is more than the system can or should support (and the worms still persist with the excess nutrients). If nutrients are limited and the pest population of worms are controlled by the predator instead, the predator at some point will reduce bristle worm levels down to a threshold where it is no longer adequately sustained and it must seek alternate foods for life. If you then provide food to the predator in the absence of bristleworms, you may have be again on the same slippery slope of net excess import of nutrients into your system. Whew! What a conundrum. You can see that any way you choose to look at it, excess bristle worm growth is largely a matter of excess food supporting them. Buying a fish or invertebrate for the sole purpose of controlling bristleworms is therefore not the solution; controlling nutrients is the answer.

There is a very long list of fishes that are reliable predators on bristleworms. Almost the entire wrasse group from the tiny six-line wrasse (*Pseudocheilinus hexataenia*) up to the beefy Bird Wrasses (*Gomphosus* species) all prey upon worms. Smaller species in this group (like

Stenopodid Boxing Shrimp (*Stenopus hispidus* and the like) are somewhat more reef-safe predators on bristleworms than the highly touted Arrow crabs (*Stenorhynchus*), although not without similar risks of aggression to small fishes, other shrimps, crabs, and mollusks. Very few predators are likely to single-handedly effect a cure for plague populations of bristleworms in a tank with nutrient accumulation problems. Aquarium Image.

Halichoeres) are very trustworthy in invertebrate displays while larger species like *Thalassoma* and *Coris* that can be a living nightmare as they mature and wreak havoc with other invertebrates and smaller fishes. One of the most effective and least harmful (to desirable sessile invertebrates) fish groups is the Pseudochromidae (Dottybacks); the longer-snouted species in this group from the Red Sea are the most effective including *Pseudochromis aldabraensis, P. flavivertex, P. fridmani,* and *P. springeri*. The Goatfishes (Mullidae) and Sandperches (Pinguipedidae) are also outstanding here but a bit "rough and tumble." This mention of fishes predaceous on bristleworms is by no means complete. Do research and consider the many other possibilities if the bio-load of your system will support them after denuding a nuisance bristle worm population.

Platyhelminthes

Flatworms

(Not sea slugs and not all really "planaria," but they truly are worms!)

Phylogenetically, flatworms are the beginning of bilateral life. Starting here, we see the replaying theme of being able to divide creatures along a midline axis to observe essentially mirror-image anatomy. This simplicity does not make them wholly primitive, though. Many are evolved parasites, displaying both direct and complex developmental lifestyles. As their namesake tells us, these creatures are dorso-ventrally flattened (back to front), worm-like animals of pronounced thinness. They lack a body cavity (coelom) where their body organs might reside and be suspended as ours are, and are thus classified as acoelomate animals. Hopefully you're not eating while reading this passage, because flatworms also only have one opening to their digestive tract. What goes in their mouth either stays inside them, or goes out the same... well, you get the point. Most are carnivorous and very specific about their prey. Few are "reef-safe" in aquaristic terms as their preferred prey includes cnidarians, bivalves, sea squirts, bryozoans, other worms, crustaceans, or snails. Some are even cannibalistic. Vegetarian flatworms are hard to come by! Taxonomic debates aside as to the number of valid classes in this phylum, most living subdivisions of flatworms are parasitic in nature like the Trematode flukes and Cestode tapeworms. Aquarists, however, are mostly familiar with the free-living group of flatworms known as Turbellarians, which are not much more welcome in captive marine systems. The larger ones tend to be predatory and carnivorous while the smaller ones are often quick-to-reproduce and become a nuisance (the acoel species popularly known as "planaria"). Muco-ciliary action, a gliding over moist surfaces of their own slime, accomplishes movement of these fascinating creatures. Some species also locomote (swimming through the water, in this case) by an undulating action. Sexually, the Platyhelminths are also true hermaphrodites... being both male and female in one individual. Cross-fertilization is the rule with a "pair" rudely stabbing one another with a hard penile **stylet** to actuate sperm transfer. As you may recall from school experiments with freshwater Planaria, they can also reproduce by splitting asexually (fission). Although there are more than a few remarkably beautiful organisms in this group, most are of little attraction to home aquarists. We suggest that you avoid most all flatworms in the aquarium and will advise you here of at least a few high-profile nuisance or predatory species that you should know.

Polyclads (meaning "multi-branched"... in reference to their digestive system) are Turbellarians that are rarely collected for the hobby, but do appear on occasion as incidentals. Many are quite beautiful and resemble the Opisthobranch sea slugs. In fact, there seems to be some degree of mimicry in more than a few species between them. The adaptation seems to favor the flatworms, which are believed to generally be not quite as noxious or toxic. Nonetheless, they can be equally poisonous to aquarium systems and inhabitants, and are no less carnivorous. Most are a nuisance if not patently predatory on desirable tank inhabitants. Unfortunately, very little is known about their collective husbandry and viability in aquaria. What we do know however indicates that few if any aquarists should deliberately seek the keeping of flatworms. If you discover one in your aquarium, you are advised to either leave it be, or remove it to a quarantine tank for observation and experimentation to determine its feeding habits. While in isolation, take the time to research your find with hopes of discovering its history and toxicity. Flatworms are typically

Nudibranch?... no, its a flatworm: *Pseudoceros dimidiatus*

Planaria?... not exactly: Acoel flatworms (*Convolutriloba retrogemma*) (A. Calfo)

Beautiful, but unwelcome...

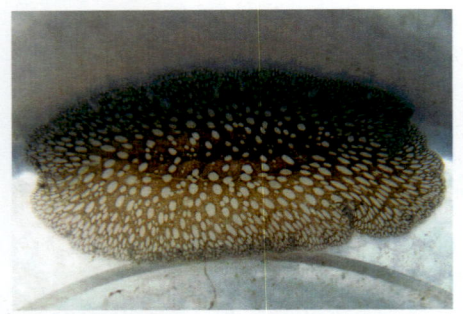

This predatory flatworm looks remarkably like the dorsally viewed mantle of a like-patterned *Tridacnid squamosa* clam: notice the ruffled edges and pattern arrangement in the center image. After reading the short introduction to this chapter, you can probably guess why this is so! This series of images also gives some indication of how very changeable in form an undulating and motile flatworm can be. (*A. Calfo*)

oval in appearance, although some are more elongate. You may notice they bear not-so-prominent head structures too. Cryptic species tend to be small at less than 10 mm, and non-descript, but some species are larger (up to 60 cm) can be gorgeously adorned.

Acoel Flatworms (*Convolutriloba, Convoluta, Waminoa,* etc): Commonly referred to by aquarists as "planaria," these include the often symbiotic red, rust, orange, brown-hued flatworms that reach infamous plague proportions in aquaria. Infections of "gut-less" acoel worms covering photosynthetic corals, plants and algae impede the ability for that organism to feed (capture plankton, conduct photosynthesis), adequately shed metabolites, derive oxygen, and more. Although they may not prey directly upon their living "perch," their stifling presence can become harmful or even fatal, nonetheless. They are considered to be generalist feeders and contain zooxanthellae for harvesting the by-products of their photosynthetic symbionts. The zooxanthellae may be acquired by digesting decaying coral tissue. They do not however seem to consumer healthy cnidarian tissues. Adult size is typically 1/4" or less (<5 mm).

Are they Predators, Are they Parasites, Are they Pests? Who Cares? The mostly reddish brown colored individuals of small ovoid pest species like *Convolutriloba retrogemma* can be a cause for concern. In brightly lit systems, utilizing their zooxanthellae, these animals can quickly reproduce, and in good numbers exude enough mucus to damage soft and hard corals. There are strategies for intervening detailed below. Are they eating the coral polyps, maybe just living on those animals mucus, perhaps stimulating them to produce more for their use? With so many species (even specimens from more than a few genera all generically called "*Convolutriloba*"), we cannot fairly comment on their risk categorically. Simply put, if they seem to be harming your livestock, consider the following possible courses of action.

Controlling Flatworms

There is no question that even the most noxious and unpalatable flatworm species on the reef still have some kind of predator that feeds upon them. Unfortunately, aquarists with plague populations of such worms in aquaria have not yet found an enthusiastic predator that is commonly available but not obligate to feed on the worms. The Cephalaspid Head Shield sea slug, *Chelidonura varians*, is touted as the quintessential Acoel flatworm eater. True as that might be, even aquarists with the worst infestation of nuisance flatworms must concede what will happen to the *Chelidonura* once the food supply runs out. It's surely irresponsible to harvest this obligate feeder with the knowledge that it will starve to death after the worm prey population has waned. *Chelidonura* are not even cultured. Fortunately, they are also hard to find in the trade. Please be sure to pass on them if you see one for sale. Aquarists must seek other means of controlling nuisance flatworms even when such methods are slower or less effective.

One of the simplest ways to distinguish an ornate flatworm like *Pseudobiceros hancockanus* from the mollusk Nudi-branchs ("naked-gills") is to verify the lack of external gill structures which are usually visible on the sea slugs. Flatworms also tend to be much thinner and faster moving.

Water flow: In some aquaria, an

increase in water flow was been demonstrated to reduce flatworm populations. For others it is weakly effective if at all. The benefits seem to be both direct and indirect. Acoels appear to dislike strong water movement and are commonly seen in the areas of weak activity. If this is your problem, it may be a matter of flow distribution or overall volume. A turnover of total tank volume at 10 to 20 times per hour is recommended. Increased water flow will also keep the periodically liberated worms (swimming or dislodged) in suspension for mechanical filters and skimmers to export them. Aquarists have often reported seeing deeply colored skimmate and filter media from aggressive skimming and filtration. Indeed, these improvements will help overall system health if nothing else, so be sure to confirm or improve performance in these areas as a first address of the problem.

Low salinity and freshwater: "Been there, tried that, bought the T-shirt," pretty much sums up this issue. It's true that low salinity impedes the growth of Acoel flatworms when done as freshwater dips or low salinity baths of longer durations. Unfortunately, hyposalinity techniques kill far more desirable life forms without any guarantee of curing the flatworm problem! Be sure, at least, to never try the low salinity gamut in the main system. This will have serious long-term consequence to system health in part by stressing or killing organisms in remote locations (deep rock or sand). Any experimentation with salinity manipulations is best done outside the aquarium. We recommend none as a control for flatworms.

Chemical Treatments: "Don't try this at home." Some folks have cited the use of commercial and DIY remedies for "selectively" poisoning flatworms in hobby aquaria. With any kind of reasonable consideration, we must concede that this is impractical if not wholly inappropriate for any

Pseudobiceros bedfordi- this species is amongst the most commonly encountered in the Indo-Pacific. It is a characteristically predatory flatworm, feeding on tunicates and micro-crustaceans. Photographed here in Indonesia.

Pseudoceros bifurcus is distinguished by its distinctive white mid-line stripe that finishes orange at the head of the worm. This species has a wide range that includes East Africa to the Philippines. Pictured here off of Heron Island feeding on common prey, the ascidians (sea squirts).

Pseudoceros sapphirinus is but one of many flatworms that looks remarkably like a Nudibranch sea slug with its prominent head structures. By mimicry, toxicity or both, such Polyclads can feed by day and night on a reef.

Thysanozoon flavomaculatum a strikingly bespeckled flatworm pictured here in the Red Sea.

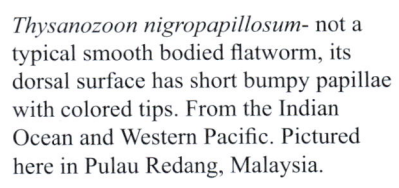

Thysanozoon nigropapillosum- not a typical smooth bodied flatworm, its dorsal surface has short bumpy papillae with colored tips. From the Indian Ocean and Western Pacific. Pictured here in Pulau Redang, Malaysia.

Worms

Potential Piscid Flatworm Predators

Synchiropus picturatus (Peters 1877), the Green, Picturesque, Target, and Psychedelic Mandarin Dragonet. To seven cm. Indo-West Pacific; Philippines, Indonesia, northwestern Australia. Aquarium photo.

Macropharyngodon meleagris (Valenciennes 1839), the most common species offered in this genus to the aquarium trade, is the Leopard or Guinea Fowl Wrasse. Indo-Pacific; Cocos Keeling to the Western Pacific. To six inches (15 cm) in length, but rare.

Anampses twistii, the Yellow-Breasted or Twist's Wrasse. From the Pacific and Indian Oceans, but best out of the Red Sea. The entire genus is sensitive and tends to do poorly in the aquarium.

Halichoeres chrysus (Randall 1981) the Golden or Canary Wrasse. A bright bold sun-yellow color and most often listed as the Yellow Coris Wrasse, though it is not a *Coris* genus member. This is an exemplary aquarium species that is suitable for peaceful fish-only and reef systems. To a mere 4 inches or so total length. (10 cm)

Valenciennea strigata (Broussonet 1782), the Blueband or Sleeper Goby. Indo-Pacific; East Africa to the Tuamotus. To a little over seven inches (17 cm+) in length. Second most popular species of the genus in the aquarium interest.

system with living rock, sand and invertebrates. There is no such wonder drug that will kill a specific variety of nuisance flatworm while leaving so many other desirable creatures of similar or sensitive anatomy unharmed (coral tissue, polychaete worms, snails, clams, etc). Take our advice that no such drug is possibly that discriminating. If someone does in time tell you that a magic drug works on just that one species of pest, ask how? It's a simple and fair question. In the meantime, take our advice and pass on the anti-biotics (anti-life) and snake oil remedies.

Manual siphoning is moderately effective but quite laborious. The truth of the matter is that such flatworms are common and present in many tanks. They will wax and wane and rarely linger in a large sustainable population. In most tanks, an infestation reaches a critical mass within months and simply crashes. Have a good skimmer working and saltwater on hand for exchanges when it does. Good husbandry is really the best long-term solution. We haven't found a "perfect" tank yet where some aspect of hardware or husbandry didn't need improved or modified. Even if such a tank were infected, the worms truly cannot last forever. Occasional siphoning will support the effort in wait for the natural population to wane. If you lack patience, simply blast the worms weekly or more often with a strong stream of water (as from a water pump or power head) and keep the worms in suspension to be exported by the skimmer. In such cases, your skimmate will look like awful red wine many hours later! Have faith; these worms really are a small concern.

Natural predation may help control flatworms, but is generally unpredictable in effectiveness. Success is often a matter of random fortune to find individual specimens of known species to eat flatworms and are not indicated by any given

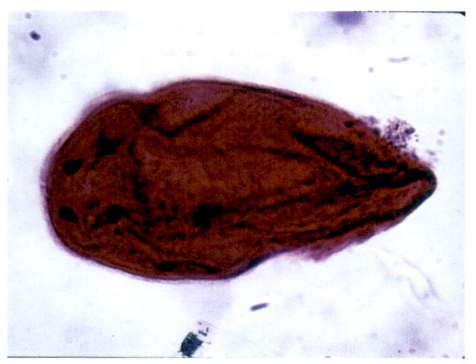

Many aquarists do not even realize that the common "Black-spot Disease" so often seen on yellow tangs (*Zebrasoma flavescens*) is a Turbellarid flatworm (a stained microscopic image of *Paravortex* pictured at left). It's strange any aquarist should find this parasitic worm persistent or difficult to eradicate. It is easily controlled and removed by routine freshwater dips. Any infected fishes held in a proper bare-bottomed quarantine tank and given daily dips for one week can be assured of a cure. Cleaner shrimp (*Lysmata*) and cleaner gobies (*Gobiosoma* and *Elacatinus*) will also help to control this parasite. (right: *J. Cross*)

species. Aquarists abroad have reported situations where a specimen completely controlled a pest population, other circumstances where they merely helped, and more often situations where the same "predator" ate little or none and became just another burden on the system.

Potential fish predators of pest flatworms:

- Dragonets (*Synchiropus*)
- Leopard wrasses (*Macropharyngodon*)
- *Anampses* (Tamarin Wrasses)
- Various *Halichoeres* Wrasse (Including *H. chrysus*)
- Various Sleeper Gobies (*Valenciennea*)

We do not recommend any of these creatures as the *primary* means of controlling pest flatworms. If you have the space and means to otherwise support such fishes, you may consider them.

Summary

Patience is a Virtue with Acoel Flatworm Infestations

"This too shall pass," might seem like a simple expression of hope and conciliation in dealing pest flatworm infestations, but in many cases (perhaps all), it's the best approach.

There really are no guaranteed flatworm predators that are reliable, hardy and commonly available to us yet. Most curious potential predators find flatworms quite distasteful. Yet, we are sure that such populations will wane naturally in weeks to months in any case. You need not wonder where they came from either. Most aquariums have flatworm species naturally and many more invertebrate species that go unnoticed. For reasons unknown to us, they simply flourish at times in some systems. Fooling with these life forms, by environmental manipulation, chemical treatment, even physical removal can prove to be an expensive and disastrous experience. Scientific literature is replete with the nasty cocktail of toxins these deceptively "easy meals" contain. Catastrophic results have been reported from massive kills or die-offs by the sudden release

Flatworm plagues in aquariums are not uncommon, though more of a nuisance rather than dangerous. Many are not feeding directly on the corals, but rather metabolites, mucus, by-products, or simply looking for a place to perch as they are mixotrophic (symbiotic with zooxanthellae they possess). Unfortunately, they deprive corals of light and water flow as they settle in plague proportions. (*L. Gonzalez*)

Worms 181

Waminoa is but one of many nuisance flatworm genera. Individually, it is distinguished by its pumpkin-like profile. The color is said to be imparted by the presence of its algal symbionts (dinoflagellates, golden brown diatoms). Infestations do not only occur in the aquarium as revealed here in the Lembeh Strait (Indonesia). Pictured here on Bubble Coral (*Plerogyra*) at **left** - Flower Pot Coral (*Goniopora*) in the **center**, and on a Plate Anemone Coral (*Heliofungia actiniformis*) at the **right**.

of toxins into the system. Proper quarantine protocol will reduce the numbers of such pests that enter your main display, and as such will reduce the chance of proliferation too. Your best cure besides patience for the spike to subside it to focus on aspects of good husbandry like the issues of water changes, manual siphoning, improved water flow and aggressive protein skimming.

Nemerteans, Sipunculids - PEANUT WORMS!

Sipunculids are wonderful little worms that cannot be bought or collected easily, mostly for their burrowing nature. Yet, they are commonly imported and cause quite a commotion on discovery (for their unique shape) if stressed out of their lair in sand or rock, as it commonly occurs with broken rock during a move or import of fresh substrate. Occasionally you can spy them in the aquarium pushing their tentacles slightly out of a hole in the rock. To see the body of the creature for its asymmetrical form leads to many common names and comparisons. Someone obviously thought it looked like a peanut. Many others have said part of it looks like an elephant's trunk. Most everybody thinks the body of the worm looks firm and a bit rubbery. There are perhaps several hundred species of Sipunculids, but aquarists commonly see the same few varieties carried in with Indonesian rock and coral, and live rock from Fiji. Rock-dwelling species have the capability to secrete chemicals to dissolve limestone and mine their residences in time. Whether they are found in rock or sand, they are particularly sessile and depend on food to be carried to them. In the dynamic of an aquarium, this does not bode well for their wild prosperity and proliferation. Most peanut worms that are not fed or kept in mature refugia will suffer attrition and die within months. They are unsegmented species, which typically only grow to about 2" (5 cm) and can move remarkably fast and retract their "tube-trunk" (the **introvert** which is crowned with feeding tentacles) into their body. The group includes both suspension feeders and "detritivores" (deposit feeders). The main part of the body that the introvert is attached to is larger, irregular and somewhat bulbous. Peanut worms as a rule are dedicated nocturnal species and shy away from light (they are photoreceptive and negatively phototropic). It is rare to find a specimen that is not nestled in rock or sand if it is healthy. Although they may reproduce asexually (fission), shedding gametes and pelagic larvae are a more common strategy. For this reason, at present, propagation of Sipunculids in the aquarium is uncommon.

Nemertean Ribbon Worms like this *Baseodiscus hemprichii* are rarely seen in the trade. Little is known about their captive requirements and what is known makes them wholly unsuitable for the average home aquarium. Among their many potential challenges is the adult size of specimens like this, pictured off of Heron Island Australia, which can reach 3 feet in length (90 cm). Unbelievable field reports of long slender specimens, particularly from temperate seas, cite them at 50 and 100 feet in length (15-30meters)!!! Most are believed to be predatory carnivores on other worms and crustaceans.

A series of photos showing a Sipunculid burrowing its way into the substrate. These unusual animals require a mature substrate to prosper. Commonly imported with Indonesian rock and coral as well as live rock from Fiji. (*R. Hilgers*)

This Terebellid "spaghetti worm" spreads across the sand of an aquarium. (*L. Gonzalez*)

The feeding tentacles (feathers) of this worm are complex structures that have evolved to act as filter-feeding aspects. This crown is an assemblage of branches that each have side branches (pinnules) in turn. This collection altogether resembles a feather but performs more like a highly intricate "sieve." (*L. Gonzalez*)

184   *Natural Marine Aquarium Volume I - Reef Invertebrates*

MOLLUSKS

There are seven extant (living) classes of Mollusks:

- Aplacophora (deep ocean... many eat cnidarians...that's about *all* we need to know as reef aquarists)
- Bivalvia (clams, oysters, scallops, mussels)
- Cephalopoda (octopus, squid, cuttlefish and nautilus)
- Gastropoda (marine snails, slugs, and hares... very popular)
- Monoplacophera (deep ocean... not seen or kept by aquarists)
- Polyplacophora (chitons)
- Scaphopoda (fascinating but obscure to aquarists, includes the mucus-feeding detritivorous Tusk Shells)

Aquarists are familiar with four classes in the phylum Mollusca: bivalves, cephalopods, gastropods and chitons. Without a good understanding of the taxonomy, it seems remarkable to think that a marine snail is more closely related to an octopus than it is to a barnacle (crustacean)! In time, mollusks have managed to occupy many different terrestrial and aquatic habitats with remarkably few changes from their progenitors (like the chambered nautilus- relatively unchanged for perhaps 400 million years). There are common traits characteristic to most. They are bilaterally symmetrical, and markedly *cephalized*- having a distinct and well-developed "head" from which a muscular "foot" is attached. In some fashion, each has a mantle which sheathes their delicate organs and also serves to secrete a shell (most species). They are dioecious, although the modes of reproduction employed are quite variable. Most animals in this group possess a rasping organ called the radula to assist with feeding. In some predatory species it it used to spear or stab prey, in others it is used to simply rasp or convey food into the mouth. In some sub-groups, the shell and radula may be secondarily modified, sometimes radically so as with Sea Slugs and the venomous Cone Shells, respectively. Mollusks have hearts, circulatory systems, "brains" (ganglia) and well-developed gills in most cases too. All told, they are highly evolved and include some of the most intelligent animals on the planet (the cephalopods).

Historically they have been crucial to human populations as a resource used for food, ornamentation, and even currency (shells)! Modern man studies this unique group for numerous known and potential compounds to utilize in medicine. Their roles in the living reef dynamic are crucial particularly with the algae-grazers for nutrient cycling. They also comprise a significant amount of bio-mass as food for animals higher up the food chain. In the modern reef aquarium, you may not readily notice their presence, but rest assured that you would notice their absence. They address one of the most fundamental problems in the aquarium: nuisance algae control. Some are acquired incidentally with living substrates (rock and sand), plants, corals and decorative algae. Others are sought deliberately for specific proclivities to control targeted nuisance organisms. Many are predatory on other invertebrates like cnidarians, echinoderms, other mollusks, and even fishes. A few can be fatally dangerous to man like the infamous textile cone snail.

Species targeted for collection and import are enjoying much better survivability, beyond improved shipping techniques, for the evolution of the natural marine aquarium with abundant live rock, plants, algae and refugiums. Nonetheless, there are still many that are wholly inappropriate for aquarium life, or at least casual aquarists, like the shell-less snails (nudibranchs, sea slugs & hares) and most cephalopods. At length, though, mollusks are a very diverse group... ranging from millimeter sized gastropods to some of the largest animals on the planet (Giant squid). Research your candidates for purchase thoroughly and identify hitchhikers quickly, as many in this phylum are parasitic, predatory or otherwise dangerous. For the sheer size of this group, we could not possibly get into comprehensive specifics for all as the nature of our content would need to shift from hobby reference to an encyclopedia-sized dissertation. The number of species in this phylum, in fact, is second only to the Arthropods with estimates near 100,000. Of a necessity, we can only cover aspects and species most frequently encountered (useful and otherwise) with perhaps a pleasant measure of photogenic eye-candy (galleries of images) thrown in for inspiration and mutual admiration. Aquarists are categorically empathetic souls, and no one wants to see the natural resources we admire negatively impacted by our participation. Take heed of our advice here, conduct research beyond this text, and use good common sense to be a truly conscientious aquarist.

Gastropods and Polyplacophorids
Snails and Chitons - the shelled grazers

Trochus is a fine grazer that eats a wide range of nuisance algae species. It is a long-lived and sometimes large species that moves fast and feeds eagerly. They culture readily in aquariums, particularly where the young are free of predators. Here, images showing one as it struggles to use an empty snail shell to right itself after falling upside down. (*L. Gonzalez*)

Overview

This chapter addresses the groups of shelled mollusks that we commonly call "snails." Taxonomically and in popular literature alike, a similar discussion might be limited to the class Gastropoda only. However, we have elected to add Chitons (class Polyplacophora) for convenience and the practical reality that most aquarists do not recognize or care to distinguish between the shelled mollusks. Chitons comprise a small passage here and share many similarities in husbandry. You may assume their inclusion in any generalized statement about Gastropods, herein.

The group of mollusks that we call snails is rather enormous, with over 50,000 living (extant) species. The stomach-footed mollusks, Gastropods, are second only to the insect class in diversity, and they comprise nearly 2/3 of all mollusks. They can be found on land and in both fresh and saline waters. In the sea, there are more species of Gastropods than there are corals, fishes and crustaceans combined! There are four living subclasses of Gastropoda: 1) Heterobranchia (parasitic species uncommon to us), 2) Pulmonata (freshwater and terrestrial species), 3) Opisthobranchia (shell-less sea slugs: the nudibranchs and sea hares), and 4) Prosobranchia (the shelled marine snails). Obviously, we are interested as marine aquarists in just the last two groups mentioned. Yet, comparing the two, their participation and viability in the hobby could not be any more different. The sea slugs (Opisthobranchs) are categorically difficult to ship, keep and feed. The Prosobranchs, however, include numerous species that are hardy, useful and popular in aquariums. With the shelled Gastropods we can discuss them in greater depth and with success as celebrated aquarium specimens. Surprisingly, for all their diversity in numbers, they have many commonalities that allow us as aquarists to make some practical comments and standardized remarks about their husbandry and behavior.

Marine snails can be found over a wide range of habitats from tropical to temperate waters and at all depths. Most are benthic, but some live free-swimming (pelagic) lives. Some of the Prosobranchs familiar to aquarists include: "Turbo snails" collectively (*Turbo*, *Astraea* and *Trochus* species), Limpets, Nerites, Ceriths, Abalone, Vermetids (worm snails), Conch, Whelks, Tulips, Cowries, Helmet, Murex, Egg, Mud and Paper Shell snails (*Stomatella*). Aquarists primarily use snails for utilitarian purposes as "clean-up" scavengers of detritus and algae. Alas, such blind faith that snails will find their

> **A summary of the key points for selecting marine snails**
>
> Good specimens:
>
> - have been held at the latest stage of import (your dealer/source) for nearly a week or more
> - have firm grip of the substrate
> - are responsive to touch (operculum closes quickly or species without one clench the substrate fast and firm)
> - do not appear to have any injuries to the foot, nor is the foot limp

own food in the aquarium can lead to attrition and death for many. For all of their hardiness, snails cannot live on water alone and will require due diligence on your part to ensure or provide adequate food and bio-minerals to thrive.

Selection

The proper selection of snails for the aquarium has as much to do with foresight and preparation as it does in finding a good specimen. Like so many invertebrates, snails are very sensitive to sudden changes in water quality (salinity, oxygen levels, temperature, etc.). As such, it is very important to know your supplier and the history of the snails that you are receiving. Most importantly, you will want to know how long the last person (your dealer) has had them in stock; hopefully, they will have been held for a week or more. It is common for marine livestock to change hands several times while navigating the chain of custody on import. From the collector, they go to a holding and acclimation station for "hardening" (purge waste and to be exposed to handling and water changes). From there they are likely conducted on the orders of transhippers to go to wholesalers or retailers. It is not uncommon for these merchants then to trade or sell to other wholesalers or resellers and thus another link in the chain is added. Under the best of circumstances, most marine livestock receives several different changes of water over a period of several days to more than a week before making it to your local pet shop or home. Much of this time is spent in transit, in the dark, and sometimes even out of water (dry shipping, in actuality damp shipping, is helpful for many species, but leaves snails vulnerable to greater temperature changes).

Most of the living mass of a snail is deliberately obscured by its shell. This can make an assessment of their health somewhat of a challenge. Nonetheless, snails that demonstrate a firm grip upon the substrate are likely a good start for candidates. In fact, if you know that the snails in your dealer's tank have been acclimated and held for even a day or more, they should have a firm grip upon the rocks, aquarium walls or sand. Untethered snails are vulnerable, and such a sight is rare with healthy specimens. Some species can pull their foot and vulnerable body parts completely within their shell. These species have an **operculum** that retracts with the foot to seal their shell and protect their living mass when disturbed. They may retreat from predation, poor water quality, or low tide (and air-exposure in shipping) in this manner for example.

Gastropods that lack a sealable shell with an operculum, like limpets and abalone (plus the Polyplacophorid chitons), will instead hunker down tightly to the substrate to wait out whatever threat it perceives. These species generally have a much stronger grasp of the substrate and occur in areas with more dynamic water flow, which brings us to the next point of concern when selecting Gastropods: foot health. The large meaty foot of a Gastropod is its lifeline to the world. Obviously it is used for locomotion - creeping over rocks, tunneling through the sand, and even swimming by undulations in some cases. Even a slight compromise to it impedes its ability to feed itself and defend against predation. Great care in handling must be exercised when pulling a snail off the substrate for fear of straining or tearing this muscle. To harm it would more than likely be fatal for the snail. Take note of a limp or weakly responsive foot on newly imported snails that would suggest mishandling in collection or shipping.

Care

Acclimation

Beyond addressing the key requirements for the selection and handling of snails mentioned above, the single biggest obstacle to their health is **osmotic shock**. For the entire captive life of Gastropods in your care, you must be mindful of their great sensitivity to changes in water quality. Even slight changes in composition can prove to be fatal as many aquarists can attest after doing a seemingly normal water exchanges that killed snails.

> **Common Gastropod sensitivities**
>
> - salinity changes (like excess freshwater evaporation top-offs, indifferent water exchanges or lack thereof)
> - aquarium supplements, additives and medications (great sensitivity to metals and organic dyes)
> - temperature shock and oxygen deprivation (improper water changes with low-oxygen warmer water, for example, and the shipping induced expressions of these problems)

Cyphoma mcgintyi Aquarium image.

Not all snails are harmless and innocuous... some are innocuous and predatory! Pictured here is the Sundial or Box snail (the Architectonicid - *Heliacus* sp.). They are decidedly predatory on zoanthids and are usually only discovered after they've eaten most of the colony that they were using for food and cover. (*A. Calfo*)

Despite the fact that many snails come from rugged marine environments (surge zones, tidal pools, and so on), such tolerance does not readily translate to aquaristics. Their squishy invertebrate tissues are thinly layered and, while sinewy and strong, are poorly adapted to suffer changes in the chemical composition of their habitat. They suffer osmotic shock and overdose easily from unnatural concentrations of various common aquarium elements. Abidingly, changes in salinity must occur very slowly over a period of many hours. We recommend **drip acclimation** of new specimens over a few to several hours if you can maintain a stable temperature in the shipping vessel (heated bath or float acclimation). In kind, changes in salinity must be avoided during water changes (adjust salinity to match display water carefully) and evaporation top-off (do not let aquarium stray far or for long, and compensate slowly if it ever does, altering specific gravity no more than .002 per day).

Know that most additives, supplements and medications are extremely toxic to Gastropods (some antibiotics excluded). Of particular danger are metal supplements (any copper, and even excess magnesium dosing, for example) and organic dyes (medications with malachite green, methylene blue, etc.). Exposure to such toxins may not cause immediate death, but rather saturates tissues and thus will overdose the animal after some days or even a couple weeks. Last of all, you will find that snails are at least as sensitive to changes in temperature and dissolved oxygen levels as much as any other invertebrate, and more than some fishes.

Difficulties with these parameters are typically encountered during shipping and hardware failures or power outages. In such cases, contingency planning will make the difference in their survival as with heat or ice in packs for shipping, and ready-made seawater and battery back-ups for power/pump failures.

After all of this, many aquarists might wonder why these common and plentiful creatures should be so sensitive or why we should even care. When regarded as mere clean-up organisms, it's easy to place little value on their lives. You might even think that they are naturally short-lived for their abundance. Nothing could be further from the truth. Gastropods include some of the longest-lived invertebrates known to man. Many can measure their lifespan in decades - they have reliably read growth rings like a tree. Some specimens have been measured at well over 100 years old. Even if they were naturally short-lived, as empathetic aquarists we place no value or limitation on their due husbandry as living creatures.

The Lamellariid *Coriocella hibyae* is one of the most unusual fleshy Prosobranch gastropods with an internal shell that leads you to want to guess it is a nudibranch (shell-less snail). A striking black specimen is pictured here in the Maldives.

Natural Marine Aquarium Volume I - Reef Invertebrates

Feeding

With estimates of more than 50,000 species in this class, you can imagine that there are more than a few feeding habits and preferences. Still, popular species in the aquarium trade have been sorted and can be fairly categorized to some extent. Most favorable species are overwhelmingly motile herbivores and prefer microalgae. A few target macroalgae or even plants. The dangerous or generally undesirable species are predatory carnivores. And we even see some unique sessile specimens that eat phytoplankton or trap micro-organisms via mucus-feeding strategies (Vermetid snails). The nuisance algae grazers are understandably the most popular with aquarists. You must realize though that each snail is highly specialized to feed on very specific types or even single species of algae. They each have highly specialized radular teeth designed for a purpose and feeding preference. Thus, physiologically, it is illogical and unreasonable to expect one species to be able to eat brown diatom slime, sinewy Bryopsis and overgrown *Caulerpa* with equal enthusiasm or ability. As with all new aquarium specimens, do enough research in advance of a purchase and confirm their feeding preferences in a proper quarantine tank before letting them get lost in a large display.

Reproduction

In contrast to the decidedly hermaphroditic shell-less Gastropods (the Opisthobranchs), Prosobranchian Gastropods are generally dioecious. Success with the captive reproduction of marine snails runs the gamut from likely impossible for years to come, to prolific without even trying. The easiest species to spawn and rear have no pelagic larval stage but brood or demonstrate direct development instead, like *Strombus*. The more challenging to rear snails disperse their gametes into the water column (like abalone) for a complex and extended larval period. Species that cannot presently be reared in captivity, however, benefit the system significantly by shedding gametes, which serve as very nutritious and unique plankton in the aquarium for filter-feeders like the now-common *Turbo* spawning events. There is a bit more information on Gastropod spawning in the dedicated chapter on reproduction. Detailed data is available for those interested by following some of the links in the bibliography. If you ever have spawning Gastropods in your system, please document and video or photograph the event and share your information; mention it at an aquarium society meeting, on the Internet message boards or write an article about it for the print hobby literature. Detailed information on spawning and rearing popular marine snails needs to be better established.

Compatibility

Most of the desirable marine snails in the trade have few, if any, compatibility issues beyond getting along with anything that won't eat them. Most are indifferent to competitors and are only limited by available food (grown or provided) in the aquarium. Nonetheless, there are some snails that have unfavorable feeding habits. Some groups and families have both good and bad members for the aquarium and are described in the gallery below.

Newly acclimated snails, if healthy, will quickly begin exploring their new environment. (*L. Gonzalez*)

Hobbyists have also become familiar with some uniquely predaceous species like the Box or Sundial snails (*Heliacus*) that feed on zoanthids, Pyramidellid snails that prey on Tridacnid clams, *Thycra cristallina* that preys on Linckia starfish, and many others targeting specific urchins, sea stars, corals and other invertebrates.

A rather large (1/2"/1.25 cm) stomatellid snail, with clearly visible shell. Many of the stomatellids you might see in aquariums have a completely hidden shell embedded within the mantle. (*H. Schultz*)

Mollusks: Snails & Chitons

Trochoids - Turban, *Turbo*, *Trochus*, Top Shell, *Astraea* and Star snails

This group includes some of the most popular, hardy and useful species of snails in the marine aquarium hobby. Most all are herbivorous and peaceful with each other and other desirable invertebrates. Some species seem more inclined to eat specific types of algae. *Turbo* snails (Turban) are better adapted to eat tougher or longer hair type nuisance algae. They have also been known to chow through some macroalgae as well. *Astraea* species (*A. tectum/tecta*, common turbo, *A. phoebia* - Star snail, and *A. americanum* - American turbo) favor finer films of green and brown (diatoms) microalgae. *Margarites* and *Tegula* species seen in the trade are temperate specimens and do not belong in the tropical marine aquarium.

Turbo fluctuosus is a very popular snail in the American hobby. Commonly collected in the Sea of Cortez, this snail can control tougher and longer filamentous algae in the aquarium than most any other snail. They readily consume other micro- and macroalgae and can be long-lived, but need large quantities of food to survive. Without target feeding (dried seaweed, for example), it is unlikely that you will want or need more than one snail per 25 US gallons in the aquarium. There are several other *Turbo* species seen in the trade, including the Chestnut Turbo, *T. castenea* (top left) from Florida. (All images *L. Gonzalez*)

Natural Marine Aquarium Volume I - Reef Invertebrates

Several *Trochus* species are available to hobbyists from both wild and aquaculture stocks. They are very fine grazers that eat a wide range of nuisance algae species. Aquarium images. (*L. Gonzalez*)

Astraea phoebia is a handsome star shell snail that can often be found in shipments of common *Astraea tecta* "turbo" snails. (*L. Gonzalez*)

Mollusks: Snails & Chitons

More than a few genera of useful Trochoid snails, AKA Turbos, are available to aquarium hobbyists. They are some of the very best herbivores for controlling various microalgae films and filaments. They are peaceful, hardy, long-lived, and commonly shed gametes in the aquarium. Some have suggested stocking at a rate of one snail per 10 US gallons, but ultimately it is your growth of suitable algae (or provision otherwise) in the aquarium that will dictate stocking densities. Pictured at left, one of the most popular aquarium snails, *Astraea*. (L. Gonzalez)

Although the "reef-safe" hermits do not actively seek to kill or consume live snails, they will do so if empty shells and adequate food are not available. (L. Gonzalez)

192 *Natural Marine Aquarium Volume I - Reef Invertebrates*

Neritids - Nerites: *Nerita, Neritina,* and *Puperita* species

Nerites have recently begun to garner recognition for their peaceful utility in refugiums and tidal displays. At least several species are commonly encountered in the trade, but aquarists must be careful to avoid inappropriate specimens. Some are strictly intertidal (they need and desire air-exposure) and will literally crawl out of the aquarium and die waiting for a tide that never comes back in. Nerites are mostly herbivores and can often be found among seagrasses and in the shallows on many shores. As such, they make fine refugium dwellers where they can tend nuisance algae growths on desirable plants and macroalgae while liberating plankton and epiphytic matter for filter-feeders by their rasping activities. They are distinguished by their handsomely rounded shells, which are sometimes striated or checkered. Imports from the Atlantic (Florida and the Caribbean) are popular in the trade. Egg laying is common but a protocol for successful captive rearing has not yet been clearly established.

Nerites are mainly intertidal, and best utilized in tidal displays and refugia where they can escape full submersion, like upon prop roots of red mangrove seedlings (*Rhizophora mangle*). If forced to live submerged full-time, many will die within about a year. Pictured here, not surprisingly, at the water's edge and above in a wholesaler's aquarium.

Mollusks: Snails & Chitons   193

Ceriths

Ceriths are some of the best, yet most underrated, snails for the marine aquarium. They are simply outstanding algae grazers, quintessentially peaceful and compatible, long-lived and prolific. They will flourish in aquariums with deep sand beds and refugiums especially, where they admirably control diatom algae and aerate the substrate. Most are acquired incidentally with a purchase of live rock or sand, but some are deliberately collected and sold. Aquarists are encouraged to share these snails amongst friends, aquarium clubs and local merchants. In some aquariums, their reproductive success and eagerness to consumer algae can displace the need for most other wild collected gastropods. Finicky aquarists may not favor their prolific nature, but rest assured that they (like any organism) are limited by available food. Thus, if you have a plague of Ceriths in your aquarium, then you obviously have nutrient control issues that afford the algae growth that feeds and supports a large population. Rather conveniently from an aesthetic point of view, most Ceriths come out and are active at night. Tiny reef hermit crab species often use their shells and will kill a living snail for their residence.

Cerith Snails (*Cerithium* and like genera) are some of the best and safest snails for marine aquariums. They are superb grazers on detritus, brown diatom algae, and like-films but cannot control filamentous varieties. Ceriths are prolific and reef-safe by any definition and recommended for deep sand beds and mud systems (to stir and agitate the substrate). Deliberate collections of this snail are made in tropical Atlantic waters.

Strombids - Conchs (*Strombus*), Vase, Harp, and Spider Shells (*Lambis*)

The few Strombids that are common and appropriate for the aquarium trade have enjoyed a somewhat to very good reputation. Despite the fact that some species in their (super) family are larger or even highly predaceous carnivores, several have been enjoyed and even aquacultured as useful herbivorous scavengers. *Strombus alatus* (Fighting conch), *S. gigas* (Queen conch) and *S. maculatus* have all been demonstrated to have merit in the aquarium, although *S. maculatus* seems to be the only one without any significant limitations in large part because of its demure size. The Fighting conch, however, commonly grows to a clumsy 2-4 inches in length (5-10 cm) and has been cited as large as 5" (12.5 cm). The Queen conch grows even larger, to 6-9" commonly (15-22.5 cm), and is cited at a respectable 12" in shell length (30 cm). If we are to be responsible aquarists, we will plan for the full lifespan of these and any living creatures we place, deliberately or not, into our aquariums. If we are to be practical here, either species will be too large or too clumsy (knocking over and perhaps injuring sessile invertebrates) for most private aquariums. They also require deep sand and soft substrates to burrow into and more food to survive than you may be willing or able to provide as they grow (importing a significant measure of nutrients to the system). In fact, they live poorly on rock and actually need large sand beds to thrive, if not at least survive. Natural diets include algae and detritus, but these snails will usually scavenge for most anything; farmers have even developed a pelleted chow to feed them. Because of the popularity of a few useful and herbivorous species, collectors have taken to sending the trade more than a few unsuitable aquarium species. A mistaken identification can be devastating, as with the use of the commonly available "Crown conch" (*Melongena corona* - Mangrove conch), which can eat several hundred turbo snails in just a couple weeks! Indeed, be aware that various kin resemble *Strombus* but prey on desirable invertebrates like other snails, sea stars, urchins, and even fishes. For the availability of desirable and known-safe aquacultured varieties, there should be no trouble avoiding a wild-caught predator. Reproduction in aquaria is not uncommon or difficult with some Strombids (such as *S. maculatus*) that exhibit direct development of (non-pelagic) larvae. This group of snails can be hardy, long-lived, and useful in marine aquaria if given a suitable large, sandy habitat along with adequate supplies of food.

Lambis truncata, Spider Shells: occasionally becomes an inevitably collected species for its beauty and unique shell. They are best kept in large aquaria (over 200 US gallons) with deep, large sand beds. Excess rockwork will impede the necessary daily habits of this ornate snail. Red Sea image.

Strombus alatus, the Florida Fighting Conch: a popular species of medium-large size (5"/125 mm) that enjoys many of the same accolades and criticisms of the herbivorous sea urchins. They are clumsy but efficient grazers that require considerable space to be useful and effective. Fighting conch, like all *Strombus* in the hobby, need large, deep sand beds. Be clear on species identification with conchs as some predatory species can be mistaken for reef-safe candidates. Pictured here losing its fight to a Queen conch (*S. gigas*).

Strombus gigas, the Queen Conch: a large herbivorous species that can grow to 12" (30 cm) and naturally requires enormous aquariums. Deep sand beds are also a must, and as such make this species a temporary or unsuitable specimen for most private aquaria. If deemed appropriate for your system, please seek some of the aquacultured individuals as populations of wild specimens are generally threatened. The genus *Strombus* has more than fifty recognized species; aquarists can likely find a smaller and more suitable species for the home aquarium.

Immediate left and below: ***Strombus sp***. Aquarium images. (*L. Gonzalez*)

196   *Natural Marine Aquarium Volume I - Reef Invertebrates*

Vermetid Snails

Worm Snails are rarely recognized for what they are: Mollusks. It's little wonder, too, for they live sessile lives by building hard calcareous tubes and sit in one place for life as mucus-net and filter-feeders. They really look nothing like our traditional definition of a snail: motile, stomach-footed and shelled. But they are no less snails, and rather common in aquariums at that. Worm snails are regularly imported with live substrates and can proliferate in many aquaria if they are afforded adequate levels of high dissolved organics and particulate matter. Overall, they may be regarded as desirable for their harmless lifestyle and filter-feeding nature. Indeed, their calcareous tubes are sharp to the touch and a bit of a surprise to an aquarists picking up heavy rocks. Yet even when they reproduce abundantly, we cannot fault them, but only our own lapse in husbandry that allowed them to flourish and grow (nutrient control issues). In essence, their presence is not at all unfavorable, and in reasonable numbers can be regarded as quite healthy and natural. Like any organism though, a sudden increase in their number and presence is an indication of water quality and available food conducive to their increased presence. In this case, plague numbers may likely reveal a flaw in nutrient export mechanisms (poor skimming, inadequate water changes, overfeeding, and so on). Frankly, a few tens or even a couple hundred of them in a larger aquarium are a very fine addition to the biodiversity. They have few natural predators in the reef-safe aquarium and must be limited by availability of food. Vermetids are completely reef-safe and pose no threat to desirable marine invertebrate life. Worm Snails occur in many varieties in the aquarium from thin to large bore, small whorled tubes to magnificent spiral shells, and ranging in size from nearly microscopic to several inches in length. They are a harmless sessile group that often trap food in a mucus net. It's quite natural to mistake these organisms for a calcareous tube-building worm.

Tropical Vermetid snails can be spotted developing in the protected areas of your reef. Their calcareous shells are a bit sharp to the touch, but otherwise these organisms are enjoyable additions to the bio-diversity of your display. (*A. Calfo*)

Cowries

Cowry snails (mostly *Cypraea* in the trade) are a large group of snails with several hundred recognized species. They are gorgeous and unique creatures that are highly prized by aquarists and shell-collectors alike. They are quite variable in suitability for aquarium life, but mostly should be avoided. Some are carnivorous and even the safe herbivorous species tend to grow large and clumsy. Hobbyists with mixed garden reef aquariums will need to find small herbivorous specimens, keep large aquaria, or forego their inclusion altogether. In fact, some cowries feed on sponge, corals, ascidians, or colonial anemones and cannot realistically be kept at all. For many others, their diet is not even known at this time. Aquarists

Cypraea caputserpentis, the Snakehead Cowry: common in regions where tropical collectors frequent (Indo-Pacific and Hawai'i), this snail is safe and suitable for the reef aquarium if it is handled and shipped considerately on import. A small herbivorous species. Aquarium photo. (*L. Gonzalez*)

seeking reliable herbivores in this category should look first to species like: *Cypraea moneta* (the Money Cowry), *C. annulus* (the Gold-Ringed Cowry) and *C. caputserpentis* (the Snakehead Cowry). It should also be mentioned that cowries tend to be rather strict in regards to water quality requirements (including high saturated oxygen levels) and as such can be difficult to ship. Be sure to quarantine all new specimens in a heavily aerated isolation tank. Even the herbivorous species appear to be relatively indiscriminate on captive diets and will scavenge for dried seaweed (nori), thawed ocean meats and even pelleted foods. Feed this large snail well and appropriately in captivity.

Cypraea annulus, the Gold Ringed Cowry: an uncommon but excellent snail for marine invertebrate aquariums with an adequate source of algae to feed upon. This species is distinct and unmistakable with a bright yellow-gold ring atop a stark white shell. The fleshy mantle is variably earthen colored and mottled. A wide distribution in the Pacific from the Red Sea through the Indian Ocean all the way to Hawai'i's leeward islands and the Cooks.

Cypraea caputserpentis, Photographed off of Maui, Hawai'i.

Not pictured but worth mentioning, *Cypraea moneta*, the Money Cowry: the famous, and one of the few, cowry species used as currency by some native peoples. Indo-Pacific; Red Sea to Polynesia.

Cypraea miliaris, (**right**) the Millet Cowry: a cryptic and nocturnal species with a dark, fleshy mantle covered with numerous branchlike papillae. Its range includes the West Pacific, Malaysia, Japan, and Philippines. North Sulawesi image

Cypraea tigris, the Tiger Cowry (**left**): the most popularly exploited species by the shell and "curio" trade, this snail is the most recognized by aquarists in kind. Sometimes precarious to keep in captivity and generally is not recommended. It feeds without much discretion upon algae, ocean meats (dead matter), and various desirable reef invertebrates like corals and bivalves. They are too large for most home reef aquariums, however well-behaved they might be, at about 4" (10 cm). South Africa, Red Sea, Hawai'i, Society Islands. Pictured here off the Gilis, Lombok, Indonesia.

Ovulids - Egg Cowries, Shuttle Shells, Flamingo Tongues

Although Ovulids are closely related to the true cowries, they are wholly unsuitable for aquarium life and care due to their predatory and parasitic nature. They feed on cnidarians (corals, gorgonians, hydroids), which most hobbyists are unlikely to want to feed to a snail, even when adequate supplies can be cultured. In many cases, Ovulids feed selectively on a single type or species of animal. For reasons beyond logical explanation, the Flamingo Tongue cowry (*Cyphoma gibbosum*) is still collected on occasion for the aquarium trade; it needs to eat gorgonians to survive (one or two species only). A notorious white egg cowry *Ovula ovum*, appears incidentally on occasion and is a parasite on *Sarcophyton* Alcyoniids (Mushroom and Toadstool Leather corals). None of the Ovulids can be recommended for casual aquarium keeping.

Cyphoma gibbosum, the Flamingo Tongue (**upper right**): is an attractive but tragic aquarium specimen. Conscientious aquarists do not buy these snails because of their obligate gorgonian diet (*Gorgonia flabellum* and *G. ventalina*). They hail from the Tropical West Atlantic and are small species about one inch in length (25 mm). Bahamas image.

Cyphoma macgintyi, the Spotted Cyphoma (**middle right**): a striking species that also is an obligate feeder on Gorgonian tissue. Aquarium image.

Dentiovula dorsuosa: (**lower right**) this cryptically colored species has evolved to resemble its polyped cnidarian prey. The mantle has polyp-like bumps (papillae) on solid colored flesh, which blends into the soft coral upon which it feeds. Pictured here in North Sulawesi, they range the Western Pacific including Indonesia, Malaysia and Japan.

Phenacovolva rosea is a remarkably well-camouflaged parasite in like visage to the gorgonian prey upon which it lives, eats and lays its eggs. Its range includes East Africa to the West Pacific. Pictured in action here in North Sulawesi.

Epitonium billeeanum is an Ovulid in true form for this group having evolved to resemble its prey, the bright orange/yellow *Tubastrea* Sun corals (ahermatypic stonies). Distributed throughout the Tropical Indo-Pacific. Pictured here in Indonesia.

Cone Snails - *Conus, Turris, Terebris*, and more

Cone snails are infamous for their wickedly predatory nature. They can catch and kill fishes, and at least one species is fatal to humans (*Conus textile*). Not all species seek piscine prey; many simply feed on worms and others on mollusks. Still, all in this family are carnivores and serve little useful purpose in the modern marine aquarium that aims to preserve and enhance bio-diversity in living substrates. Cone snails seize their prey by spearing them with a modified radula through which venom (conotoxin) is injected.

Muricid snails are commonly acquired as incidentals with live rock. They are predatory and may grow to become a serious problem in a mixed invertebrate aquarium. Keep murex in fish only aquaria and offer them meaty foods. (A. Calfo)

Conus textile - The infamous Textile Cone snail by nature feeds on other mollusks, but can in fact kill a human with its potent conotoxin. The drowning deaths of some swimmers are theorized to be from contact with Textile species. Indo-Pacific; Red Sea, much of the rest of the tropical Indo-Pacific. Pictured here in the Red Sea.

Murex Snails

Muricids are a group sporting remarkably beautiful shells that sadly have little place in a mixed invertebrate marine aquarium. They are quite predatory like so many snails in this order. Although they often seek very specific prey in the wild like bivalves, barnacles and other snails, most can readily be acclimated to eat thawed ocean meats like one would offer to captive fishes (shrimp, krill, clam, and so on). There are members of this family that eat coral (like *Drupella cornus*) or clams such as the infamous Oyster Drills (*Urosalpinx*). Attempts at keeping most any snail in this family will be more successful in aquariums without sessile invertebrates.

Natural Marine Aquarium Volume I - Reef Invertebrates

Whelks, Tulips and Mud Snails

This group includes a range of species that run the gamut of aquarium suitability from outstanding to dangerous and everything in between. They also range wildly in adult sizes from a fraction of an inch to several inches in length, even approaching a foot (30 cm). One very exceptional species of Fasciolariid, the Florida Horse Conch (*Pleuroploca gigantea*), is recorded at 30" (75 cm), and is surely one of the largest snails in the world. Make no mistake that all snails in this group are essentially carnivorous - eating fare that includes worms, dead organic matter, other mollusks, and even fishes. Please research these and all aquarium candidates before you take them home to be certain that you understand their exact dietary needs and can provide them. Some familiar snails from this group include:

Family Fasciolariidae: Tulip Snails

These snails are common and hardy, but not trustworthy and should be regarded as unsuitable like whelks and other predatory snails in a mixed invertebrate aquarium. Among the many desirable creatures that will be eaten in your display, other snails are a favorite, and Tulips can eat small Turbos and like-species at an alarming rate.

Family Buccinidae: Whelks

Most of the snails in this family are so decidedly predatory that short of a species tank or large fish-only (FO) display, they have no practical application in the reef invertebrate aquarium. Ignorant or unscrupulous collectors have sent whelk species under the guise of "Conch" snail (*Strombus*) with devastating results on the other invertebrate life in the dealer's or customer's system. Whelks are some of the most predatory snails to be found. Very few suitable species are imported for aquarium use. We mention an attractive and fairly harmless relative here - the Bumble Bee snail.

Engina (*Pusiostoma*) *mendicaria*, the Bumble Bee Snail: this species is often cited as being beneficial for aquariums, but that is doubtful for live sand, refugium and live rock bio-diversity. We'll concede that these whelks are simply beautiful and fairly harmless. But to be clear, like the other members of this family, they are physically unable to eat algae as they lack the necessary feeding aspects to accomplish this. They are predatory snails that eat worms and perhaps other desirable fauna in the sand. For their very small adult size, however, they likely do little harm in a healthy system. All things considered, we recommend that you avoid keeping them in small aquaria and refugiums, but enjoy them in larger tanks or very deep sand beds instead. No doubt, their activity on and in the sand as agitators has some consolatory benefit.

Fasciolaria tulipa, the True Tulip: a large adult species at 6-10" (15-25 cm). These snails commonly are imported incidentally at small sizes with live rock, rubble and other substrates. Although hardy and easy to care for in captivity, they have no place in mixed invertebrate reefs and should be removed for their predatory nature. Cozumel image.

Family Nassariidae: Nassa or Mud Snails (*Nassarius*)

Nassarius species are useful, safe and remarkably uncharacteristic of this predatory snail family. They are superb detritivores for reef aquaria. *Nassarius* actively burrow and agitate live sand (fine grain is best) and feed voraciously on dead (meaty) organic matter. They seem to place no significant burden on desirable infauna by predation. (*A. Calfo*)

Limpets - *Diodora, Lucapina, Scutus* and more

Numerous species occur in the marine aquarium by way of collection and incidental acquisition with hard substrates. They are voracious algae grazers that are not always useful for the mixed invertebrate aquarium. Although most are herbivorous, some are predatory like the coral-grazing Caribbean Orange Limpet, (*Lucapina*) and the fleshy mantled Black Limpet (*Scutus*). Even among the algae grazing species, they can be devastating to desirable algae forms like corallines, as well as nuisance forms. Some also reproduce prolifically in the aquarium for better or for worse. While we would not dissuade you from keeping an active algae-grazing limpet, we do suggest that you restrict them to mature systems with reliable growth of algae to sustain them. There is some concern with limpets denuding young or weak batches of live rock of vital corallines, which could invite undesirable algae species to settle instead. Legend has it that their very dense radula and strong rasping action can abrade the interior surface of acrylic aquariums too. As some consolation, many limpet species will stay in one spot (like one rock) for most of their lives if sufficient algae can be grown there.

Acmaea sp. A dorsal view of a Limpet in the Caribbean. Perhaps you can see why they are sometimes called "Keyhole shells" or "China Man's Hats."

Some limpet species are intertidal and as such are unsuitable for aquarium life with regard for their need to be exposed to air and their inclination to crawl out of the aquarium. Some temperate species also enter the trade and should be avoided for tropical aquariums. Even the tropical, subtidal (appropriate) species can only be found in areas of dynamic water flow. All limpets have evolved to tolerate remarkable water flow and surge. While such wave action is not possible in the aquarium, their muscular foot is a weakness with ill-advised aquarists that try to pull these snails from the glass or rocks. Great care must be exercised when moving these snails with concern for tearing or injuring a limpet. It's best to move the snail while attached to a small rock if possible. Although a few species in large aquaria are likely to be innocuous, we suggest that you avoid the deliberate acquisition of most limpets when so many less troublesome gastropods are available. *Lottia gigantea*, the Owl Limpet (not shown) grows 3-4" (75-100 mm) is but one of the intertidal species occasionally seen in the trade. Be certain to avoid any intertidal or temperate species for use in a display of typically subtidal, tropical invertebrates.

Stomatella varia, the Paper Shell snail: is not deliberately collected for the aquarium trade but is exceedingly common as an incidental import from the Indo-Pacific. They are conspicuous and familiar to many aquarists for their unique abalone-like shape, very fast mobility (a speedy snail!) and their prolific nature in aquaria. A harmless, nocturnal herbivore to be shared among aquarists. (*A. Calfo*)

Scutus cf. *unguis*, the Black limpet (left and above): an incidental acquisition from common Indonesian imports of live base rock and live rock with coral. It's unfortunate that we cannot trust this snail, like most limpets, as they are inclined to graze live coral and other desirable invertebrates. They are otherwise excellent algae-grazers for non-reef aquaria (*sans* cnidarians) and breed prolifically in captivity. A handsome nocturnal species with a large, dark fleshy mantle that obscures most of its telltale limpet shaped shell. (*H. Schultz II*)

Chitons - Polyplacophorids

Chitons comprise a small group of snails that are dedicated herbivores in rocky and high flow environments. They are reliably peaceful and useful algae-grazers for reef and mixed invertebrate aquariums. Most are acquired incidentally with living substrates (live rock, coral base). They tend to be remarkably inactive, favoring very short travel (moving mere feet in one year) and hiding in the dark crevices of the seascape. Chitons can simply be allowed to stay, if not be encouraged to breed, for their innocuous nature and proclivity for eating unsightly brown diatom algae. They are almost wholly reef safe as a group.

Chitons can often be spied by day resting below the substrate and clinging to the glass. They are harmless if not useful algae grazers that reproduce readily in the aquarium. (*L. Gonzalez*)

Mollusks: Snails & Chitons

A brief mention about Non-Tropical (cool-water/temperate) snails

Unfortunately, we do occasionally see species of non-tropical animals in the tropical aquarium trade and it just serves to remind us to encourage you to research the needs of *all* of your charges before taking them into your care. This group is no different, and we have made mention earlier in this chapter about intertidal and temperate snails that you may find for sale. In many cases, the animals may be very hardy and well suited to aquarium life, but only within the safe parameters of natural habitat- namely, cool water. Most will suffer accelerated metabolisms and die prematurely at best. Since we are coral reef enthusiasts keeping tropical species only, these snails have no place in our aquariums. Below we list just a few familiar specimens, which is by no means even remotely inclusive of all likely or possible cool-water species in the trade.

Norrisia norrisi, Norris' Top Shell (Mexican Red-Footed snail): by nature these are peaceful herbivorous snails. However, they must be kept under 70F to fare well in captivity. This temperate species can be found in nature grazing on Giant Brown Kelp (*Macrocystis*) off of the California coast. Photographed here in a marine livestock wholesaler's tank.

Temperate snails (not pictured)

Cypraea spadicea, the Chestnut Cowry: collected in California. This snail is not suitable for the tropical reef aquarium by any means. Beyond its temperate needs, they prey on desirable invertebrates like Ascidians (tunicates and sea squirts), Porifera (sponges), anemones, and snail eggs.

Astraea (*Lithopoma*) *undosum*, the Wavy Top Turban Snail: a cool water Turbo snail ranging from the California coast down to Baja, Mexico. Some specimens may take weeks or even months to succumb to warmer temperatures, but be assured they have no place in the tropical reef aquarium.

Megathura crenulata, the Keyhole Limpet: a sub-tidal but temperate snail. They grow too large to be useful for most home aquariums at any rate (to 4"/10 cm).

Opposite above image, the endearing face of an inquisitive *Strombus* snail with conspicuous eyestalks. Opposite below, the mixed company of nicely encrusted *Trochus* and *Turbo* snails. (Upper: H. Schultz) (Lower: L. Gonzalez)

Mollusks: Snails & Chitons 205

Nudibranchs (Opisthobranchs)
Sea Slugs

Dendrodoris tuberculosa is one of the largest Dorid sea slugs found throughout the Indo-Pacific. The elaborate compound tubercles of this species almost permit it to be mistaken for an ornate anemone! They can grow to 6" (15 cm) in length.

Various common names are exchanged freely with the shell-less Opisthobranch "snails" that we most often call- Nudibranchs [*pronounced* (noo-dee-branks)]. Also known as sea slugs, they are mesmerizing "butterflies of the sea" by any name. This group of mollusks includes a greater number of colorful and exotic creatures than perhaps any other we know. As aquarium specimens they are categorically inappropriate for casual keeping. As aquarists, however, we cannot help but be drawn to their fantastic shapes, colors, patterns and behaviors. An investigation into their feeding habits and biology is even more fascinating. While we do not encourage the aquarium care of most Opisthobranchs, we offer this collection of images and a bit of guidance on their husbandry in shared admiration. A few species are actually very hardy and suitable for captivity (they feed, grow and reproduce). For the rest, we have hope that specialized aquarists will prudently experiment in endeavors to discover possible and practical regimes of husbandry for future aquarists to keep and reproduce these remarkable creatures.

As you wade into this chapter and gallery of images, please note that at times we may refer to this subclass (Opisthobranchia) of organisms collectively as "sea slugs" for sake of a convenient common name. There are very specific differences between the groups commonly imported.

Sea slugs are mollusks, which not surprisingly, are related to the shelled terrestrial snails that we see on land. They are found worldwide in both tropical and temperate seas. Most are rather small at 1-3" (25-75 mm), but some are microscopic and others approach 12" long (300 mm). As they are predominantly shell-less, daytime active organisms, it's little wonder that they have evolved some seriously potent and complex anti-predatory defenses. Toxic species are most always remarkably colorful and noticeable to advertise their noxious flavor (**aposematic coloration**). Less toxic species usually match the substrate that they live or feed upon (**cryptic coloration**). For all of the above-listed reasons and more, we regret to say that most sea slugs are either difficult or impossible to keep in small private aquariums. Their diets are too specialized, they are delicate to ship and sensitive to acclimation, and there is a very real danger with some that on their demise (even if kept healthy for their short natural life) the decomposing release of their toxic composition into the aquarium can disturb or kill other organisms. If you must keep them, we recommend quarantine and then limit their number and mass in the system (few slugs per volume of water) and please research their prospective prey *before* you buy them! Very large and mature aquaria (several hundred gallons in size and aged over 1 year) may improve your chances of having and culturing suitable prey incidentally. Ultimately, though, there are many reasons for leaving these beauties in the sea.

Selection

If we haven't frightened you away from keeping sea slugs yet... read on. You are either a committed aquarist, or should be (a many *entendre* statement!). You don't have to worry about company for your sea slug... they are not gregarious and your system will be safer from their toxicity with as few as possible. Having a single specimen or only two will also increase the likelihood for some species to find adequate food in the aquarium. As hermaphrodites, you don't need to worry about finding the right mate. Mixing species may or may not be permissible; some species are "cannibals," but most are so specific in diet as to be safely mixed inter-species. When you seek to buy a specimen, please do yourself a favor and honor the creature at the same time- know its needs (especially diet) before you buy it! The very best way to buy a sea slug is on or with the food it eats. That may be algae, sponge, anemone (hopefully not!), coral, worms, or much more.

For their characteristically toxic nature, you can expect a prospect to be active by day and crawling around your dealer's aquarium. If your dealer cannot identify or advise you on a specimen's care, they can at least tell you where it was collected from (geographic regions provided by their collectors) and should be willing to hold the specimen for you while you take some time to research it. Take notice in our bibliography (plus the Internet and from other printed sources), that there is a tremendous amount of documentation in galleries and identification references for these high-profile animals. You should be able to place your candidate within a genus or group to extrapolate possible husbandry required based on what is known about like species.

When selecting for a healthy specimen, there are two kinds of sea slugs to choose from: 1) viable, and 2) "soon-to-be toxic goo." The difference is fairly unmistakable: healthy sea slugs are active and responsive. Sick or dying specimens (unresponsive) will die and dissolve quickly for their lack of dense muscle or skeletal mass. They are comprised mostly of water, have naturally short lifespans and are less evolved to be durable organisms. Such animals don't need to be too tough or long-lived with spawns of eggs that number in the millions for some. This sort of M.A.S.H. mentality with Opisthobranch health and morbidity is rather akin to anemones; they heal or die quickly with a very narrow margin in between. Assuming the candidate that you seek appears healthy, you must establish that it has been soundly acclimated to the dealer's tank and has been undisturbed ideally for a week or more. Else, offer

> **What makes a sea slug?**
> Some commonalities in the group:
>
> - Most are **shell-less**, others have only reduced shells
> - All are **hermaphrodites**
> - Most have a **radula** (feeding aspect)
> - Most are **toxic** in some form or another (elaborate chemical defenses)
> - Most are bilaterally symmetrical (mantle cavity and viscera are on the right side of body)
> - Very **short natural lifespans** are the rule with most living months to not much more than a year
> - Most have internal gill structures that have regressed, leading to secondary external aspects (**pseudobranchia**)
> - Other features include a forward pair(s) of cephalic tentacles/chemo-sensory **rhinophores**
> - Their specialized diets are often limited to one group of organisms or even a single species

Gymnodoris ceylonica: a subtly colored beauty ranging from the Red Sea over to Japan and Australia. Color scheme and pattern is fairly consistent for the species. To two inches in length (5 cm). Pictured here in the Maldives.

a deposit to the dealer to hold the specimen for at least a week's time from their acquisition. Even then you will want to quarantine this animal like all livestock at home for a couple of weeks to one month. In isolation, you will be able to verify your beliefs or estimations on the animal's dietary needs. Be warned: if you cannot confirm that the animal is eating in QT, *do not* send the specimen into the display. A weak and starving sea slug poses more threat to other livestock than you have risk for losing its life. It's a sacrifice that you should not have to make at any rate with good species selection and proper QT protocol.

Many aquarists on seacoasts may be able to collect their own Opisthobranchs due to their cosmopolitan distribution. You can find sea slugs on both coasts of the Americas... throughout tropical waters... in cooler Japanese waters... and even in downright cold waters off the coast of Scotland! Sea slugs are everywhere... like Elvis. As with collecting any creature, investigate local regulations first. You might be surprised to find sea slugs in waters remarkably close to home. For many reasons that you can imagine, collecting your own is far better than purchasing air-shipped (stressed) animals. Above all, you have the opportunity to observe and verify the specific food(s) they eat. You can collect them together, and likely collect more food later if it cannot be cultured at home.

Care

Feeding

For how specific Opisthobranch diets generally are, and for how little is known about the necessary diet of so many, we have little to share beyond the obvious and what has been already stated: only buy sea slugs with their prey (or when prey is known), or don't buy them at all. For example, some Aeolid species feed only on *one* species of hydroid, and some Dorid species limit themselves to one food type, like sponge. If you cannot culture their prey, you likely should not keep the predator. In recent years, we have discovered species that eat common and plentiful nuisance organisms in the aquarium like pest anemone munchers (*Aiptasia*-eating *Berghia* nudibranchs), nuisance flatworm eaters (The Head Shield *Chelidonura* sp) and simple, hardy algae eaters (Lettuce slugs- *Elysia* sp).

The phylogeny of Opisthobranchs can grossly be categorized, in part, by their feeding habits:

- Sea hares (**Anaspidea**) are traditional algae grazers... rasping away at gross algal matter

- **Saccoglossa** are highly specialized algae-eaters that rasp algal cells to suck out the contents

This predatory sea slug, often cited as a *Tritoniopsis* sp., is a regular import with the soft corals upon which it feeds. Remove promptly if observed in the reef aquarium. (*L. Gonzalez*)

208   Natural Marine Aquarium Volume I - Reef Invertebrates

- **Cephalaspidea:** Head Shield Slugs mainly eat "worms" (polychaetes, flatworms, etc), foraminiferans, and other mollusks including sea slugs. Bubble Shells eat a significant amount of algae
- **Nudibranch** are all carnivores- consuming a wide range of animals (sponges, hydroids, anemones, coral, etc)
- **Doridacea** and **Notaspidea** mostly eat sponges, bryozoans and sea squirts (tunicates)... some coral too
- **Aeolidacea** and **Dendronotacea** principally eat Cnidarians (stinging organisms like corals and anemone)

If you are interested into delving further into the feeding habits of sea slugs, there is an amazing (and huge!) compilation of references on the "Worldwide Food Habits of Nudibranchs" by Gary R. McDonald and James W. Nybakken at their website... (see bibliography)

Defensive Strategies of Sea slugs

So what's the big deal with sea slug toxicity? Hmmm. let's see? If you had the size, strength and/or consistency of a marshmallow and you were creeping around a crowded reef of hungry animals in plain view, your evolutionary choices would be limited. You would either develop strategies of defense, or... you'd become a salty *bon-bon*. As aquarists that admire these beautiful creatures, we are very pleased to see that they have succeeded in the former strategy. It's like the joke about the amazing cliff-divers of Acapulco- there are no mediocre or bad cliff-divers. There are only two kinds: *Great* cliff-divers... and, "stuff on a rock." Failure is not an option. In the same spirit, Opisthobranchs have evolved some incredible strategies of chemical defense to overcome their vulnerability.

Having read this far into the text and perhaps from reading other hobby literature, you realize that many sea slugs have toxic flesh. To be specific, you likely understand that some slugs eat stinging animals like corals and then use their stinging components (cnidocysts) in their tissue to be likewise dangerous. But sea slug defenses go far beyond this fascinating adaptation. Among Opisthobranchs, some of the most formidable are the Sidegill sea slugs, which can secrete self-produced sulfuric acid! Yes... that's an effective anti-predatory strategy. Lets examine some of the many ways sea slugs have evolved to defend themselves:

- Toxic secretions like poisons or sulfuric acid
- Physical deterrents like spiny spicules embedded in the fleshy mantle
- Assimilation of a foreign defense like the ingestion of cnidocysts (cells which contain stinging nematocysts)
- Cryptic coloration (matching habitat or prey as with yellow slugs eating yellow sponge and laying yellow eggs)
- Aposematic (warning) coloration- a conspicuous display of toxic potency

Beyond any issues of stress from feeding, compatibility, acclimation or husbandry at large, we know that most every sea slug you are likely to keep in your aquarium will die within months naturally. They are generally collected as adults and few live more than a year. With their sudden death, the substance of their composition, however noxious or toxic that may be, will be released into the system. To reduce the potential for disaster, you will need very good chemical filtration (carbon changed weekly instead of monthly is a good habit in general), adequate water flow, and a skimmer or ozonation to help manipulate and/or export some of these elements. When such creatures die in the aquarium or are unaccounted for and presumed dead, be prepared for significant water changes to be safe as well. You may be wondering, just how serious is sea slug toxicity in the aquarium? Just the very touch, mucus or slime of some species is potent!

For perspective, let us recall a pet department in a Sears & Roebucks stores the early seventies, when they sold live pets. There was a problem tank in which various marine fishes would perish within days for no apparent reason (even by standards of the day). They tried all the usual remedies: massive water changes, draining and refilling the tank... even throwing away the gravel and decorations. In time, they realized that the Spanish Dancer (swimming *Hexabranchus* nudibranch) that they

Hexabranchus sanguineus, the Spanish Dancer: a large sea slug that approaches 40 cm in length (16") with a conspicuous presence on the reef. This animal earns its namesake for the graceful undulations of "flight" when inclined to swim through the water. They are not reef safe or hardy, but eat a number of desirable aquarium denizens like sponges, worms, echinoderms and even other mollusks! They are distinguished by laminar rhinophores, partially retractable secondary gills, and large flap-like oral tentacles. They are highly variable in color and pattern- smaller specimens of unique or uncommon color are often mistaken for *Chromodoris* sp. (a difference - branching gills here on the Spanish Dancer). Shown here, an adult in the Cooks Islands (left) and ribbon egg mass in Malaysia (right).

had kept in the tank had bumped against the walls depositing stinging cells all over. In effect, the imparted "nettles" were debilitating everything they came in contact with. The tank was acid and bleach washed and salt-scoured and the curious dilemma was solved. Such potency, of course, wanes in time with carnivorous sea slugs that are starving or cannot otherwise find cnidarians to feed on. On the contrary, freshly fed and imported slugs can be dangerous to even touch. Some of these snails even feed on the Portuguese-man-o-war and Jellyfish and there have been reports of divers and swimmers getting injured in kind for touching such nudibranchs. The commute of stinging cnidarian cells is remarkably effective in Opisthobranchs.

The process of assimilating "live" nematocysts is a natural wonder. Nudibranchs are able to consume the tissue of cnidarians without the stinging cells being tripped or fired. These nematocysts are then carried through the convoluted digestive tract on the back of the sea slug to reside in the tasseled cerata (usually concentrated in the tips). Cerata are essentially specialized extensions of the digestive tract. They have very thin tissue not only to impart these stinging cells in defense, but also to better derive oxygen from seawater over the extended surface area. The skin there is so thin that you can often see packets of zooxanthellae or the very solid matter of a recent meal inside. Cerata are one of the most visible and perhaps remarkable aspects of sea slug evolution.

Water Quality

For their extraordinary diversity and distribution in habitats of the world's seas, you will have to really do your homework to discern suitable water quality parameters for your specimen(s). Opisthobranchs can be found from the depths to the shallows and from some of the coldest seas to the equator. There are no standardized parameters for sea slugs. Most in the trade (except many sea hares) are tropical and are likely to favor traditional reef aquarium water quality. Pay special attention to maintaining high and stable Redox levels (350 mv-425 mv), high saturated oxygen levels, and very stable salinity (evaporation top-offs should always be gradual, for example). Medications in the form of metals and organics dyes are not safe to use. Some antibiotics may be safe, but all are still unlikely to be necessary for the volatility of the group regarding injury and disease. In ways natural, as well as in relation to aquaristics, sea slugs lead rather ethereal lives.

Reproduction

Opisthobranchs are true hermaphrodites- simultaneously both male and female, although self-fertilization is very rare. Spawning can occur between any two specimens and the commencement often stimulates many individuals to spawn *en masse* by chemical cue. Reproduction usually occurs in reciprocal pairs, but sea hares will reproduce as a group forming "chains" of adults, readily visible to people, in the shallows. Events of reproduction occur sea slug-slow and may take many hours to complete. The demersal egg cases are truly things of beauty and nearly so recognized and admired as the sea slugs themselves. Picturesque casings are laid in a spiral-ribbon like fashion, which can number from a few tens of eggs to nearly countless batches (a million or more have been cited in number). The eggs are often the same color as their typical prey (colored sponge, for example). You might also find their eggs in strands, strings or sheets. In time, sea slug eggs hatch as free-swimming larvae (**veliger**). Most larvae have a vestigial shell that will be lost during metamorphosis. The extended period of development and the various specific plankton needed at various stages or species of development make captive rearing for most very difficult. Nonetheless, progress can be made as we have seen in the cases of the commonly cultured *Berghia* "*Aiptasia*-eaters." We hope to have made it apparent that most endeavors with sea slug care and husbandry, including reproduction, is precarious at best and likely must be conducted in species tanks where their needs are the primary focus of the

Reproduction in sea slugs generally occurs in pairs of hermaphrodites that cross-fertilize. Self-fertilization is rare. This handsome, unidentified pair was photographed in the Red Sea.

Sea Slug Gallery

With compatibility issues and the essence of a summary contained in the brief offerings on husbandry above, we now proffer a bit of information and enticing imagery on some popular groups and orders.

Anaspidea

The Sea hares

Sea hares are unique among Opisthobranchs, most notably for having internal gills. It is indeed a stark contrast to the "naked gill" Nudibranchs. They are mostly herbivorous creatures eating all manners of plant and algae (albeit with some species-specific preferences). They are motile and can swim in the water column with sometimes graceful undulations of the body and mantle. Aquarists must take heed, though, that many sea hares found in the trade are temperate species. There is also cause for concern that they will pollute water quality if disturbed by releasing an anti-predatory "ink" in the water (**toxic** to fishes). Many have cryptic coloration to match their plant and algae "prey" and as such are not as highly regarded by aquarists and divers for their natural beauty. In truth, some sea hares are rather repulsive to look at for their appearance as a big slimy green slug. Despite their larger size, they are consistently short-lived hermaphrodites which spawn in unique groups in a chain. Some small specimens may be suitable for controlling nuisance microalgae.

Family Aplysiidae:

Aplysia californica, the California Sea Hare are some of the largest sea slugs to be found- growth to fifteen inches (375 mm) and several pounds in weight. They are temperate herbivorous species and not at all suitable for the tropical marine aquarium.

Aplysia dactylomela, the Spotted Sea Hare ranges wide from the shallows to below 100 feet (30 m +) and perhaps more suitable for the aquarium as a tropical West Atlantic species. Commonly collected at 3-6" (75-150 mm) and said to reach 12" (300 mm). Handle with caution and regard for their noxious "ink" secretions.

Saccoglossa

Photosynthetic herbivorous sea slugs from the genus *Elysia* have enjoyed great success and popularity in the aquarium trade. Many are easy to keep and reproduce, and they are quite utilitarian for their decidedly algae-grazing diets. They have been reported to eat a wide variety of desirable and undesirable algae in the aquarium including *Caulerpa, Codium, Ulva, Bryopsis, Cladophora, Enteromorpha* and numerous other nuisance microalgae. Most, in fact, are colored green for their diets as they consume and store living chloroplasts from algae to use in like symbiosis (the newly hosted algae produce some food for the sea slug). You will need to research any species you seek for such control as they tend to be species-specific on dietary preferences. *Elysia subornata* is inclined to eat *Caulerpa* for example. In addition to their highly specialized diet we should mention their notable feeding strategy. Some have given these slugs the moniker, "cell-suckers," for their unique feeding style of piercing algal cells and sucking the contents out. Hmmm... sort of a vegetarian vampire!

* Elysia (=Tridachia) hails from the Atlantic and Pacific.

Cephalaspidea

Head Shield slugs & Bubble Shell Cephalaspids are regarded as the most primitive of the Opisthobranch orders in part for their remnant shell and similarities to the Prosobranchs (shelled snails). One group in the order, the "Head Shield" snails, are so-called for the large fleshy shield on their head that they use to prevent matter from entering the mantle cavity as they bulldoze through substrates including sand. Members of this group include species with the most highly evolved sensory aspects: they are carnivorous hunters that can sense and track the mucus trails of their prey! Burrowing varieties mainly eat bristleworms, other worms, foraminiferans, bivalves, and snails. *Chelidonura* species are famous and sought after by aquarists for their voracious appetite for nuisance flatworms. *Navanax* is the cannibal of the sea slug class and one of the very few creatures that can safely prey on Opisthobranchs. The shells of Head Shields are generally not at all obvious in plain view. Another group in this order has quite distinct shells like *Bulla striata.* Their shells also lack significant pigment, as observed empty, which has given rise to their common name ("Bubble Shells") for the likeness to large foamy bubbles on the waves and shore. In contrast to Head Shields, popular Bubble Shells are herbivorous.

Chelidonura varians (AKA *C. varia*): this species is coveted for its obligate diet on flatworms which include the sometimes plague and pest populations of acoel rust-brown flatworm species in aquaria (AKA "Red planaria"). From the Tropical Indo-West Pacific.

Bulla striata, the Striate Bubble snail - circumtropical distribution. Grows 1- 1 1/2" (25-37 mm) This species is a wonderful specimen for the aquarium. They grow and reproduce readily in captivity and eat rather indiscriminately for a sea slug (taking diatom and nuisance algae, e.g.). They are also useful agitators in deep sand beds for their burrowing activities.

Chelidonura inermis, the Striped Sea Slug. They are sometimes a temperate species.

Elysia crispata (= *Tridachia crispata*), the Lettuce Sea slug hails from the Tropical West Atlantic. This highly variable species feeds on various micro- and macroalgae. They grow 2-4" (5-10 cm) and their colors vary considerably including a remarkable teal-blue variety. The frilly convolutions of the body become exaggerated with light exposure. This sea slug is photosynthetic, using the chloroplasts of ingested algae for photosynthesis.

Elysia ornata, the Green Sea slug is one of the most sought after by aquarists for its obligate drive to eat the pest algae *Bryopsis*. Circumtropical distribution.

Notaspidea
Sidegill Slugs
The Sidegill slugs have little appeal, or many challenges- depending on your perspective, as aquarium specimens. They are decidedly carnivorous and require prey items that are desirable or difficult to culture in aquariums. Sponges, bryozoans and ascidians (tunicates and sea squirts) are typical prey. This group also includes members that can secrete acid when threatened or harassed. Again- not for casual aquarium keeping!

Pleurobranchus grandis. Like many sea slugs, this species is highly variable in color; typically it has three red hued bands, at least. It likes to forage on and in the sandy seafloor for prey. From the Indo-West Pacific, Red Sea, Philippines, and Australia, it grows to about 8" (20 cm).

Dendronotacea
Family Bornellidae
Dendronotid sea slugs are comparatively rare in the aquarium trade and likely have little place or purpose as casually kept captive specimens. They are occasionally imported incidentally with rock, plants or algae and they are specialized carnivores that favor cnidarians (hydroids are common prey). Aquarists may recognize *Bornella*

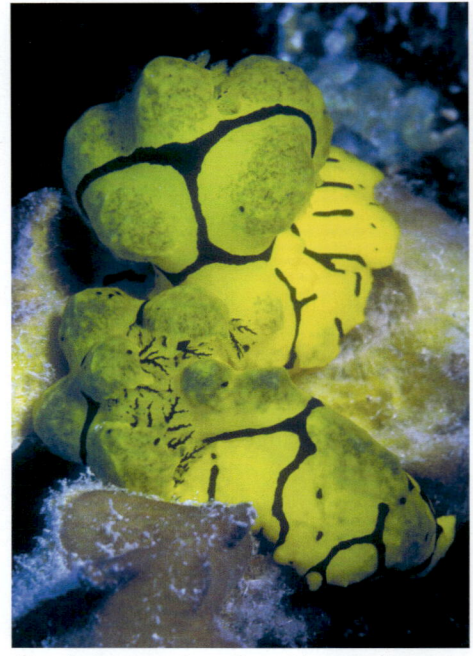

Notodoris minor. A sea slug with a penchant for Porifera. A fine example of conspicuous aposematic (warning) coloration in this sponge feeder. Pictured off of Queensland, Australia.

calcarata, the Tasseled Nudibranch, from the Caribbean.

Doridacea
The Dorids
Quite a few groups and species are recognized in this suborder. They often can be found in rocky intertidal zones circumtropically, and sometimes in large numbers. Dorids are characteristically oval or elliptical in shape with a large fleshy mantle that drapes the foot. Their posterior gills, or gill plumes, and cephalic tentacles (rhinophores) can be retracted

Elysia crispata. Aquarium image. (*E. Kruzel*)

fully. These exposed secondary gill structures are the aspect that gave rise to the namesake "Naked Gill" Sea slugs. The subtleties of color, shape, and size range wildly within the order, but we can say that they are carnivores with most preferring to eat sponge. As such, many of the assimilated sponge toxins make these nudibranchs highly toxic. Thus, the combination of high toxicity and unfavorable or rare prey (in aquaria) make this beautiful group of sea slugs almost wholly undesirable to casual aquarists.

Family Chromodorididae: There are several hundred species in this genus with almost half in the genus *Chromodoris* alone. They are widely distributed throughout the worlds oceans.

Chromodoris annae. Indo-Malaysian Peninsula- to less than 1" (25 mm). (*D. Fenner*)

Genus *Chromodoris*: The largest genus of Nudibranchs, by species, recognized.

Chromodoris kunlei: found in the central Indian Ocean to the southwestern Pacific. Pictured here in the KBR covered dorsally with many hypnotic eye-spots. (*D. Fenner*)

Chromodoris fidelis: ranging from the Red Sea to the Marshall Islands- grows to 1.5" (3-4 cm) and is said to feed exclusively on the Purple Sponge, *Aplysilla violacea*. North Sulawesi image

Chromodoris willani feeds on sponges- this one in Indo. Like most members of this group, it is not suitable for aquarium life. (*D. Fenner*)

Ceratosoma tenue: a widely distributed from the Red Sea, Tanzania, Norfolk Island, Australia, New Caledonia, Indonesia, Philippines and Hawai'i - to approximately 4" (10 cm). North Sulawesi image.

Chromodoris lochi: a striking variety with a pale blue body and black lines. Its small gills, tentacles and rhinophores are somewhat variable in color. They feed on Porifera (sponges). North Sulawesi image.

Genus *Glossodoris*:

Chromodoris quadricolor is one of several remarkably similar looking Pacific species in this genus which includes the complex of *C. annae* (yellowish appendages), *C. elizabethina* (orange appendages), and *C. magnifica* (reddish appendages). Growth to about 2" (5 cm).

Glossodoris cincta - Red Sea, Eastern Africa, Japan, Fiji. To just over 2" (5-6 cm). Complex and cryptic coloration is the standard for this species.

Mollusks: Nudibranchs

Genus *Hypselodoris*: These sea slugs are grossly similar to *Chromodoris*, but have longer, deeper bodies, wider heads and more conspicuous rhinophores.

Glossodoris rufomarginata: Red Sea, Eastern Africa, Fiji, Hawai'i (to 2"/5 cm). As its namesake suggests, an orange-brown border is consistent for this species. North Sulawesi image

Hypselodoris bullockii: too often seen in the trade, this sea slug has little chance for survival in the aquarium beyond a few weeks or couple months at best. A highly specialized feeder. Egg-laying before death is common in aquaria.

Hypselodoris infucata: Eastern Africa to Australia and the Philippines. Growth to 1.5" (3-4 cm). Highly variable. North Sulawesi image.

Glossodoris stellatus: New Guinea, Indonesia- to 5" (125 mm). A very uniquely patterned sea slug with a distinctive white-speckled black body. An absolutely "stellar" visage like the night sky. They are commonly found on flexible sponges. Sulawesi, Indonesia image here.

Hypselodoris nigrostriata: a unique sea slug- yellow bodied with dark purple lines. Its rhinophores and gills are generally red-orange. Growth to 1.5" (3-4 cm). North Sulawesi image.

Family Halgeridae: Genera *Halgerda, Sclerodoris, Aphelodoris, Asteronotus, Artachaea.*

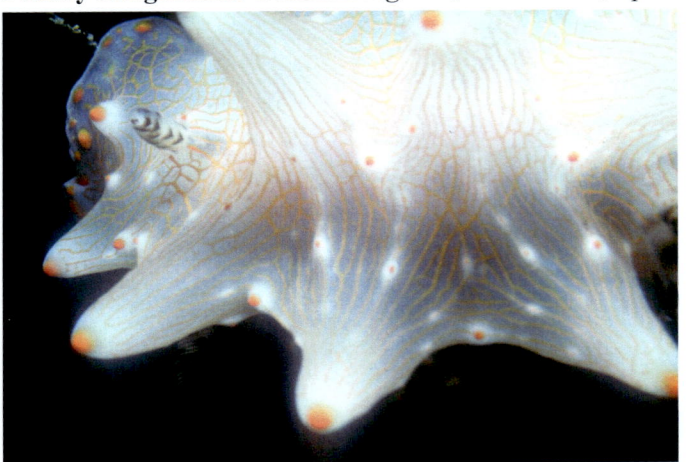

Halgerda malesso: Philippines, Indonesia, Guam, Mariana and Marshall Islands... growth to nearly 3" (75 mm) and typically feeds on Porifera. North Sulawesi image. (*D. Fenner*)

Halgerda tesselata: Image made off of Pulau Redang, Malaysia. Range includes Madagascar, Thailand, Micronesia, and Australia. Growth to an inch and a half in length (3-4 cm). This sea slug has a rather unique and consistent pattern among species in this family.

Family Hexabranchidae: Monotypic - the Spanish Dancer.

Family Phyllidiidae: distinguished by their warty bodies and unique placement of gills between their mantle and foot (not the dorsum). These sea slugs are tough-bodied, toxic sponge eaters that are not suitable for aquarium life. They are infamous for releasing toxins in aquariums that kill fishes and some invertebrates.

Hexabranchus sanguineus is a large, highly specialized and toxic nudibranch that is wholly unsuitable for casual aquarium keeping. Admire this beauty in the sea. Cook Islands image.

Phyllidia arabica, photographed in the Red Sea.

Phyllidia coelestis: South Africa, China Sea, Australia, Fiji. To 6 cm (~2.5"). Similar to *P. verrucosa*, but distinguishable by *P. coelistis* divided black markings. Sulawesi image. (*D. Fenner*)

Phyllidia elegans is widely distributed, including Red Sea to Maldives, Fiji to Australia. Pictured here in Indonesia. (*D. Fenner*)

Mollusks: Nudibranchs

Phyllidia ocellata: a highly variable species over a wide range in the Indo-Pacific. (*D. Fenner, left*)

Phyllidia varicosa: wide distribution from the Red Sea to Hawai'i with variable color. They grow to nearly three inches in length (75 mm). (*D. Fenner*)

Phyllidiella pustulosa: another beautiful species of black body color with striking pink tuberculations. Common from the the Red Sea to Hawai'i. Pictured here in North Sulawesi. (*D. Fenner*)

Phyllidiopsis annae: (not pictured) distinguished blue and black color with three longitudinal body ridges. A small species to 1.5 cm in length (1/2-3/4").

Reticulidia halgerda: Advertising its toxicity with glorious color contrasts, this sea slug is similar to *Halgerda* spp., but lacks the circum-anal gills of the latter. A Fiji image.

216   *Natural Marine Aquarium Volume I - Reef Invertebrates*

Family Polyceridae

This family of sea slugs uses narrow, extendable throats as a suction pump in feeding rather than as a scraping radula (they are "slurpers"). As such, they may be easier than other nudibranchs to keep in captivity. They take a wide range of animal prey which may include pest anemones like *Aiptasia*.

Nembrotha rutilans ranges Indonesia to Australia and grows to 2" (5 cm). A species of extremely variable color and pattern, some with almost "clown-like" combinations. Feeds in part on ascidians.

Roboastra tigris: the Sea Tiger. Tropical Eastern Pacific; Sea of Cortez. This is a large species that grows to 12" (30 cm) and feeds on other nudibranchs.

Nembrotha lineolata (top right) hails from the Western Indian Ocean and Pacific; Seychelles, Philippines, Western Australia, Fiji. It is a fast-moving species that feeds on ascidians (tunicates and sea squirts). North Sulawesi image. Growth to about 2" (5 cm).

Nembrotha megalocera (middle right) is a Red Sea endemic that is inclined to "swim" by undulations of the body if disturbed.

Roboastra arika: (bottom right) a bryozoan feeder of highly variable color, including rhinophores, and pattern.

Halgerda malesso, Sulawesi.

Mollusks: Nudibranchs 217

Tambja morosa is commonly found near the base of reef slopes from East Africa through Indonesia and the Philippines to Hawai'i.

Tambja sp. North Sulawesi.

Tambja kushimotoensis: Mauritius to Japan

Family Facelinidae: includes notably photosynthetic species. They are still predominantly carnivorous on cnidarians.

Aeolidacea
Aeolids (*pronounced* 'Eye-Oh-Lidz' and sometimes spelled "Eolids") This group of sea slugs is likewise carnivorous, toxic and generally inappropriate for casual aquarium keeping. They eat cnidarians (corals, anemones, jellyfish, etc). Aeolids can be distinguished from the popular Dorids by their extra pair of oral tentacles and parapodial tentacles. Most notably, however, they are distinguished by often dense clusters of cerata which perform duties of respiration, digestion and defense. Their very physical assemblage is intended to look like a cnidarian cluster of stinging polyps or tentacles. Aeolids are well defended with assimilated undischarged nematocysts from consumed prey, poisonous glands, noxious mucus secretions and calcareous "spines" (spicules) embedded into the mantle.

Phyllodesmium longicirra - a larger sea slug from Indonesia, New Guinea, East Australia (to nearly 6"/15 cm). Heads-up reef-keepers! This beauty, like so many nudibranchs, is advertising its potency and proclivity - the long "tassels" (cerata) on its back are filled with symbiotic zooxanthellae and defensive nematocysts acquired by *consuming* soft coral. In very large, specialized soft coral displays (over 200 gallons and mature), however, this species may be kept with little harm to its hosts for their fast growth. Alcyoniid "Leather Corals" can be suitable fodder.

Mollusks: Nudibranchs

Pteraeolidia ianthina: the Sea Dragon from Mauritius, East Africa to Australia, Japan and over to Hawai'i. They are partly photosynthetic via endosymbiotic zooxanthellae, harvested by eating hydroids. Grows to 6" (15 cm). North Sulawesi image.

Flabellina exoptata: from the Indo-West Pacific with growth to over one inch (3 cm). Indonesia image. (*D. Fenner*)

Family Flabellinidae:
Distinguished in this order in part by their long oral tentacles that resemble "handle-bar mustaches."

Flabellinopsis iodinea: the Spanish Shawl... a temperate species unsuitable for the tropical reef aquarium. Photographed here off of Submarine Rock, Catalina Island (Southern California).

Family Tergipedidae

Phestilla melanobrachia: (not shown) Reef-keepers will have little trouble guessing what coral this sea slug eats! It bears a striking resemblance to the Dendrophylliid- *Tubastrea* (AKA Orange Sun coral) upon which it feeds and lays its eggs. Distributed in the Red Sea, East Africa to Australia, Japan, and Hawai'i with growth to about 1" (25 mm).

Bornella calcarata, the Tasseled nudibranch, is quite variable in color. Somewhat commonly found in the Tropical West Atlantic, this species is wholly inappropriate for the home aquarium for its carnivorous and highly specialized cnidarian diet.

One of the handsome and hardy Lettuce-type nudibranchs, *Elysia diodomeda*, which feeds on nuisance hair algae species.

Mollusks: Nudibranchs

Clams, Scallops, Mussels and Oysters
Class Pelecypoda (Bivalvia)

Spondylus americanus, the Atlantic Thorny Oyster: a common import from the Caribbean that is challenging to keep for its size and nature. Collected at depth (40-100 ft), supplemental plankton via live drips or passively from large refugiums will almost certainly be necessary. Pictured here in the Bahamas.

Bivalves are a class of organisms of enormous ecological importance to man and on the reef. Also known as Pelecypods, there are about 15,000 species in this group, which are predominantly marine with a fraction occurring in freshwater habitats. They comprise a significant measure of the biomass on and *in* a reef (bored in carbonate rocks and coral, and nestled down in soft substrates). Most are filter-feeders and work with tremendous efficiency scrubbing water clean as attested by many aquarists that favor Tridacnid clams in their reefs. In a much more poignant example, consider how the zebra mussel (*Dreissena polymorpha*) has impacted the ecology of the Great Lakes of America in slightly more than a decade by accidental introduction from the Caspian Sea via transoceanic vessels. They spread rapidly and at first appeared to do a wondrous job of cleaning up said waters from a previously "unenviable" state. Very soon, however, it was realized that they had inadequate natural predation, not only becoming fouling organisms, but also massively reducing natural levels of plankton, which in turn harmed many higher animals up the food chain. Their accidental introduction has been an ecological disaster and a lesson in the fragile and complex dynamics of the food web. But bivalves feed in many different ways, as you will discover, and range from harmless filter-feeders to nearly autotrophic sun-soakers to obligate parasites (as boring organisms or as larvae on hosts).

As their name implies, bivalves are mollusks with paired shell-halves that enclose the body. They include the commonly recognized (and eaten!) clams, scallops, mussels and oysters. All have been kept in aquaria with variable success. Some groups, like the Family Tridacnidae, are wonderful specimens for aquaculture and aquarium life. They are as attractive as they are utilitarian. Most bivalves, however, are not at all well suited for aquarium life for their filter-feeding habits and the inadequacies of natural plankton in the aquarium. The modern evolution of aquaristics, however, is making it possible with refugiums and live food reactors (drips) to keep many species considered difficult to date.

The definitions of Bivalvia taxa are not clearly understood, although five subclasses seem to be legitimate: Anomalodesmata, Heterodonta, Palaeotaxodonta, Protobranchia, and Pteriomorpha. Ultimately, the physiological and various differences in taxonomic dissection of the class have little bearing on the practical aspects and understandings of aquarium life for these specimens. Thus, we'll forego an investment in detailing their classification and briefly describe unique aspects and features of species within these groups for aquarists to recognize.

Selection

Even with recent improvements in natural aquarium methodologies, aquarists have a long way to go before we can successfully keep many of the filter-feeding bivalves. Of course, "success" needs to be defined as more than just keeping them alive for short periods of time. Too often we find an incidental filter-feeding bivalve as a hitchhiker on rock, or even buy one of the challenging species like Flame Scallops (*Lima scabra*), only to look back after six months or a year and claim success at having kept it. Yet, 6-12 months is not the natural lifespan of such creatures. Good husbandry for any captive animal often extends their life beyond what we know or perceive to be a natural lifespan in the wild. Thus, if we can say that many bivalves will live 3, 5 or more than 7 years in the wild, then we can agree that specimens living less than 12 months in the aquarium do not qualify as success stories. Tridacnid clams are known to live many decades with some of the largest (*Tridacna gigas*) having a potential well beyond 100 years. Being mindful of any such potential, it should be apparent that life in the aquarium of even a year or two for many bivalves (particularly the filter-feeders) could very well have been an act of slow attrition. Success instead must be defined by growth and vigor, if not by observed acts of reproduction in the aquarium, which is regarded by many as the ultimate compliment to husbandry (most animals need to be well fed and in good condition

What makes a bivalve?

- They form two bilaterally symmetrical calcium carbonate based shells
- Their shell is exterior (secreted by the mantle)
- The are the only mollusk that lacks a radula (grinding "tongue" feeding aspect)
- They have well-developed gills (ctenidia) for filter-feeding and respiration
- Most are filter-feeders
- They are motile, but live fairly sedentary/benthic lives anchored by a muscular "foot"
- Unlike most mollusks, they are as not conspicuously cephalized ("headed")

to spawn). We challenge you to ask yourself with each of the species you keep, "what exactly does it eat?" Be specific beyond "filter-feeder" or "mixotrophic." For all of the unknowns in husbandry with bivalves, an inability to answer that question need not preclude your possession or the responsible care or purchase of the animal. But, it does demand prudence and due diligence in husbandry; seek to discover how to make that specimen flourish in your care and be sure to share that information and opinion with others. If not, you may simply be a shell collector.

Specific tips for selecting bivalves for aquarium use are few, in large part for the reality that most of their living tissue is obscured by non-living shell. Tridacnids with large fleshy mantles offer greater opportunities to evaluate health and vigor, and these are addressed in a dedicated

Fileclams, AKA "Scallops" are delicate and demanding beauties for the aquarium. They are naturally found lodged in crevices in the reef. Make no attempt at keeping these creatures without being mindful of their obligate need for a lot of very fine plankton. Natural plankton-generating methodologies will be helpful or necessary for success.

Tridacna clams are some of the very best bivalves for the aquarium. Their needs are well known, they are aesthetically some of the prettiest to be found, they exist as superb filter-feeders, and they are available aquacultured and in good health all year around. Among the most forgiving of this popular group, *T. squamosa* is pictured here at a retailer. (*L. Gonzalez*)

section of this reference. For all bivalves, though, a period of extended observation in quarantine and holding is called for. Collection often requires the separation or severing of byssus/muscle tissue that tethers a bivalve to the substrate. Improper collection, which includes outright pulling or tearing, can cause damage to a specimen that is not readily apparent but nonetheless harmful or fatal in time (days to weeks). The isolation of new bivalves from the main display is crucial for even small specimens have a considerable amount of flesh (bio-mass) that can degrade quickly if a new or stressed animal dies, possibly leading to a catastrophe in the main/display system. Indeed, the rotting flesh of a dead clam overnight is no different than the equivalent of flake food or thawed frozen meats overfed and left in the system. Such large portions or organic matter invites the rapid proliferation of (mostly) non-pathogenic bacteria, which can become pathogenic in large numbers or simply deplete the system of oxygen in a short period of time for the activities of decomposition. Short of gaping open unnaturally, little can be said for evaluating a new bivalve behaviorally. Most are sedentary and purse open only enough to filter-feed. They are naturally "closed" for much of any given day with only a siphon in evidence to the keen-eyed aquarist. For attached species, solid anchoring to a bit of hard substrate is a very good sign, but the lack of substrate means nothing to the contrary for how some species naturally occur or can safely be collected. Rest assured that observation in a proper quarantine tank for a few weeks is the best course of action for evaluating the readiness and suitability for putting a bivalve in your display.

One thing that you can do to hedge your bets for good species selection is to start with specimens known to be hardy in captivity, and avoid those that do not (unless you are a specialist). Of the incidental species acquired with live rock, most of the hardy ones that will survive long-term are the ones that simply live through importation, while most others are dead and even cleaned out before arrival to you. Thus, your choices as a hobbyist are rather straightforward if not simple.

Good selection suggestions:
Tridacnid and Turkey Wing (*Arca*) clams. These two bivalves are very different from each other yet either group has been demonstrated to live for many years in the aquarium. Tridacnids are cultured and well documented, and success for an aquarist requires very little effort to research the way. *Arca sp.* "clams" are usually acquired incidentally but occasionally are collected. They have proven to be forgiving to a wide range of physical parameters (temperature, oxygen levels, etc) as well as indiscriminate about suitable fare and feeding options. Aquarists often do not need to target feed living Turkey Wing shells with an adequate bio-load of display animals that are fed well and generate sufficient nutrition for this filter-feeder, incidentally. Flat-tree oysters (*Isognomon*) are also observed in the hobby and have a similar history and husbandry as Turkey Wing clams.

Dubious or challenging selections:
most oysters, scallops and file clams, and mussels. Thorny Oysters (*Spondylus*), so-called Flame Scallops (Limids, AKA Rough Fileclam), and Green Mussel clusters (*Perna* sp) are common imports that fit this category. By any definition they are challenging to keep and recommended only for advanced aquarists and specialists. Throwing them into the average reef aquarium is not only a poor way to keep them, but it is a liability and undue risk to the whole system in the sudden and likely event of their death. They are heavily heterotrophic and require measures and matters of food, including the finest plankton that

we cannot yet produce easily in the aquarium. Bottled supplements to date have proven inadequate for feeding bivalves and are more likely to corrupt water quality in time. Unfortunately, these clams are not only common imports but inexpensive, and as such tempt too many aquarists to make unwise or impulsive purchases. A conscientious aquarist will likely avoid these specimens for some time until the specialists advise us of sound and reliable methods of husbandry to ensure their prosperity in aquaria.

Care

Water quality

Bivalves need the same considerations of water quality as other reef invertebrates. Maintain high water quality conditions (Near Sea Water) with special attention to monitoring bio-minerals to support these calcifying organisms. As with reef aquaria, remember that stability and consistency in trace element and bio-mineral levels are more important than occasional spikes toward "ideal" maximums. Take heed therefore of careful dosing techniques. Any sudden influx of concentrated iodine, calcium hydroxide, etc., can prove to be stressful or fatal to a bivalve that draws the temporary concentration in with the water. The dilution of supplements in appropriate measures of fresh or saltwater is always recommended. One of the most surprising permutations of this sensitivity is when supplements are dosed in a sump with the belief that they will be adequately dissipated only to discover bivalves living under or near the return lines in the tank above suffer. Some bivalves have conspicuous inhalant and exhalent siphons (modified folds of the mantle). In others there is but a single siphon that is partitioned to keep opposing streams separate. Take notice of the sight and response of mantle and siphon activity as an indictor to bivalve health.

Hyotissa (*Pycnodonta*) *hyotis*, the Honeycomb Oyster: just one of the many unique and ornate-shelled bivalves. It is a heavily obligate filter-feeder with a wide natural range that includes South Africa to the Red Sea, and through to New Guinea, Indonesia and the Philippines. They require large and mature aquaria and are not recommended for beginners. Pictured here in the Red Sea.

Handling

Another unique condition of bivalve handling and acclimation is a consideration for gentle physical manipulations. One of the most sensitive aspects of bivalves in the aquarium, beyond issues of water quality (including salinity shock), is the protection of the **adductor muscle**. The adductor muscles hold the shells of a bivalve together. They facilitate the opening and closing of the animal for normal daily routines as well as "seal" its shells tightly closed to prevent predation or shock from water quality (low tide, temporary pollution like heavy

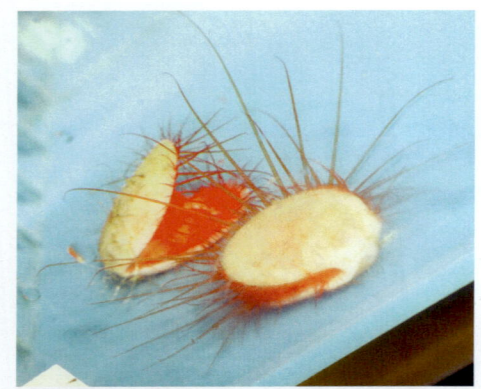

Lima (*Limaria*) *scabra*, the Fileclams sold as "Flame Scallops" in the aquarium trade. They are one of the most difficult bivalves to keep successfully for more than a year. They can demonstrate remarkable and rapid motility if threatened. Bottled supplements are not likely to keep this animal fed well. Most starve in captivity in time.

Aquarium images of unhealthy (top) and healthier (bottom) specimens. Notice the wide and vulnerable gaping of the unhealthy specimen.

sediments as with storm activity, etc). Most bivalves have two muscles, but some groups, like Scallops, have only one (centralized). Closely tied to this aspect is the foot of the animal. The tactile foot is used for motility (some can crawl along a substrate), burrowing in soft substrates, or holding fast to a hard substrate. To support settlement, many species secrete byssus threads, which are issued as a viscous liquid that becomes fibrous and quite strong. For some species of dedicated benthic lifestyles, these byssal threads cannot be sheared from hard substrate without harming or killing the animal (as with many mussels). For other, like the Tridacnids, set byssus material can be gently cut away to collect and move a clam. The giant clams will often abort the old material and secrete new byssus in time. However, all bivalves that make an effort with their foot and/or byssus to attach to a hard substrate must be treated with extraordinary care not to pull or tear the adductor muscle in the process of harvest. To do so would impair its ability to open and close its shell and subsequently feed or protect itself. Under no circumstance should you attempt to pull a bivalve from a rock. Approach it giving advanced warning and allow itself to close tightly and retract. You may then gently and slowly rock the animal to a side and sever byssus threads with a razor or scalpel. Better yet, transport the animal undisturbed and still attached to the substrate if at all possible.

Placement

In the display, most filter-feeding bivalves will need to be kept as they naturally occur, in crevices & caves, under overhanging rocks and always with plenty of current. Although some can move if necessary, most are dedicated to a benthic lifestyle and, like corals and anemones, they are critically dependent on the currents bringing food and oxygen to them and carrying waste matter away. Non-photosynthetic species often have photocells ("eyes") that sense light and instigate them to move away from it. Light does not harm these bivalves but makes them "realize" they are vulnerable and in plain view. They may simply stick their foot out and crawl away (or pulse their shells and swim like "scallops") to a dark nook in the tank. To prevent them from finding their way to an unseen, inaccessible or inappropriate place in the aquarium, be sure to place the shady species in a proper spot at the start and not perched on a rock in full light exposure. Of course, for the mixotrophic species that have zooxanthellae, the quality of light is crucial and a flat rock under the "sun" is close to ideal. Tridacnids are far and away the most common group in this category and have been addressed in a full chapter in this reference. Like you'll read so many times in this reference, researching an animal's natural history before bringing it home will spare life-threatening situations and prepare you to serve your charge satisfactorily.

Delicious cooked in so many ways, the Green Lip Mussel, *Perna canaliculus* is sometimes offered as a reef invertebrate to aquarists. This is a cool to very coldwater species however (it hails from New Zealand) that requires cold water conditions.

Pedum spongyloideum, the Iridescent Scallop: this little bivalve lives embedded in corals (often *Porites* species). They are not a boring species but rather incidental and impose themselves by settlement. The coral is unharmed and merely grows around them while they enjoy the benefit of a living stony home. A wide distribution including the Red Sea (above) and Indonesia (bottom-right, this page).

Feeding

Most bivalves are attached to rock or found within the substrates (in sand or bored into rock or coral). There are but few varieties that are free roaming. Almost all, however, are filter-feeders to some extent, including the oft-mentioned "Giant" Clams (*Tridacna* and *Hippopus*). The nature of necessary food matter varies considerably with the many thousands of species of bivalve recognized. There are four types of feeders within the group, essentially categorized by the application of their gill structures. Some feeders are ciliary while others are macrophagous.(1) Ciliary feeders (usually non-photosynthetic) purse their shells open to reveal "fine-haired" structures composed of cilia and coated with mucus. Nanoplankton and like-matter gets trapped in the mucus and carried by cilia to the mouth. (2) Macrophagous feeders may use large gills (ctenida) instead to rake plankton from the water. (3) Other bivalves only use their gills for respiration and catch food in labial palps instead. And lastly, some bivalves employ a septum across the mantle cavity to pump food-laden water into the mouth (4).

All are very specific about particle size, however, and this has been one of the biggest obstacles to successful long term care of bivalves (and aposymbiotic corals in kind). Too many species at present are not well suited for aquarium life for their particular feeding habits and the insufficient levels of the right kind of plankton in the aquarium. We believe or know that most need very fine plankton that cannot be produced or provided easily if at all from prepared foods like bottled food supplements. The few such supplements to date that show promise still have their limitations (shelf life, dosing protocol, particle size, etc). Thus, we must depend on evolving technologies for producing plankton like live phytoplankton drips (AKA "Phytoplankton Reactor"), seagrass refugia (shedding "phyto" and epiphytic matter) and deep sand beds for bacteria, microscopic plankton and other nutrients.

Reproduction

Most Bivalvia are dioecious (separate sexes) and reproduce by shedding gametes. For most, fertilization is external and larvae either develop to be free-swimming or parasitic. Almost none can be raised in the confines of small private aquariums for practical concerns. They often have long or complicated larval cycles that place impossible demands on filtration and hardware aspects of captive systems. In some cases, brooding occurs and these specimens include the best or most likely candidates for

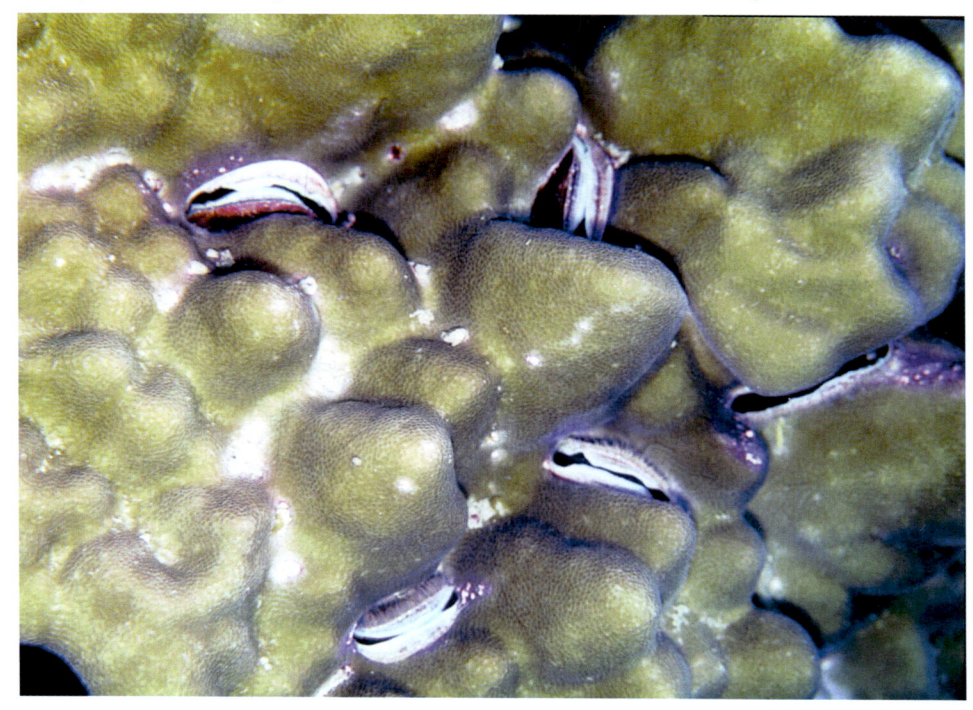

Mollusks: Bivalves

Atrina pectinata, the Pen Shell (Family Pinnidae): a few of these bivalves make their way into aquaria. They have a distinct and memorable orientation nestled vertically in the soft seafloor. Echinoids are common predators on bivalves as pictured here above. The remarkable strength of the adductor muscles (keeping their shells closed tight under attack) is rivaled by the sea star's extraordinary strength and tenacity to squeeze the bivalve until it fatigues and opens wide enough for the sea star to invert its stomach and mount a digestive assault. Pen Shells are circumtropical.

Dendostrea frons, the Frond Oyster: this handsome-shelled species hails from Tropical West Atlantic and commonly grows to two and a half inches in length (62 mm). Pictured here in the Caribbean sharing the denuded gorgonin stem of an octocoral with a small colony of ascidians.

Pteria aegyptica, Egyptian Wing-Oyster: most always found attached to gorgonians (pictured here), stony corals, soft corals, hydrozoans, etc. This organism needs strong current, moderate light and microscopic plankton (perhaps less than 15 milli-micron). Grows to about 3" (7-8 cm) in length. Indo-Pacific and Red Sea.

Limaria orientalis, the Eastern Fileclam: a solid red variety that is consistent in habitat and perceived needs for husbandry like most "Flame scallops." This bivalve has a broad distribution in the Western Pacific; New Guinea, Indonesia, Philippines. Photographed in North Sulawesi.

Limaria fragilis, the Fragile Fileclam is widely distributed in the Indo-Pacific; Red Sea to Japan, Australia. Currently, this is a very poor candidate for captivity, best reserved for specialized aquarists to discover practical husbandry guidelines. Like most "scallops," they need to be secured in rock or rubble and provided strong water movement. Queensland, Australia image.

228   Natural Marine Aquarium Volume I - Reef Invertebrates

replication in aquaria. The broadcast spawning events, however, can be a blessing in the aquarium as the gametes of bivalves are crucial plankton for many filter-feeding reef invertebrates. Marine species go through a planktonic **trochophore** stage before developing into a tasty (free-swimming) **veliger**. Parasitic species forego becoming veligers and instead latch onto fishes for food and to be dispersed. Information about reproduction of bivalves in the aquarium is sparse although sure to be expanded in due time. Commercial industries have been culturing numerous varieties for many years and interested aquarists should seek their fisheries data to see what if any methodologies can be extrapolated and applied to aquaristics for success at home.

Lopha cristagalli, the Cock's Comb Oyster: a commonly recognized species by divers and aquarists for its distinct "zig-zag" shells. A heavily dependent filter-feeder in aquaria. It lives attached to hard substrates and is pictured here in the Red Sea.

Compatibility

Most issues with bivalve compatibility revolve around discovering what won't eat them. Worms, crabs, snails, echinoderms (starfish and urchins), many other mollusks (especially gastropods like whelks) and numerous fishes all prey on bivalves. They are a fundamental element in the food web on a reef. Beyond the careful selection of suitable macro-organisms, guarding against predation from tiny worms, snails and boring organisms can be difficult with the incredible diversity of life often carried in with live rock and live sand. The very best defense against such predation is good placement of the shell. Locate the byssus of the clam (opening that byssus threads are issued from to attach the living shell to the substrate). Be sure that clam is given a sound and solid rock upon which to attach. Tridacnids placed in the sand for example should still have a small flat rock buried underneath, upon which they can set and attach squarely. Solid attachment to hard substrates prevents small predatory creatures from crawling under the bivalve and literally "waltzing" into the mantle cavity through the byssal port to eat its host from the inside out. Research the natural habitat of each bivalve in your charge to be certain that you have commuted it to the best possible place and position for healthy display.

Spondylus varians, the Variable Thorny Oyster: This oyster is the most common bivalve in its Western Pacific range. They are striking specimens with orange mantles, blue "eyes," white scutes adorning the shell which invite the settlement of a variety of cryptic fauna (sponges, bryozoans, algae, etc). This specimens photographed off the Gilis - Lombok, Indonesia.

Mollusks: Bivalves

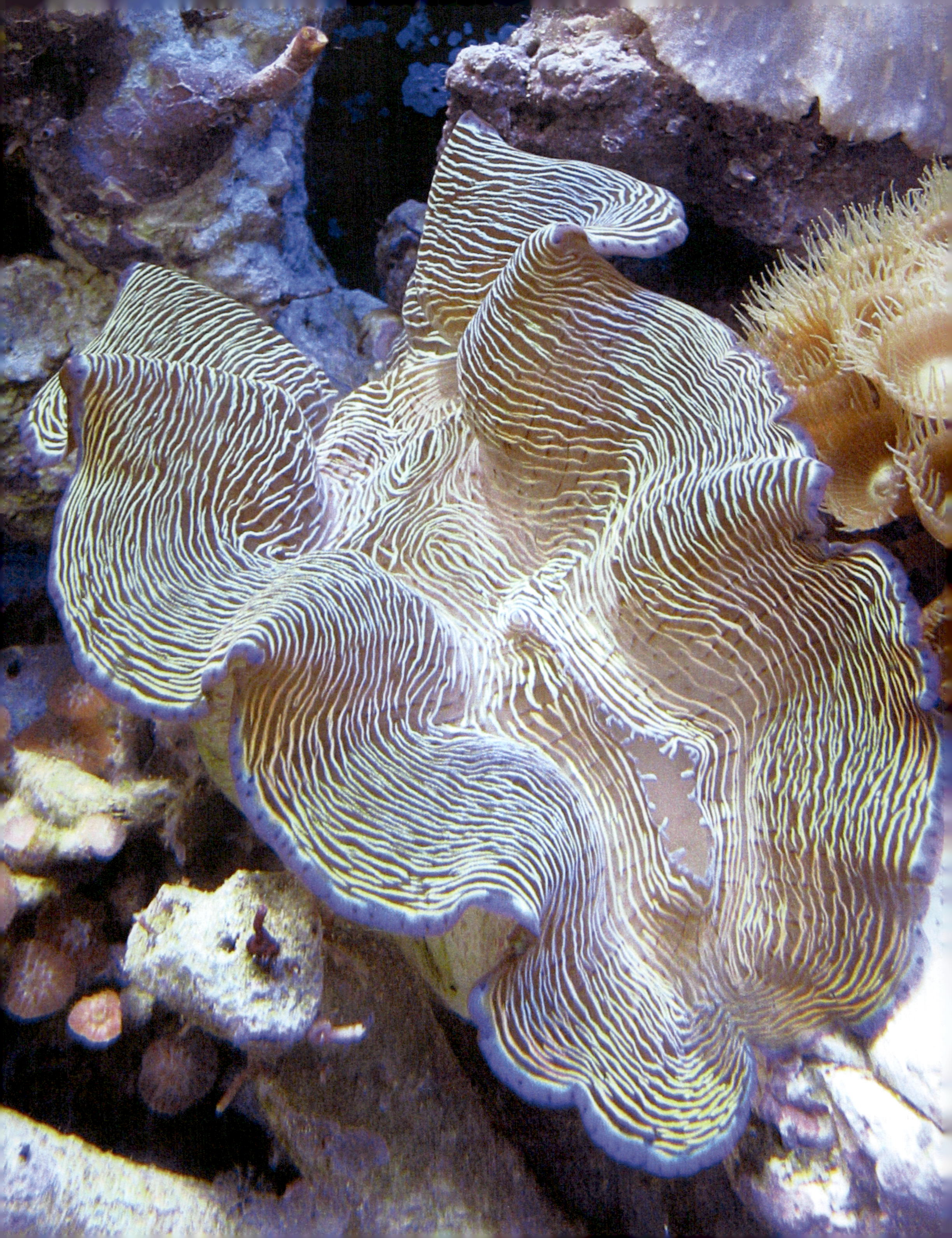

Giant Clams
Sub-family *Tridacnidae*

Tridacna crocea - The smallest Tridacnid species. One of the most colorful, variable and sought after Tridacnids, *T. crocea* rarely exceed 6" (15 cm) in length and is commonly known as the "Boring Clam" for the ability (and preference) to mine a residence in hard carbonate reef structures. (*L. Gonzalez*)

Tridacna gigas - The largest Tridacnid species. Legendary sightings peg *T. gigas* at nearly six feet in length, although 4 feet is quite a remarkable specimen and not entirely uncommon. The age of such living wonders is awesome… perhaps well over 100 years old. This one photographed off Queensland, Australia.

There are at least seven *Tridacna* species and two *Hippopus* species in the sub-family Tridacnidae. In the genus *Tridacna*, five are offered in the aquarium trade: *T. crocea*, *T. maxima*, *T. derasa*, *T. squamosa* and *T. gigas*. Among the species of interest to aquarists, Tridacnid clams are found over a very wide range of depth - from lagoon-shallow water to more than 45 feet (the rare *T. tevoroa*, however, is found at amazing depths approaching the limits for zooxanthellate-symbiotic animals). Species are also found in niches running the gamut of water turbidity from muddy to pristine clarities. As such, the dependency of these animals in captivity on lighting and supplemental feeding to algal-symbiotic activity (feeding organismally and by "absorption") ranges accordingly. With further consideration for differences in hardiness and sensitivity due to shipping and acclimation, it becomes apparent that a categorical assessment of husbandry for the subfamily as a whole in aquaria cannot be proffered.

Selection: Choose Captive Produced vs. Wild-Collected

Most of the Tridacnid clams found in the ornamental trade are cultured, although some wild-harvested specimens are unfortunately still collected from reefs. Aquarists will want to make a concerted effort to acquire cultured specimens preferentially. These animals are much more likely to excel; they are easier to acclimate for having been farmed in captive conditions and transported with fewer traumas (e.g. handling, transit time, water quality, etc). Contrarily, progress from the point of collection through the chain of custody on import for wild-harvested animals is often burdened measurably. This leads to stressed clams offered at inflated prices to compensate for the categorically higher mortality

Left: *Tridacna derasa*, at a retailer in San Diego. This 10" individual might be only a few years old! (*L. Gonzalez*)

Mollusks: Giant Clams   231

T. maxima - Tridacnids develop a fabulous range of colors and patterns, even within the same species. These are small maximas at a retailer. (*L. Gonzalez*)

imposed upon them. Some of the problems with wild-harvested clams include: increased likelihood of damage to sensitive adductor muscle tissue from improper collection, increased likelihood of parasitic organisms (like "pyram" snails… the Pryamidellids), and weakened condition on import from poor water quality due to fouling organisms that commonly travel attached to clams (many sponges and other sessile invertebrate growth forms do not weather transit well). Some aquarists feel as though wild clams are prettier, but this is not true. Wild clams are less likely to be graded and cherry picked on import in contrast to cultured clams that are necessarily sorted and graded. The fact of the matter is that aquarists who are willing to pay a premium for the more desirable mantle colors drive some markets. As such, an American aquarist is less likely to see or pay such a premium for the highest grade clams while Japanese aquarists, for example, are willing to pay handsomely for these favored varieties.

Selection: Criteria

When selecting Tridacnid clams, many of the above considerations come into play and our recommendations reflect the realities of importing live clams that may not be readily apparent to an aquarist. One of the first things that a buyer should look for in a healthy Tridacnid is alert and responsive behavior. While some species and larger specimens in general can be somewhat slower to respond to a passing shadow overhead, all should respond promptly to such disturbances. Lazy or non-responsive behaviors often indicate a stressed animal. In a similar vein, a "gaping" clam indicates a serious and often fatal condition. **"Gaping"** is indicated by a stretched appearance or splaying of the animal's shell halves, as well as the inhalant siphon (the larger opening on the mantle of this filter-feeder) appearing to be open particularly wide and non-responsive. Quite

T. maxima - with internal organs visible, an example of stress-induced gaping (*L. Gonzalez*)

frankly, it is symptomatic of the last stage before death for many specimens. It does not, however, indicate any specific condition and can be caused by prolonged siege by predators like pyram snails, saline or luminary shock, severe trauma to adductor muscle and/or internal organs, and more.

Additionally, when selecting a Tridacnid, its **mantle** (the colored fleshy lobes extending to and beyond the top of the shell) should be fully extended and its color should be rich, dark and crisp indicating a healthy population of resident zooxanthellae (symbiotic algae). Individual "bleached spots" in the color of the mantle, or pale

Hippopus sp. - Photographed in the aquarium of hobbyist Jim Nastulski. (*L. Gonzalez*)

color overall, indicates problems with the life-supporting symbiotic algae. Clams, like coral, that have expelled their zooxanthellae are stressed and struggling animals. For such specimens to reach their compensation point for survival and ultimately recover, regular feedings may be necessary. Clam feeding is best and most likely done with the finest food suspensions if not exclusively with a dissolved nutrient (more about this below in the *care* section).

Another unusual symptom to look for in Giant Clams is **gas bubble disease**. It is amazing just how easily this deadly condition can be induced. Gas bubble disease is indicated by hidden air bubbles trapped in the tissue of a clam. This gas can even make small specimens buoyant! Similar to the dangerous nitrogen bubbles sometimes experienced by rapidly surfacing SCUBA divers, this problematic condition is caused by supersaturated atmospheric gasses escaping solution within the very tissue of the animal. In aquaria, this is most often induced by pouring hot water into cool water, sometimes done when making buckets and baths for acclimating animals (a very dangerous way to quickly make warm water). The condition is also created by the pinhole aspiration of air (as with a leaky fitting) into a water pump or plumbing, causing the supersaturated condition. These causative agents are easily prevented, identified and remedied, however the condition in the living

T. maxima. - 5" (125 mm) specimen in captivity. (*L. Gonzalez*)

Mollusks: Giant Clams   233

Tridacna squamosa. - 3" (75 mm) juvenile, in captivity. *Valonia* and coralline algaes share space on this individual. The exaggerated scutes of *T. squamosa* encourage colonization by such "hitchhikers." (L. Gonzalez)

animal is serious, and there is no therapy available to the average hobbyist.

Scrutinizing the shell and mantle of a Tridacnid is also crucial when selecting a specimen. The very structure and texture of many bivalve shells is intended to attract incidental organisms to settle and live as camouflage. In shipping, hitchhiking organisms like algae, sponges and other invertebrates can easily foul, leading to stress or death of a clam. Some Tridacnids are more inclined than others to attract such growths by virtue of their attractively exaggerated **scutes**. *T. squamosa* is perhaps the most interesting if not attractive Tridacnid in this regard. In many cases, when such co-imported sponge growth is allowed to remain on the clamshell, the sponge can, over time, grow to the point of impeding available water and light to the clam's mantle. Some algae and sponges are in fact boring (as in drilling, not yawning) species that can be quite destructive - mining fatal holes straight through to the clam's body cavity. Be alert of suspicious pinhole formations on the shell that may indicate such harmful species. Once again, for these reasons and more, cultured clams with clean shells are better choices for aquarium display.

Clam Pests & Predators

Many pests and predators are attracted to Tridacnids. Boring sponges, segmented worms (errantiate polychaetes), crabs, predatory mollusks (snails) and flatworms are commonly caught preying on *Tridacna* and *Hippopus*. Most can be observed through careful observation and removed

2-3 mm shells from parasitic snails of the family Pyramidellidae (AKA "pyrams"). (B. Neigut)

manually. Many can be prevented or protected against by founding Tridacnids on rock to protect their vulnerable byssal port (the opening on their underside through which byssal threads attach and secure the specimen). For clams placed upon a sandy bottom, a solid flat rock buried underneath is strongly

Pseudocheilinus hexataenia - The "Six-line Wrasse." A favorite predator of small snails and worms.

Halichoeres chrysus - The "Yellow Coris Wrasse." Also family *Coridae* (but not genus *Coris*!). Another great snail hunter, up to 5" (125 mm) fully grown.

recommended. Pyram, Murex and Costellarid snails often prey on such clams. The Pryamidellids are perhaps the most common, prolific and tedious, if not difficult, to eradicate due to their small size and camouflage coloration. Natural predators, such as smaller species of wrasses, can be helpful for reducing the population of such predators, but none can be assured of providing complete protection.

The gelatinous and, at times, numerous egg masses of predatory snails require frequent re-inspections by an aquarist to detect and remove in order to maintain satisfactorily low populations. Large clams can safely sustain siege from a small amount of pyram snails for many months, but smaller clams can succumb in just a few days to weeks. Routine inspection of clams for some pests is *de rigueur* in systems that otherwise properly quarantine all new rock, coral and sand. It is absolutely required for all other systems run with less discipline. Monthly inspections for the first 3-5 months are suggested and quarterly after that. All things considered, the aquarist should allow four weeks for proper quarantine and screening of Tridacnid species. Many pests and predators of clams can be baited in a bare bottomed quarantine tank with pieces of food-clam from the local grocer or pet store freezer.

Bigger Isn't Always Better: Size Recommendations

Beyond aesthetics, clam size is an important consideration in price and selection. The potential adult size of a clam is often overlooked at the time of aqcuisition. *Tridacna gigas* can grow several feet in length and attains an extraordinary weight of many hundreds of pounds in as little as ten years. A responsible aquarist plans for the adult size of this magnificent animal just as one would do for a large family dog. It is not sensible or appropriate to keep the true Giant Clam in tanks of even a couple hundred gallons in size on the hopes and dreams of getting a bigger tank one day. *T. derasa, T. squamosa* and *H. hippopus* in suit reach impressive sizes approaching two feet in length. Due respect and consideration is required for keeping such animals. However, even with a suitable sized display tank, one must also be selective about the size upon acquisition for various species of clams. As a rule, very small and very

Tridacna derasa - This 12"/30 cm monster in a retailer's aquarium is still only half grown! (*L. Gonzalez*)

large clams tend to ship, handle and acclimate poorly relative to smallish to intermediate sized specimens for a given species. For *T. crocea* and *T. maxima*, 2 1/2" – 4" (62-100 mm) specimens are generally best, stronger for shipping and acclimation. For all other Tridacnids - 3"-6" (75-150 mm) specimens tend to handle best upon import. Let these guidelines serve the aquarist well during information gathering as he/she attempts to make an informed buying decision based on an intelligent (hopefully!) consensus.

Clam Care: The A, B, C's

Care of Tridacnid species in captivity is rather straightforward once a healthy specimen has been selected and established. These clams are quite long-lived with natural life spans of decades. Tridacnids are largely symbiotic animals that can secure much of their sustenance from photosynthetic activity. The quality and intensity of artificial light provided them is therefore a matter of great importance. Numerous developments in lighting technology employed

Summary of Selection tips for Tridacnid clams:

- Check for alert and responsive behavior to changes in light

- "Gaping" inhalant siphons indicate stressed or dying animals

- Mantle should be extended fully

- Pigmentation should be rich, dark and crisp indicating healthy symbiotic algae

- Bleached spots or overall pale color indicates stressed/ starving animals

- Beware of gas bubbles in tissue

- Look for holes in shell from boring organisms

- Look for pests and predators, their egg masses, or signs of damage particularly under mantle, between shell scutes and around byssal port (underneath clam)

- Intermediate sized clams tend to acclimate best to captivity

Clockwise from top-left, *T. maxima*, *H. hippopus*, *T. crocea*, and *T. squamosa* in captivity. (*B. Neigut*)

in keeping photosynthetically symbiotic animals has evolved rapidly in recent years and is sure to continue. Nonetheless, time-tested technologies are available now to recommend to aquarists intent on keeping Tridacnids. All symbiotic animals must acclimate to changes in light on arrival and clams have proven to be one of the most adaptable groups of reef creatures here. The very notion of a homogenized lighting recommendation for successfully keeping a subfamily of symbiotic creatures that runs such a wide gamut of exposures to light at depth and turbidity in the wild is testimony to their adaptability indeed. Lighting requirements for Tridacnids is rather akin to lighting necessary for popular species of shallow water so-called SPS corals (small polyped stony scleractinians). Their needs may be regarded as moderate to bright... leaning decidedly towards bright (intense). Under-illuminated clams will change color: often darkening at first in an attempt to cultivate more zooxanthellae with the purpose of trying to capture more of the diminished available light energy. In advanced stages, such clam's color will pale as zooxanthellae are expelled under duress by the starving animal. Supplemental feeding of dissolved nutrients (sources of ammonia/nitrate) may stave off or delay the inevitable. Under-illuminated clams may survive in captivity for many months or even more than a year before perishing "mysteriously" (a mystery only to the aquarist that does not realize that illumination was waning or inadequate). Most clams are best kept under high intensity lamps like metal halide lighting. Bulb

New shell growth is obvious on this juvenile (2.5"/63 mm) *T. derasa* in an aquarium (*L. Gonzalez*)

temperatures in the range of 6500K to 10,000K are ideal for Tridacnids and most popular corals in the ornamental trade. A 175-watt lamp per four square feet (2' X 2') will serve the purpose nicely in water 20-30" deep. Fluorescent lighting of various formats (PC, VHO) is fine and aesthetically attractive for shallow water environments less than 24" deep. When fluorescent lighting is utilized, clams should usually be placed in the top 18" of water. Acclimation of clams in very shallow water or under brighter lights (250 watt MH and higher) must be done with great caution and consideration, including starting the specimens at depth, and initially shielding/shading the specimens. Maintenance of good water clarity through consistent use of chemical filtrants and water changes, along with clean light intensity (lamps/covers free of debris and salt creep, and fresh lamps of high PAR delivery rotated regularly) is necessary for clams and all symbiotic reef animals.

Placement

Placement of clams is another simple but crucial matter of importance. As with hermatypic corals and like symbiotic animals, Tridacnid clams should be placed in a good spot the first time and left in place! Moving healthy and established animals throughout a display can impose tremendous stress as they try to compensate for the changes in lighting, circulation and proximity to other life. Repetitive moves of newly acquired animals within a matter of days or weeks can prove to be fatal. Aquarists are advised to research the clam's needs as best as possible, make informed decisions, and place them permanently the first time. Clams should be set upon a hard flat rock (burying the rock in the sand if necessary or desired) to reduce the risk of siege by predators through the byssal port and damage to vital organs. Water flow is a matter of somewhat lesser importance for Tridacnids so long as a few basic needs are met. As a filter-feeder, Tridacnids are dependant on adequate water flow bringing nutrients to them and carrying waste products away. Unlike corals though, they are not so directionally discriminating about the dynamics of water movement for growth or their very morphology. A moderate to high turnover of system water in random turbulent flow serves Tridacnid clams very well. Laminar (one direction) flow or affronts with direct streams of water is to be avoided. Aside from being unnatural, a directed stream of water might carry temporarily spiked additives or supplements in a stressful or fatal concentration

Large *Tridacna maxima* in the Red Sea.

Mollusks: Giant Clams

T. crocea off Queensland, Australia.

around this filter-feeder. Clams are especially sensitive to metals in the water and they have been reported anecdotally to suffer from accidental local concentrations of spiked iodine when water movement in the tank was inadequate.

In summary, placement of Tridacnid clams is rather like the handling of most reef corals in captivity; they require confident and secure placement (on a stable flat rock) under moderate to strong random turbulent water flow. Another consideration when planning a display that will house Tridacnid clams is that they can shoot a stream of water up and out of the aquarium. You need to take into consideration the depth of your tank, the depth at which you will place the specimen, its adult size, and the distance between your lights and the surface of the water. These estimates will play an important role in several lighting decisions such as the intensity of the lamps you use, whether you place a cover on the aquarium, and whether or not you use a shield on the light fixtures themselves. Of course, you should always use fixtures that are designed for marine aquarium use. It would also be prudent, in the event that your fixtures do come into contact with saltwater (risking a short-circuit and/or shattered lamp), that the lighting system operate on a separate circuit than the pumps and other vital equipment. Careful planning will protect you and your livestock from a complete system failure.

Feeding

Feeding Tridacnid clams is an often-overlooked aspect of their husbandry. They are filter-feeding animals that do indeed have a fully functional digestive system. As form follows function, it stands to reason that they must feed on something. Beyond nutrition derived from the products of photosynthesis from their hosted symbiotic algae, Tridacnid clams feed on very minute ("nano") plankton. **Nanoplankton** particle sizes cannot be easily provided by aquarists, through traditional prepared food suspensions. In

Left, *T. maxima,* middle a brown/black *T. maxima,* and *T. derasa* at right. (*L. Gonzalez*)

Top, ***T. maxima***, bottom, ***T. squamosa*** in a "look-down." (*L. Gonzalez*)

Mollusks: Giant Clams 239

T. squamosa - Red Sea.

fact, target feeding of any kind is very difficult to do with clams (especially smaller specimens). Larger particles of food will be collected and rejected from the clam with a sudden and disapproving expulsion like a cough! Clams instead need dissolved organics and the finest plankton (microscopic) on which to feed. Cultured specimens held in raceways are fed ammonium chloride just to maintain their active and growing metabolisms, a precarious and potentially toxic practice especially when employed on smaller systems. Clams that pale in color overall (uniformly) after an extended period of time in captivity may well be starving for a source of nitrogen. In aquariums that are aggressively filtered and skimmed, nitrate levels can be so low that endosymbiotic zooxanthellae in clams and corals begin to suffer, with waning color of the host as testimony to it. Some aquarists have tried to correct this deficiency by adding sodium nitrate in very small concentrations (.1% Knop, 1996). Adding ammonia and

T. maxima. (*L. Gonzalez*)

nitrate however can be a difficult or even dangerous endeavor for the casual reef aquarist and is not recommended for most.

Other alternatives for nanoplankton in aquaria include yeast and cultured phytoplankton (greenwater). Bottled greenwater is often touted as ideal clam food. However, even if Tridacnids feed on such products from a bottle, the necessary limitations of the food and its delivery are so strict that it is unlikely that much of it is useful. The reality of it all is that few aquarists are aware of the proper and necessary protocol for preparing and offering bottled greenwater supplements to insure a fine and usable particle size. Bottled phytoplankton must be packaged, shipped and held under constant refrigeration: it should be purchased from a refrigerator and kept in a refrigerator. Ideally, the bottles will be dated and used within 6 months of production. Greenwater cultures clot with age and thus particle size can increase dramatically. Whisking with an electric blender may be necessary to reduce particle size to a hopefully usable measure for clams and corals; simply hand shaken greenwater is not remotely adequate for this purpose. Such restrictions to the successful application of bottled "greenwater" products are indeed tedious. Nevertheless, such products may still serve a useful purpose for aquarists that cannot or will not take steps to culture live phytoplankton. Live phytoplankton cultures that were formerly tedious to produce are nowadays simple to culture and harvest (see index of contacts for plankton culture kits/supplies). Recent developments with so-called **phytoplankton**

Unusual gold *T. maxima* - aquarium photo (*A. Calfo*)

Large 18" (46 cm) *T. derasa* at the Shedd Aquarium in Chicago (*L. Gonzalez*)

reactors show great promise for the convenient culture and delivery of this elemental foodstuff to reef invertebrates and will likely help to unlock the mysteries of husbandry of many captive animals. Live baker's yeast suspensions are also nutritively dense (rich in vitamins and proteins) and likely beneficial to clams and some corals. However, they must be prepared with the same tedious methods as bottled greenwater described above. Regarding the applications of feeding, all such foods mentioned here are best delivered through a slow and extended drip, as clams do not respond favorably to sudden attempts at target feeding with a pipette or baster.

Tridacnid clams in aquariums with high bio-loads from heavy feedings to various animals, including fishes, will undoubtedly benefit from available dissolved nutrients. Perhaps the best source of nutrition for clams is a large, mature inline fishless refugium. Refugia with seagrasses, containing rasping snails/mollusks, worms and more, with strong water movement can generate significant and useful amounts of epiphytic material and plankton. Refugium technology in its many forms and styles has been fully embraced by modern aquarists that are just beginning to understand and define the merits of such applications. And remember that there are *many* useful variations on refugium technology beyond the limited abilities of *Caulerpa* filled vessels. Judicious experimentation with a combination of the above listed feeding options will easily support thriving symbiotic clams under appropriate illumination.

Summary of Care for Tridacnid clams

- Provide moderate to bright lighting over shallow water for most Tridacnids. Power Compact (PC) or Very High Output (VHO) flourescent for water under 24" deep. Use Metal Halide or HQI for water more than 24."

- Put the animal in a good place the first time and let it acclimate (Leave it alone!)

- Clams should be set upon a hard flat rock (or in sand on buried rock)

- Avoid placement or exposure of clams to direct streams of water that may carry temporary spikes of additives or supplements (iodine, metals)

- Any feeding of clams is best conducted with a slow drip/continuous feed

- Fishless refugiums and phytoplankton reactors are natural food sources

- Ample & stable supplies of calcium and carbonates are necessary for growth

---

Besides quality light and available nutrients, an aquarist must also realize the importance of maintaining adequate and consistent supplies of bio-minerals for calcification in Tridacnids. Indeed, consistency of water quality is of greater importance than reaching the high end of some idealized water chemistry. In this case, an alkalinity of 8-12 dKH and calcium ranging 350-425 ppm would be quite satisfactory to maintain Tridacnids.

Compatibility

A discussion concerning the compatibility of Tridacnids with each other or any other reef animal is delightfully simple. Essentially, any tank big enough to house Tridacnids with anything that won't eat them or overgrow them is a compatibility success! Unlike many other marine invertebrate groups, they exude no significant noxious compounds or chemically defensive elements. With enough space to grow and mature under an unobstructed view of appropriate light, Tridacnid clams will prosper. Beware, though, that some so-called "reef safe" creatures find such clams to be a delicacy. More than a few species of shrimp, crabs, and fishes may prey on clams but not necessarily corals. And once a clam is established in a safe residence, aquarists must of course be diligent in preventing sponge, coral, algae and other invertebrates from growing, overshadowing or stifling their passive Tridacnid clams. While Tridacnid clams seem to be somewhat indifferent to the sting of many reef corals

Mantis Shrimps   Triggerfishes   Cleaner Wrasses

Crabs   Angelfishes   *Lysmata* Shrimps

These reef creatures are somewhat to very likely to prey on Tridacnid clams.

H. hippopus (B. Neigut)    *T. squamosa* - aquarium photo (B. Neigut)

9" (22.5 cm) ***T. derasa*** - at a US retailer. (L. Gonzalez)

and invertebrates, the simple obstruction of light to the symbiotic clam by the overgrowth of a close neighbor can prove to be fatal. Aggression by cnidarian neighbors is surely inevitable with some species. As with coral, it is best to prevent stinging animals from growing upon or contacting Tridacnid species.

Special Notes

Categorically, the best clam species for display by casual aquarists are *Tridacna derasa, Tridacna squamosa* and *Hippopus hippopus*. These three Tridacnids are the hardiest members of their subfamily with all things considered. They ship well and acclimate to a wide range of light and water quality conditions. Aquarists commonly find that they are the least demanding Tridacnids with regards to light intensity and spectrum, and that they often fare better in aquariums that are overstocked, overfed and have a higher level of dissolved organics. Again, if any of the Tridacnids could fairly be called "starter" clams for

Tridacna maxima   ***Tridacna crocea***

These two species are frequently confused. *Tridacna maxima* has a more oblong shell with deeper grooves and very pronounced scutes. (*Top row: B. Neigut, Bottom row: L. Gonzalez*)

beginners it would be *Tridacna derasa*, *Tridacna squamosa* and *Hippopus hippopus*. The only significant consideration for keeping these three species is their adult size at over 18." As with all reef creatures, long term planning for growth is quite sensible and necessary.

Tridacna crocea and *T. maxima* are indeed the most popular and sought after clams for their extraordinary mantle colors in combinations too numerous to mention. Alas, they are also more likely than most Tridacnids to suffer from shipping-induced duress. Furthermore, they can be quite sensitive to waning or inadequate light over aquaria. For these reasons and

T. maxima is commonly imported at 2-3" (5-7.5 cm) but is not the best "starter" clam for new aquarists due to its strict demands for light and other sensitivities. (*L. Gonzalez*)

Mollusks: Giant Clams   243

more, the crocea and maxima clams are not to be recommended to new aquarists and beginners.

Tridacna gigas is somewhat of an uncommon import into America. The limited availability is just as well, since this true "Giant Clam" seems to suffer from handling and shipping trauma easily. It is also the fastest growing species (1/2" per month is easily possible). Few privately held aquariums can adequately house such a specimen in progress through adulthood to an extraordinarily large weight and length!

The huge *Tridacna gigas* at the Waikiki Aquarium, weighing several hundred pounds and captive grown for over ten years.

Tridacna crocea (L. Gonzalez)

Tridacna maxima (L. Gonzalez)

Hippopus hippopus

T. squamosa unusual blue, in captivity

Tridacna derasa (L. Gonzalez)

Tridacna gigas in Fiji

Right, *Tridacna maxima*. The *Acropora* coral seen here can quickly over-shadow and light-starve a clam if allowed to grow unchecked. (L. Gonzalez)

244 *Natural Marine Aquarium Volume I - Reef Invertebrates*

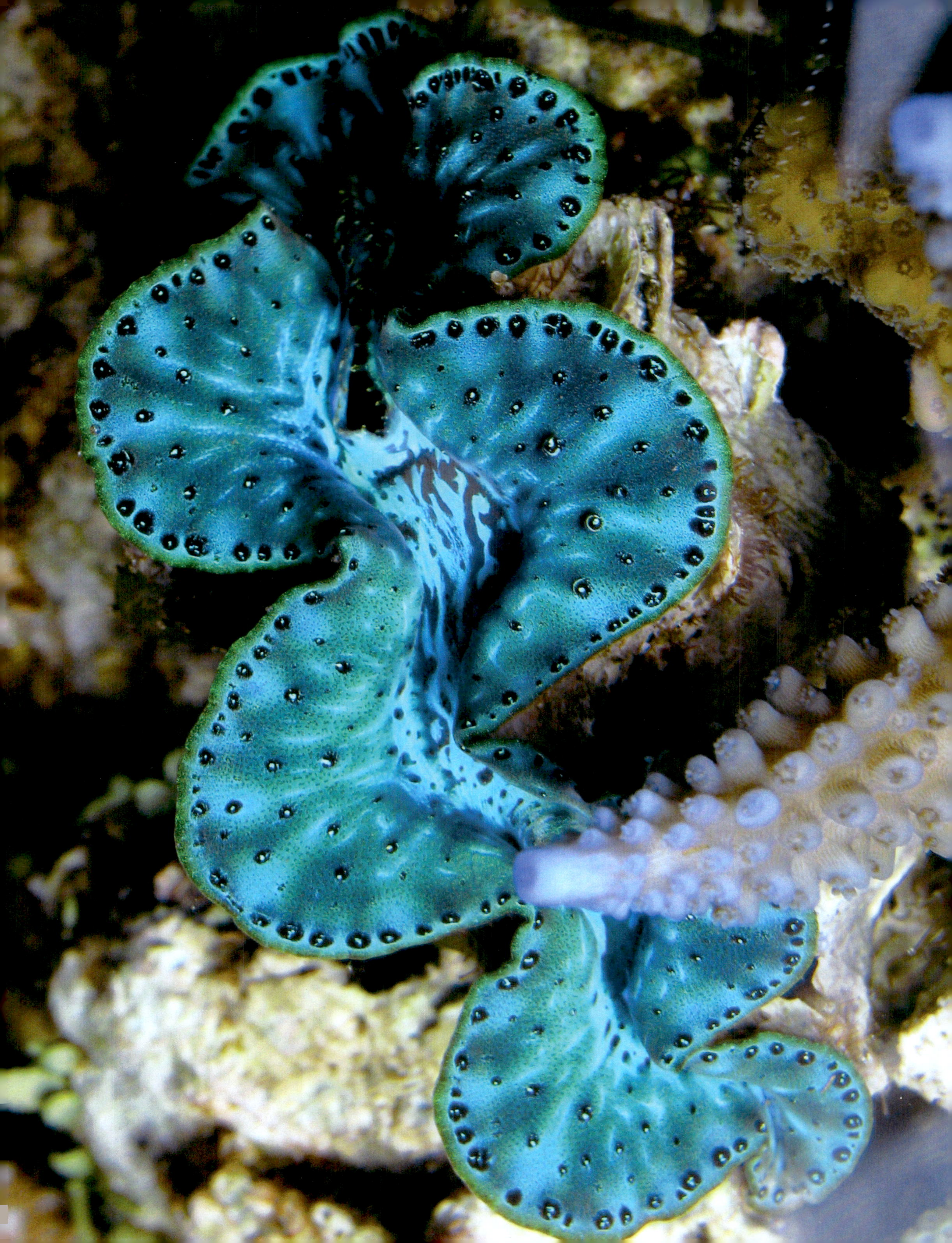

Octopus, Cuttlefish, Squid and Nautilus
The Cephalopods

It's difficult to even begin to find the words to describe such beautiful and fascinating creatures in the sea. Pictured here is a Flamboyant Cuttlefish, *Metasepia pfefferi*.

Class Overview - Cephalopoda

It is quite natural as aquarists to want to hear guidelines and "rules of thumb" regarding husbandry for various groups of animals among diverse choices in the hobby. With the entire class, Cephalopoda, we can fairly say that *every* specimen you will encounter requires specialized aquaria. In most cases, even then you will be limited to a single specimen per tank for their notable aggression and predatory nature. Here's your rule of thumb: one specimen of any species per tank unless specifically noted otherwise (*Nautilus* and a few cuttlefish, well fed, barely being exceptions).

In all likelihood, octopuses are the only cephalopod that aquarists will or should see in the hobby. Quite frankly, they are challenging enough for casual aquarists to keep and we rarely encourage it. Cuttlefish and *Nautilus* are imported on occasion but are impractical for more than a few good reasons. Most cephalopods are best left in the ocean for all but the most dedicated and specialized aquarists. Still, we write at some length here about each of these groups for both edification and inspiration.

Cephalopods are found throughout the world's oceans, from the shallows to the depths of the abyss, and in cold and tropical waters alike (marine only). Adult sizes range from a few centimeters to the giant squid, which is cited at more than 60 feet long (near 20 m). All are carnivorous and most are even cannibalistic.

Taxonomy

The taxonomy of this class of organisms, like so many, is somewhat unclear... or rather, it is at least in dispute. A strong camp suggests a five-order (extant species) system with Nautiloids in a subclass. Aquarists may recognize 3 groups: the orders Octopoda (octopus) & Sepiida (cuttlefish), and the *subclass* Nautiloidea (nautilus). Unfamiliar to hobbyists will be Teuthida (squid) & Sepiolida (cuttlefish-like cephalopods) which are not likely to find their way into the pet trade, and Vampyromorphida (deep-sea vampire squid)... as cool as they are (!)... are not mentioned here either for obvious reasons. All told, there are somewhere near 700 species of cephalopods. It's little surprise to the amateur naturalist in kind, for the privilege of so much good nature programming on television, that cephalopods are regarded as perhaps the most intelligent of the invertebrates. They are highly evolved organisms, indeed; they possess eyes with focusing lenses, and an efficient closed circulatory system, relatively large "brains," and are capable of demonstrable learning. In fact, they can be found on the top-10 lists of animal intelligence, out-ranking numerous mammalian vertebrates and other common household pets.

Anatomy

Some common traits among most or all cephalopods include their namesake "head-footedness." Their hallmark tentacles are attached directly to the head in some fashion or another that is common to all. **Tentacles** range in number from less than a dozen to as many as ninety by species. There are numerous subtleties and variations to these "feet" or arms within the group. Some are for grasping prey and others are for reproduction. Another very distinguished aspect of the foot-tentacles has evolved to form a siphon, the **hyponome**, which craftily allows them to move about by drawing water into the body (**mantle**) cavity and then expelling it forcibly out through the siphon- jet propulsion! A cephalopod can manipulate the direction of its siphon any which way it chooses to motor about. Another common trait shared by all cephalopods, save for the octopuses, is a chambered

246 *Natural Marine Aquarium Volume I - Reef Invertebrates*

shell. For most it is internal like with cuttlefish and squid, whereas the *Nautilus* displays their modified shell externally.

One of the most remarkable and formidable aspects of a cephalopod is its mouthparts. Nestled inside their tangled mass of tentacles is a hard, chitinous beak that resembles and performs rather like a bird's beak. Animals that get close enough to actually see this beak are in *dire straights* (the state of being... not the band). This powerful structure quickly and easily punctures and mauls even the toughest shelled crustaceans and bivalve shells, which comprise a significant part of cephalopod diets. A characteristically Molluscan **radula** acts like a rough textured tongue, ready to rasp and convey food in the mouth. With the familiar exception of deepwater species like *Nautilus* that have refined other senses for life in the dark, most cephalopods have very acute eyesight to support their decidedly predatory lifestyle.

Delving further into the anatomy of this group we find some very fascinating details of these highly developed invertebrates. In evolutionary stride, some are the largest in the sea. For their size and motile lifestyle, they have several hearts in an essentially closed circulatory system. That places them more than a few rungs up the evolutionary ladder! They also have an exceptionally advanced nervous system, and a brain-like assemblage of nerve cells that is strikingly reminiscent of a vertebrate brain mass. This brain-cluster (**ganglia**) actually controls many complex functions like the tactile manipulations of the suckers and tentacles. We are not just talking about the rudimentary instinct to grasp something and eat it, either... but rather thoughtful tasks like the legendary events of trained octopuses learning to open a closed-lid jar for food. At any rate, you can be sure that these extraordinary invertebrates present special challenges in husbandry, some of which you will read below, in part due to their intelligence.

Cephalopod Defense Strategies

Its like the old saying goes, "The best offense, is a good defense... so locate what offends you and kill it." Or something like that. There surely is a political joke in there somewhere too. And all is in jest, though, as most cephalopods are pacifists when it comes to combat. The primary strategy of defense with most every cephalopod is **evasion** for their relatively soft-bodied "defenseless" invertebrate physique. There are exceptions, of course, with a few frightfully aggressive species, and some fascinating mimics that will bluff aggression. For most cephalopods, however, their first response is to run like water through a first-time tourist in the Third-world... er, that is to say... they simply swim away fast.

There are indeed other strategies of defense among cephalopods. Passively, many exploit their remarkable ability to change color, texture, and pattern through cryptic and highly variable camouflage with **chromatophores.** Unwelcome eyes laid upon them are usually hard pressed to distinguish their visage from the surrounding seascape. In the aquarium, most will blend in quite well with the various rocks, ornaments and substrates offered. Aquarists can enjoy experimenting with different fixtures in the display to observe the wondrous color changes and camouflage. Mimicry in some species is an astounding refinement of this strategy. For some, it is a simple matter of having one or more pairs of large eyespots (being ocellated) used to confuse or dissuade a would-be attacker as to which end is the "business end," or just how big it is with these "eyes." In a few notable species, imitation is taken to another level by posturing in shapes that resemble other reef animals altogether.

"Mimic" and "Wonderpus" octopuses are the epitome of this behavior with a still fully unrevealed repertoire of imitations that can be shuffled rapidly as needed. One of the primary reasons suspected for this evolution is the need for protection and safe conduct for these uniquely diurnal species. Unlike most cephalopods, which are active at night, the mimics can exploit the daytime niche through

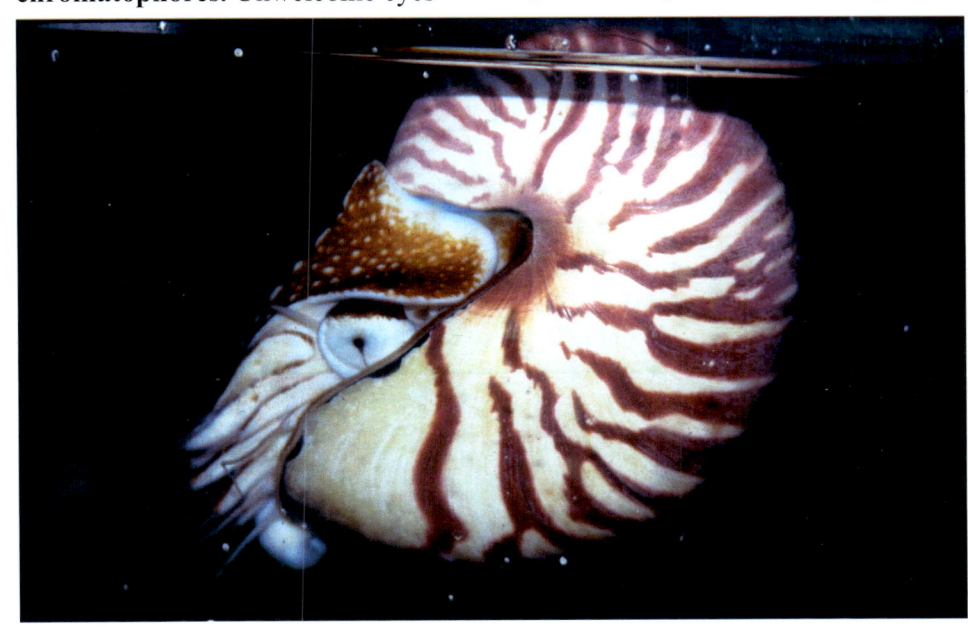

A very common and clearly stressful problem with captive *Nautilus* is trouble with buoyancy. Specimens like this will bob and float at the aquarium surface while struggling to descend. From their natural habitat at depths beyond that which any man can dive freely, to their cool water needs and great sensitivities to water quality, the Nautiloids are best left to expert aquarists. (A. Calfo)

This series of images depicts the same specimen photographed in sequence, illustrating the amazing ability of cephalopods to qui change their appearance by **chromatophores.** *Octopus cyanea,* The common Day Octopus is so named for its unique diurnal behavior. Indo-Pacific distribution: here in Queensland, Australia.

their adaptation. When a territorial damsel threatens to bite and nip the octopus out of business in the area, the mimic can rise up to imitate a sea snake- a natural predator on damsels. If a rowdy crab or similar benthic bully threatens the peace, the mimic octopus can emulate a wicked Stomatopod (the Mantis Shrimp). For bigger bullies and would-be predators, the mimic octopus has even been observed to rise up off the ocean floor in an extremely vulnerable position to mimic a venomous lionfish (like *Pterois sp.*). Talents with mimicry in cephalopods, however, are not only used in defense. They may imitate schooling prey to swim amongst candidates from which to make a meal, or they may simply imitate a less-threatened animal like flatfish to cruise about by day innocuously. Modes of mimicry are remarkable and one of the most intriguing reasons for keeping and studying cephalopods at large in aquaria. Populations of mimics specifically, however, are still uncertain and their appearance in the wild is rare. Limited experience with imported specimens has not been promising as the Mimic Octopus has been sensitive to handle, slow to feed and suffers a poor rate of survivability on import. It is also suspected that they need extremely deep, fine sand beds (over 8"/20 cm) to bury in for periods of time. Aquarists are advised to leave the Mimic and Wonderpus in the sea until we have a better indication of the potential for sustainable harvest and successful shipping techniques.

Inky-do

One of the most notorious means of cephalopod defense is **sepia**, the ink-like substance used to create a confusing cloud for evasive maneuvers. Ask most folks what color they think the ink is and you'll usually hear black, brown, or perhaps even blue or purple. Red probably would not rank high on the list of guesses. Nevertheless, red wins a prize too (deeply concentrated). Sepia is primarily a highly concentrated solution of the pigment **melanin** and mucus, plus some **tyrosinase** (used to irritate eyes and perhaps temporarily afflict the sense of smell). It is produced in a little sac that most cephalopods have functional from birth. The mucus helps to coagulate the melanin into globules, which may act as a decoy (bait). Else, less mucus in the mix creates a smoky screen. Some deep-sea species do not appear to have or need an ink sac. Similarly, older octopuses are less inclined than youngsters to issue it. Theories abound for the reasons why, but we can only guess at this point. "Inking" in the wild or in the aquarium is a sign of severe fright or stress and is used as a final means to avoid predation or harassment. Part of standard protocol for good cephalopod husbandry in aquaria is having enough conditioned and ready-made fresh seawater on hand, plus quality chemical filtration media, in case the emergent need for large or complete water changes arises. Not all cephalopod inking events in aquaria require such drastic measures, but that does not preclude the safety and sensibility of being prepared.

Cephalopod Compatibility

Cephalopods are a truly challenging class of reef invertebrate for aquarium life. Considering their need for space, constant live foods, oversized filtration to handle messy feeding habits, demands on superb water quality (including high oxygen levels), hostility to most every other living animal including each other, and their proclivity to escape and become carpet fodder... ask yourself if you up to the challenge? Most species will probably have to be kept them by themselves... and most species are nocturnal. Short of a few brief-shining moments each week at feeding time, most cephalopods are quite boring for the casual aquarist. Make no mistake, we have seen too many kept by the ill-prepared and strongly encourage most aquarists to admire these beauties from afar.

In passing, we'll mention that natural predators on cephalopods are many. This is not surprising for how soft bodied and tasty they are (did someone say, *tako*?). Eels and triggers rank high on the list of willing predators. Numerous other indiscriminate and opportunistic fishes and crustaceans will steal and arm or more if they can. The tables are turned for many small reef animals, though, as cephalopods are dedicated carnivores. The greatest of them all, if legends are true, is the giant squid (*Architeuthis*): alleged to attack sperm whales. As the largest

248 *Natural Marine Aquarium Volume I - Reef Invertebrates*

creature in the animal kingdom, this may well be true.

For additional information on cephalopods, consider reading, "Cephalopods - A World Guide" by Mark Norman and Helmut Debelius, and be sure to visit Dr. James B. Wood's "The Cephalopod Page" on the world wide web (see resources).

Order Octopoda: Octopuses

Characterized by having no shell (unique in the class, on whole), a sharp beak, and **eight arms.**

Aquarium use - Octopus:

Of the various divisions in the order Octopoda, hobbyists will only find members of the family Octopodidae in the aquarium trade. We'll spare you any taxonomy beyond this point and leave you to your vices with science television programming and Hollywood deep-sea movies for the rest. Of all cephalopods, octopuses are the only members of this class that aquarists have any realistic chance of finding and being able to keep healthy with strict husbandry and preparation well in advance. By no definition are these animals likely to survive an impulse purchase from an overly enthusiastic aquarist. Beyond issues of aggression and the need to be kept in isolation, many are so temperature sensitive that they need a chiller on top of an already high-end filtration system for their voracious diets and messy feeding habits. If you are willing to make the investment in hardware and research needed to keep an octopus, there are a few species suitable for aquarium life. Consider the following, though. They are escape artists that can climb out of water and squeeze through an opening that is only a fraction of the diameter of their head. They are sensitive to temperature (it needs to be stable and slightly cool or cold for most species offered for sale), oxygen levels must be high (near saturation), and trace contamination must be strictly guarded against (they are intolerant of metals, medications, and organic dyes especially). Even common sea salts and reef supplements rich in metals (magnesium, for example) can prove fatal. They are easily disturbed and readily ink, especially when small, from even the occasional routine and necessary tasks of aquarium maintenance. The natural lifespan of tropical species is categorically *less* than coldwater denizens with most living a mere 12 to 18 months total. The unknown or mature age of specimens acquired for the hobby does not bode well for prospects of long-term enjoyment. The investment of more than a thousand dollars in hardware, including a chiller, for some just doesn't seem like such a great idea for an animal that might only live a few months. Even the largish darling of public aquaria, *Octopus dofleini,* the Giant Octopus only lives 3-5 years if fed well and kept at a chilly 10 degrees Celsius. Do you still want an octopus? Consider *Octopus bimaculoides* as one of the very best cephalopods for captivity if you are inspired to keep one. We don't disagree that octopus at large are fascinating and wondrous. They are smart to the point of being able to use tools, learn and remember complex tasks, and have focusable, color vision. But they simply aren't practical for casual aquarium keeping.

Should you choose to specialize and endeavor to keep cephalopods, look to academic and scientific literature for the species you pursue. Many have been kept, studied and even cultured by science, and information on specifics of husbandry may be available to you. Seek fisheries data too if available, although such will be focused on larger and likely cooler water species harvested for food. Feeding of most cephalopods is fairly easy. They are not very discriminating. Fishes, crustaceans and some bivalves are favored food items. Octopuses usually ignore gastropods, although cuttlefishes relish them. A common staple in captivity is the crayfish (*Procambarus* and like genera) or grass shrimp (*Palaemonetes pugio* and like species). Most any shrimp or crab-like arthropod will be taken. Be sure to gut-load prey with nutritious foods (spirulina-based dry foods, algae, krill, plankton, etc) before feeding them to the octopus, especially if variety in the

The "Mimic Octopus" (*Octopus sp. 19*) is a celebrated cephalopod for its highly evolved state of mimicry. Mimicry is used at times for defensive posturing in imitation of more dangerous reef denizens like Mantis "Shrimp," Lionfish, and Sea snakes. It is also used to stalk prey or simply feed by day in the guise of a less-threatened organism like the banded flatfish (Sole). The Wonderpus *Octopus sp. 5* (not pictured) is a similarly celebrated mimic. Pictured here in North Sulawesi

Octopus vulgaris, the Common Octopus: a relatively large species ranging worldwide in tropical seas. A commercially important fisheries species for human consumption and a common import into the aquarium trade. This species is occasionally active by day and lives 12-18 months. Its natural diet includes bivalves and crustaceans. This one photographed in Cancun, Mexico.

Octopus macropus, The White-Spotted octopus: a common species observed by divers in a wide distribution around the globe (tropical waters). This reef denizen preys mostly on hermit crabs. A small individual pictured here in Cozumel.

captive diet is limited. When feeding crustaceans, you may notice that your octopus leaves piles of husked prey (bivalve shells, bits of crab and shrimp exoskeleton, etc) outside of their lair in **midden piles**. These piles in the wild are useful, in fact, to scientists and students for discovering not only the captive diet of a given octopus species, but the diversity of life that might otherwise go undiscovered if not for the cephalopod's far-reaching hunt for prey, which includes the depths and crevice of the living reef.

Be careful when feeding cephalopods. **Never feed these animals by hand** but rather use tongs or skewers if necessary to target feed. Their beaks possess **cephalotoxins**, which largely do not affect humans, but can severely afflict us in rare cases. Beyond the infamous, and potentially fatal, Blue Ring octopus, *Hapalochlaena lunulata*, other species have been known to have potent bites. Reports have suggested that symptoms may be like the sting of "mildly" venomous scorpion fishes like a *Pterois* Lionfish: a severe burning sensation starts at the puncture wound and progresses, swelling ensues followed in time by blisters and possible necrosis. Treating the local area of the bite within minutes with hot water has been said to help, anecdotally. Any bite or wound from a marine organism should be treated immediately by a doctor, as there are septic concerns beyond issues of potential envenomation.

Order Sepiida: Cuttlefishes
Characterized by having **ten arms** (two are extended for food gathering), and a porous internal shell.

Aquarium use - Cuttlefish

Cuttlefish are every bit as fascinating, albeit less common, to aquarists among the cephalopods. They are most famously recognized for their hypnotic color changes that at times occur at duration and in such a myriad of colors that you can't help but liken them to a living kaleidoscope. This adaptation is theorized, in fact, to literally be confusing if not hypnotic to prey, affording a hesitation that proves to be fatal for sighted animals sought as food. Although still uncommon in the aquarium hobby, several species of cuttlefishes are cared for and cultured, not the least of which is the well-documented variety, *Sepia officinalis*. Surplus stock and availability has been limited mostly to zoological and academic (research) organizations. It is only a matter of time before enough broodstock makes it into the hands of dedicated aquarists that forge the industry in a waiting and willing market for the hobby. Be sure to follow the leads to individuals and research cited in this chapter and in the bibliography for further information on the specifics of husbandry including successful captive breeding efforts.

Like most cephalopods, one of the greatest challenges to keeping cuttlefish is their extreme sensitivity to shipping and handling. The thought of air shipping any is daunting, and you are strongly encouraged to find an outlet regionally. They suffer all the same concerns in husbandry as the above-mentioned octopuses regarding inking, water quality and conspecific aggression. Importing smaller specimens is logically better for the higher ratio of water and oxygen per animal... not to mention consideration for their very short lifespan. You may even be able to acquire eggs, which are much easier to hatch than octopus (requires brooding).

Cuttlefish occur in a wide distribution across the globe in temperate and tropical seas alike. If you find a source for live cuttlefish, be sure that you are correct on the identification of the species and its needs. Even tropical species barely tolerate 77F (25C) at the highest end of their range. High temperatures markedly reduce the lifespan of even tropical species. As with most cephalopods, a chiller will almost certainly be necessary. At any rate, temperature stability must be ensured. They also are far more active than octopuses and require larger and longer aquaria to thrive. The size of the display will depends on the needs of a given species, but suffice it to say that vessels 4 foot long (120 cm) or 40 gallons (150L) are a fair minimum. There is a chance of sociability with some cuttlefish species, although we strongly recommend against it in practical aquariology without a very large aquarium. One specimen per tank is always best and safest with these carnivores and "cannibals." Almost anything else in the aquarium, short of Cnidarians (corals, anemones, jellyfish, etc) is likely to be killed if not eaten. Favorite foods include most of the Arthropods (shrimps, crabs, and the like) and fishes, as well as many mollusks. Through bony fishes and chitinous invertebrates, cuttlefish derive a substantial amount of bio-minerals to form their internal shell (AKA "cuttlebone" familiar to many bird keepers as a calcium supplement). Although live food is preferred, cuttlefish take thawed frozen meats more readily than octopus. Review the notes on feeding octopods above and take heed of concerns with gut-loading prey, never-hand feeding, and treating bites should they occur. At length, cuttlefish present similar challenges to aquarists as the keeping of octopus, but ultimately must be regarded as somewhat more challenging. They are more active, require more food and subsequently need larger aquaria, better filtration, and better husbandry (water changes, attention to filter maintenance, etc). We do not encourage the keeping of any wild imported specimens by casual aquarists and would guide specialists to seek domestically cultured broodstock at universities and large private aquaria.

Subclass Nautiloidea: Chambered and Pearly Nautilus

Aquarium use - Nautilus:

Throughout this text and our other works we have used the phrase "casual aquarist." It has never been used with a negative connotation and simply is a moniker to sincerely describe what most of us are: hobbyists. We have great admiration for the sea and aquatic life forms and we endeavor to keep them to the best of our abilities.

Most of us are somewhat limited from delicate or challenging species by our time, space or disposable funds available. As such, we are "casual aquarists" with a noble and realistic purpose in our endeavors to study and enjoy hardy reef invertebrates. In contrast, "dedicated" aquarists, or perhaps better-termed "specialists," do not necessarily have more passion to keep aquatics, but rather the *means* to do so. Such means might entail the purchase of a full reef lighting system to keep a single anemone properly in an aquarium without any other cnidarians (anemones, corals...). Closer to the heart in this chapter, it would entail the purchase of a lot of expensive hardware including massive filtration and water chilling units to keep a single cephalopod. Even then, a challenging and potentially laborious journey with much more time invested to successful husbandry awaits. It is beyond casual aquarium keeping and is the realm and reality of keeping most cephalopods, and certainly all Nautiloids in captivity.

Importing live and healthy nautilus is an extraordinary feat. Like most cephalopods, they are extremely sensitive to low dissolved oxygen, increased water temperatures and the inevitable rigors of transit. Let's be very clear here, chambered nautilus have been tagged and measured to descend daily to an average depth of nearly 500 meters! Can you even begin to conceive of what the physical

Sepia pharaonis, The Pharaoh Cuttlefish: a fast-growing and warm-water species that is cultured in Asia and occasionally found in the aquarium trade. This cuttlefish grows to about eighteen inches (45 cm) and may live 2-3 years in captivity. They are very aggressive - keep one specimen per tank! Photographed here off Gili Air, Lombok, Indonesia.

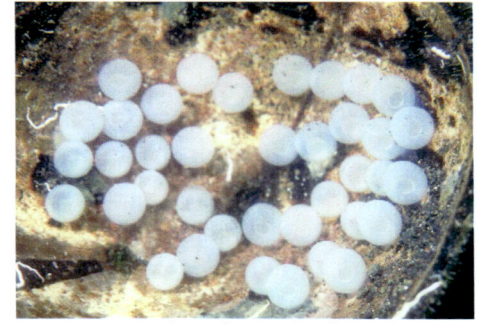

Metasepia pfefferi, The Flamboyant Cuttlefish: this species grows to about four inches (10 cm) in length. A photogenic cephalopod, it is almost always photographed "disturbed" in its brilliant coloration. This species can be observed hunting actively by day, searching with little discretion for fishes, crustaceans, and gastropods. Its broad range includes the Western Pacific: Australia, Indonesia, and the Philippines. Pictured here: An "alarmed" specimen (left), and a batch of eggs (right) laid inside of a coconut shell. The eggs of some cuttlefish can be shipped and hatched, unlike those of octopuses which require deliberate brooding.

Sepia latimanus, The Broadclub Cuttlefish: A large species that grows to about twenty inches in length (50 cm!)... it can be found in the Red Sea and widely throughout the Indo-Pacific. Prawn (shrimps) make up a significant part of its diet. N. Sulawesi photo here.

dynamics of the environment that they spend much or most of their life in is like at 1500 feet below the surface of the ocean? Cold and dark does not even begin to describe what must really be an awesome environment. The pressure alone is testimony to the integrity and natural wonder of their chambered shell, which they use for buoyancy by regulating air and water between the perforated segments. Thus, shipping in a sealed and un-pressurized bag of warm seawater from the Philippines, for example, to a wholesaler anywhere in the world must exact considerable duress on the animal. Perhaps related to life in atmospheres above sea level, many nautiluses have great difficulties with buoyancy in captivity. Few private aquarists can keep these cephalopods alive for very long and fewer anywhere in the world can raise their offspring successfully. They require cool water temperatures unconditionally. Captive lifespans will likely be cut to less than 6 months if held over 70F (~20C) [pers. comm. Dr. Bruce Carlson]. Ideal temperatures for captive specimens should be around 64 F (18C) or slightly cooler. Their natural lifespan is unknown but we do know that they are much longer lived than most any other cephalopod at perhaps 20 or more years to senescence. Feeding is one of the few easy aspects to nautilus husbandry- they eat most any meaty fare live or dead that they can get their tentacles on, including each other at times. Shelled crustaceans and bony fishes are necessary to wear beaks and for providing the considerable bio-minerals needed to manufacture shell. In fact, their propensity to gnaw is legendary and aquarists have learned that they must keep Nautiloids in specialized aquaria. This nippy mollusk will scratch and gouge acrylic, and will chew away the silicone in standard glass aquariums! Insulated glass aquaria with guarded seams have been employed successfully. For the necessarily cool temperature of their water, single pane glass would otherwise form unsightly condensation and obscure viewing. Very large aquaria are standard for keeping this animal healthy too. Vessels of several hundred gallons or more are recommended for just a few individuals. For these reasons and more, let us persuade you to enjoy this ancient marvel through aquatic science books and film instead.

Unique among Cephalopods

In gross form, nautilus share common traits will other members of their class (head-footedness, beak, tentacles, hyponome), but differ in some significant ways from cephalopod brethren. As a deep-water species, they lack the acute vision and focusing lens of squid, octopus and cuttlefish. Instead, the nautilus has a primitive eyehole almost literally like a pinhole camera. Of course, feeding in the shallows at night and living at depths of several hundred *meters* certainly dispenses the need for precise vision as we understand it. They have a much more heightened sense of smell instead. Another notable distinction is their external shell, which serves as protection from both predators and pressure at depth, at the expense of agility and speed.

Sepia sp. Dwarf Cuttlefish in N. Sulawesi

Sepia sp. pair of Dwarf Cuttlefish, N. Sulawesi

Nautilus also lack the ability to color change conspicuously like most other cephalopods, but this is in stride with their lack of diurnal habits. Their tentacles lack suckers but number more than most any other member of their class with perhaps as many as eighty or ninety. And alas... as cool as they look, they are not as "intelligent" as other cephalopods and behave nearly so primitively as they first appear to be.

Classification

Nautilus are strictly marine animals and believed to be limited to the tropical reefs of the Western Pacific, although this may not true given how very little is know about their habits and distribution. There are seven extant species recognized: *Allonautilus perforatus, A. scrobiculatus, Nautilus belauensis, N. macromphalus, N. pompilius pompilius, N. pompilius suluensis, N. stenomphalus.* In time, we are likely to discover new species or subspecies, as this subclass is better understood. Time and technology will surely reveal new information about this living fossil. In fact, Nautiloids (extant) and Ammonoids (extinct) once were dominant groups in the sea with about 3000 fossil species recognized to date and a presence in the record for roughly 500 million years. Some scientists believe that the modern Nautiloids have lived relatively unchanged for about 400 million years, and that makes them a *bona fide* living fossil by any definition!

Nautilus are victims of their own natural beauty and uniqueness; they are collected in great quantity for their shells as ornaments. Thousands are taken from seas each year for the curio trade, with a small number traded live for hobby and zoological collections. We know from science that they have extremely slow reproductive rates, thanks in large part to the studies of Japanese researchers and the work done by the Waikiki aquarium in Hawaii with Dr. Bruce Carlson. *Nautilus* mature sexually at 5 to 10 years of age but only lay about 12 eggs annually. Not all hatch, and not all can be expected to survive, naturally. To say that this animal could be in peril of over fishing would be a gross understatement. They grow slow, breed slow, and we really have no idea how large or small their natural populations really are. If history and human impact on such animals is any lesson (whales, sharks, rhinos, etc), then we need to be soberly realistic about the impact of any human activity, be it shell collecting or aquarium collections, on this living treasure. While we would not want to discourage any serious aquarist with resources and skills from keeping live Nautiloids, we are saying that even buying *Nautilus* shells for merely curios or fancy is perhaps unnecessary.

Anatomy

The shell of a *Nautilus* is divided into chambers (**phragmocone**), which are further divided by "partitions" (**septa**). The newest chamber, or last chamber formed, is the living chamber, which is the largest and contains the hood, head, sheathed tentacles, etc. All chambers are pierced with a tube (the siphuncle), which controls the flow of "air" and water between the chambers to maintain neutral buoyancy. Although cited at 750 meters deep (nearly 2500 feet!), few are suspected of venturing beyond a "mere" 300 meters (~ 1000 feet). Like other cephalopods, they use a hyponome to channel pumped water drawn into the mantle. All are dioecious (separate sexes) and oviparous (egg-laying). We are not aware of any private aquarists successfully rearing spawned *Nautilus*, whose eggs can take a full year to hatch.

Nautilus belauensis, the Pearl Nautilus. Showing the characteristics of all extant species: the spiraled shell containing the chambered (air/water) sections utilized in buoyancy control. This species and *N. pompilius* comprise what little representation we see in the aquarium trade.

254 *Natural Marine Aquarium Volume I - Reef Invertebrates*

ARTHROPODS

Arthropods (Arthro- "joint"; Pod- "leg"): we all know these animals, the insects, crustaceans, spiders... and well we should, as there are more of them on the planet than any other phylum. They have infiltrated the land, sea and air around the globe. Insects, mites, spiders, millipedes and caterpillars surround most of us every day, even when we cannot see them. Aquarists have admired and enjoyed quite a few arthropods from the earliest days of the hobby. In this group, there is something for everyone: hardy, long-lived, useful, and highly ornamental species abound. With perhaps more than a million species in this phylum, however, there are also plenty of undesirable members that grow too large, too demanding, or simply predatory. There are species that prey on parasites and live as useful cleaner-animals, and there are members of this group that *are* parasites (isopods). We find arthropods in our aquariums brought in as incidentals with live substrates, and we see numerous species deliberately collected and offered for sale. Their presence in the modern marine aquarium is necessary and inevitable, from the microcrustaceans (plankton) that flourish in refugia, rock, sand and elsewhere in the system, to the large ornamental reef shrimp and crabs that grace the reefscape.

It would be a veritable understatement to say that the taxonomy of the subphylum Crustacea is both complex and unsettled. For those interested in the most strict interpretations and expressions of modern nomenclature, we ask for your indulgence and understanding of the enormity of the issue. For casual aquarists, we encourage you to study and recall the scientific names of organisms with, if not in preference to, the common names of animals for a standardized language with which to exchange information globally in the Information Age. With that said, the basic groups or categories of crustaceans familiar to aquarists include: Microcrustaceans (isopods, amphipods, and other zooplankton), Brachyura (true crabs), Anomura (false crabs), Natantia (shrimp), Merostomata (horseshoe crabs), Palinura (Lobsters), and Stomatopoda (mantis shrimp).

To make some generalizations about crustaceans at large: **crabs** are likely the hardiest of the useful species in this group (hermits, porcelains, and true crabs), **shrimp** are better behaved but more sensitive, stomatopods (**mantis shrimp**) are predatory but make great feature display animals, **isopods** are generally parasitic, and **lobsters** have little place in private systems for their aggression or adult size. All arthropods are somewhat sensitive to water quality and rapid or sudden changes. Osmotic shock is readily induced with simple changes in salinity as with a rushed or improper acclimation, evaporation top-off with freshwater, or an aggressive water change. Iodine is needed for their exoskeleton and molts, and it can be provided through food or liquid supplementation. The frequency of molts is highly variable by species and gender. Nonetheless, frequent occurrences are a good indicator of health and growth. In preparation for a molt, an arthropod creates a new but soft exoskeleton beneath the old hardened one. In due time, the old shell splits (usually down the back or at the base of the tail) and the "bug" (shrimp, crab, lobster, etc) walks out of its old skin, so to speak. The new soft animal swells up measurably (25% in size for many) before hardening to accomplish a successful event of growth. Exoskeletons are made of proteinaceous chitin and are often consumed by the former owner. If not, many other creatures seek to eat it for the same reason. While the freshly molted and vulnerable arthropod is still soft, it will go into hiding for several days.

Keeping and even breeding many of the joint-legged animals is fairly well-defined. Most need very stable and high water quality near natural seawater levels. While dilute seawater can be stimulating for some fishes, it is very stressful for most arthropods who depend on full-strength seawater for delivering crucial bio-minerals and to prevent osmotic stress. Even the intertidal species appreciate stable water quality.

Most arthropods are generalists when it comes to feeding. They usually are flexible omnivores that will eat both green and meaty matter. There are some notable exceptions such as the Porcelain crabs which have evolved to consume fine particulate matter (filter-feeding). Success in keeping arthropods healthy, as well as at peace in the aquarium (avoiding aggression), is naturally a matter of research and good husbandry. Once you've identified candidates you'd like to keep, you must the determine the size, number and gender (where applicable) of the creatures that strike your fancy. If you have the means to satisfy their needs in the aquarium with regard to their husbandry issues, you must then navigate your new specimens through the acclimation process. Be mindful of the sensitivities of these shelled creatures and ask specific questions of your dealer about the prospects you seek or see. Ideally, new imports will be allowed to stabilize in your dealers tank for at least a week or more before being sent into a proper quarantine (QT) tank at home for another few weeks. Please note the advice on QT protocol herein and do not let the lessons there and beyond be lost on you.

Whether in quarantine or the display proper, know that compatibility issues with arthropods are many. Territorial disputes are not uncommon from small pistol shrimps that think that they can chase everyone away, all the way up to the 24" (60 cm) Lobsters in the sea that *know* they can! Explore their history and husbandry and discover the many fascinating behaviors of arthropods beginning here.

Stomatopods
The Mantis Shrimp

Large or small, the stomatopods (AKA Mantis shrimps) can be formdiable predators and are highly intelligent creatures. Pictured here is the "business end" of *Odontodactylus scyllarus* - magnificently colored and one of the so-called "Peacock species." (*K. Gosinski*)

Traditional reef aquarists (keepers of cnidarian displays) often cringe at the thought of mantis shrimp in their aquarium. Stomatopods are certainly efficient and highly evolved predators upon many desirable reef invertebrates from the smallest snails and bivalves to some surprisingly large fishes. But this book is a guide to all reef aquarium life, embracing available livestock for community displays as well as species-specific aquariums. It is a celebration actually of all reef invertebrates, and in that spirit we've included the often beautiful and always fascinating stomatopods. Mantis "shrimp" are only distantly related to those other crustaceans that we call true shrimp. They get their insectorial moniker ("mantis") from their specialized forelimbs (raptorial appendages) that resemble the praying mantis insect (*Stagmomantis*). Stomatopods can be found worldwide in both tropical and temperate seas commonly ranging in size from less than 1" (2.5 cm) to 12" (30 cm) in length with citations even larger. They tend to live solitarily, though monogamous pair bonds are formed in the wild. While their young still go through planktonic stages, such pairs settle out of the plankton together and create burrows where they may live together for the rest of their lives. Life expectancy is over five years for most species and longer for the larger varieties.

Selection

Getting past the stigma, legends and lore, there are several hundred species of stomatopod among which some are attractive, entertaining and welcome guests in (some) aquaria. Many of the small incidental species acquired with live rock can be quite harmless and easily kept with most desirable reef invertebrates (corals, anemones, echinoids, etc) and all but the tiniest or slowest fishes. For the more aggressive or larger mantis, a species tank is well worth the time to set up for study and enjoyment. Small or large, peaceful or aggressive, it is a safe assumption nonetheless that mantis shrimp should be kept as one species and one specimen per tank. They are fairly intolerant of each other, but opposite sexes in large quarters are advised if attempts are made at mixing individuals. There is always some chance that individuals will take advantage of the other when it molts and is vulnerable.

Habits and activity vary greatly by stomatopod species. Many are diurnal, yet you will find nocturnal and even crepuscular (dusk-dawn) species as well. Stomatopods are divided into two broad categories by feeding mode: the "**smashers**" possess a club-like appendage used to crush prey, and "**spearers**" utilize a set of spiny forelimbs to impale and capture soft-bodied prey. Spearers have been recorded as striking prey with one of the fastest known movements in the animal kingdom, up to 10 meters per second! In reef and community aquariums, the smashers tend to be more destructive in aquaria than the spearers for the higher abundance of suitable prey for the former. Both types tend to favor some type of burrow although the smashers are more inclined to stay visible or in the open. Some industrious individuals will carve burrows out of carbonate rock if there are no acceptable bolt holes in the aquarium. A 1.25" (30 mm) *Gonodactylus smithii* was observed to bore a lair approximately 6 cm X 10 cm X 2 cm into a piece of aquacultured Florida live rock in less than one week (Mike Bloss, pers. comm.).

Smashers occupy open areas of the reef and more often can be seen on hard substrates. These are almost always the incidentally imported species with live rock, corals, etc. They typically feed on hard-bodied creatures like true and false (hermit) crabs, and numerous mollusks (snails and bivalves are favorites), which they stun and break open with a sudden strike. Gonodactyloids (mostly smashers) are perhaps the most familiar and loathsome members of this phylum. Not only are they the most common and destructive stomatopods in reef tanks, but they're highly adaptable in captivity. Some are very capable of cutting your finger open or stabbing a fish... after they've smashed and killed all of your snails!

Spearers are far more secretive in the marine environment and prefer to live in burrows in soft substrate like sand and mud as dedicated ambush predators. They are rarely imported incidentally and must be deliberately collected for the aquarium trade. Squilloids are some of the few members of this group familiar to aquarists or divers. Lysiosquilloids are mostly larger and less common in the trade. Their spearing aspects bear anywhere from three to seventeen sharp spines that are used to literally impale soft creatures like shrimp and fish.

Selecting mantis shrimp specifically for the aquarium requires the same attention to fundamental signs of good health and behavior listed for other crustaceans. Avoid buying specimens with missing appendages. Although they can and will regenerate missing "parts," they are not legendary healers

Large mantis species are rarely acquired incidentally with live rock, often for their sandy habitat preferences, and are usually collected and traded deliberately. (*L. Gonzalez*)

Arthropods: Stomatopods

like some other crustaceans and, more importantly, absent limbs may indicate a severely stressed animal. A healthy individual is always alert and never seen sitting sluggishly in the open flat of a dealer's tank. The merchant will hopefully have provided a bolt hole, piece of pipe, or crevice for the mantis to hide in and reduce stress during holding and quarantine. Oddly enough, these wicked predators can be slow to feed at first and may be hesitant to eat dead foods (more about this below). Be sure to see any candidate feed in the aquarium before transferring them to another display. A potentially good specimen may need a little more time to regain an appetite after the stress of import. A premature move of a newly imported specimen could doom an otherwise viable candidate.

Care

Husbandry for mantis shrimp is relatively simple and straightforward. They have no special sensitivities or needs that differ from most any other crustacean. Once established, they are fairly hardy with regard to water quality and survive in any reasonable kept system in the range of natural seawater conditions. As with all invertebrates, you will want to avoid low salinity, metals and organic dyes. As with all invertebrates, also exercise caution using antibiotics and other medicants.

Ideally, stomatopods should be kept in a display tank of their own, with adequate filtration to handle the metabolic wastes produced by heavy and messy feedings (rather standard protocol for predatory animal care). There is a legend about glass aquarium breaking potential with larger individuals. We recommend that you avoid keeping big mantis shrimp (over 4"/10 cm) in thin glass aquariums. On occasion, they have been said to crack or break the walls of their containment, though neither of the authors has personally seen this first-hand. When selecting an aquarium, be mindful of the adult size of your targeted species. Although they are not as active or demanding as fishes for tank size and length, mantis need a comfortable aquarium to insure good steady water quality if nothing else. While there may not be a rule of thumb on proper stocking densities for stomatopods, we should all be able to agree that a species that potentially grows to be 8-12"

Odontodactylus scyllarus, the Peacock Mantis Shrimp (AKA Clown Mantis), is a beautiful and legendary "smashing" species. The multitude of colors are almost hypnotic and have made this stomatopod popular with aquarists that have a flair for the wild and dangerous side of life. Peacock mantis are a large species that approach 8" (20 cm) with a range that includes the Indo-Pacific through to Hawai'i. Pictured here on sand in the open in Indonesia.

as an adult (20-30 cm) is wholly inappropriate for a tank merely three times as long as its body. Provide a suitable habitat for your mantis type: soft substrates for spearers and rocky seascapes for smashers. It is sensible and strategic to install planned bolt holes or crevices to better contain and enjoy your stomatopod's visible residency in the aquarium. Capped plastic pipes buried at the base of rockwork will often make a fast and inviting home for your mantis shrimp. Let the open end project just within view and place several around the aquarium for easy feeding and photographic opportunities. Burrowing species also appreciate small rubble or shells which, they'll utilize for a stable burrow if left to their own devices in natural sand and mud substrates. In such circumstances, be sure to provide adequate substrate at depth: more than 3" (75 mm) is likely necessary. For some species, a large ornamental shell like a Conch may be the best decorative home you can provide. From just these few suggestions, we are sure you can envision other creative possibilities for stomatopod habitat and housing.

Handling

After all of the admonitions you've read here on mantis shrimp "weaponry," it should go without saying that extra care in handling is called for. These animals have truly earned their nickname "Thumb Splitters" by divers, fisherfolk and collectors worldwide. There are even some wild legends about large mantis shrimp severing digits from people. The latter claims are falsehoods, but what *can* happen from an injury by mantis shrimp, as with any wound in marine ecosystems, is a chance for infection. Smacks, taps and slashes from small mantis in aquaria will hurt, and some will draw blood. If it happens, rinse in hot water, disinfect the wound and be mindful of the possibility of needing routine medical attention and antibiotics. Prudence and

Lysiosquilla is a "spearing" genus of Stomatopod. They prefer to lay in wait at the mouth of their burrow for the chance to strike out at a soft victim that happens by. The wide range of this genus includes the Philippines and Indonesia.

prevention are the best strategy for safe handling and possible removal of mantis shrimp by traps, nets and bags.

Feeding

Feeding mantis shrimp is not so much an issue of "what" to feed, but rather "how." Some like to be teased with dead prey on a stick (thawed frozen meats), other find this irritating and just want you to leave it outside of their lair. Some, especially newly imported specimens, will wait for live prey to be offered. All are carnivores that can ultimately be weaned onto meaty dead fare. Just be sure whichever mode of feeding you choose to keep your hands far away! Feeding sticks and tongs have the added benefit of conditioning your mantis shrimp to become familiar and comfortable with the sight of them and train them to come out on command expecting food. Meats of marine origin from the pet store or grocery store freezer will be fine. Thaw all meats slowly in cold water to protect their nutritive value (never hot or room temperature water). Popular food items of stomatopods include krill, cocktail shrimp, clams, and squid. Spearers especially enjoy occasional sacrifices of live "feeder" grass

Arthropods: Stomatopods 259

Mantis Shrimp and Live Rock
Much ado about nothing sometimes...

While some larger species of mantis shrimp pose a legitimate threat to desirable livestock in an aquarium, some incidentally acquired species that "show up" with live rock are really little trouble or cause for concern. Oftentimes, the drab little brown or green species like *Gonodactylaceus falcatus*, and *Pseudosquilla ciliata* pose little likelihood of ruining your day, with an adult size of 2" (5 cm). Some small invertebrates may indeed be at risk (like snails), but there is no dire need to call off work, hide the children, or call in troops to seek and destroy the invader. It's just a little stomatopod after all. Trap it in due time if you like, or simply confirm its identification as a small and relatively innocuous species. Rest assured that such species are less of a risk to your fishes than a common carpet anemone (*Stichodactyla*) or elephant ear corallimorph (*Amplexidiscus*).

shrimp (*Palaeomonetes*) or crayfish (*Cambarus* and like genera). We recommend feeding these crustaceans specifically not only because they are nutritious, but also because they are aquacultured unlike some other popularly touted prey items like wild-harvested fiddler crabs. Live steamer clams, if acclimated slowly, can also be a very good food for stomatopods, and can live in the system for a week or more. Feeding freshwater fishes, like goldfish, as a staple is discouraged as an inferior diet for any marine predator due to their inherent nutritive deficiencies. Be aware that unwanted live prey may simply be killed in a territorial display of aggression and not consumed. Live food is not necessary at all, but it is fascinating to observe this quintessential hunter's skill. Mantis are well-studied in science in large part for their well-developed eyesight. They are said to have vision that is orders of magnitude more evolved than human eyesight, with the ability to see ten times as many colors as we do! The wonder of stomatopod vision aside, we recommend small daily feedings, with occasional fasts of a few days.

Reproduction

There have been reports and allusions to breeding attempts and successes with stomatopods. For their gross similarities in modes of reproduction to other easily propagated crustaceans, we have good reason to hope that captive propagation is possible in home aquariums. Mantis shrimp lay eggs that the females carries on her "abdomen" *ala* shrimp and crab-style. Larvae go through a series of molts in non-feeding stages before maturing to a point where they live free and feed. In the confines of an aquarium, and especially small breeder tanks and partitions, the young larvae are highly cannibalistic. Rearing mantis shrimp babies has very real practical challenges to space, time, and husbandry for an aquarists. Still, dedicated enthusiasts are encouraged to continue work in this area.
Molting occurs more frequently with young and varies by species. For young adults and mature specimens in general, you can expect cycles ranging

Not all stomatopods are bold and beautiful. Most nuisance species acquired with live rock and hard substrates are well-camouflaged ambush predators.

from twice monthly to once quarterly. Good water quality, adequate nutrition and available iodine are necessary as with other crustaceans for successful **ecdysis** (shedding the exoskeleton) and its replacement.

Compatibility

Reef keepers can relax knowing that stomatopods are not interested in stinging animals (corals or anemone) as prey. Mantis are essentially only a risk to other motile reef invertebrates and sessile mollusks (especially bivalves). Regarding the control of unwanted stomatopods in the aquarium, baiting and trapping such hitchhikers is best through properly quarantined live rock and like vehicles (attached corals, live sand, etc). Few natural predators will safely control mantis shrimp and not eat the very things you were trying to protect from the shrimp (snails, small fishes, etc). The usual predators have been implicated: triggers, puffers, some eels, and large wrasses. More likely, you end up with a large specimen in your display system and some form of store-bought or homemade trap will be necessary to remove pest mantis shrimp. One of the simplest and most successful traps is to sink a wide-mouthed jar into the aquarium with meaty bait in it. In the mouth of this jar you need to install a funnel pointed inward. The mantis will smell and seek the food and, if they find their way down through the funnel, are rather slow to discover the suspended funnel mouth as a way out. Instead, trapped mantis usually try to pursue futile escapes on a endless loop around the inside perimeter of the jar-trap. Another common DIY trap is to tie meaty bait in a nylon satchel and tether it on a string that is run outside of the aquarium. Lay the nylon satchel along the rockwork or sand bottom near a lair-opening or the regular haunts of the offender. Some mantis will get inextricably snarled by their spiny appendages in the material and can be dragged out from the rockwork and hiding by the tethered string. Many other tips and tricks abound on the Internet. Do be careful of the more extreme chemical measures like magnesium chloride or carbonate soda baths to dip suspects hiding in rocks. Although such chemical treatments will often draw a mantis out from hiding, they also kill or stress many desirable creatures. Our advice is to not suffer or punish beneficial creatures (microcrustaceans, polychaete worms, etc) for lack of good screening in a QT tank. Take the extra time instead to trap them passively, as there is usually no imminent danger necessitating their removal. Commercially made traps are also available for this duty. Bear in mind too that the "hallmark" clicking sounds ascribed to mantis shrimp hunting and feeding is not so very distinguished and indicative of their presence in a tank. More often than not, it is simply an Alpheid/pistol shrimp and no worse for wear.

Summary

If you'd like to keep these animals, we advise you to set up a species tank with a single specimen and appropriate decor. While there certainly are manageable species and specimens that can be kept in mixed company, the overwhelming majority are too aggressive, territorial or indiscriminately hungry to trust otherwise. Mantis shrimp really need to be the focus of a given display and most other motile tankmates will just be temptation. Be prepared for their reclusive tendencies; much like octopuses and other interesting predators, mantis shrimp are only dramatic at feeding time. If you follow this advice and choose to keep a mantis shrimp featured tank, you might well enjoy the company of this very fascinating arthropod for many years to come.

Like other sensitive crustaceans, stomatopods should be carefully acclimated to the water chemistry of any new environment. This precious time in the specimen cup or other transparent container also provides an excellent opportunity to photograph your adorable new pet! (*K. Gosinksi*)

Brackish shrimps (and crayfish) are good "Live Feeder" options for large and small Stomatopods.

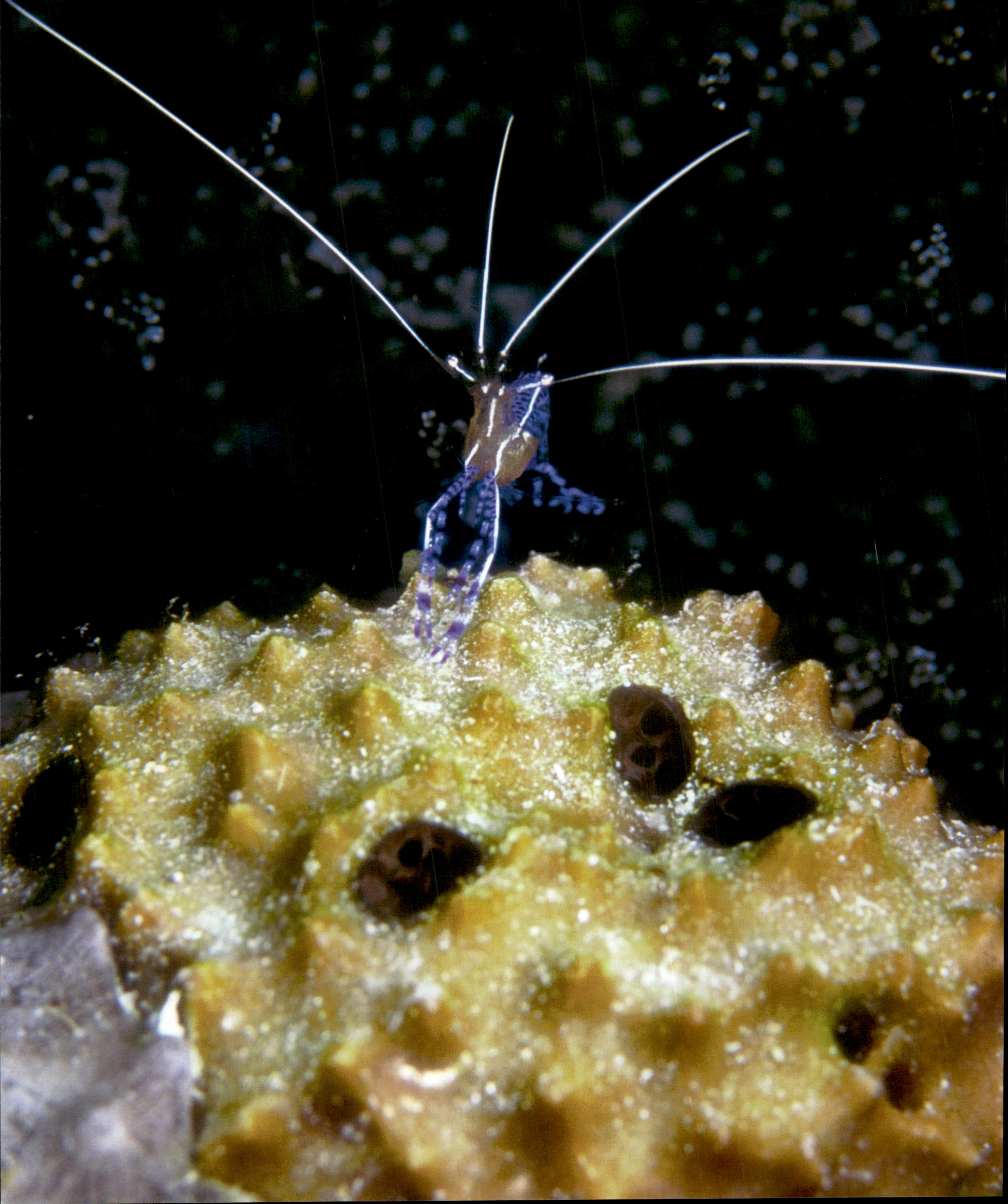

Shrimp

Cleaners, Dancers, Harlequins, and Pistols

Summarizing the groups of reef shrimp that aquarists utilize is quite a task because of their wide array of habits and habitats. Some are useful cleaner organisms that set up stations offering to harvest parasites and unwanted material from larger reef creatures. Others have unique commensal symbiotic relationships with various reef invertebrates like the Anemone Shrimp with cnidarians and echinoderms. And others, like the *Hymenocera* Harlequin Shrimp, actually prey on echinoderms. One common characteristic of shrimps is their fascinating relationships with each other and other reef creatures that we study and enjoy in aquaria.

Regardless of how ideal a given species may be for use in our aquariums - collection, shipping, and acclimation are significant obstacles for crustaceans in general to overcome. Shrimp are especially sensitive to handling and abrupt changes in water chemistry. Low pH and low oxygen levels in shipping containers take a heavy toll on new specimens. Rapid and aggressive acclimation procedures impose immediate and sometimes irreversible osmotic shock on their weakened constitutions. Frequent moves during various exchanges in the chain of distribution also amplify risks and burdens upon shrimps. Shrimps will require at least as much, if not more, time for acclimation to aquarium life, as well as allowing a suitable recovery period post-import. Again, follow the same rules we've outlined for all other specimens for acclimation and quarantine. In QT, provide segments of PVC pipe for functional but sterile refuge to your shrimps. In the display aquarium, a rocky habitat will be necessary for most. Researching each of your specimens' needs *before* you acquire them will help ensure good buying decisions as an educated consumer.

Rhynchocinetes durbanensis, the Camel Shrimp (the Durban Dancing Shrimp), is perhaps the best-known and loved species from this family. They live in large populations in the wild and will fare better if kept in a group in the aquarium. This species is quite attractive and hardy for aquarium life but may not be kept in aquariums with mixed reef invertebrates. They have stronger carnivorous inclinations than the other reef-safe shrimps and can be harsh on microfauna in the display or refugiums.

When moving shrimps between aquarium systems, drip acclimation is the rule. Intervals of 20 minutes or longer will be fine, as long as the drip rate is slow and the temperature in the acclimation vessel is maintained steadily. Once shrimp have been acclimated to a system, pay special attention to evaporation top-off and water changes, as abrupt moves in salinity can be stressful or fatal to them. As mentioned with crabs and other crustaceans, iodine is necessary for most shelled invertebrates to achieve successful molting. Although iodine does not stay in solution very long (weekly water changes cannot come close to keeping a stable level), many foods are rich in this element, and small daily feedings of krill and the like will adequately support **ecdysis** (shedding the exoskeleton). It's recommended that you leave the shed, nutritious molts of shrimp in the aquarium for them or other creatures to consume. Freshly molted shrimp are quite soft bodied and vulnerable to predation for several days while they swell with water and their new shell hardens. During these brief reclusive periods, remarkably, missing or damaged parts of a shrimp can be regenerated with enough successive molts.

Most shrimps are carnivorous as well as scavengers, with noted exceptions. The hardier species readily adapt to dry and thawed frozen meaty foods. Those with special needs are described and generally discouraged for aquarium keeping. Inter- and intra-specific compatibility issues range wildly and are detailed below by family or genera. If you have doubts

Periclimenes pedersoni, Pederson's Shrimp, is the most commonly imported and recognized Caribbean anemone shrimp. It is an excellent cleaner of parasites from fishes, but is, like most shrimp in this family, very difficult to keep alive in aquaria. Pederson's Shrimp must be kept with its anemone host and requires a species tank. Most aquarists are not willing or able to set up a dedicated display for such a tiny and cryptic animal, however beautiful it may be on close examination. Beyond husbandry issues, this creature is a poor shipper. Pictured here on a sponge in the Bahamas.

Arthropods: Shrimps

There remain numerous unknown and undescribed species of Rhynchocinetid, mainly due to their cryptic habits. Collections of aquarium fish and invertebrates at night are less common and more specialized to find high-dollar species. Thus, unusual *Rhynchocinetes* specimens are unlikely to find their way into the hobby on a regular basis. These specimens were photographed at night in Fiji (left), and the Red Sea (right).

about a particular specimen's feeding habits or social behavior, testing in quarantine with a clear divider is a sensible, safe manner of finding an experimental solution. For their size and agility among the rockscape, capturing rogue shrimp in a fully decorated display can be a nightmare. If you should need tips on doing so, please refer to the crab or mantis shrimp sections of this book. There are quite a few trap designs that are ingenious and will work given time and patience. For marine aquarium shrimp, we divide our discussion into the following groups: dancers, cleaners, harlequins and pistols.

Dancers

Rhynchocinetids are commonly known as Camel, Camelback, Dancing, Candy, or Hingeback Shrimp. This family is distinguished by a conspicuous camel-like hump on the back and their possession of a moveable rostrum (beak). These shrimps are hardy and peaceful towards most tank mates, including conspecifics, although they cannot be trusted with cnidarians (zoanthids, corallimorphs, corals, anemones, etc.). Some aquarists have had good fortune with this family of shrimps in reef aquariums for years, but there are many more reports of Rhynchocinetids eating coral tissue in time. While not reef-safe species, they are a wonderful regular offering to the marine aquarium hobby and can be recommended for all aquarists, including beginners, to keep in a proper display.

Most camel shrimp are small and reclusive and can be found in large groups on the reef, especially at night. Without any significant means of defense, they have evolved large, reflective eyes to serve them keenly. They also conduct themselves in a peculiar and sometimes hypnotic group dancing, which surely challenges the tracking abilities of an ambush predator. We recommend that you keep these shrimps in small groups for their benefit, as well as natural presentation. Eggs or ovaries can often be seen in females, and males generally have larger chelipeds and claws than the females. Provide a rocky habitat for the group to retreat to. Avoid obviously predatory fishes and be mindful that larger clawed shrimp and crabs like pistol shrimp (Alpheids), Boxer Shrimp (*Stenopus*) and Arrow crabs (*Stenorhynchus*) can easily kill these small shrimps. It may be safe to mix other small-clawed crustaceans with camel shrimp, however.

They are often mislabeled or mistaken for Peppermint shrimp (*Lysmata*) with which they share some marked differences in husbandry and habits, not the least of which is the aforementioned risk to corals in the reef aquarium. Ironically, despite their seemingly greater inclination to nip cnidarians, they are not reliable predators on *Aiptasia* pest anemones like the true Peppermint shrimp (*Lysmata wurdemanni*). Camel shrimp are carnivores, and very good ones at that. This is one of the very things that make them such delightful and hardy aquarium guests. They will consume almost any thawed/defrosted-frozen or prepared meaty food item you care to offer, as well as flakes, pellets, tablets or whole prey like shrimp and krill.

Cleaners

Cleaner Shrimps

There are numerous species of cleaner shrimps, so-called for their inclinations to service fish or other reef creatures in therapeutic efforts removing parasites and necrotic tissue. In exchange, cleaner organisms are usually spared from predation by a significant number of potential foes. Alas, this agreement does not always hold up in the confines of an aquarium with frequent, repetitive and tempting contact between natural predators and prey (groupers and shrimps for example).

264   *Natural Marine Aquarium Volume I - Reef Invertebrates*

Lysmata amboinensis is the renowned White-Striped Cleaner Shrimp of the Indo-Pacific (Eel-, Skunk-, Scarlet-, or Ambon Cleaner Shrimp). This species is very widespread throughout the tropical Pacific and Red Sea. They are hardy and well behaved in aquaria, and may be kept singly or in groups. This species, along with its Atlantic kin *Lysmata grabhami*, are perhaps the most active and effective cleaner shrimp for hobby use. Do not rely upon cleaner shrimp to find their own scraps of food among detritus, but rather feed them deliberately and nutritiously with meaty fare of marine origin. **Note:** although considered reef-safe, cleaner shrimp have been known to nip and harass desirable bivalves in aquaria like Tridacnids.

Lysmata grabhami is known as the Atlantic White-Striped Cleaner Shrimp and looks fundamentally similar to *Lysmata amboinensis* from the Pacific. They are all hermaphrodites, and as such can be kept and bred successfully in pairs of any two. Provide rocky habitats for hiding and cleaning stations. Be mindful of their great sensitivity to contaminants in the water as well as abrupt changes in salinity.

Lysmata californica, the Catalina Cleaner Shrimp, is often sold as a "Peppermint Shrimp," but is more than twice the size of a true Peppermint Shrimp, *Lysmata wurdemanni*, as adults. They are also a darker and duller red color and more stoutly figured in general. This species is temperate, however. They require temperatures between 50F-68 F (10-20 C). **Note:** this shrimp will not control pest *Aiptasia* anemones like *L. wurdemanni*.

Lysmata wurdemanni is known as the true Peppermint Shrimp (Caribbean Cleaner Shrimp). In community fish and invertebrate aquariums it is celebrated for its hardiness and peaceful disposition. In reef aquariums it is heralded for its proclivity to control pest *Aiptasia* anemones, which it does very well. Be sure to purchase *this* species if you seek Peppermint Shrimp for *Aiptasia* control. There are more than a few *Lysmata* that resemble this creature, none control cnidarians as well, and some do not eat *Aiptasia* at all. In turn, it must be conceded that its willingness to eat one cnidarian (the anemones) can be directed towards other desirable species (various corals). Although considered to be reef-safe, keep a close eye on corals and clams with all *Lysmata* in the aquarium. This species is commonly and commercially cultured, and can be bred in home aquaria.

Arthropods: Shrimps 265

Lysmata debelius is one of the most expensive and highly prized cleaner shrimp because of its stunning color and rarity. Known as the Blood Shrimp (Fire Shrimp), this species must be collected singly or in pairs at depth, which drives market prices up for specimens. It is found throughout the Indo-Pacific and is remarkably peaceful to most fish and invertebrates. Protect this species from larger-clawed shrimps and crabs like **Stenopus** (Coral Shrimp), **Stenorhynchus** (Arrow crab) and Alpheid Pistol Shrimps. Also avoid bright or excessive direct light like that associated with stony coral and clam-featured reef aquaria. *Lysmata debelius* is an excellent scavenger and feature animal for a deepwater or twilight biotope display.

Lysmata prima - Yellow bodied with distinctive red lining. East Indo-West Pacific, Maldives, Andaman Sea, Indonesia. N. Sulawesi picture. Found living on sand and mud bottoms.

Saron marmoratus (above) is commonly known as the Marble or Saron Shrimp, and is a typically nocturnal member of this family. They are collected in Hawai'i for the American trade and in the Red Sea for European hobbyists. Red and green color varieties are seen and may indicate the depth of collection, with red hailing from deeper waters. For all of the nice things said about the cleaner family of shrimps, *Saron* are neither nice nor cleaners. On the contrary, they are destructive, carnivorous and aggressive. They eat all types of meaty foods and are thorough scavengers in aquariums with sturdy fishes and few invertebrates. They cannot be trusted in traditional reef aquaria, however, as they will eat cnidarians and fight with other crustaceans. *Saron* are usually found in pairs in the wild, but must be purchased as such if you have any hope for a pair in your aquarium. Same sex specimens and incompatible pairs will fight to the death. Males are distinguished from females by their very elongated first pair of walking/fighting legs (female pictured here).

Thor amboinensis (above) is a quiet and polite member of this family, famously known as the Sexy Shrimp. It is common in all tropical seas and is also known as the Squat Anemone or Broken Back Shrimp. They are quite small at ¼"- ¾" (6-18 mm) long. Due consideration in the aquarium design must be given for this shrimp's small size. Pump and filter intakes and overflows are common everyday risks. This shrimp is much better suited for small stable aquaria without complicated or dangerous plumbing aspects. They are usually found in association with Giant, Sun, or Elegant Anemones. For small home aquariums, a common *Condylactis* pink-tip anemone from Florida may be suitable if given enough light and food. Sexy shrimp are hardy once established and will eat most any meaty fare. Offer *Thor* minced krill, pacifica plankton and mysids as food for starters. The specimen pictured here was photographed on a *Condylactis* in the Caribbean.

"Welcome to The Three Amigos Cleaning Service" Aquarium image. (*L. Gonzalez*)

The most commonly available or recognized cleaner shrimp species belong to three genera (there are many more): *Periclimenes*, *Lysmata*, and *Stenopus* - all have long, white signaling antennae. Their efficacy as cleaners is highly variable by species and by individual specimens. As cleaners go, *Lysmata amboinensis* and *L. grabhami* (the Eel or Skunk Cleaner Shrimp) are the most effective and reliable. To the chagrin of many itchy fishes, cleaners can be quite selective about which species they will service.

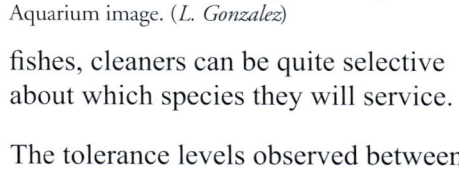

The tolerance levels observed between species and genera of cleaners are quite variable, ranging from the extremely tolerant *Lysmata*, to the extremely intolerant, *Stenopus*. In a large enough aquarium, however, even the aggressive species can cohabitate. Adequate food is necessary, though, as hungry cleaner shrimp are still dedicated carnivores and may become cannibalistic. Most species are reef-safe, but have sometimes been known to nibble upon desirable cnidarians or clams. The Peppermint Shrimp (*L. wurdemanni*) is one of the riskier species with corals, and has been known to eat large polyped fleshy corals (LPS), polyps, and corallimorphs as readily as it eats

Stenopus pyrsonotus, the Ghost Boxing Shrimp, is more expensive, less common, and more delicate than *Stenopus hispidus*, largely because of its deeper water residency.

Arthropods: Shrimps

nuisance anemones. White-Striped Cleaner shrimps (*L. amboinensis* and *L. grabhami*) have been known to harass bivalves such as Tridacnid clams. Our advice is to simply keep an eye on all or any members of this group with your fleshy sessile invertebrates. Be sure not to confuse the excited pounce of a shrimp upon recently fed coral as evidence of a carnivorous attack. Most often, they are simply bold and enthusiastic feeders that seem to think they are immune from cnidarian stings.

All cleaner shrimp species prefer subdued lighting, plenty of rocky cover, and nooks and crannies. Healthy specimens will molt generally on a monthly basis. Most will also breed readily, and some can be raised in captivity. Search the Internet and printed works for data on the

Stenopus hispidus, the Coral Banded Boxing Shrimp, is known and loved worldwide by aquarists. Circumtropical in distribution, they are hardy, inexpensive, and fairly well behaved. It's best to only keep one per tank, unless you are confident in your dealer's means to procure a compatible pair. They are generally reef-safe and will appreciate a large cave or opening in the rocks to hide under by day.

aquaculture of this group, like, "How To Raise and Train Your Peppermint Shrimp," by April Kirkendoll. One of the best things about *Lysmata* in the aquarium for breeding is that they are hermaphrodites. You might literally select any two specimens and have yourself a viable breeding pair. That is, without a doubt, bolstering for those considering the expensive purchase of *Lysmata debelius* (Blood or Fire Shrimp). They simply take turns in reproductive roles during spawning events with one fertilizing the other, and then switching roles with the next spawning. It is even possible to see both of your shrimp carrying eggs at the same time.

The Peppermint Shrimp (*Lysmata wurdemanni*) is not especially active as a cleaner shrimp, but it is a popular and reliable predator on *Aiptasia* pest anemones. Several similar shrimps are marketed for the same purpose. We may find the *Rhynchocinetes durbanensis* (Camel Shrimp) marketed for this purpose, or the **coldwater species**—from the West coast, we find the California Peppermint Shrimp, *Lysmata* (*Hippolysmata*) *californica*, and from the East coast

Dasycaris zanzibarica is as unusual as it is attractive, and clearly demonstrates the skills of anemone shrimp to change to match their specific host (a gorgonian sea whip pictured here). Indonesian image.

Lysmata rathbunae, the Hidden Cleaner Shrimp (north to Virginia, uncommon in Florida). Please be sure to avoid purchasing the coldwater peppermint shrimps, as they not only do *not* prey upon pest anemones as well as *L. wurdemanni* (the so-called true Peppermint Shrimp), but also are wholly inappropriate for a tropical aquarium being coldwater species. All are carnivorous, however peaceful they may seem. The popularized recommendation of peppermint shrimp for refugiums, based on their propensity to breed and shed gametes, is rather strange to us. As carnivores, and especially if left unfed by an aquarist, they are likely to be a greater burden on desirable microfauna in refugia, rather than a benefit for the occasional spawn thrown back into the bio-mass. Do consider the focus of your refugium and weigh the burden of this otherwise delightful species in it.

In summary of the group:

- *Lysmata* and *Stenopus* are easy to feed, hardy and peaceful enough for most aquariums.
- *Periclimenes* (Anemone Cleaner Shrimp) are difficult to ship and keep and not suitable for any but the most specialized aquaria.

Periclimenes amboinensis (left) is yet another strikingly colored and patterned anemone shrimp that has evolved to cryptically blend into camouflage with its obligate host - a feather star (Crinoid). The wide range for this species includes the Western Pacific, Indonesia (pictured here), the Marshall Islands, and Australia.

Periclimenes brevicarpalis (right), the photogenic Pacific Clown Anemone Shrimp, is widely distributed throughout the Indo-Pacific. Males are conspicuously smaller than females, with both maturing to less than one inch in length (<25 mm). Although they must be kept with a host, they are far less discriminating as a commensal than their kin, and can be kept and fed reasonably well in captivity. Offer finely minced meaty foods of marine origin.

Hippolytids are the most commonly recognized and popular Cleaner Shrimp species. Most are peaceful and useful in mixed invertebrate aquariums. Some species in this family can be quite dangerous or predatory in a community tank, like Saron.

Stenopodids are long-time hobby favorites driven by the industry staple, *Stenopus hispidus* (Banded Coral Shrimp or Boxing Shrimp). This group is also known as Scissor Shrimps and

Periclimenes holthuisi, (left) Holthuis' Cleaner Shrimp, is a commensal with a wide array of cnidarians, including jellyfish. For your best success at keeping them in aquaria, they should be purchased on, and along with, the exact host with which they associate. In captivity, they will eat most meaty foods offered (*gammurus* and mysids are nutritious fare). Specimens can be found in association with both soft and hard corals in the coral seas. This one is commensal with a *Heliofungia* plate coral.

Periclimenes magnificus (right) is reportedly found in close association with Elegance Coral (*Catalaphyllia*), the anemone, *Dolfleina armata*, and even with snake eel species. Commonly found on soft substrates and muddy niches.

Periclimenes yucantanicus, (above) the Spotted Cleaner Shrimp, is a familiar Caribbean denizen in *Condylactis gigantea* (Giant Pink tip Anemone) and other tropical West Atlantic anemone species.

Arthropods: Shrimps   269

Periclimenes imperator is a species of highly variable color known to associate with less common hosts like Spanish Dancers (the nudibranch *Hexabranchus*), Sea Cucumbers (*Stichopus, Bohadschia, Synapta*), and Sea Stars (*Gomophia*). They have a very wide range from the Red Sea and East Africa to Japan. Pictured here in Indonesia on a holothuroid.

can be found in great abundance in the shallows and depths alike. They are exceedingly common in some natural habitats, and can usually be spied clinging underneath a cave or overhanging rock. They are generally hardy to keep and easy to feed, taking a wide variety of fare but preferring meaty foods. If left to forage naturally in a large reef aquarium, *Stenopus* seek polychaete worms and can help control nuisance, errant bristleworms. They are also fairly well behaved, although small fishes and other shrimp are sometimes at risk.

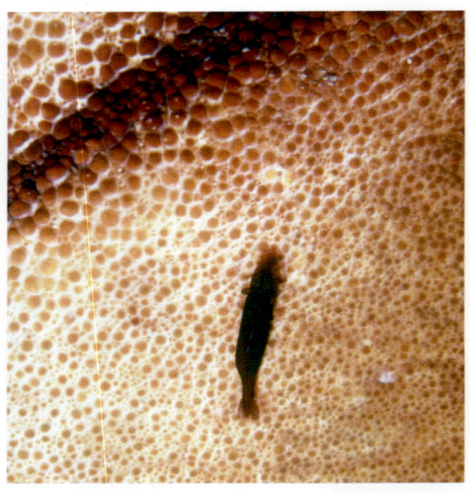

Sexed pairs can often live together, but same sex mixes will elicit a violent territorial response. We recommend only one per tank unless you can buy a pair that is known to be living together peacefully. Unrelated species of especially small-clawed shrimps (like *Lysmata*) may be pulled limb from limb if introduced to their territory. In large uncrowded tanks, however, they can be quite tolerant of all, and reef-safe with almost anything that won't eat them. They may also perform as cleaners, although this behavior is usually not observed with the same fervor as observed with *Lysmata* species.

They are commonly found in mated pairs in the wild, at cleaning stations they set up to service visiting parasite-infested fishes. Mature females are distinctly larger than males, and often

(Left) *Periclimenes soror* is similar in gross form to *P. imperator*, but is more slender. It demonstrates great skill at changing its color to match its host. Viable sea star hosts include *Linckia, Culcita, Acanthaster*, and *Choriaster*. This species has an enormous distribution from the Red Sea all the way to the Sea of Cortez.

carry eggs or reveal bright blue-green ovaries through their bodies. Natural lifespan has been recorded at over 5 years, and spawnings in captivity have often been observed. Frankly, this is almost inevitable in a healthy system. Data and details on breeding attempts and successes abound on the Internet in places like the Breeder's Registry (see *resources*). Much insightful information can also be gleaned from the now commonplace aquaculture of *Lysmata* peppermint shrimp.

A special mention about handling *Stenopus*: the body of this shrimp is covered with spiny "hairs" which can become easily tangled in coarse netting material. Furthermore, frightened *Stenopus* will execute autotomy and abandon a large claw. Although the lost appendage will grow back after successive molts, it leaves the shrimp vulnerable in the interim. These shrimp are very predictable swimmers, however (backward by propulsive flips of the tail), making the procedure of netting a relatively simple affair. Catch them in an open bag or submerged clear plastic cup

Periclimenes longicarpus is a delightfully elegant endemic of the Red Sea and Arabian Sea, which can often be found in association with the Bubble Tip Anemone (BTA), *Entacmaea quadricolor*. It is pictured here instead on a Bubble Coral, *Plerogyra sinousa*.

A very well-camouflaged *Periclimenes* shrimp on a Crinoid feather star in Indonesia. Some shrimp have evolved to blend remarkably well in camouflage with their hosts...

... while other shrimps seem almost bold if not contrasting in color to their hosts. They clearly rely on such natural cnidarian defense mechanisms to dissuade would-be predators.

by coaxing them with a stick or net handle into the waiting container. Then, they will predictably swim away from the affront (backwards into the waiting cup or bag).

Palaemonids are commonly known as Anemone Shrimp for their commensal associations with cnidarians. Most are small (less than 1"/25 mm), very cryptic, and easily overlooked in their natural habitat. They tend to be delicately colored, often translucent and usually quite beautiful. Aquarists are attracted to these shrimps for their lovely visage and unique commensal relationships with anemones, corals, large nudibranchs, sea cucumbers, and urchins. Unfortunately, they are usually very difficult to keep, even under the best of circumstances. Anemone shrimp are delicate to collect and ship, are challenging to feed, and are easily predated by even the best-behaved aquarium fishes. Extremely consistent and high quality aquarium parameters are crucial for their success in captivity. Some clownfish and damsels will tolerate a shared space with Palaemonids in a large host anemone. Others will simply eat a new shrimp dropped into the aquarium before it even sinks to the aquarium floor. Hawkfish and small wrasses are also notoriously predatory on shrimps among the reef-safe fishes. Even once acclimated successfully to a peaceful and specialized system, it is still difficult to protect these small shrimp from pump and filter intakes, overflows, and other dangerous aspects of aquarium hardware. Most simply starve to death in the aquarium for a lack of natural plankton and mucus from a suitable host. Because of the great challenges to keeping most host anemones in aquaria, it is recommended that casual aquarists and beginners avoid trying to keep these creatures, especially in community aquariums with mixed species of fishes and invertebrates. This is an easy group to make an argument for leaving in the sea. Should you not be dissuaded, do research extensively and provide a species-specific biotopic display for these shrimp, along with providing them their symbiont anemones. For advanced aquarists and specialists only.

Hymenocera picta is known as the (eastern) Harlequin Shrimp. Another species perhaps, *H. elegans* (debated as a variety of *H. picta*), is familiar to divers and aquarists as the Hawai'ian Clown Shrimp, and is distinguished by an even brighter medley of color, with strong yellow tones to the white areas. All live in pairs and should be kept as such in species-specific tanks only. They are shy and peaceful to most all reef life except their echinoderm prey. Mature females may be larger than males, and spawning of this shrimp has been observed in captivity. They have been successfully reared in public aquaria. (*J. Chodakowski*)

Harlequins

Gnathophyllids - Harlequin Shrimp (Clown or Painted Shrimp) are infamous shrimp that have gained notoriety for their macabre culinary delight in eating the tube feet of echinoderms. Members of this group are unmistakable in pattern and form, with their stout bodies and large colorful spots. They are gaudy in a most attractive fashion, with pink or purple polka dots on a stark white body. Unfortunately, they are seen too often in the trade and, worse still, offered to aquarists who are ill prepared to feed such delicate and demanding species. They are obligate feeders on the ambulacral system of spiny-skinned animals, and will starve to death without tube feet to eat. Some aquarists have been successful in culturing rearing tank-raised sea stars to feed their Harlequin shrimp. Some of the shrimp have responded well to cultured food species, while others opt to starve in wait for a natural prey item. Chocolate chip

A close-up of the tube feet of a sea star. Harlequin shrimp are obligate feeders on these tube feet, and will attack sea cucumbers, urchins, and sea stars. Outside of home-breeding echinoids to feed such species, we cannot recommend this species for aquarium life.

Hymenocera elegans variety (Clown Harlequin Shrimp): For the dedicated few aquarists willing to set up species-specific tanks just for this amazing shrimp, we recommend the further exploration of cultured echinoderm prey, nuisance *Asterina* sp., instead of feeding wild harvested sea stars. (*G. Rothschild*)

starfish (*Protoreaster nodosus*) and Blue Linckia starfish (*Linckia laevigata*) are commonly used wild "feeder" sea stars for Harlequin shrimp. Responsible and conscientious aquarists will not buy any of these sea star species, however, for their dismal survivability on import. Other genera are accepted, but none are sensible prey for an empathetic aquarist to use. The shrimp, and feeding of wild prey really, are incongruous with considerate aquarium keeping.

Beyond ethical issues of sustainable harvest and best use of resources, buying wild-harvested sea stars to feed a captive Harlequin shrimp is impractical and unreliable. Too many ship poorly or die in transit, and without regard for the dynamics of the impact of each purchase, we must consider how assured or consistent the food supply is. Wild sea star availability is subject to seasonal shifts, variable mortality rates on import, as well as the buying preferences of your dealer. In turn, the available food supply and subsequent survival of a given Harlequin shrimp hinges on these factors. It is very hard to rationalize the responsible keeping of this shrimp in captivity without a home-cultured supply of acceptable sea stars. The easily propagated (cutting and simple fission) *Asterina* sea star has been suggested as a possible species to culture and feed *Hymenocera picta*. Because of the aforementioned difficulties associated with the husbandry of such obligate feeding strategies, we strongly discourage the conscientious aquarist from purchasing these small lovelies.

Pistols

The Alpheids

Pistol shrimp (Symbiotic or Snapping Shrimp) often go unnoticed, or at least unidentified, by marine aquarists. These largely nocturnal crustaceans are seldom seen, but often heard with their distinct snapping sound, which may crescendo at night. They can be found circumtropically and in temperate seas alike. Divers and aquarists have described their sound production in many similar ways: a pistol crack, cracking glass, or a marble striking glass. They are often mistaken for mantis shrimp that make a similar sound, which ironically brings some aquarists much fright. We say ironic because the relief they feel once they discover the noise-maker is not a stomatopod (mantis shrimp), but an alpheid, is sometimes short-lived for the latter's rival aggression or predatory nature. This is not to say that all snapping shrimp are dangerous - quite the contrary. They are well equipped and territorial. They use their single, enlarged claw to shock or stun prey, and sometimes to literally punch a hole in the exoskeleton. The loud sound is also simply used as an alarm at times to ward off intruders. Their small size minimizes their overall threat; most snapping shrimp only grow 1"-2" in length (25-50 mm) with few exceeding 3" (75 mm). If kept with appropriate tankmates and given space, they can be peaceful residents and a benefit to live sand methodologies.

Symbiotic Associations

Almost every aquarist at some point in time has been inspired by the sight of the symbiotic relationship between Clownfish and anemones in the wild or a home aquarium. Their pairings are one of the most symbolic and familiar relationships in the sea. Unfortunately, this is one of the more difficult challenges because of the inherent difficulties of keeping anemones, not to mention the question of their dubious sustainable harvest from the wild. Thankfully, there are many other interesting symbiotic relationships that we can enjoy in our aquaria. Xanthid crabs in Acroporids (*Acropora* and *Montipora*), Trapezid crabs in Pocilloporids (*Pocillopora*, *Stylophora* and *Seriatopora*), Cardinalfish with

Harlequin shrimp are best kept in small, specialized aquaria, in part, to concentrate feeding opportunities and ensure health. (*G. Rothschild*)

Although hardly as glamorous as clownfish symbiosis with anemones, the pairing of pistol shrimp and shrimp gobies is a far more realistic and practical relationship to pursue in aquarium keeping. These creatures are much hardier in captivity than the categorically difficult husbandry of anemones at large. Purchase pairs together whenever possible for best success. (*G. Rothschild*)

Arthropods: Shrimps   273

Synalpheus stimpsonii is a very unique symbiotic alpheid species that often exhibits contrasting body color to its crinoid hosts. Not all Pistol shrimps have mutualistic partnerships, but some are merely harmless and commensal as pictured here.

Diadema urchins, and of course, our subject at hand, Shrimp Gobies and Alpheid Shrimp.

Alpheus and *Synalpheus* shrimp are two notable symbiont genera, with shrimp gobies of the genera *Amblyeleotris*, *Cryptocentrus*, *Ctenogobiops*, *Istigobius*, and *Stonogobiops*. In their **mutualistic relationships**, both participants benefit; the shrimp constructs the shared burrow-home, while the goby keeps a sharp vigil against predators with its keen eyesight. Some of these alpheid shrimp species are considered functionally blind and need the goby to watch for protection while they build and maintain the burrow. The partners stay in constant touch via the shrimp's long antennae, and a retreat is immediately sensed from the goby when identified. As you can imagine, these partners work out a refined body language of communication, from twitches to panic thrashes, to indicate the need and nature of movement in and out of the burrow. Research has shown that some of these partnerships are formed by chemical attraction, which brings the pair together as juveniles. Pairs are thought to be formed for life. Alpheid shrimp species also have commensal relationships with other reef organisms like sponges, crinoids, corals, or anemones. These various partnerships make for fascinating presentations.

Selection

Not all alpheids live or need to live in symbiotic relationships. Those that are active in such relationships are highly selective in regards to acceptable hosts. In some cases, the shrimp must be collected with its partner or host or it will not re-pair with another in captivity. Indeed, the best way to ensure you obtain a compatible pair is to purchase them together. It is very difficult to collect and navigate a specific pair of creatures together through the chain of custody on import. Pairs get mixed up, or even misrepresented. A true bonded pair is a rarity in the trade and hobby. Again, research is your best hope of success. Only then should you seek them on special order from you dealer, and understand that patience and time will be necessary for you to secure a healthy and compatible pair. We strongly advise you to avoid alpheid shrimp that have relationships with challenging invertebrates like crinoids and Poriferans (Sponges).

Care

Husbandry for pistol shrimp is like that of most other shrimp and crabs. They are slightly more tolerant of water quality issues in comparison to other crustaceans, as evidenced by their propensity to survive import with live rock out of water for days at a time. In fact, most aquarists receive their pistol shrimp as incidentals this way. Further evidence to their toughness - they have to survive the sometimes extraordinarily foul curing process for live rock. This reality should not give aquarists a license to abuse them, of course, but merely illustrates the considerable hardiness of the group. Full strength seawater at NSW values is recommended, and you should always avoid low salinity with most marine crustaceans unless they are known to be intertidal and tolerant. Regular water changes, available iodine, and stable water quality will insure success. As with others of this group, feed staple foods rich in protein and iodine, such as krill or other crustaceans. Partner gobies will often forage for other fare near their burrow to share with their den mate. As with any aquarium creature, offer a wide variety of foods and do not limit them to one or two prey items. In captivity, alpheid shrimps will take both algal and meaty foods and benefit from a mixed omnivorous diet, although they do lean decidedly towards being carnivorous. **Note:** please be sure to avoid feeding adult frozen brine shrimp as a staple (more than 20%) to this or any aquatic organism, as it lacks sufficient nutrients; indeed, animals can literally starve to death if forced to subsist on it.

Habitat

One key thing to remember when planning a display that houses alpheids is that they *will* dig. The ramifications of this are two-fold. First, you must be certain that all rockwork is firmly anchored and secure. The possibility of an avalanche from the tunneling activity of a pistol shrimp is very real, and tumbling rocks inside the aquarium can easily break glass or a seam of the aquarium. To ensure this will be prevented, we recommend the use of plastic cable ties (tethering drilled rocks) or a PVC substructure to aid you in creating an interesting and structurally sound display. Secondly, there is the practical matter of supporting the shrimp's efforts to build a burrow with coarse sand and crushed shell or rock in the substrate. Pure sugar-fine sand makes for miserable engineering of an Alpheid burrow and will instead instigate the shrimp to tunnel under rocks instead. One of the best and safest ways to provide for a pistol shrimp is to install capped (the buried end) tubes or pipes in strategic

The long, sensitive antennae are the hallmark of this gregarious cleaner shrimp. Although not as active at cleaning in the aquarium as *Lysmata* species, *Stenopus* is generally collected in the wild at their cleaning stations, often in mated pairs.

Thor amboinensis - the "Sexy Shrimp" in Sulawesi.

Arthropods: Shrimps 275

An ich-infested *Zebrasoma veliferum* (*desjardinii*) Sailfin Tang receives some welcome dentistry from *Lysmata amboinensis*. (*L. Gonzalez*)

areas before the shrimp is added to the tank. The mouths of such tubes should be placed naturally in convenient places to view the alpheid's behavior. This arrangement makes target feeding a lot easier, too. Inventive, handy aquarists might even plan for residency in advance by sealing in clear substratum tubes against the viewing panes of the aquarium so that the activity of the shrimp and its partner if any can be clearly observed, even when they travel below the sand.

Compatibility

Pistol shrimp are somewhat misunderstood with regard for their aggression. Outside of mates, they are very intolerant of each other, as well as most other shrimps and crabs. They will even catch and kill small fishes. Mandarin fish, small sleeping wrasses, and gobies, for example, are somewhat vulnerable to a prowling pistol shrimp at night. Attacks on fishes or other invertebrates are usually not of a predatory nature, however, but more borne of territorial aggression, as when an intruder tries to commandeer its territory or bolt hole. Most at risk from this sort of aggression are new and confused fishes that have not yet learned the layout or social dynamics of the aquarium. In large aquaria with active, sturdy fishes, alpheid shrimp present little danger to other desirable residents, and can be enjoyed for their functional abilities and unique behavior.

Tiny *Mysis* shrimps can multiply rapidly in a fishless refugium, providing plankton to the filter-feeders in the main display. (*H. Schultz*)

Here's an unidentified Alpheid beautifully camouflaged in a Crinoid feather star in North Sulawesi, Indonesia.

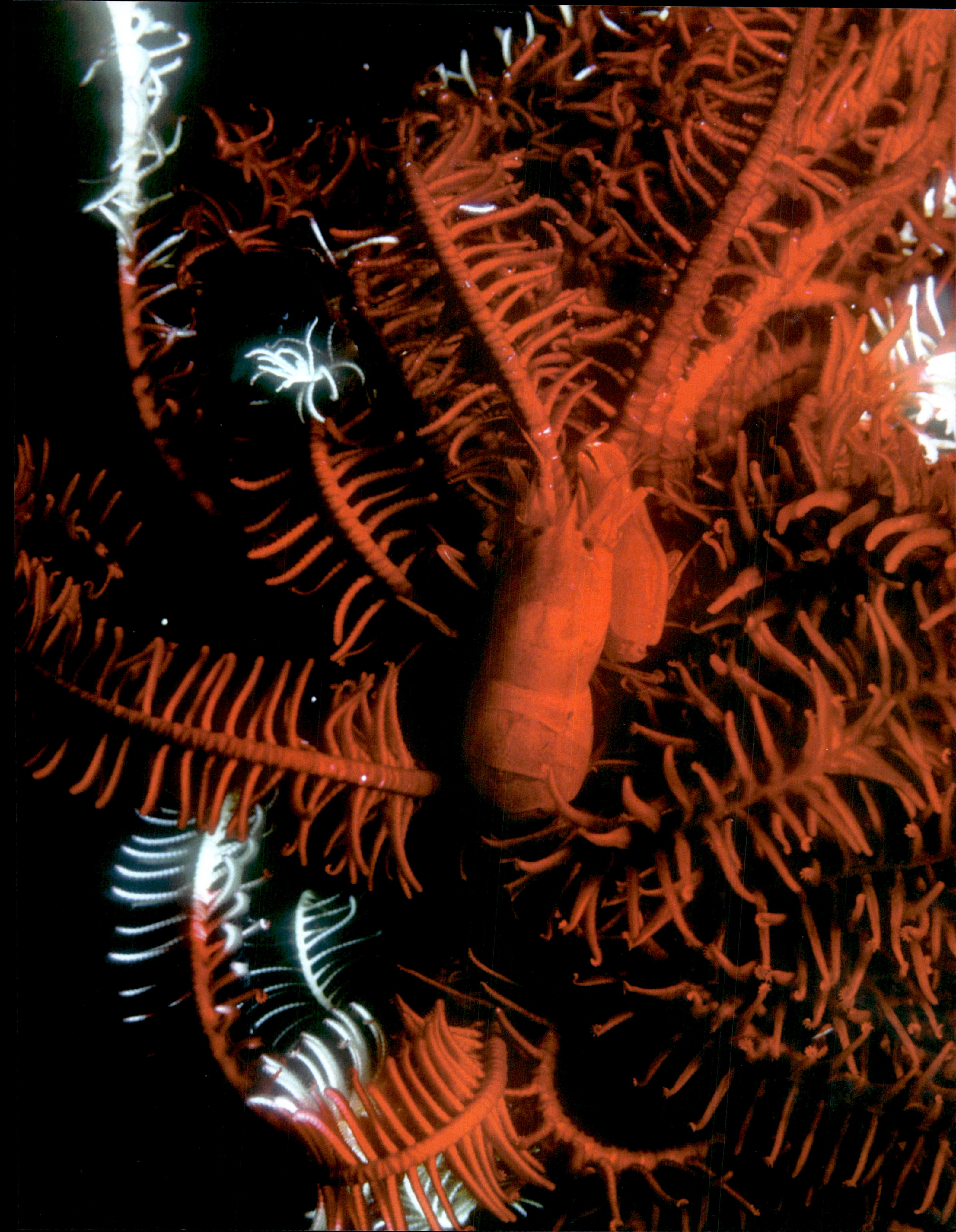

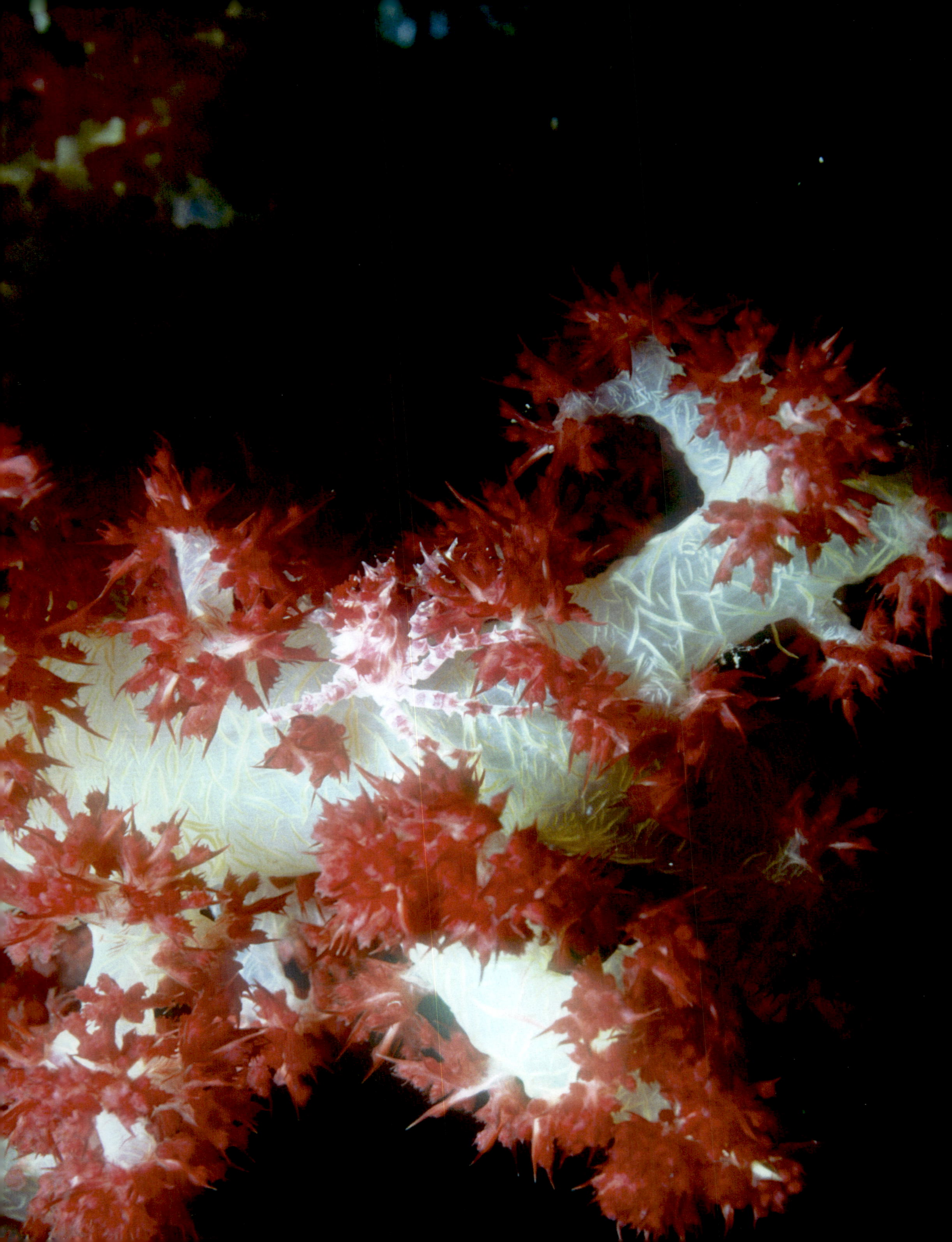

Crabs
Briny Bugs? Or rather... Salty Arthropods!

Crabs are a popular and well-represented category of reef denizens in the aquarium trade. Some species are not true crabs at all like the hermit crabs (Paguroidea), horseshoe crabs (Merostomata) and sea spiders (Pycnogonida), but the number of species and genera collectively seen by aquarists is extraordinary. There are over 4,000 described species of true crabs alone (the infraorder Brachyura, distinguished in part by skeletal plates/epistomes fused to the "body"/carapace). Their suitability and husbandry in aquaria ranges widely from blessing to curse. With an extended Arthropod family of about 42,000 species including shrimp, lobsters, crayfish and an abundance of terrestrial insects, you can imagine what a tremendous contrast there must be between members overall. As such, it is clearly impractical to make any categorical generalizations about the group in aquaria. However they are commonly opportunistic feeders, which often leads to compatibility problems in the narrow confines of an aquarium. Most crabs are omnivorous and many are quite predatory. As a result, nearly all crabs need to be watched closely in marine aquariums.

In this introduction to crabs, we hope to acquaint you with the many fascinating possibilities for keeping this diverse group of crustaceans. But we must also warn you that these sturdy and well-designed survivors are resourceful and may be risky in a collection of fish or mixed invertebrates.

Selection

A discussion of crab *selection* assumes that the debated specimens were or will be acquired deliberately. However, that is not always the case, as you may already know, for numerous species in the aquarium trade. Many crabs are accidentally imported as "hitchhikers" with live rock or by hiding in coral and amongst other invertebrates. It is a testament to their extraordinary hardiness that they can survive the process of collecting, shipping, and curing live rock. Hitchhiking crabs must live for days out of

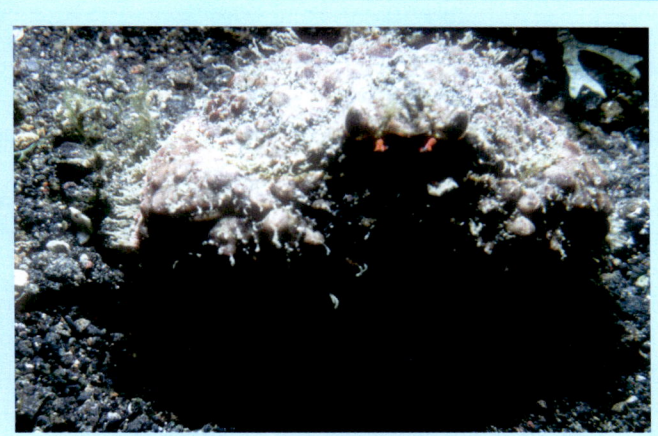

Calappa heptatica - Shame-faced Crab

Achaeus japonicus - Decorator Crab

Aniculus aniculus (G. Rothschild)

Hyastenus bispinosus - a decorator sea spider

Left: *Hoplophrys oatesii*

A Xanthid crab in residence. (*L. Gonzalez*)

Plagusia depressa a Grapsid rock crab.

Glyptoxanthus erosus the Eroded Mud Crab..

Neopetrolisthes maculata, the Anemone crab. (*G. Rothschild*)

Mithraculus sculptus, emerald Mithrax crab. (*L. Gonzalez*)

Lissocarcinus orbicularis, the Harlequin Swimming Crab

The variety of sizes, feeding behaviors, and forms of crabs observed is rivaled by few other groups of animals in the hobby. They run the gamut from highly specialized to staggeringly indiscriminate in their selection of foods and habitats. Be observant when evaluating species for the invertebrate aquarium; form follows function, and large destructive claws are rarely if ever ornamental.

---

water (sometimes more than a week with many shipments of live rock!), wrapped in damp newspaper if they are lucky, and then suffer through the osmotic shock of being dropped into saltwater without any acclimation. They must then suffer extremes of water quality during the curing process, which can involve levels of nitrogenous decay and mineralization that would kill most any other aquatic organism. The marine crab must truly be appreciated for its toughness!

Some crabs are commensal or symbiotic, others are predatory, and others still are purely accidental in residence. Ultimately, a general discussion about crab selection, whether deliberately or accidentally acquired, in marine aquariums must be reduced to *suitability* for captive study. The variety of sizes, feeding behaviors and forms of crabs observed is rivaled by few other groups of animals in the hobby. They run the gamut from niche-specialized to amazingly indiscriminate. Some species, for example, have specific commensal relationships with cnidarians. In such cases, the benevolent and desirable crab protects the coral or anemone in which they reside from cnidarian predators while being protected by the coral from crustacean predators. Other species, however, *feed* on cnidarian tissue and some are even able to store the stinging nematocysts of the prey much like predatory nudibranchs (sea slugs). It is a very interesting adaptation that affords these crabs the noxious protection of their prey's former defensive mechanisms.

Coral crabs are commonly found in the branches of scleractinians and are often observed with trepidation by reef-keeping aquarists. More often than not though, the hitchhiker is a commensal species: Xanthids in Acroporids (*Acropora* and *Montipora*) and Trapezids in Pocilloporids (*Pocillopora, Stylophora* and *Seriatopora*). Some species are rather mundane in color while others are quite dramatic... veritable living treasures for the underwater photographer! The very presence of a crab within a colony of coral, however, does not automatically qualify the specimen as a benevolent commensal. In fact, a sessile invertebrate prey item is the ideal place to discover a cnidarian predator! Thus, an aquarist must keep a watchful eye on such incidental organisms for damage to the host. There is no better place for observation than in quarantine where all livestock should be confined on import or acquisition. Screening or trapping undesirable organisms is straightforward (quite simple) in a proper (bare bottomed) quarantine tank. However, the removal of a small predatory crab from a fully rockscaped display can be an extraordinary challenge if not downright impossible. The accurate identification of an incidental species is recommended whenever possible and a mindful watch of behavior and husbandry for other specimens when identification is not possible. Please study, learn and share your knowledge freely.

Decorator crabs are another fascinating group of crabs favoring

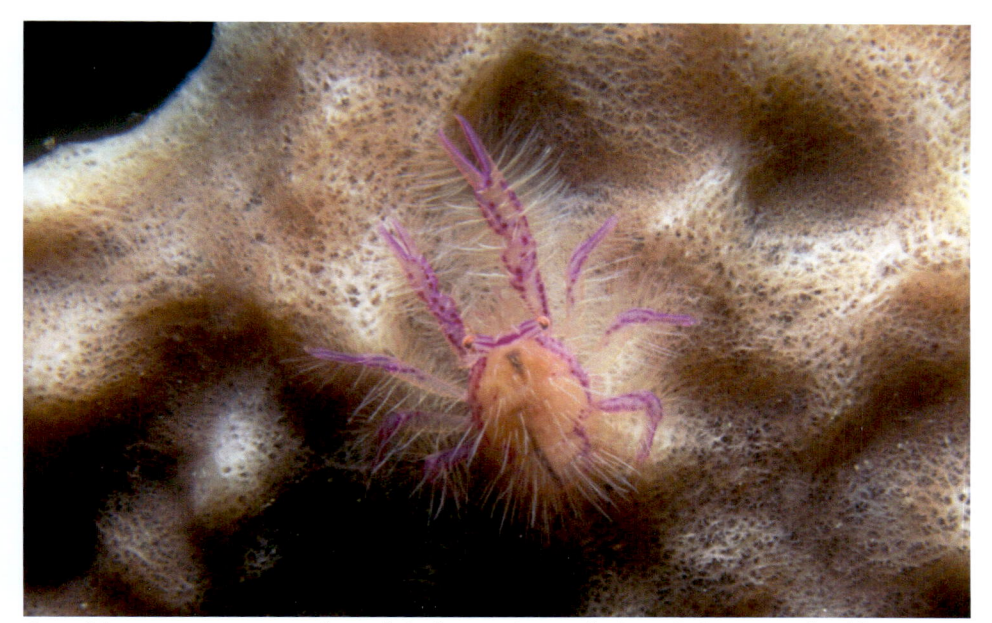

One of the more uncommon arthropods in the aquarium trade, Squat Lobsters are related by superfamily to the popularly recognized Porcelain crabs. Pictured here is the exquisite Galatheid, **Lauriea siagiani** (AKA "Xeno crab").

various forms of natural camouflage, which is at least irritating if not harmful to the other life forms in an aquarium that are pillaged for such adornment. This can become quite problematic as they pickup, move, and modify (read: molest!) your prized specimens in the tank. Bits of algae (micro and macro), polyps, zoanthids, sponge, and mushrooms are commonly accepted ornamentation for these crabs. Most decorator crabs will also eat small fishes. As a rule, they are decidedly omnivorous with a palate leaning towards carnivorous. Decorator crabs are very interesting and generally hardy crustaceans for fish-only aquariums, but mostly inappropriate for mixed invertebrate displays.

Decorator *Sponge* Crab (*Schizophrys dama* and the like) from the tropical Atlantic usually reach 1-2" (2-5 cm) as adults for most species. They are categorically less hardy than other decorator crabs. The very noxious nature of sponge (compositionally) and its tendency to ship poorly may be a mitigating factor in higher mortality for these crabs on import.

Decorator *Arrow* Crabs come from the Atlantic and although more slender in stature is no less formidable than other decorator crabs lacking discrimination in their diet. It is an opportunistic feeder that is likely to wreak havoc on microfauna in live rock and sand like most crabs.

Sea Spiders (example *Anoplodactylus sp.*) are common imports with gorgonians, live rock and invertebrates. These arthropods are generally small (<5 cm), fragile, poorly understood, and almost always acquired incidentally. They are perhaps best remitted to fishless refugiums upon discovery (with the "host" invertebrate if possible) and studied closely for insight to feeding and general husbandry requirements. Some are known to have very specialized diets.

Decorator *Spider* Crab (example: *Acheaus japonicus*) from Indonesia is a small and very innocuous species. Many such sea spider crabs are rather fragile by design and strategic as decorators employing live cnidarian and hydrozoan animals for camouflage. Tube anemones, hydroids and small polyps are popular epaulettes. They are omnivorous by nature.

Decorator *Spider* Crab (*Camposcia retusa*) Indo-Pacific: One of the most common decorator species, this creature reaches an adult size of about 8 inches (20 cm) and has a very indiscriminate appetite! Although a risk in mixed invertebrate "reef" displays, this hardy species is best kept in aquariums with fast, restless fishes.

Decorator *Spider* Crab (*Xenocarcinus sp*) Micronesia: Inclination towards algae grazing. A well-behaved decorator with fishes and invertebrates. Adult sizes are well under 7 cm for most species in this group.

Neopetrolisthes maculata is commensal with anenomes. This handsome crab possesses flattened, large chelipeds and a white body with red spotting. Note the extended maxillipeds, its feeding mechanism. Aquarium photo. (*G. Rothschild*)

Arthropods: Crabs

Decorator *Coral/Sponge* Crab (includes Dromid species): A stout bodied group, these destructive decorator crabs remove large, whole specimens of sponge and octocoral (including impressive chunks of Leather Coral/Alcyoniids!) from hard substrates and fix them to their carapace. Often acquired incidentally with live rock, such specimens have no place with a casual marine aquarist or in mixed invertebrate reef aquariums.

The most popular crabs in the aquarium trade are known or believed to be algae eating species. They are found in all shapes and sizes and are popularized commercially, regardless of their actual efficiency or suitability for the modern marine aquarium. Many such crabs are marketed in "clean up crew" packages, but some can be quite problematic. While it is true that they can be excellent at nuisance algae control, even safe species can become a burden to the biodiversity of a system in large "herds" or populations (as they are sometimes encouraged to be kept). Most all are omnivorous to some degree and may eat other desirable plants and animals, including each other! Common prey for crabs include (but certainly are not limited to): snails and other mollusks, algae, other crustaceans, fishes, and worms. It would be fair to say that many species are a burden to the microfauna of live rock and live sand in natural marine aquaria. Accepted food items may be dead or live matter as most crabs are quintessentially opportunistic feeders. Hermit crabs are the most popular of all arthropods in the aquarium trade. Many of the rumors and misconceptions about species popularly believed to be "reef safe" becoming dangerous could be credited to a predatory species (or one in like appearance to a known "safe" species) slipping unnoticed into a good group upon collection.

Clibanarius tricolor, the Blue-Legged

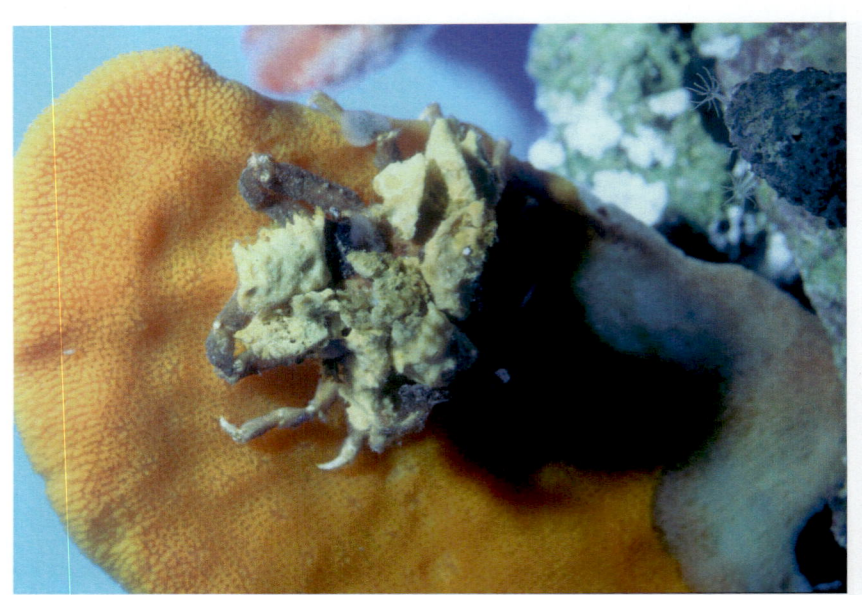

Schizophrys dama - Yellow Decorator Crab

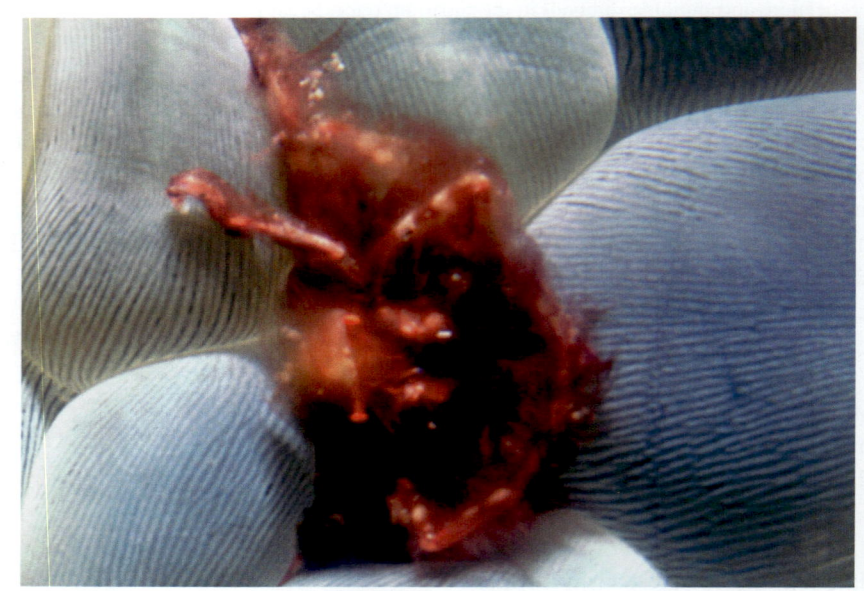

Achaeus japonicus - Decorator Spider Crab

Xenocarcinus conicus - Decorator Spider Crab

Hermit Crab. A typically passive and "reef-safe" species, this crab can be helpful in marine aquaria for controlling some nuisance algae species. A benefit to reef aquaria if not kept in large numbers (with concern for decimating microfauna in living sand products). With an adult

Clibanarius tricolor, Blue-Legged Hermit Crab. (*L. Gonzalez*)

size of less than 1" (2.5 cm), this species is generally compatible with anything that won't eat it. Be careful upon acquisition that incidental and potentially predatory species do not enter the aquarium unnoticed with the guise of a "typical" Blue-Legged Hermit (cerith) snail shell.

Paguristes cadenati (Forest 1954), the Scarlet Reef Hermits are slightly larger than and have similar merit to the Blue-Legged Hermit Crab (*Clibanarius tricolor*) in reef aquaria. This species is sought after for its striking color and functional nature. Generally inoffensive, this raven beauty has an inclination towards climbing live coral and irritating polyps (especially at night) but causes little or no harm.

Polypagurus (*Manucomplanus*) *varians*, the Staghorn/Antler Hermit Crab. A dramatic and unusual species defined by its relationship with the hydroid, *Janaria mirabilis*, which encrusts the hermit's shell and forms magnificent antler-like extensions. It seems to be an invertebrate safe aquarium resident but not exactly hardy or easy to care for. It is inclined towards very small bits of meaty food (marine origin) while its commensal hydroid feeds by absorption and likely nanoplankton. A challenging species

> **Not all species of crabs are decorators, and not all decorating crabs are true crabs! Some of these fascinating species are, in fact, sea spiders.**

best left in the sea or for species-specific displays with advanced aquarists.

Dardanus peduculatus, the Anemone Decorator Hermit. This fascinating species is renown for adorning its shell with living anemones (some with *Calliactis* species). The anemones serve an apparently clever purpose of defense with their stinging tentacles for the traveling hermit. Said anemones are carefully transferred when the hermit changes shells. The many perils in the confines of an aquarium (unnaturally repetitive and close brushes with other cnidarians, filter intakes, etcetera) make the anemones on the shell of this hermit crab inappropriate for all but a species tank. Not recommended for the casual marine aquarist.

Trizopagurus strigatus, the Striped or Halloween Hermit Crab. One of the most strikingly colored and patterned hermit crabs in the aquarium trade, this species attains a moderate adult size of several inches. It is an excellent scavenger for safe, fish-only displays, but is unsuitable for aquariums with most other small invertebrates by virtue of its clumsy and brusque size if not its predatory inclinations.

Aniculus maximus, Yellow Hairy Hermit Crab. An amazingly beautiful specimen for which pictures simply do no justice, this hermit crab will eat anything in its path: a true omnivore. They are hardy and long-lived in aquaria once established but are notoriously sensitive to shipping. Requires higher water quality to survive in captivity than most crabs.

Petrochirus diogenes, the Giant Hermit Crab. This remarkable species grows to an amazing adult size of 12"/30 cm (this is not a misprint!). Needless to say, it is not a reef safe creature by aquarium standards and is really best kept in a species tank. At this size, few fishes are safe and the family cat should probably stay off of the top of the aquarium.

Criteria for Suitability...

So-called "Safe" versus "Unsafe" Crabs:

The description of any reef animal as "safe" or "unsafe" must be accepted as an arbitrary assignment. Such valuations can still have merit and are offered here to guide aquarists to make successful choices. However, inevitably all reef animals must eat something else on a reef and in the strictest definition are not "safe" in marine aquaria. To be specific, though, we shall describe "safe" species here in relative scope with their likelihood to eat other popular and commonly kept fishes and invertebrates. For species that can be identified accurately, we generally have good to very excellent information about their behavior and husbandry. Rest assured, however, that reliable deductions of such information

Portunus sp.

are also possible with unidentified species by careful observation of physiological characteristics. We can tell a lot about a crab's behavior by studying the form of key features and appendages. Form almost certainly follows function, as they say. As such, a specimen with large, sturdy pincers like a *Calappa* species, the Box or

Arthropods: Crabs   283

Dardanus megistos (Herbst 1789), Shell-Breaking Reef Hermit Crab, often sold as the White Spotted. Members of this genus are predaceous and will gladly consume any fishes they can get their claws on. To six inches (15 cm). Place with large, aware fishes only.

Shame-faced crabs, is likely to use them for similar purpose: chipping and crushing pieces of thick mollusk shells to reach living snails, hermit crabs and bivalves (clams and the like). The large and powerful claws indicate a formidable predator not to be trusted in a tank with mixed invertebrates. In suit, an unknown species of crab with feathery appendages (maxillipeds) like the Porcelain crab species (*Neopetrolisthes* and *Petrolisthes*) is certainly some kind of filter-feeder and likely quite safe with most other fishes and invertebrates. Diminished claws and filter-feeding appendages indicate a very compatible crab in mixed species displays. In between these two extremes of "safe" and "unsafe," most crabs fall. It is no surprise then, as most are very opportunistic feeders, that this large group of crabs in the middle has mediated appendages that serve a true omnivore well. A sleuthing aquarist will evaluate a new crab carefully and can be assured that larger claws are for crushing (prey capture), and smaller claws are for cutting (algae and invertebrate grazing). Thus, armed with this information, you can make an informed decision with reasonably good confidence as to the suitability of a crab for your marine aquarium.

De-Selection: Removing Unwanted Crabs

Many folks discover resident crabs that seem to appear from nowhere. Indeed, they most often arrive as hitchhikers on live rock or with other hard substrates including corals and other invertebrates. Aquarists are strongly encouraged to isolate all new rock and invertebrates (and fishes of course) in a proper quarantine tank where such crabs can be discovered, observed and removed if necessary. With proper quarantine protocol for all livestock, new imports are acclimated, addressed and evaluated with due process if not proper respect for their very lives. As responsible aquarists, we must also protect our investment and the many lives of our animals in the full aquarium display from the introduction of a parasite, pest or disease. To place a new piece of live rock or livestock into an aquarium

Clibanarius tricolor, the ubiquitous Blue-Legged Hermit Crab. (*K. Gosinski*)

Paguristes cadenati the very common and popular Scarlet Reef Hermit. (*L. Gonzalez*)

Clibanarius seurati. (*L. Gonzalez*)

without quarantine is literally playing a game of chance with the lives of your charges.

Proper quarantine is to last 4 weeks in a bare bottomed aquarium, with 2 weeks being a bare minimum for reasonable assuredness of health. Live rock and any other new substrate (coral, shells, plants, etcetera) is to be propped off of the bare bottom of the QT (quarantine) vessel as if placed on an open rack. Improvisation with a plastic grid (like common "eggcrate" light diffuser material used for dropped ceilings) and short PVC legs makes a suitable table to perch a specimen on.

Trapping in Quarantine...

Baiting and trapping predatory crab species is simple enough. If an undesirable crab resides with the QT specimen, it will be easily attracted from high upon its perch to the bare bottom of the QT vessel with an enticing piece of meat (of marine origin). Reef keepers often use minced clam as bait, fearing a predator likely to target Tridacnids. Truth be told, most predatory crabs are so indiscriminate that a leftover piece of ham sandwich would likely work as well! Trapping generally must be done in the dark of night, as many undesirable species are shy and nocturnal. Some species seem to be indifferent to the presence of red light and an incandescent, red-colored "party bulb" may remain lit over the QT aquarium for periodic inspection by the aquarist on nights when trapping is conducted. Most crabs are to slow or unable to scale the smooth legs of the makeshift PVC rack easily and upon discovery are easy for the aquarist to net and remove. For agile crabs, the bait may be tied in a satchel of nylon (aquarium netting bought for the purpose, or a boiled piece of ladies nylon stocking). Predatory crustaceans like crabs and mantis shrimp (Stomatopods) often are snared easily in nylon netting; a small length of fishing line can be tied to the satchel of bait for extra security. Whenever possible, spare an unidentified or offending crab's

> **Hermit crabs run the gamut in size and "politeness" ranging from the demure Blue-Legged Hermit crab (upper left) at less than 1" (<2.5 cm) to the rogue Giant Hermit crab (lower-right) at a meal-sized 12" (30 cm)!**

life by allowing it to live in a fish-only or otherwise appropriate display, refugium, or sump where they can be quite useful scavengers. Otherwise, donate your trophy to a local aquarium store or aquarium society.

Trapping in Displays...

Capturing predatory or nuisance crustaceans in a fully rockscaped display is sometimes not as difficult as it may first seem. Crabs are particularly clumsy and not as skittish about migrating to bait as are Stomatopods. Using the above technique of a satchel of fragrant meat (silversides, clam, squid, and krill, for example), take the bait tethered by a long piece of nylon (clean fishing line or clear sewing thread) and gather a

Calcinus tibicen, the Orangeclaw Hermit Crab (*L. Gonzalez*)

Unidentified sp., red-striped hermit. (*L. Gonzalez*)

Petrochirus diogenes, the Giant Hermit Crab. (*G. Rothschild*)

Arthropods: Crabs

An irresistably photogenic and interesting aquarium denizen, *Neopetrolisthes maculata* "Anemone crab." (*G. Rothschild*)

A wee, tiny *Zebrida adamsii* safe among the spines of a much larger urchin, *Astropyga radiata*. Sulawesi photograph

Hairy versus Smooth Carapace Legend

In popular aquarium lore, a legend exists about crabs with hairy carapaces. Although it often proves to be true, there are more reliable methods of evaluating suitability. It is said that an unidentified crab species can be assessed as "reef-safe" by the texture of its carapace (back/body). Hairy crabs are believed by some to be categorically dangerous while smooth shelled species are thought "safe" with mixed invertebrates. Furthermore, it is said that species with dark or black tipped claws are inclined to prey on fishes or other invertebrates. The truth of the matter is that such distinctions are neither practical nor reliable. There are some notorious hairy bandits that have made their way into reef aquaria and wrought havoc like the *Dardanus* species (lower right). However, quite a few crabs without hairy carapaces or appendages can be equally destructive and predatory like the box or shame-faced crabs. These *Calappa* species (bald as a pool cue ball) seem like they will eat anything including Volkswagens given the chance to (and using their powerful claws like a can opener!). As a rule, the shape (form) and function of a crab's claws are a better indication of likely behavior as described above (crushing versus cutting pincers). Generally, your best bets for safe selections are species that get no larger than a coin as adults (<3 cm) and have diminished pincers. Smaller species pose less risk to fishes and a modest risk to invertebrates. Beware, however, of hearsay opinions regarding "average" adult sizes of popular crabs. Commonly imported at 1-2" (2.5-5 cm) in diameter, we have seen *Mithraculus* (*Mithrax*) *spinosissimus*, the Red Channel crab, that grew to nearly 6" (15 cm) in diameter and were quite aggressive to fishes!

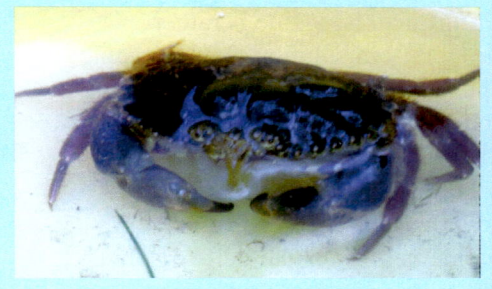

Smooth exoskeletons are not indicative of behavior in reef crabs. Many "hairless" species are quite destructive in a reef aquarium. This *Cancer antennarius* is one example.

This *Dardanus megistos* hermit crab is hardy and beautiful - but would certainly wreak havoc in a reef aquarium - even the fishes would be at risk. Those hefty claws are made for crushing!

small to medium sized wide-mouthed glass jar. Pickle and mayonnaise jars work nicely for this purpose: smaller jars for smaller game may work fine. The wide-mouthed jar can be buried almost entirely in the sand. If you prefer, the jar can also be leaned against the rockscape. The tethered bait, of course, is placed into the jar while you wait for the targeted scavenger to come out at night. Capture may simply be a matter of chuckling diabolically as you watch the trapped crab try to escape up the slippery glass walled prison of the sunken jar. Indeed, for most crabs it is easier to slide down into a glass jar after food than to climb back out with or without a meal. If necessary, the bait can instead be weighted in the jar and the nylon leash can be fashioned to pull a trap lid over it. As you might imagine, there are many variations on this theme... be resourceful.

Care for Crabs

Although crabs are usually hardy and tolerant of a wide range of water quality, they should be afforded the same care and handling as any invertebrate. Good invertebrate husbandry includes full strength seawater (a specific gravity of 1.024-1.026), stable pH (averaging ~ 8.3), zero ammonia or nitrite, and little nitrate (<10 ppm.). Special care should be exercised during all acclimation events as crabs, like most crustaceans, can be quite sensitive to osmotic shock from differentials in salinity. Research all specimens thoroughly for feeding requirements and consider all of the above possibilities about behavior and physiology for unknown species. Be assured, again, that most are opportunistic, omnivorous feeders and that many can become predatory.

Regeneration

It is quite fascinating to discover how effectively crabs and crustaceans regenerate and repair lost and damaged appendages! Entire legs can be regenerated after a series of molts. As a point of note: **Iodine** has been recognized as a necessary element for health and growth in many crustaceans. A deficiency of this element has been implicated in incomplete molts and a higher mortality of crustaceans in aquariums, particularly during molting periods. Unfortunately, iodine is rather temporary in solution due to organic uptake, oxidation, protein skimming and chemical adsorption. It has a lifespan of mere hours in aquarium water, which lends support to the argument for dosing such useful supplements in small daily amounts rather than larger weekly spikes. Regular water changes are also a highly recommended way to replenish minerals and nutritive elements such as iodine while diluting contaminants and other undesirable components of aged seawater.

Compatibility of Crabby Denizens

Even when tolerant of fishes and other invertebrates, many crabs are quite territorial towards one another. Most are best kept in small numbers if not singly. Progress in breeding attempts with several species has been made and interested aquarists are encouraged to set up species-specific systems for the purpose. Compatibility issues are more likely to arise with benthic fishes and motile invertebrates rather than open-water fishes or sessile invertebrates.

Best Bets with Crabs and Why

Petrolisthes galathinus, the Lined or Banded Porcelain Crab, is a striking species collected in the Caribbean. It is small in stature, typically reaching not more than 1" (25 mm).

Advice on choosing the best and worst candidates among crabs is truly an ambitious endeavor. As described here and beyond, the scope of crab size and behavior is extraordinary and diverse. To give such advice fairly, we must assume that aquarists will heed fair warnings about the inherent risk of keeping these omnivorous scavengers in mixed species displays.

Hermit Crab general overview (many genera):

Schizophrys aspera, a species unlikely to behave for long in an aquarium with mixed reef invertebrates.

Far and away, as the most popular crab in the aquarium trade, some species of hermit crab are inevitably going to make it on a list of "Best of Crabs" from the sheer number of represented individuals in the hobby. Hermit crabs are categorically more tolerant of fluctuating and less-than-ideal water quality than most crabs. Very few are specialized feeders or require any significant attention in husbandry. Most only need a regular supply of food, reasonably good water quality and a supply of empty shells to shuffle between with growth (and for amusement at times it seems!). The provision of extra shells is overlooked by aquarists more often than not and forces many hermit crabs to kill live snails in the display for necessary housing to grow into. Many such attacks could easily be prevented (and the subsequent bad reputation such crabs get for it) if this sensible consideration was addressed. Despite the sales pitch, most hermits ultimately cannot be recommended in any significant number for mixed invertebrate displays. Safer specimens generally do not exceed coin-size as adults (under 2"/5 cm) and have slender claws indicating a greater dependence on cutting and grazing. Beware of hermit crabs with large, stout claws for crushing and chipping: an indication of a predatory nature.

"Dwarf" Reef Hermit crabs: Regardless of the best marketing efforts to the contrary, none of the following "dwarf" hermit crabs excel beyond another to substantiate any claims to being the "best" detritivore, scavenger or algae grazer. Moderation of stocking numbers is recommended to reduce the burden of these opportunistic feeders on live sand microfauna. They are generally hardy and innocuous: Blue-Legged Hermit (*Clibanarius tricolor*) Atlantic, Scarlet Reef Hermit (*Paguristes cadenati*) Atlantic, Zebra Hermit (*Calcinus laevimanus*) Pacific, Red Tip Hermit (*Clibanarius sp.*) Mexico, "Neon" blue striped/knuckled (*Calcinus elegans*)

Common (Green) Hermit (*Clibanarius vittatus*) East Atlantic: It is the most common of the larger hermit crab species kept in aquariums and it is highly recommended to those with active, sturdy fishes in display. This species in particular has demonstrated a time-tested and extraordinary hardiness and resilience in aquariums. They are very tolerant and efficient scavengers that will eat most anything. Never trust this species with invertebrates and exercise due caution when mixing them in like kind with concern for territorial aggression.

Sally Lightfoot Crab (*Percnon gibbesi*): This popular and attractive crab has been an aquarium favorite for decades. A categorically hardy crab, the Sally Lightfoot can be recommended to most new aquarists as a durable and relatively peaceful species. Small specimens feed predominantly on nuisance algae and scavenged bits of meaty food, which makes them quite useful and desirable for aquariums. Large specimens have been known to attack small fishes. *Percnon sp.* live on the reef (often hiding under urchins), and in rocky areas. They will climb out of the water given the chance

Uca crassipes - Fiddler Crabs

and due consideration must be given to openings for a fatal escape in aquarium systems for Sally Lightfoots. They cannot be fully regarded as "reef-safe" crabs by aquarists, as they have been known to occasionally molest and consume polyps and cnidarian tissue. They make great denizens for most FOWLR (Fish Only With Live Rock) displays.

Emerald Crab - *Mithraculus* (*Mithrax*) *sculptus*: This crab is heralded for eating the dreaded nuisance Bubble Algae (*Valonia* and like species), but it is also quite capable of killing small fish. Some *Mithraculus* species are cited as reaching adult sizes ranging from 2.5" to well over 6." Larger specimens around 6" (15 cm) are in fact not at all uncommon and are rather intimidating to small fishes at that size. The Green Emerald crab is a useful but overrated algae scavenger in the authors' opinion. It is not to be purchased with the exclusive hope of grazing nuisance algae. While *Mithraculus* have been frequently observed to eat various species of nuisance algae with vigor, they eat none to exclusion and are omnivores in the truest definition of the word, although still an altogether useful and worthwhile aquarium inhabitant.

Arrow Crab (*Stenorhynchus seticornis*): This popular aquarium specimen is renowned for eating bristleworms (errantiate polychaetes) but also for killing small fishes and other crustaceans. It is a true carnivorous species and perhaps a measurable burden on other desirable microfauna in live sand and live rock products. Arrow crabs are generally safe with hardy corals, anemones, and larger fishes. They fare well and are likely to serve a greater good in larger "reef" aquariums. Egg-laden females are often observed with a full brood held conspicuously in a "trap door" under the thorax. Rearing larval crabs is still challenging on a hobbyist level, but these frequent products of reproduction provide excellent food for small fishes, corals and other invertebrates.

Porcelain Crab (Porcellanids including *Petrolisthes*): feed themselves by extending their specially adapted, feathery appendages (maxillipeds) into the water column. A good flow of water is necessary for this filter-feeder to survive in captivity. Unlike the anemone crab (*Neopetrolisthes*), however, this animal is not as dependant on a cnidarian host and tends to fare better in captivity. They use comb-like mouthparts to sieve tiny plankton from the water column for food and may benefit from supplemental feedings of newly hatched brine shrimp and rotifers. Natural plankton from a fishless refugium is likely much better.

Worst Bets with Crabs and Why

Decorator Crab (various species): Simply destructive (and repetitively so) to desirable benthic and sessile organisms. Most are not trustworthy with invertebrates or small fishes. Some species have specialized and nearly impossible diets to support, while many other species

Percnon gibbesi the Sally Lightfoot Crab (*G. Rothschild*)

have categorically carnivorous tendencies. The Decorator *Spider Crab* (*Camposcia retusa*) may be the hardiest and most suitable of these fascinating crabs for captive study and display (AKA the Micronesian Spider Decorator Crab).

Sea Spiders (the Pycnogonids): These are fascinating arthropods that occur over an extraordinary range from the abyssal deep ocean floor to shallow tropical waters. Several hundred species have been identified but little is known about their captive care. Diets vary but mostly include some sort of "meat"; they are generally predatory and carnivorous with some species known to have very specialized diets. Tiny hydroids, cnidarians, zooplankton, pieces of fish, nudibranchs and gastropods are some of the many prey animals favored by sea spiders.

Box/Shame-faced Crabs (*Calappa sp.*): These are formidable predators, as mentioned above, on mollusks and other crustaceans. Many get large, can dishevel a rockscape with their stout bodies, and generally wreak havoc on invertebrate tank mates. They are known to feed on mollusks and crustaceans. Study of such crabs may best be conducted in a dedicated species tank.

Pom-Pom Crab AKA Boxer Crabs (*Lybia sp.*): These are fascinating, tiny crabs that hold small anemones in their pincers and use them to ward off attackers. The anemones are also used to swab the substrate as a feeding tool from which collected matter is harvested. Aquarium success with this crab has generally been a challenge. Duress in shipping seems to be a significant problem with this crustacean. The imposed stress of unnatural, repetitive contact with fishes and other invertebrates upon this nervous little animal in the confines of an aquarium is also suspected as a limiting factor to the survivability of *Lybia* in captivity. The study of Pom-Pom/Boxer Crabs is best conducted in a dedicated species tank.

Mangrove Fiddler Crab (*Uca crassipes*): Males grow one very large claw. This is the common "fiddler" crab species used in the freshwater and marine aquarium trade. They can be kept with fishes, but must have a place to climb and dry out of water. They favor clean, consistent high quality water and are essentially inconvenient for a casual marine aquarist to accommodate without a dedicated refugium or similarly specialized niche aquarium. Males are decidedly territorial.

Molting

Periodically, all crustaceans molt (or undergo **ecdysis**, if you prefer) to allow for growth. It is highly recommended that you leave molted exoskeletons in the system for the same or other animals to consume. A crustacean molt is rather nutritious matter (high protein). In some species of arthropod, the frequency of molting is a sex/gender indicator. For several days after a molt, crustaceans are vulnerable (soft bodied until they harden) and require safe haven. During this time, it is imperative that you provide sufficient hiding spaces and food for your crustaceans to survive. Oily meats of marine origin, HUFA-rich matter (highly unsaturated fatty acids like the supplement, Selcon), and high protein fare are all important staples in the diet of crabs and many marine aquarium specimens.

Anemone Crab (*Neopetrolisthes sp.*): A filter-feeder that usually requires intermediate to expert care. They are almost always found in pairs in the wild (and should be purchased as such if keeping two). Anemone crabs often conflict with clown fishes over host anemones. Plankton generating refugiums seem to contribute significantly to success with this challenging creature; fishless refugia will contribute zooplankton and high-flow seagrass systems may impart nutritious epiphytic material and phytoplankton of use to anemone crabs. Due to the equally challenging nature of most anemones, this crab is relegated to the category of "avoid" for all casual aquarists. A thoughtful and well-designed aquarium system will be necessary to successfully keep anemone crabs and their commensal host, similar to Porcelain crab (*Petrolisthes*).

Horseshoe Crab (*Limulus polyphemus*): This living relic has fascinated mankind for time untold. Alas, by virtue of its adult size (easily approaching 12 inches/30 cm) and feeding habits, it is unsuitable for all but the largest aquariums. Horseshoe crabs need vast planes of deep, fine sand to adequately thrive. An aquarium of several hundred-gallon capacity is unlikely to sustain a single specimen with enough food to survive long term. Even if alternate food sources could be supplied, the restless foraging habits of this creature every night agitates sand and sediments and can cause serious problems with water clarity. The frequent turbidity will impact filter performance, and systems with photosynthetic invertebrates will suffer markedly by the reduced light. To deny a horseshoe crab deep fine sand and sediments will impose a duress that is likely to impact its survivability significantly. Another species-specific display will best serve *Limulus*.

Giant Hermit Crab (*Petrochirus diogenes*): This large hermit crab and like species of size are best limited to large aquariums with select fishes that can easily reckon with their predatory nature and formidable weapons. Casual aquarists with mixed fish and/or invertebrate displays under 200 gallons will likely suffer the company of this species and its destructive habits.

Anemone/Coral Hermit Crab (*Dardanus sp.*): The precarious anemones placed upon this hermit crab's shell make this species inappropriate for all but a species tank. The anemones are likely to suffer the rigors of aquarium life and an abbreviated lifespan for the dangers therein. The repetitive contact of these anemones with the abrasive rockscape, other cnidarians, pump and filter components, and more require that a responsible aquarist provide a species tank for this hermit crab and its captives. Anemone Hermit Crabs are not recommended for casual aquarists.

Galatheid - The Squat Lobsters are an interesting group of arthropods. They range from tiny ornamental species in tropical waters, rarely seen by divers let alone aquarists, to rather large edible species in temperate waters with fisheries industry built around them. More closely related to hermit crabs, they are called "lobsters" because of their long chelipeds. The more photogenic and interesting species are found as commensals on corals and Crinoid feather stars. Little is known about their captive husbandry but their small size and often commensal nature makes them unlikely to be suitable for newer aquarists or generalized aquarium systems like typical garden reef aquaria.

Porcellanella picta is naturally found in association with Sea Pens (*Pteroeides*, *Veritellum spp.*). Females of this species are larger than males.

Summary

We generally do not recommend crabs for mixed invertebrate systems. They are typically better suited to fish-only or species-specific displays. Whatever scavenging benefits most have are outweighed by their potential to kill desirable fish and invertebrates. For aquarists seeking nuisance algae control, it is likely better to focus on improving nutrient export mechanisms and employing alternate herbivorous livestock like snails. The risk of most crabs in time to demonstrate their true omnivorous or even carnivorous tendencies is simply too great. It is especially ironic when an "algae-grazing" crab is routinely inclined to kill and consume other algae-grazing invertebrates!

Stenorhynchus seticornis - Arrow Crab, photographed in Cozumel. (*D. Fenner*)

Limulus polyphemus - Horseshoe Crab. This is one of the "not a crab" crabs. It's big, bulky and hungry, with a burrowing habit that can topple rocks and corals in the aquarium. These animals require special consideration during aquarium planning. Casual aquarists should avoid them altogether.

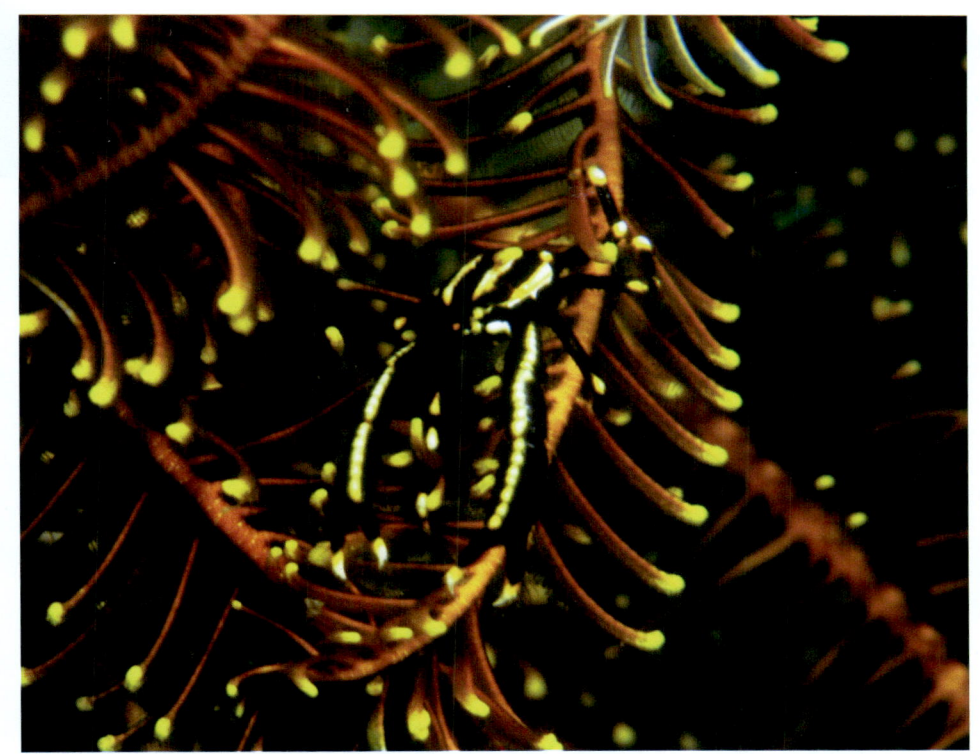

Allgalathea elegans is a striking example of commensal Galatheid with color that varies by host species of Crinoid feather star. They seem to be dependant on their specific host for feeding opportunities and as such are doubly poor specimens for care in captivity due to the very challenging nature of keeping feather stars alive in aquaria.

The Banded Clinging Crab, *Mithraculus* (*Mithrax*) *cinctimanus*, is a familiar commensal with *Condylactis gigantea*. They are somewhat to very difficult to keep in captivity. Cozumel image.

Lissocarcinus laevis, N. Sulawesi

Arthropods: Crabs   293

Lobsters
Suborder Pleocyemata, infraorders Astacides & Palinura

There are several small, colorful species of lobsters that can be quite tempting to marine aquarists with mixed fish and invertebrate displays. Unfortunately, few if any are suitable for such company due to their indiscriminate and opportunistic feeding habits and aggression. Small fishes are especially at risk as well as shrimp, crabs and even other lobsters. If kept in a species tank, however, they can be quite hardy and long-lived. *Enoplometopus* is a commonly imported genus of "Reef Lobster." They are circumtropical in distribution and rowdy carnivores to be sure. Fishes are pursued and sessile invertebrates like coral are often simply disturbed or dislodged by their activities. Pictured here, ***Enoplometopus antellensis*** (AKA Flaming Reef Lobster), one of the smallest species in the trade at 4" or less (<10 cm).

Lobsters in the marine aquarium have been somewhat of a novelty. Although they are fascinating and handsome creatures in their own right, they present some significant challenges to husbandry for aquarists. Central issues of difficulty with keeping lobsters include messy feeding habits, copious waste produced, and destructive behavior to animate and inanimate objects. To generalize, they occur in two sizes for the aquarium: large & rowdy, and small and rowdy. Exaggerations aside, while there are a few better-behaved species, most lobsters are clumsy if not patently destructive. They all lean toward indiscriminate "omnivorism," eating everything in their path that doesn't move fast enough away. Many are simply dedicated carnivores, and all in the hobby are opportunistic predators in some fashion, which makes them potentially dangerous and unreliable in mixed company. We can fairly say that all specimens over about four inches in length (>10 cm) are unsuitable for anything but "rough and tumble" systems with tank mates of appreciable size (outside of a dedicated species tank). All lobsters should be considered intensely aggressive inter-specifically (with same/like-species), and few fish are truly safe in their company. It is advised to keep just one lobster per tank, and only with hardy, "aware" tank mates that are fast swimmers and light sleepers.

All these frightful warnings aside, we can tell you that given a suitable aquarium and habitat, lobsters are generally hardy, long-lived and easy to care for. Most are not too terribly active, but food will coax all but the shyest from hiding. They are nocturnal creatures as a rule, and with the common "reef lobsters" (*Enoplometopus*) you will find that they are extremely secretive...

preferring the deep dark crevices of the rockscape. But with display arrangement, it takes little effort to make an attractive and focal hiding spot for lobsters in plain view so that you can enjoy seeing them in the aquarium most any time. A rock cave or clear section of tubing can provide an interesting view; the strategic construction of an open cave or overhanging ledge facing the viewing pane will almost certainly be adopted if no better cave is constructed elsewhere in the habitat. Most lobsters like to burrow at least a little and favor a more coarse media rather than the sugar-fine sand that has become so very popular and useful in reef aquaria. Such gravel, however, is challenging to keep clean with a normal fish load and even harder with sloppy "bugs" in residence. Feeding preferences for lobsters vary somewhat by species, but most found in the trade will eat most any meaty foods proffered. Include algae-based foods (spirulina or nori) in the diet too to enhance their colors. A wide variety of meats of marine origin are recommended as the staple. Offer lobsters a selection of at least 4-6 different foods as you would fishes to improve the overall nutritive quality of their captive diet. Selection and husbandry is essentially the same with lobsters as for crabs and other large crustaceans. Although they are hardy in many ways, they are still invertebrates with the same inalienable sensitivities to water quality (low oxygen levels, changes in salinity, etc) and toxins like metals and organic dyes. Realize too that the large appetites and messy feeding habits of most species require extra attention to water quality. Heavy biological, mechanical and chemical filtration is necessary for keeping lobsters. Aggressive protein skimming can relieve some of the burden on these filtration aspects.

Lobsters are most sensitive, perhaps, to rapid changes in salinity; be mindful of this when conducting water changes, acclimating them to new systems, or adding freshwater for evaporation top-off as most are subtidal species. Like many Arthropods, they also seem to need or at least favor iodine in the diet to facilitate healthy growth and successful molts (**ecdysis**). Iodine can be added as a supplement to the water or it can be provided simply in rich foods like shrimp and krill. Good feeding and frequent molts in fact are all it takes from these crustaceans to demonstrate their amazing regenerative abilities. Whole arms and legs once missing can re-grow within a short series (months). It's quite remarkable to see the process executed with the delicate finesse of a molt around eyes, antennae and even the individual **sensory hairs** (principally on the legs and used mainly for taste) and **proprioceptors** (used for touch). Regular and complete molts are a sign of good feeding and water quality. Incomplete molts, which leave parts of the body fragmented inside, indicate the contrary. Frequency of ecdysis varies by species and with environmental conditions. Nonetheless, a molt once every two months is a reasonably good range of activity. There can be differences in frequency between the sexes too. Lobsters are dioecious and practice sexual reproduction as a rule. Several varieties are commercially aquacultured, but their complex larval cycles make reproduction in the home aquarium difficult. It should go without saying that the temperate species so common for the dinner plate are wholly inappropriate for tropical aquarium life for their size and water quality needs.

Despite their reputations in the aquarium as tough and aggressive, lobsters are not wholly "untouchable." Common predators of them include other lobsters, large crabs, octopuses, and the usual piscine suspects: triggers, puffers, large angels, and eels. These will all go after lobsters with little provocation. On the other side of the coin, some favorite lobster prey

Enoplometopus daumi, one of the clawed lobsters, is quite common in the aquarium trade and goes by several familiar names including The Purple, or Purple & Orange lobster, Daum's, or the Violet Reef Lobster. *E. daumi* is one of the quieter and more reclusive species. They are often kept in reef aquaria with sometimes good results. It is still a calculated risk that we do not recommend. A better solution might be to keep your lobster in an ancillary refugium-like vessel to protect the main display inhabitants. This species is commonly imported from the Philippines and Indonesia.

Enoplometopus occidentalis: AKA the Hairy or Red Reef Lobster, this species has a wide distribution throughout the tropical Indo-Pacific. It is distinguished by white ocellated spots all over the body. Like most lobsters, this reef denizen is nocturnal.

Homarus gammarus, the European Lobster (above) is a cool water species from the Mediterranean sea and the Atlantic ocean to Norway and quite similar to the American Lobster of culinary fame. (below)

Homarus americanus, like *H. gammarus* above, sports two powerful claws, with one generally larger than the other (bigger in *H. americanus*, though). The oversized claw is used as a crusher for smashing shells. The smaller claw is used like a knife with its serrated edge to tear flesh. *H. americanus* "Maine Lobster" is known in cuisine by many names: iseebi (Japan), bogavante (Spain), homard (France), and hummer (Germany). (*J. Chodakowski*)

Panulirus argus, the Caribbean Spiny Lobster is primarily a nocturnal feeder but will take opportunities to dine by day especially as large, mature specimens. Pictured a night image in the Bahamas. Growth to 24" (60 cm)

Panulirus gracilis, the Green Spiny Lobster (right) is a species of interest to science, fisheries and hungry folk. This lobster grows to over 12" in length (>30 cm) and is of little to no interest in private aquaria. Its range includes the East Pacific.

296 *Natural Marine Aquarium Volume I - Reef Invertebrates*

includes sleeping fishes of all sorts and sizes, other lobsters, marine plants and algae, shrimp, crabs, mollusks, worms, and other sessile non-stinging invertebrates like bivalves (they can easily cut the adductor muscle of anchored clams and make a meal of them). Cnidarians are generally safe from being eaten, although unstably perched specimens in the display are at risk of injury from being knocked over by clumsy lobsters. Ultimately, like anything else we've covered in this reference, with a properly designed aquarium system and habitat, these organisms can live full and healthy lifespans in captivity. We recommend a species tank for most lobsters to be safe.

Gallery

Species that are commonly found in the aquarium trade can be described in three major groups:

- **Astacoidea** (Crayfish and Clawed Lobsters)
- **Palinura** (Spiny Lobsters)
- **Thalassinidea** (Ghost Shrimps and Clawed Lobsters)

Astacoidean (Nephropids): Clawed Lobsters... this is the most popular group for tropical aquaria. Commonly referred to as "reef lobsters," they are some of the smallest and most colorful species available. They are distinguished in part, by a single, conspicuous pair of antennae. *Enoplometopus* is commonly found in deep rock crevices on the reef and is strictly nocturnal.

Palinurids: Spiny Lobsters (AKA Rock lobster or Langoste/Langouste/Languste). The first glance of this group reveals that they lack the large tasty claws of the aforementioned Astacoideans. In fact, more than 40 species of Palinurid are instead harvested commercially for lobster tail-meat. Their first large paired appendages are walking legs and not claws or pincers, and paired legs may be **chelate** or **subchelate**. Body form is somewhat variable in the group, but a stout, cylindrical cephalothorax is typical. Some also commonly can be heard to make a rasping sound which has been described like the sound cicada (locust) make. The process is called **stridulation** and is created when they rub the base of their antennae against serrated ridges on their head. This aspect is hardly the most fascinating or unusual for the group. The most unique attribute of Palinurids must surely be the "March of Spiny Lobster." Each year in the Bahamas, hundreds to a couple thousand lobsters line up in columns and march across the sand floor triggered by early autumn storms. The behavior is remarkable, bizarre and complex. This often solitary "bug" begins to collect in large numbers under ledges in shallow water around

Panulirus inflatus, A Blue Spiny Lobster: a familiar denizen of the East Pacific. Pictured here in Magdalena Bay, Baja.

Panulirus marginatus, the Banded Spiny Lobster is a Hawaiian endemic ("ula") of tremendous commercial value. It has been well studied and protected by strict fisheries protocol. Recreational collection of the species is limited to hand-catching (no nets or traps). Juveniles and reproductive females are protected year around, and all are protected through the breeding season. They are not "reef-safe" and eat both carrion and reef invertebrates indiscriminately, playing a crucial role as both predator and scavenger in the reef community. This species lives a dedicated nocturnal lifestyle in the deep crevices of the reef and will try to ward off intruder with loud rasping sound (stridulation). Mature specimens top off at 12-16" (30-40 cm). This one in a public aquarium, at about 12:44, according to the reflection of Bob's old Seiko dive watch.

Panulirus guttatus, the Spotted Spiny Lobster (AKA Spanish Lobster) is an ornate species with über-specimens cited at 18" (45 cm). These lobsters actually live on the reef proper, spending their time under corals and ledges by day and then traveling to seagrass beds and lagoons at night to feed and forage. Pictured here in Cozumel.

Panulirus ornatus, the Ornate Spiny Lobster with a wide range that includes the Red Sea, and East Africa to Fiji Islands. This species is large (20"/50 cm) and aggressive. Posturing here in defiance of Bob and his camera! Photographed in the Seychelles.

the time late seasonal storms are due to roll in. The single file columns are a defensive strategy we think which helps to improve their chances of safe passage as they march down to the deeper water *en masse*. If the lead lobster is knocked out, the next in line carries on the command. Studies have been done on this natural phenomenon that make interesting reading for aquarists and divers alike. Fishermen with big nets and short sight have over-exploited this phenomenon to the injury of their own livelihood and natural populations alike. Protection of these lobsters in some areas has been imposed.

For all their fascinating history and behavior, Palinurids are still best suited for life in the sea, or the dinner table, before captivity in a mixed aquarium community. There are many species within this order that are temperate or grow very large. A species-specific tank is best if lobsters are to be displayed, and chillers may be necessary. Aquacultured tropical specimens can be much easier to keep. Some familiar genera include *Jasus, Linuparus, Palinurus, Panulirus,* and *Puerulus*.

Scyllarids: the Slipper, Spanish, Locust, Flat-Head and Shovel-Nosed Lobsters. Genera *Arctides, Ibacus, Parribacus, Scyllarus, Scyllarides,* and *Thenus*. The Scyllarids are seen regularly in the aquarium trade and can fare well in captivity in a proper display with deep substrates and segregation from most other reef invertebrates. They are best kept in fish-only aquaria with sturdy tank mates. They obviously lack large pincer claws like Astacoideans, and they also do not have long sensory antennae like the Palinurid spiny lobsters. Interestingly, the large conspicuous bony plates at the front of Scyllarids are the modified antennae of these burrowing species. Although they may not look very menacing, they are dedicated carnivores and cannot be trusted with snails, mollusks, echinoids, and other crustaceans. Feed Scyllarids a variety of meaty foods of marine origin. Be mindful of their tendencies to rocket-swim when disturbed (tight aquarium covers).

Panulirus versicolor is known by various permutations of the names Painted Rock lobster (science) and Blue Spiny Lobster (hobby). It is distributed widely throughout the tropical Indo-Pacific and can grow to an impressive 20" (50 cm) length! Despite being a real bruiser at maturity, aquarists still seek and buy this species for the aquarium. Admittedly, they are rather complacent, compared to other lobsters, with regard for the tolerance of other tankmates and even others of their kind if given enough space. Large aquariums are still required (minimum 100 gallons). Meats of marine origin are the expected fare for this carnivorous scavenger. In the aquarium it will take residence under an overhang or simply burrow a strategic pit backed up against one of the corners of the aquarium. Pictured here is a juvenile in Bunaken (Indonesia) at left, and adults in Fiji (center) and North Sulawesi, Indonesia (right).

Panulirus penicillatus is one of the few spiny lobsters seen occasionally in the aquarium trade. Commonly called the Tufted or "Hawaiian" Blue Spiny Lobster, imports are generally collected outside of Hawai'i. They are protected elsewhere from overfishing too, like Vanuatu. Growth to about sixteen inches in length (40 cm).

Arctides regalis, the Red-Banded Slipper Lobster has a wide range in the tropical Indo-Pacific. They are the quintessentially opportunist, catching and killing most anything they can find including conspecifics (yes, they are "cannibals"). This beautiful creature can be kept easily in a species tank and is long lived, but is not "reef-safe" by any definition when it comes to mixed invertebrate aquarium communities.

Scyllarides latus, the Cape Town Lobster, is a large species (to 12"/30 cm) ranging from the East Atlantic to the Mediterranean. Keep only in large aquariums with sturdy fishes and *sans* other invertebrates.

Scyllarides squamosus, the Scaly Slipper Lobster, is a cryptic Scyllarid commonly covered in tubercles and camouflaging growths. Adult size is reported at 16 " (40 cm). They are imported on occasion from Hawai'i. This pair was photographed off Roca Partida (a small volcanic island southwest of Baja, Mexico).

Parribacus antarcticus, the Sculptured Slipper Lobster, is a circumtropical species but is harvested for the aquarium trade from Florida. A nocturnal creature of little appreciable activity in the marine display aquarium. Pictured here in St. Lucia (Caribbean).

Arthropods: Lobsters

Crustacean Microfauna

The shrimp-like Amphipods are some of the most common and important microcrustaceans in the marine aquarium... serving as efficient scavengers and as nutritious prey. (H. Schultz III)

Bugs, critters, scuds, fleas, zooplankton, unknown "things"... they are called by many names and misnamed by many more. By any name, however, these tiny crustaceans are abundant, most always beneficial, and an integral component of a healthy aquatic ecosystem in the wild and at home. Systematically, micro-crustaceans encompass a diverse and vast array of organisms, making up nine suborders of the order Pericarida, plus two subclasses of the class Maxillopoda. We will briefly cover the four most important groups to aquarists here: the Amphipoda, Mysidacea, Isopoda, and the Subclasses that include ostracods and copepods. While there is little practical information on their care and culture, the proliferation of desirable microcrustaceans is inevitable in healthy marine aquariums that employ natural strategies including live rock, live sand and refugiums.

Perhaps "micro" isn't the best term for describing all these other non-ornamental crustaceans with regard to size of some of their members... like the celebrated deepwater isopods that exceed 12" (30 cm) in length. However, the vast majority of the crustaceans mentioned here are decidedly very small in size at a fraction of an inch or mere millimeters in length. Don't let their tiny size lead you to believe that they are unimportant, though. These groups are essential organisms in marine food chains with many of them consuming planktonic algae, others eating zooplankters, and all converting such matter to a larger biomass, which in turn becomes available as food for fishes and other larger invertebrates. There are also a few species in these groups that are debilitating parasites, though the vast majority are truly aquarium-desirable.

Amphipods: AKA Gammurus (**Scuds**), Skeleton Shrimps, Whale Lice, Beach Fleas, and the list goes on. Crustaceans of the suborder Amphipoda are similar in body plan to isopods, but are more "shrimp-like" with an arched back, laterally compressed body, and legs tucked underneath. Taxonomic validity aside, you can find more information on amphipods in the public body of information by referencing the names of the following four suborders: Caprellidea, Gammaridea, Hyperiidea (planktonic), and Ingolfiellidea (perhaps simply Gammarideans). Aquarists and divers will mostly recognize the Caprellids and Gammarids from shallow marine environments. The Caprellids are long and skinny critters commonly known as skeleton shrimp. The Gammarids are the most recognizable member of this group in reef aquariums by far, AKA scuds or gammurus. They are poor swimmers and move about by crawling or scurrying. Typically ranging in size from 5-10 mm, they are widespread on the seafloor from pole to pole and occupy numerous freshwater, brackish and terrestrial environments as well. Some of the deep-sea species are even bioluminescent. In aquaria, they are often bold and plentiful and will feed in the open by day or night, although most prefer to forage nocturnally. They are very efficient scavengers as a rule with carnivorous tendencies. Dead and dying meaty tissue are quickly consumed while living tissue is left unharmed by most species. In a group with many thousands of described species, its no surprise that a few are parasitic. One of the most familiar examples of this ectoparasitic lifestyle is the whale-lice (*Cyamus* and the like) looking like whitish fingers on the surface of these oceanic mammals. In aquariums, swarming amphipods are often blamed by the ill-advised or mistaken for attacking live corals and other reef creatures. This is highly unlikely, as the amphipods are almost certainly simply scavenging about or consuming dead and dying material. Amphipods are in fact tremendously beneficial to aquariums, supporting the stability of a system by quickly and efficiently consuming and converting rotting organic matter. They eat detritus, algae, bacteria, fungus, plants and animal remains.

Mysids are beneficial and important prey in marine food chains. (*H. Schultz III*)

They are one of the most successful and likely microcrustaceans to develop in the home aquarium and should be encouraged to increase in population by keeping natural live rock and sand substrates, sumps and fishless refugiums.

Aquarists seeking to cultivate this nutritious planktonic groups members *en masse* can install a simple, upstream fishless refugium kept dim or dark and packed with coarse filter media like bonded filter pads, algae scrubbing pads or polyester filter-fiber. Given adequate water flow and a daily food source, they will massively colonize this "plankton reactor" and overflow good numbers of live plankton on a continual basis to the fish, coral and other invertebrates' mouths below. Amphipods reproduce easily and can often be seen carrying eggs in a ventral brood chamber. Remarkably, some species demonstrate parental care after the young leave the female. There are almost no practical disadvantages to keeping and encouraging amphipods to develop in your aquarium. Have faith that they are limited wholly by the food source that is entirely in your control ("excess" populations mean excess available food in the system).

Mysids: AKA **Opossum Shrimp** are members of the suborder Mysidacea (a commonly encountered species is *Mysidopsis bahia*). They are larger than amphipods at 10-20 mm (30 mm maximum) and look very much like miniature shrimp. Mysids possess a little pouch (**marsupium**) underneath their bodies, which lends them their common namesake. The majority are free-living and feed on zooplankton (mostly), phytoplankton and detritus. A few species live in association with corals, hermit crabs, etc. Like krill, copepods and other micro-crustaceans, mysids are significant items in marine food chains. The wild-harvest and processing of food quality mysids for the aquatics industry has single handedly made possible the aquaculture of a few commercially important species as well as the keeping of many ornamentals like the Syngnathidae (pipefishes, seahorses, sea dragons, etc).

Mysids are actually quite easy to breed at home and aquarists at the Shedd Aquarium and elsewhere have devised systems to culture hundreds of these nutritious "shrimp" daily. This kind of production alone can easily sustain a small to medium sized home aquarium stocked with planktivorous fishes and corals. In many ways, the nutritive value of these crustaceans far exceeds the popular brine shrimp (especially comparing adults of both species). Some processed mysids provide the highest available protein of any like prey in their group. They are dense and nutritious by many measures. The passive culture of mysids in fishless refugiums is also a likely, even inevitable reality. In such places as refugia, it is important to exclude predaceous tankmates that might well consume the brood population (hence the frequent admonition to keep "fishless" sumps and refugia... but also *sans* coral and like predators too). We strongly encourage you to identify and cultivate mysids in the natural marine aquarium. Live seed cultures are available from some local aquarium stores, most aquarium societies and several mail order supply companies.

Ostracods: AKA Mussel, or Seed Shrimps and living fossils dating back perhaps more than 100 million years. They resemble tiny (0.3-5 mm.) clams or seeds swimming about with jerky motions via their paddling antennae near or on the bottom of the sea. Their bodies are enclosed within a hinged bivalve casing, like a shellfish of the same name, which gives them security when they are not feeling their way about. There are several thousand species of ostracods found all over the world in fresh and marine waters, including many that are readily consumed by filter-feeding and planktivorous predators. Ostracods are very beneficial organisms that predominantly feed on detritus and algae.

Copepods: are known to science as the Oar-footed Crustaceans from the subclass Copepoda. An old grad-student roommate once told Bob

Bodianus pulchellus (the Spotfin or Cuban Hogfish) and *B. rufus* (the Spanish Hogfish) are two very large wrasse species whose juveniles are cleaner fish that have been suggested as possible predators on isopods. Most aquarists will not be able to utilize these species even if *Bodianus* would control the parasite. These fishes grow large (well over 12"/30 cm for both), require very big aquaria (long and over 100 gallons) and are rather disturbing or predatory on many other desirable reef invertebrates. In a large, active fish-only display, however, these wrasses are quite hardy and ideal aquarium specimens.

Arthropods: Crustacean Microfauna

Fenner he was going to make his mark in the world by devising the means of making pressed-formed "copepod steaks!" Nutritious, yes.. very! Palatable? That's up for debate. Nonetheless, copepods are fundamental links in aquatic food webs. They comprise the majority of plankton in most or at least many areas and are the staple food item for species ranging from tiny reef invertebrates to the largest filter-feeding fishes. Most are free-living planters principally feeding on diatom and dinoflagellate algae, but there are scrapers of algae on hard substrates, predaceous species and some parasitic members amongst the group. Due to their small size,

> **Summary of Options for Controlling Parasitic Isopods in the Aquarium:**
>
> - Strict quarantine and screening of *everything* entering the tank (especially live rock and live sand)
> - Bait with food or fish
> - Natural control
> - Manual removal (tweezers or forceps)
> - Running fallow
> - Freshwater dips
> - Medication

identifying most of the copepods requires microscopic work. Their first and sometimes second thoracic segments (of six total) are fused with the head. In free-living forms, the five pair of thoracic legs "paddles" (hence the name "Oar-Footed") the animals about, along with their two sets of antennae. The tail section is bifurcate with two **rami** (tail branches) sticking out. Copepods lack compound eyes, yet all but the parasitic species have a median eye. These are the tiniest microcrustaceans that aquarists are likely to see range from 6 mm (1/4") to much smaller. They are almost always present in reef aquaria, "hitchhiking" in on living substrates (live rock, sand, hard-base with sessile invertebrates, etc.). Like most microcrustaceans, they are desirable and readily cultivated in places of refugia like deep sand and mud beds, live rock and remote in-line refugium vessels ("plankton reactors").

Please be sure to read or review the *refugium* section in this book for more information on these plankton and all such beneficial planters in aquaristic husbandry. Aspects of refugium culture like the keeping of filamentous and unicellular algae are necessary for "breeding" many microcrustaceans that live on and in such matter (deriving food directly or incidentally as with epiphytic matter on the structures of plants and algae). Aquarists seeking additional information and supplies for phytoplankton and zooplankton culture may start by contacting specialized merchants like Florida Aqua Farms (see Resources).

Parasitic Isopods: AKA Sea Lice, **Isopods** are an order of crustacean organisms that have been well studied scientifically though not well understood by aquarists. These "bugs" are related to some readily familiar species of Arthropods like shrimps, crabs and spiders. Most folks have had experience with this group's terrestrial kin known as pill-bugs, wood lice, sow bugs or rollie-pollies... ahhh, childhood memories. The marine species can be found in all saline environments from shallows to abyssal depths and range from equatorial regions to the most frigid of waters. They are known by more than a few common names: sea lice, sea slaters, fish lice, kelp lice, and beach lice. Most isopods are small: just a few millimeters in size. But some notorious species, like the deep sea *Bathynomus giganteus*, gets to about half a meter in length! Their carnivorous inclinations run the gamut from free-living and harmless scavengers to predatory and parasitic. Although they likely will not harm your reef invertebrates, they can be quite dangerous to your fishes and even you (yes... some aquarists have been bitten or nipped by parasitic isopods. Ewwww!). Most species will take carrion and a some eat algae. They naturally are more abundant in polluted, sick and stressed environments where carrion and weakened fishes are more plentiful like habitats near heavily populated coastlines and areas where destructive fishing practices are common. Judging by hobby literature, aquarists are most familiar with the Cirolanid family of isopods. Many of these are opportunists waiting for a chance to feed on fishes. The legendary ones are aggressive parasites. The parasitic forms are also referred to as tongue-biters in reference to their habit of attacking the mouth of many fishes, and as "fish doctors," as they leech blood.

The specific modes and mechanisms of parasitism vary by species. Most parasitic isopods latch onto a swimming fish anywhere they can. Gills, eyes, fins and tail are popular sites for the ease through which blood is drawn through the host's thin membranes. The parasite may choose to leave after some time or stay with the fish until it dies and scavenging the flesh as well. Is the story all-bad about isopods? That depends on your perspective. In the ocean, not all isopods are parasitic (hardly!) and even among the carnivores, they perform a crucial role as scavengers of the dead and dying. In the confines of an aquarium, however, they must be extracted if parasitic.

Parasitic species can be challenging to eradicate from the main display and can require extraordinary measure to do so. Aquarists have reported

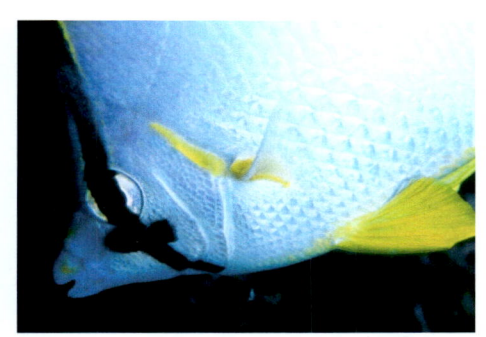

Ouch! That's got to hurt. Parasitic isopods may attach anywhere to a host, but they are commonly observed on or near the fins, gills and eyes.

removing all fish completely from their display for six months or more only to discover that the isopods survived and attacked new fishes on re-entry! Others have resorted to using a sacrificial fish to lure isopods to attack and then catch the fish repeatedly to remove isopods with tweezers. This is unnecessary for how easily it can be avoided altogether by proper husbandry and quarantining (QT) all candidates (live rock, corals, sand, invertebrates, fishes, etc). Such aggressive parasites can usually be lured with meaty bait (chunk fish meat) easily in a bare-bottomed QT just like predatory crabs, fireworms, mantis "shrimp" (stomatopods), and other undesirables. We simply cannot stress the importance of QT strongly enough, and if the risk of an eyeball-latching parasites that might seek your fishes is not sufficient motivation, perhaps you should consider another hobby! To make the task of contending with sea lice in the display that much more daunting, many such isopods suck blood *only at night* and drop off the fish to hide in the rocks by day. Thus, most folks are really put out of their way to hunt these parasites if allowed into a rockscaped aquarium.

Although isopods are circumtropical, you can expect a greater incidence of them in live products from the nearest seas to you. They also are seasonal and wax with increases in water temperatures. In America, the fast and efficient delivery of live rock, sand and animals from the Gulf of Mexico and the Atlantic carries with it a somewhat higher occurrence of parasitic isopods. Wild harvested live rock from the Gulf of Mexico, despite its extraordinary beauty (some of the prettiest live rock in the world!), commonly harbors parasitic isopods in the summertime. Current stocks of aquacultured rock from Florida suffer this same condition occasionally. This is not to say that Pacific rock is less vulnerable to infestations. It's a simple matter of processing time. For the unavoidable extended transit, Pacific rock must be aggressively stripped of sponge, algae and other soft fouling growths. The process usually involves high pressure washing which carries many of the larger crustaceans away with it. Add to that the duress of desiccation from several days to more than a week out of water with live rock imports to the US from the South Pacific and you can understand that isopods are reduced in these rock shipments. None of the above is relevant or anything to be concerned about when you properly cure all live rock and sand in isolation before adding it to the main display.

Other attempts at control and eradication usually meet with mediocre or dismal results. Freshwater dips are a very useful strategy for controlling many fish parasites. Unfortunately, isopods are not one of them. Parasitic isopods can often live longer in freshwater than their stressed hosts can. This is not to discourage you from employing the procedure, but you should have tweezers ready after the dip to extract the parasites manually if necessary. Methods of control by natural predation are even more dubious. Fishes like juvenile hogfish (as biological cleaners) and canary blennies *Meiacanthus ovalaunensis* have been employed as predators of isopod parasites. They should not be relied on as definitive controls however. The last desperate strategy that many aquarists resort to is running the system fallow (without fish hosts). For this to work, however, none of the invertebrates in the tank can be fed meaty foods, which could be scavenged or stolen by the isopods for sustenance. Six months fallow is a minimum period here, and baiting should be attempted in the interim on infested tanks. Medication used to drop off or kill isopods on fish has been attempted in hospital tanks with variable results. Afflicted fish can be captured and swabbed when tweezers seem too frightful or not possible. Iodine, merthiolate, mercurochrome, and merbromin have all been used. They are to be applied with a cotton swab but take care that no such swabs or stains approach sensitive eyes or gill tissues. Long baths in medicated water including treatments with organophosphate based medication (like DTHP; Neguvon, Masoten, Dylox) have been tried. These are chemically aggressive medications though, and proper research into specific fish species' likely sensitivity is crucial. We do not recommend chemical treatments for most infestations. Prevention is still the best treatment and less invasive measures like meaty food bait is a safer first course. All but the smallest fishes can sustain attacks by small isopods for days or weeks.

Isopod Reproduction:

After hearing all of the warnings about the devastating effects of parasitic isopods and how difficult they can be to eradicate, what could be worse? Reproducing isopods in the aquarium... yeah, that's bad. The good news is that these parasites are not hermaphrodites, and they do not reproduce asexually by fission; they are dioecious (separate sexes). Unfortunately for us, many brood their young and it is possible to bring a gravid female into the aquarium, which can give birth to 20 or more bouncing baby biters. The production of a few large, brood spawned young here makes survivability high for such a situation. It is a common theme that pests, predators and diseases that have direct development of young are challenging to control.

304 *Natural Marine Aquarium Volume I - Reef Invertebrates*

ECHINODERMS

Members of the Spiny-Skinned Animals - Or members of the phylum Echinodermata have fascinated and intrigued humankind for time untold. Their namesake means "spiny-skinned" and aptly describes the categorically rough-textured husk of most any species in the group. They have an extraordinary fossil record dating back to perhaps more than 500 million years. Indeed, they have been around quite a bit longer than men have walked along shorelines and discovered the denuded tests ("skeletons") of the departed, or spied the still-living in shallow waters.

Aquarists will recognize the three main groups: Starfish (Asteroids, Ophiuroids and Crinoids), Sea Cucumbers (Holothuroids), and Sea Urchins/Sand Dollars (Echinoids). Many animals in this group make fabulous aquarium specimens for their utility and hardiness. Inevitably, though, among the thousands of living species there are some that are wholly inappropriate and entirely unsuitable for aquarium life. In the following chapters, we will delve deeper into the differences and distinctions of the various classes and make recommendations, of course, on which specimens are best to keep and which others are best to avoid.

Despite the sometimes dramatically contrasting shapes of some relatives, echinoderms are commonly recognized for sharing an almost exclusively marine and benthic lifestyle, and for having a consistent five-part radial symmetry. A cross section of a worm-like sea cucumber, and a view of the underside of a bouldersque cushion sea star and spiny sea urchin in kind, will reveal this aspect of their nature. They all employ unique tube feet in some fashion for locomotion, feeding and respiration. The evolution of this system is quite remarkable among members of this phylum. They range from barely noticeable and innocuous to pronounced and venomous! You will want to learn such distinctions among echinoderm candidates for the aquarium when making selections; it may surprise you to discover some popular species are potentially dangerous.

Take heed of our special recommendations for their feeding and husbandry. They are often "sold-short" on their needs to thrive when regarded merely as scavengers. Be assured that most all require target feeding just like featured fishes and other macro-organisms for the marine aquarium. Some species, like the Cucumarid sea apples and Crinoid feather stars, are very sensitive to the nature of water flow (volume and direction) to even feed properly. Others, like the famous sand dollars, are such voracious feeders that almost no traditional home aquarium can provide enough food to keep one alive for even a year. Plenty of other examples in this group, however, are marvelously indiscriminate of physical parameters and feeding options as opportunistic omnivores. Some very specific individuals have become quite popular for their diligent and earnest consumption of nuisance algae. The natural and unnatural decline of *Diadema* sea urchins in the Atlantic, for example, has caused significant problems on coral reefs with such encroaching nuisance algae. At length, much is known about the proper husbandry of many echinoderm species. You can surely find interesting and useful, if not beautiful, echinoderms for your aquarium with the guidelines we have provided here. The rest can sincerely be admired from the pages of this book and visits to the sea.

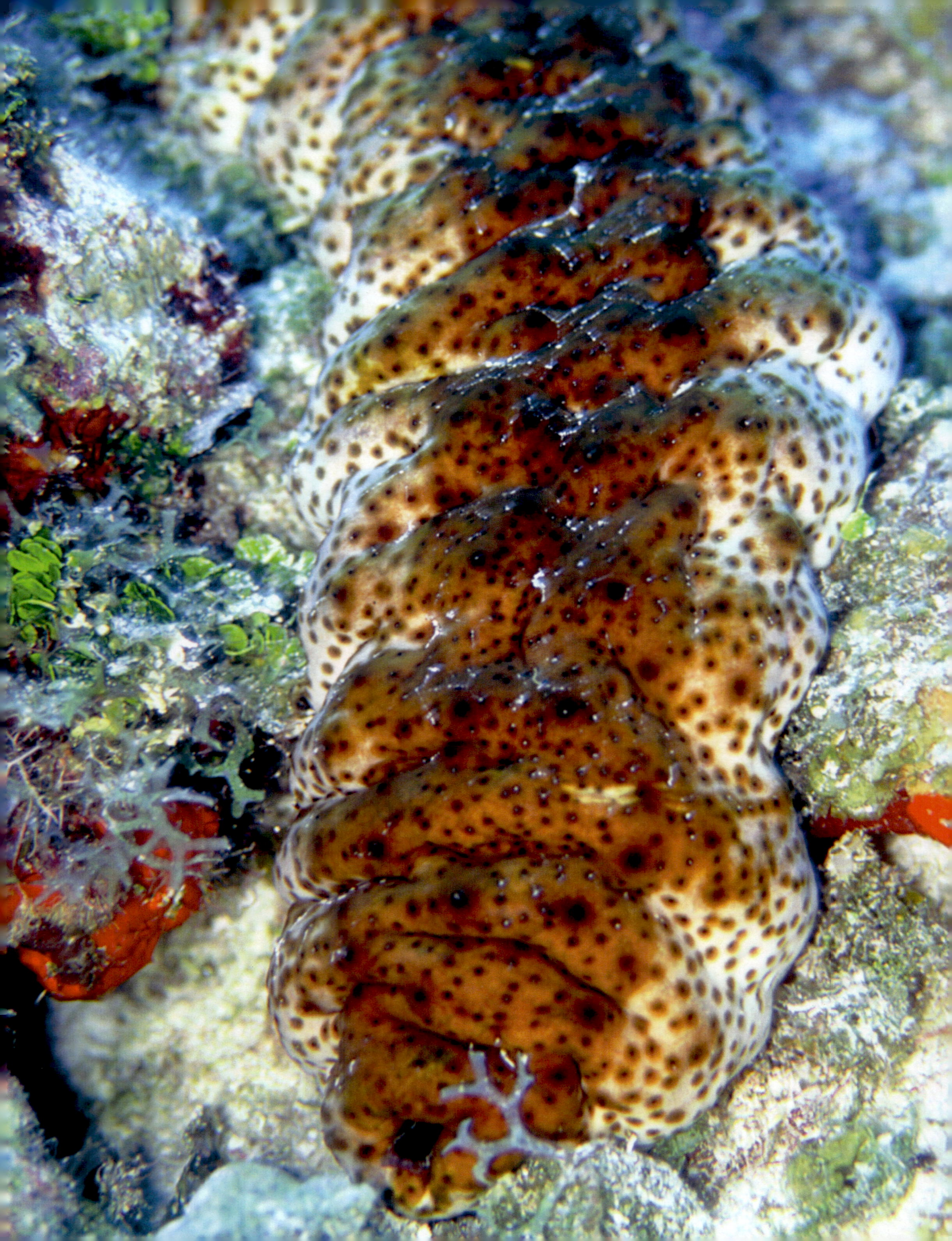

Holothuroids

Sea cucumbers (Holothurids) Sea Apples (Cucumarids) and Medusa "worms" (Synaptids)

The Holothuroids are a unique class of animals in an already unique phylum of Echinoderms. The varied morphologies and behaviors are every bit as fascinating as their often colorful or enigmatic appearance. In the aquarium trade, we see body shapes ranging from spherical to worm-like and some with bizarre rings of the sticky tentacles. The very activity of filter-feeding strategy by some Holothuroids is nothing short of mesmerizing as feathery tentacles methodically curl down to the mouth while others exit and unfurl, cleaned of savory trapped food particles. Other sea cucumbers feed altogether discreetly and "head down" as hardy and useful detritavores.

There are about 1000 recognized species of sea cucumbers with many growing to suitable sizes for care in aquariums at approximately 2-6" (5-15 cm). Members of this group, however, can vary in adult size from rather small species less than ½" (10 mm) buried in rock or sand to stunning beasts more than 8 feet long (on record at 3.3 meters)! Beyond the long and short extent of their extremes, some members of this class attain sizable mass like the *Stichopus* Sea Cucumbers which have been recorded at 11 pounds (5 kg) drained wet weight after reaching an impressive length and girth (40 X 8 inches)! Sea Cucumbers are found around the globe from tropical to temperate waters and from the shallows to the deep sea where they are a significant if not prominent element of the ecosystem. Ultimately, most Holothuroids offered in the trade are downright challenging if not outright impossible to maintain in aquariums. Large and mature aquariums with deep sand beds in refugiums or displays are crucial for supporting detritavore species (with organic sediments) and suspension feeders alike (via natural plankton).

Sea Apple vs. Sea Cucumber vs. "What the heck is that?!?"

All sea apples are sea cucumbers, but not all sea cucumbers are sea apples… as the saying goes. The point, however, is altogether moot. Common names used alone are limited in the ability to convey reliable information about species and are useful only regionally at best. With so many common names for popular creatures in the aquarium trade, and in the face of ever-refined global communications, it is important that we become equally familiar

Some Holothuroids like this Phyllophorid (above) are acquired incidentally with live rock or live sand. They reside discreetly buried in the substrate and extend branchial feeding tentacles to filter-feed upon plankton much like the photogenic Cucumarid *Pseudocolochirus* "Sea Apples" (right).

Left: *Actinopyga agassizii*, the Five-Toothed Sea Cucumber. This species has five square teeth surrounding the anus and short, knobby podia on its back and sides. Occurring in dappled tan, brown and/or yellow color on top with lighter colored sides fading to white/cream. This specimen was photographed in the Bahamas.

with scientific names for animals to clarify the everyday use of common names. While it is good practice to know scientific names in general, it is particualrly imperative in a group of animals such as the Holothuroids to clearly differentiate between the mostly non-toxic and the few toxic species in this class.

In the aquarium trade, we commonly see three groups of Holothuroids collected deliberately: The **Holothurids** (detritavorous sea cucumbers), The **Cucumarids** (the planktivorous "sea apples") and the **Synaptids** (so-called "medusa worms"). To generalize, the Holothurids are better suited for captivity as they are mostly detritavores, living in nutrient/sediment rich environments. The Cucumarids are largely inappropriate for aquarium life for their obligate dependency as filter-feeders in aquariums that are usually deficient in natural plankton. Although several Cucumarids fare quite well in captivity and all can be target fed in species specific tanks, few if any can be recommended for mixed garden reef aquariums (casual aquariology). The highly mobile Synaptids have an altogether different set of challenges (more below) and are truly best suited for dedicated refugia or species-specific displays, if kept at all.

Thelenota anax. One of the largest Holothuroids at up to a meter in length. Creamy colored with orange blotches on back. From northern Madagascar to Australia, Fiji, Guam. (*R. Fenner*)

Holothurids

By and large, sea cucumbers in the aquarium trade have muted colors and subdued markings to compliment their benthic lifestyle. A few are colorful or nicely patterned while others look like someone beat them indiscriminately with a club. It is likely that the more conspicuously colored specimens will have greater concentrations of toxins although the issue of Holothurid toxicity is somewhat exaggerated in aquarium legend. At any rate, a great many sea cucumber species can be useful and long-lived in aquariums unlike their Cucumarid and Synaptid brethren.

Tasty (Aspidochirote) Cucumbers?

In Asian culture and cuisine, *bêche-de-mer* (the name for processed sea cucumber) is known as *iriko* in

Common *Holothuria* Sea Cucumbers are popular and hardy specimens in aquariums with enough fine sand for their scavenging habits. As detritavores, they ingest significant quantities of the substrate and digest much of the organic material therein. Avoid keeping Holothurids on course substrates or shallow beds for fear of insufficient feeding opportunities. (*R. Fenner*)

Japanese, *trepang* in Indonesian, and *hai-som* ("sea-ginseng") in Chinese. It is highly regarded in medicine and haute cuisine, often as a celebratory food. In Chinese medicine, sea cucumber is used to treat a wide range of ailments in people and animals including high blood pressure, building and repairing healthy joints, and improving blood circulation. For consumption, essentially all parts of the animal are usable. *Bêche-de-mer* is a flavorless food of gelatinous texture once boiled and is often used to make soup. There are, however, many culinary uses for this animal. The sea cucumber's relation to other radially symmetrical echinoderms becomes quite apparent upon dissection for the animal has five long radiating muscles inside, which are said to taste like clam. The body wall is served pickled or raw in food preparation or may be dried out and proffered in tablet form for various remedies. In some species the skin is boiled to make a tonic. Fermented sea cucumber "guts" (*konowata*) and dried gonads (*kuchiko*) are an expensive delicacy in Japan and Korea. Although not necessarily palatable to you, the utility of a Sea Cucumber's many parts must at least be admired!

The process of preparing sea cucumber is rather complicated. It can take several days and a series of repetitive rinses, soakings and boiling. The industry of collecting *bêche-de-mer* for food use often involves drying the animals outdoors in the tropical sun on racks before export. These dried animals will require a good soaking for most culinary applications. Some Holothurian species have been threatened by over-collection for the food industry. One report cites the export of more than 4 million sea cucumbers in less than ten years from Ecuador alone! Fishery endeavors to culture this animal are well under way internationally with hopes to relieve pressure on the finite natural resource. The primary importers of sea cucumber are Hong Kong, Singapore and Taiwan with a global demand for well over 100,000 tons of this product annually. Numerous species have been targeted for food including *Holothuria edulis, Cucumaria frondosa* and quite a few *Stichopus* species.

Stichopus parvimensis, the Southern California Sea Cucumber, a temperate species. Photographed off of the San Diego coast. (*R. Fenner*)

So much for sea cucumber toxicity. Waitress… bring me another *bêche-de-mer!*

Cucumarids

The filter-feeding Cucumarid sea cucumbers, including the so-called "**Sea Apples**," are definitely more difficult to keep in captivity. They also include some of the few species in this class actually toxic enough under duress in practical applications to kill fishes (the toxin is harmful to few if any invertebrates). The very dramatic coloration of toxic Cucumarids is conspicuous and representative of the classic **aposematic** (warning) coloration. Please note as further evidence that such toxins are specialized to reduce predation by *fishes*. The warning coloration exists as a *visual* cue for sighted animals and not sightless ones (cnidarians) for which chemical cues would be in order. Nearly all toxic events can be prevented through sensible system designs and responsible husbandry: compatible tankmates, good water quality, guarded pump intakes, etcetera. It is very important to note that if a Cucumarid is ultimately stimulated to exude toxin, all fishes in the system are at risk of sudden death.

Sea Apples commonly occur in high current areas of the reef and require comparable flow in aquariums. They have a ventral side, which will always be fixed to a hard substrate in healthy animals. Dislodged specimens will promptly (well, as fast as they can move, which isn't very fast!) orient their dorsal side to life-giving currents above and re-attach ventrally. Like most Holothuroids, Sea Apples have abbreviated, non-feeding larval stages that last only minutes before settling (less than ten for many). Alas… the eggs of *Pseudocolochirus* are quite toxic to fishes and attempts at captive culture necessarily prevent the keeping of fishes in concert.

Other less toxic and arguably hardier Cucumarid species do presently exist in the aquarium trade. The Yellow

> **Stichopus** and **Parastichopus** *species* are seen occasionally in the ornamental trade and include many of the edible (human consumption) species. They occur over a wide geographic region and over a wide range of habitats including rocky hard substrates. Many are suitable reef aquarium denizens when small (avoid temperate species).

(*Colochirus robustus*) and the Pink and Yellow (*Pentacta anceps* if smooth, or *Colochirus* if spiny) sea cucumbers are commonly imported and fare reasonably well to very well in captivity.

Feeding efforts for Cucumarids should include both zooplankton and phytoplankton, and like substitutes.

Synaptids

The keeping of Synaptid sea cucumbers, so-called "Medusa Worms," is quite a challenge in every aspect of husbandry, and aquarists are strongly discouraged from keeping these Holothuroids in anything but a specialized aquarium system with dedicated care. Quite frankly, **most aquarists should not keep this animal at all**.

Synaptid sea cucumbers are extremely delicate to handle and may essentially be as toxic as Cucumarid Sea Apples if molested. Collection and transportation is hampered by their fragility, and aquarium care is challenged by their specialized feeding requirements. Rarely seen by day, the "Medusa Worm" comes out at night and is commonly observed in association with living sponges upon which they feed from the surface in the fashion of a detritavore. Some are believed to feed on the very metabolites of the sponges. Since the care and reliable culture of sponges in aquariums is difficult if possible at all for most aquarists, the keeping of Synaptids is likewise a great challenge. Few specimens survive more than a few months for casual aquarists. Species tanks are ideal if not necessary for keeping this animal successfully.

Synapta maculata is one of the longest known Holothuroids. They have been reliably recorded at 3.3 meters in length with legends of specimens in excess of 5 meters (over 14 feet).

Toxicity

The truth about "poisonous" Sea Apples and Sea Cucumbers

An Overview...

The legend of Holothuroid toxicity in aquarium kept specimens is a matter much in need of perspective. First and foremost, a few popular species of Cucumarids commonly known as Sea *Apples* **are indeed potently toxic** and potentially fatal to fishes in mixed marine aquariums. However, the overwhelming majority of Holothurids (Sea *Cucumbers*) seen in the aquarium trade present no significant if at all measurable risk of toxicity with reasonable aquarium husbandry. They present no greater danger than the keeping of many other discreetly potent invertebrates like sponges or zoantharians ("button polyps"). Aquarists will easily recognize the few significantly dangerous species as colorful Sea Apples or Violet Sea Cucumbers. Aside from these *Pseudocolochirus* and like species, most sea cucumbers are less toxic than many popular fishes, corals or other invertebrates. Indeed, zoanthids contain some of the most potent toxins in the world (palytoxin). Mandarinfish, boxfish and Gold Stripe/Six-line Groupers (soapfish) also have toxic flesh/skin secretions, for example, and are rarely if ever vilified so dramatically as sea cucumbers.

There are several ways that the cucumber toxin, **holothurin** (a saponin), can be delivered. In some cases, a simple mucous exudation of the toxin is issued under stress or attack. Under great duress, a traumatized cucumber may further compound the toxic event with complete voluntary evisceration (ejecting part or most of is internal organs). The intrinsically toxic nature of Holothuroid "innards" (**Cuvierian tubules**) is a matter of concern in closed systems and shipping containers. Evisceration, direct exudation of *holothurin,* and the release of eggs are three commonly recognized dangers of Holothuroid toxicity. The body wall of most if not all sea cucumbers is also believed to be somewhat toxic but seems to be

Colochirus robustus (Ostergren 1898) The Yellow Sea Cucumber: Eastern Indian Ocean and Western Pacific, Sri Lanka to the Philippines to Japan, Indonesia. Grows to four inches (10 cm) in length. A delightfulky attractive species that has proven durable in captivity and frequently reproduces asexually by simple fission (splitting in two). This filter-feeder moves and demands little beyond good water flow and a source of plankton. Mature aquariums with fishless refugiums to generate plankton are highly recommended for long-term success with this species.

Synapta maculata (Chamisso & Eysenhardt 1821). This slender, typically nocturnal sea cucumber can grow to length over 3 m. They are found in sandy, shallow areas… often in fouled communities. A toxic and delicate aquarium specimen: aquarists are categorically discouraged from keeping this animal in mixed fish and invertebrate displays. Members of this family are often associated with sponges and are specialized feeders. Caribbean, Indo-Pacific, East Africa, Red Sea to Society Islands. This one in Fiji at night.

of little threat in aquariums (short of a senseless fish eating the toxic flesh while embracing Darwin's theory of evolution in motion by ridding itself from the gene pool for its ignorance). Nonetheless, with well over one thousand recognized species of sea cucumber, only a handful of commonly imported species have ultimately been reported as significantly dangerous in aquariology. From misunderstanding and misinformation, many aquarists avoid the keeping of some useful species of sea cucumber for unfounded reasons.

Keep in mind that a dead, "safe" Holuthurian can indeed wipe out a tank of fishes and perhaps some invertebrates under some circumstances. This is not because it is toxic, however, or exuded some mysterious poison, but rather because of the burden that any rotting corpse does to a relatively small volume of aquarium water. Consider what a 4 oz can of flake food or a hamburger left to decay in a tank overnight does! Just the same, it could be that hamburgers are toxic by that same line of logic (insert your own joke here).

Instead, they are simply a sudden and significant influx of dead organic matter that fuels a microbial bloom and subsequent pollution from the by-products of bio-mineralization. Warm tropical aquarium water with a significant mass of decaying matter can realize an amazingly sudden deprivation of oxygen from the rapid proliferation of bacteria, thereby causing an anoxic condition that kills fishes. This can easily occur without any toxin and before the products of nitrification are evidenced, if at all, on water quality tests (ammonia and nitrite spikes). Mortalities caused by the event are amplified if it occurs over night in systems with heavy respiration (large amounts of photosynthetic corals and plants reducing oxygen levels). Indeed, this explains some of the fatalities inappropriately ascribed to perceived Holothuroid toxic events.

If you spend enough years in the aquarium industry, you will find catastrophic events that are actually caused by holothurin, and just as many or more that are not. The difference often seems to relate to the nature of the stress or death of a toxic species. The imposed stress of extremes of water quality (inhospitable temperature, salinity or other parameter of water chemistry) and imposed damage (attack by fishes, damage by mishandling or aquarium mishap like maceration through unguarded pump intakes) can trigger the apparent exudation of a potent fish toxin. However, the natural expiration of a toxic species due to senescence (old age), attrition (starvation) or extended deprivation of current (to stimulate filter-feeding), seems to present no measurable danger to an aquarium system short of burden of the decomposing mass on nitrifying faculties of biological filtration.

The key to successfully avoiding a fatal problem with toxic sea cucumbers is to simply research the potency and sensitivity of each species individually. In very gross terms, the non-descript and cryptic detritavores like the common *Holuthuria* species that come out at night (when there is less risk of predation by reef fishes) are usually "non-toxic." The colorful, filter-feeding Sea Apples that perch

Pentacta anceps, a suspension feeder among sea cucumbers. Aquarium image. (*R. Fenner*)

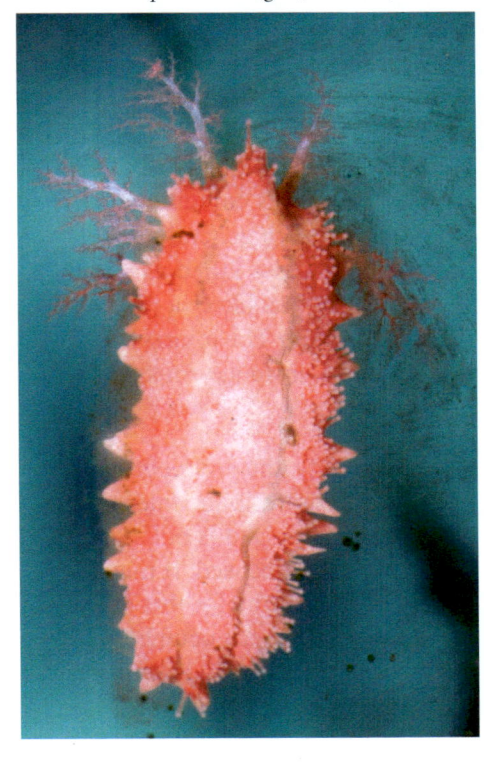

Echinoderms: Cucumbers (Holothuroids)

brazenly and conspicuously on a reef among grazing reef fishes by day are generally "toxic."

The Science of Sea Cucumber Toxicity

Holothurin, by definition, is a toxic anionic surfactant (a soapy saponin, if you will). It kills by hemolyzing (the destruction or dissolution of red blood cells, with subsequent release of hemoglobin) red blood corpuscles and causes irreversible damage. In the wild, it is very unlikely that *holothurin* kills many fishes but rather is geared towards sickening and dissuading them from attack as it is theorized with their poisonous eggs (more on this below). In layman's terms, *holothurin* affects the membranes of fishes' gills and compromises their ability to breathe. This attribute has not been lost on indigenous peoples in some areas who have learned to mash some Holothuroid species to liberate their toxin for use in lagoons to capture fishes. Most invertebrates in closed systems and certainly at large seem relatively unaffected if the condition is remedied soon enough (within hours). One theory for this difference in sensitivity is that some invertebrates can temporarily "shut down" respiration until the tides change, while fishes cannot. Ultimately though, in healthy aquariums, holothurin poisoning takes several to many hours to kill fishes. Sudden fish deaths coinciding with a Holothuroid death generally indicate that something else killed all and the sea cucumber was likely a victim and not a participant. The toxin simply may not work instantaneously even in aquaria.

Physiologically, the production of *holothurin* requires high amounts biological energy and is not a toxin that is readily dispensed. It may largely or entirely be for this reason that known toxic sea cucumber species die without "fanfare" (a catastrophic poisoning of fishes) in an aquarium system at times. Death by attrition or prolonged deprivation by other combinations (of inadequate water flow, food/particle size, etcetera) may leave an animal ill-disposed or completely unable to produce adequate toxin. If the issuance of holothurin is instead deliberate, perhaps toxic species are truly evolved to rely on other "less expensive" methods of defense like cryptic or warning coloration, strategic feeding cycles (nocturnal), and sacrificial evisceration (expelling "innards" as a distraction) in some events.

Toxicity to Humans?

Little definitive information exists concerning the toxicity of sea cucumbers to humans. Most species are regarded as non-venomous, although some are poisonous if eaten or prepared incorrectly (hence the boiling process to induce the evisceration of poisonous Cuvierian tubules). Of course, how toxic does a species like *Holothuria edulis* (AKA "the Edible Sea Cucumber") sound to you?

The concern regarding *holothurin* toxicity to human's is directed more at badly behaved divers and industry professionals handling concentrations of stressed Holothuroids in transit from small volumes of (occasionally) noxious water. Of specific concern is the handling of noxious expelled "guts" (Cuverian tubules) before touching the eyes, or a SCUBA diver clearing their mask in close proximity to a recent event of Holothuroid evisceration. Reactions may range from mild to severe inflammations of the ocular membranes as in cases of allergic conjunctivitis (symptomatically like "Pink Eye" with redness, tearing and itching) or keratitis (inflammation/infection of the cornea evidenced by pain, redness, irritation, blurry vision and light sensitivity). Keratitis should be considered a medical emergency and receive prompt medical attention since it can lead to blindness.

Aquarists are strongly advised to avoid any contact with the sticky thread-like Cuvierian "guts" of a Holothuroid. The expelled organs can be quite

Pseudocolochirus violaceus (Theel 1886), The Sea Apple or Violet Sea Cucumber... Eastern Indian Ocean and Western Pacific. Their eggs are toxic to fishes, and they may exude deadly toxins if molested/stressed in closed systems. Avoid Sea Urchins, Sea Stars (e.g. *Protoreaster, Asthenosoma*) and curious or predatory fishes.

The most widely recognized Cucumarids are the Indo-Pacific *Pseudocolochirus* species.

- ❖ *Pseudocolochirus tricolor*: body color reddish, tube foot and tubercle rows yellow, feeding tentacles yellow.

- ❖ *Pseudocolochirus violaceus*: body color variable red, blue and/or violet, tube feet and tubercle rows red on white stripe, feeding tentacles white through violet.

- ❖ *Pseudocolochirus axiologus*: the "Australian Sea Apple" has a striking mixed body color of blue or purple/violet. The oral arms range from white through pink to red, and tube feet and tubercle rows are red (sometimes with yellow stripes). This species has been reported to have an even higher concentration of fish toxins than the common Indo-Pacific species.

- ❖ *Pseudocolochirus* sp.: the "Royal Sea Apple," has a rich violet body color, bright yellow tube feet and tubercle rows: oral arms match in bicolor. Collected specimens are often unwieldy for average aquariums in this large species of sea apple. Uncommon appearances of The Royal Sea Apple in the American market insure a exhorbitant price.

adhesive (an obvious deterrent to would-be predators) and the sticky "glue" on the innards does not come off the skin easily. Indigenous peoples have skillfully used this substance to bind wounds. Clearly, toxicity of the Holothuroid viscera is of diminished risk to the skin.

Physiology

One of the great distinguishing characteristics of Echinoderms is their *penta*-meral (five) symmetry, however, sea cucumbers unlike most other Echinoderms are essentially cylindrical. Upon dissection however, their radial symmetry becomes apparent, as previously mentioned regarding culinary interests, with five strips of muscle running the length of the animal. The dermis of this "spiny-skinned" animal is embedded with **ossicles** (small skeletal plates) lending to the tough and muscular, though slimy, feel.

Thanks to the rich and albeit conspicuous slimy presence of **collagen** in the tissue of sea cucumbers (like other Echinoderms), they are amazingly capable of changing from a "fluid" to solid form very quickly. This extraordinary attribute allows them to figuratively pour themselves into tight crevices in the rocks and then tighten their skin in a firming action to prevent extraction. A remarkable adaptation.

Nervous System

Sea cucumbers have no brain, but they are highly sensory creatures with great concentrations of nerve endings in the epidermis, which is apparently enough sense to make them quite successful in the animal kingdom. Concentrations of nerves are most abundant at the ends of the animal… particularly the circumoral ("around the mouth") nerve ring near the base of the feeding tentacles.

Mobility

Although not always as obvious as among starfish or sea urchin relatives, most sea cucumbers have some form or fashion of tube feet in evidence. Occasionally they are scattered randomly over the body, often they are arranged in rows, and sometimes they are nearly or entirely absent altogether. Species with few or no apparent tube feet may be filter-feeders having less need for mobility when their planktonic food is carried to them with the currents or they will simply burrow into a substrate by alternate contractions of longitudinal and circular muscles like earthworms. The ultra-strange deep water Order Elaspodida "walk" on extended tube feet and constrictions of the body and there are even swimming sea cucumbers!

The morphology and orientation of the tube feet on a given species in fact are taxonomically revealing. The very tentacles around the mouth of some species are essentially modified tube feet. Be they detritavores or filter-feeders, all sea cucumbers have at least some modified tube feet associated with the mouth to facilitate

Below: a pair of pink and green sea cucumbers, probably *Pentacta anceps*. To 3" (~8 cm). Eastern Indo-Pacific, Indo-Australian archipelago, islands of the South Pacific. Aquarium image.

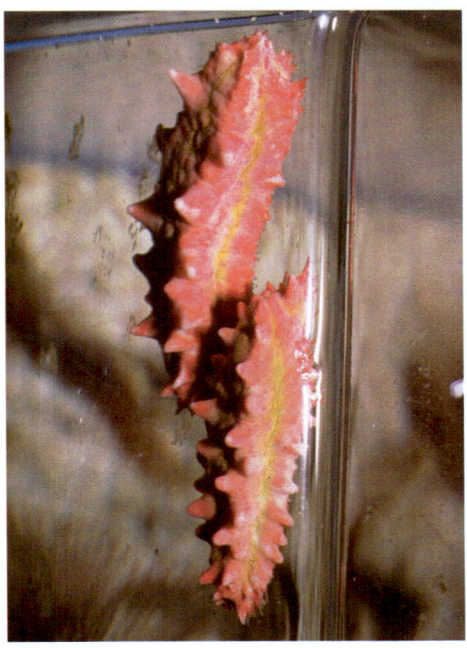

Echinoderms: Cucumbers (Holothuroids)

> **What to do if you suspect Sea Cucumber (holothurin) toxicity in the aquarium...**
>
> As with any biological or chemical contamination in aquarium systems, "dilution is the solution to pollution." Suspected events of holothurin poisoning should be promptly addressed with a fast but proper water exchange (new seawater that has been aerated, pH, temperature and salinity adjusted). Small frequent changes of chemical filtration media (activated carbon, Poly Filter, etcetera) in the ensuing days are also a good if not crucial additional strategy. The size and percentage of scheduled water changed will have to be judiciously weighed against the visible state of the system and the potential shock of the activity.

feeding. There are often ten tentacles for this purpose, but species may have from five to thirty tentacles and various numbers in between. The filter-feeding species generally have more elaborate branching tentacles than the detritavores. Beyond the basic morphology, strategies for feeding sea cucumbers in captivity are detailed later in this chapter.

Breathing and... other functions

Inside of a sea cucumber, located in the **coelom** ([*see-lum*]= body cavity), there is a specialized respiratory structure called the **respiratory tree**, which is a unique paired organ. Both sides of the "tree" merge and, by way of a common trunk, pass into a **cloaca**, which acts as a pump. Respiratory trees are rhythmically fed (inflated) and purged of water by muscular contractions for the purpose of gas exchange and excretion as most ammonia exits the body through these structures. Hmmm... this paired tree is almost like a lung; it seems like it would be more appropriate in the urochordate invertebrates (sea squirts) that are squatting in our own phylum, Chordata!

At the base of the respiratory tree is a system of tubes (the aforementioned **Cuvierian tubules** of toxic fame). The Cuvierian tubules are a series of thin, sticky tubes stemming from the hindgut. These are the most common viscera ejected by a sea cucumber under duress and often with the

Here's a *Bohadschia argus* (Jager 1833), the Ocellated Sea Cucumber, exhibiting a classic eversion of Cuvierian Tubules display. This very sticky mass is noxious and potentially deadly to fishes. Remove as soon as possible and heed the treatment advice in this passage for holothurin toxicity.

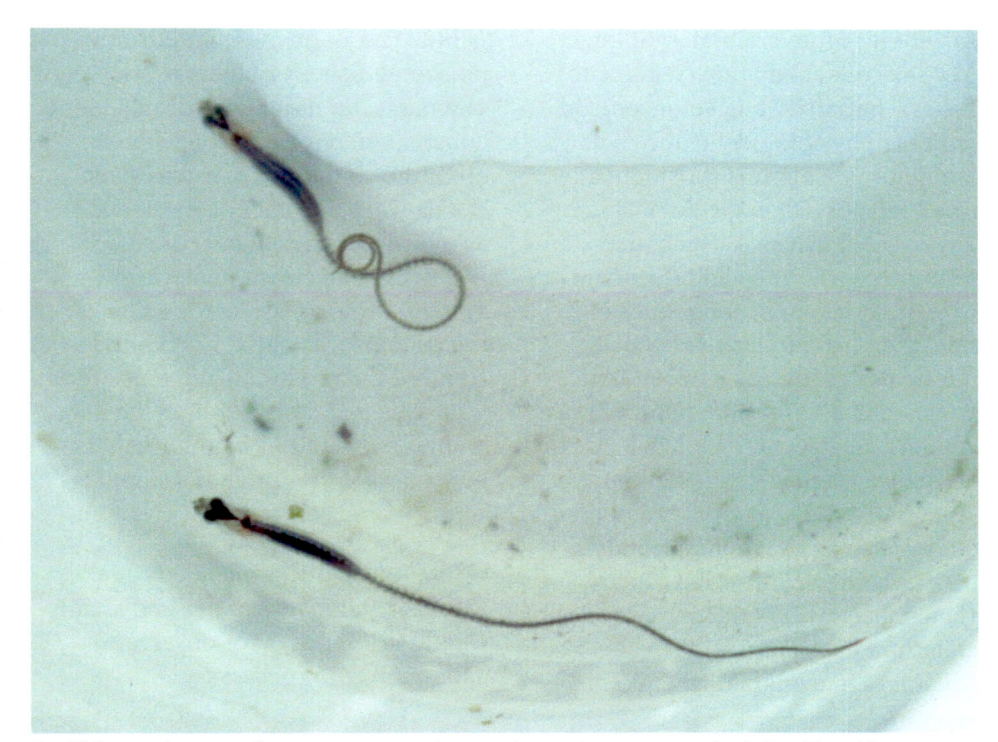

Pearlfishes are but one of many creatures to have an interesting commensal relationship with Holothrurids. As many as several at once can live in the coelom (anus) and graze the respiratory trees of its sea cucumber host. They also emerge at night for supplemental feeding activity. (A. Calfo)

toxin holothurin. Indigenous peoples often use them to protect their feet when walking on the reef. The sea cucumbers are squeezed until they squirts out their sticky "guts" which the natives put on their feet. The strong toxin in the tubules is surely quite dissuasive to unseen fishes and other injurious creatures.

Beyond the digestive viscera (not all of which is necessarily released when stressed), a sea cucumber may also expel the respiratory tree and/or gonads. While all such "guts" may be regenerated, ideal conditions are necessary and such parameters are not insured in closed captive systems. This may be the reason for some deaths of sea cucumbers in aquariums with aquarists that can have no idea that a recently acquired specimen may in fact have been eviscerated in transit and now struggles to recover. In the natural state, with assumedly ideal reef conditions, some pearlfish have learned to exploit this regenerative attribute. They have the well-known notoriety of living in the anus of a sea cumber and feeding, or rather farming, on the regenerable digestive organs. If you discover that you have obtained a pearlfish (a charming name for the fish, *pearl,* but hardly an enchanting "oyster!"), know that it must live with a Holothuroid to survive and that said sea cucumber must be kept in optimal health to support both.

Solid waste exits a sea cucumber in the form of conspicuous sandy pellets (sometimes in large amounts!) and are referred to as **casts**. Some aquarists may choose to siphon these pellets in systems with precarious nutrient export mechanisms, but most often they can be ignored, or rather… left to the mechanical mechanisms of nutrient export in a well planned system (living detritavores and good water movement to keep waste in suspension for protein skimming and filtration dynamics).

Selection

So far in our discussion of cucumbers we have already covered gut eating fishes that live in the anus of a sea cucumber… and sea cucumbers that toss their internal organs including their digestive tract and gonads at a would-be predator. These are tough acts to follow! Its like Pavarotti's fourth encore … or Bob's fourth pitcher of beer, which are one in the same thing if you've had enough beers yourself.

Sea cucumbers, once established in an appropriate aquarium system, can live for many years. In large, healthy reef displays with mature live sand, sea cucumbers commonly live more than five years with many approaching ten years. Selecting a healthy specimen is the first step to such success with these fascinating animals.

Like the rigid sea stars and urchins that are vulnerable out of water for concern with trapped air in the body (leading to infections and complications), sea cucumbers are equally vulnerable out of water for their relative lack of rigidity as their bodies collapse upon themselves. These soft invertebrates require the support of a three dimensional aquatic environment.

Because of the natural and relative inactivity of sea cucumbers, it is somewhat more difficult to assess their health when selecting a specimen. How does one differentiate between a stressed animal and its naturally sluggish behavior? This serves to complicate the keeping of some species in captivity as their lack of activity demands little energy or effort for survival and subsequently makes them seem to last longer when neglected (or starve slower than other animals!). Rather like underfed corals that are mostly but not entirely supported by symbiotic algae, the slow starvation of a sea cucumber by a net daily deficit of a mere 5% is simply not apparent to aquarists over a year's time or more. It leads folks to make statements like, "I don't know… it was fine for 14 months and then just died." In such cases, while growth or attrition may not be apparent, daily or at least weekly metabolic activity is; look for evidence of excrement! If the sea cucumber is not passing regular and conspicuous fecal pellets then it stands to reason that it is not eating enough to produce them! Hmmm… it is a gross generalization for sure but there is merit to it. You can be certain that if you notice one day that your sea cucumber is all shriveled up (speaking literally and not figuratively here) it has suffered from an extended period of nutritional neglect. Alas, it is too late in many such cases to save the animal. Please be prepared to address an animal's needs with good husbandry and placement before acquiring them.

Acclimation

Sea cucumbers underscore the sensitivity of invertebrates, at large, to issues of acclimation (relative to fishes). Initially we have to contend with the physiological reality that a Holothuroid's dermis is osmotically more sensitive than the thick skin of large scaled fishes. As such, they are far more sensitive to and disturbed by differences in water quality in transition (salinity and pH shock, nitrate poisoning, etc). Secondarily, we have that nasty little habit of sea cucumber evisceration when a stressed animal sends its guts, gonads and potentially every part of its entrails out of its anus or mouth depending upon which way the wind blows. If an aquarist is conducting proper aquarium husbandry and isolates a sea cucumber in a quarantine tank, there is little risk of an eviscerated animal poisoning fishes or other invertebrates. Of course, an ill-advised or impatient aquarists stands to poison some or all of the animals in a display tank if a new sea cucumber eviscerates as a result of stressful acclimation and direct placement in the main system. Protocol aside, a "healthy" sea cucumber is unlikely to die as a result of the mixing of shipping water with system water, although the fishes would as a result of holothurin, shed under stress, present in the shipping water.

Quarantine is recommended for all new fish and invertebrates and ideally should last 30 days for observation and treatment if necessary.

Pentacta anceps, the Red Sea Cucumber is also known as the Pink and Yellow, and the Spiny Sea Cucumber. Eastern Indo-Pacific; South Sea Islands, Indo-Australian Archipelago. To a maximum length of about three inches (7.5 cm), this suspension feeder is a hardy aquarium denizen and perhaps only slightly less hardy than the Yellow Sea Cucumber, *Colochirus robustus*. Strong water flow is necessary to stimulate a response with feeding tentacles. Aquarium photo.

How to Evaluate and Select a Sea Cucumber

✓ Check specimens carefully for discontinuities like tears, holes, or sores in the body wall. Breaches in the mucous soft body of a sea cucumber can become quickly and fatally infectious. Look for unblemished animals without indication of unnatural or odd coloration.

✓ Sea cucumbers are known to cling tightly to hard substrates with their tube feet or by collagenous contortion into crevices. Please remove them carefully with as little disturbance as possible for fear of tearing tube feet, tissue or causing abrasions of the dermis against surfaces like live rock.

✓ You need not overlook a specimen that does not respond promptly to food like a sea star or urchin. Even healthy cucumbers are not dramatically stimulated by the sudden presence of food (the detritavores more than the filter-feeders should be more alert). If feeding tentacles are only partially extended or not at all, rely on other more reliable measures of health to make a final decision.

✓ A shriveled or shrinking appearance is generally quite bad and may indicate the advanced stages of attrition (prolonged starvation) or the recent eviction of its innards under duress. It is never a good or allowable symptom.

✓ All sea cucumbers have a dorsal and ventral side. A specimen rolled over upon its back should right itself in a relatively short period of time (minutes) if in good health. Be sure to observe this behavior at least. The activity does not ensure good health, but belies poor health in a potential candidate.

Behavior

Describing what is "normal" behavior in an animal as peculiar as a sea cucumber seems somewhat ironic. Still… beyond the oft-witnessed stress-induced behaviors, there are various activities that an aquarist will need to recognize in the everyday.

Regeneration

Like relative echinoderms, sea cucumbers are capable of varying degrees of regeneration. When sliced and diced, the terminal piece containing the cloaca can survive and regenerate a body and vital organs. The act of evisceration is subsequently followed up with regeneration of internal organs.

This is also evidenced in the peculiar relationship some sea cucumbers have with the afore-mentioned pearlfishes (Fierasferidae/Carapodidae). The relationship is not clearly understood but, apparently, some pearlfishes (including several individuals at once) can live in the **coelom** (anus) of a sea cucumber. They reside inside and to some extent consume the respiratory trees of the Holothuroid by day, and emerge at night for supplemental feeding activity.

Care

Habitat

The necessary habitat for sea cucumbers can be categorized by their feeding strategies. Most detritavores need very fine sand (sugar-like grains) and sediment rich (aged/muddy) substrates. Keeping Holothurids on coarse sand or gravel is problematic if not deadly to long-term success. The filter-feeding species (like the sea apples, pink and yellow, and yellow cucumbers) live instead on hard substrates. Moderate to strong water flow is especially important with such species to stimulate and support their feeding tentacles. Without proper water flow, sea cucumbers will not extend their filter-feeding appendages and will often "travel" at risk of bodily harm. The movement of all sea cucumbers is a precarious matter as they can succumb to the many dangers that abound in the tank including overflows, unguarded pump intakes, stinging cnidarians (anemones and coral), unprotected heaters, etcetera. The keeping of Holothuroids demands consideration of their requirements in relation to aquarium hardware and design. For toxic species, damage and/or death in the mouth of a pump intake or overflow is commonly the catalyst for the exudation of toxin and thereby serious troubles in the system. Be considerate of such pitfalls in the system that could befall these slow-moving, sightless invertebrates. Also know that activity in Sea Cucumbers typically occurs at night. They do not perform on command and generally cannot be enjoyed by day like other common reef tank inhabitants.

Physical Parameters

For the reasons outlined herein, consistent, high water quality is a must for these sensitive invertebrates. Although inter-tidal species seem to

Echinoderms: Cucumbers (Holothuroids)

be more tolerant of fluctuations in specific gravity and temperature, most are surprisingly intolerant of aged water in aquariums (high dissolved organics from messy feeders, inadequate filtration, etcetera). We recommend full salinity (avoid low salinity as with fish-only displays) and well-filtered water benefiting from regular partial water changes and chemical media. Chemical media can be used in small amounts and changed regularly in good husbandry as a special consideration for this potentially toxic group of invertebrates (1 oz of carbon weekly instead of 4 oz monthly, for example). Any concerns for limiting the availability of food with aggressive filtration are secondary to water quality. Dietary needs can be addressed with fishless refugiums, deep sand beds and target feedings when possible. Strong water flow is a boon to overall water quality and filtration dynamics and will be necessary when keeping Sea Cucumbers of all kinds.

Acclimation Protocol for Sea Cucumbers

> It's a good habit to prevent exposing a sea cucumber to air.

> Always quarantine new animals in isolation. Do not add an un-quarantined Holothuroid to an established system.

> On arrival begin mixing or dripping water from the isolation tank into the shipping container to ameliorate the original volume four-fold over a roughly 20-30 minute period of time. (eg.- a cucumber that arrives in a bag with 12 oz of water will be sitting in 48 oz of mixed water after 20-30 minutes).

> Maintain stable water temperature during acclimation with insulated boxes and a thermostatic heater if necessary.

> After the acclimation period, decant most of the water in the acclimation container and transport the Holothuroid under water in the reduced bag or vessel to the holding tank. Release the specimen under water while attempting to contain as much residual acclimation water as possible to be discarded.

Feeding

A better understanding of the feeding habits of Sea Cucumbers is necessary to realize long-term success with these grotesque yet fascinating animals. Very few can live wholly unsupported for more than a year without directly addressing their specific feeding needs. For some detritavore species it is a simple matter of patience in stocking a system with a deep sand bed and waiting for it to mature, thus providing the necessary accumulation of adequate organic matter in the substrate. Taking care not to overstock with competitive species is always an important consideration and deserves equal attention in regard to sea cucumbers. For filter-feeding species, however, a finely tuned plankton generating refugium (likely fishless and without other predators on plankton, like coral) will be necessary if one does not target feed with live cultured plankton. Misinformation and misleading marketing on prepared food products is a common explanatory factor in the rates of survival, or lack thereof, of filter-feeders in captivity.

The Filter-Feeders

Target feeding Sea Apples is a challenging endeavor. They are suspension feeders that are believed to be heavily dependent on the finest plankton (nanoplankton) with perhaps a strong phytoplankton component ("algae/plant matter"). As such, standard prepared foods (like bottled invertebrate supplements) are much too large to have even a remote chance of being accepted. Even tiny fresh-

Holothuria hilla, the Pacific Tiger tail cucumber. Its rather similar to *H. thomasi* found in the Atlantic, both having the same common name. Sulawesi image.

hatched brine shrimp nauplii are too large for most. References to offering finely minced meaty foods like krill, shrimp, and fish are erroneous as no process convenient to aquarists can reduce the prey/particles to an acceptable size even if the matter were suitable. Target feeding with prepared foods almost certainly does not work. At best, live cultured rotifers may be a viable source of zooplankton. Live phytoplankton is likely a better component as a staple (not to be fed to exclusion though). Live food culture of any kind is tedious, however, for many aquarists. Perhaps the best recommendation for the masses that find themselves keeping filter-feeding sea cucumbers is to employ large planted refugiums on large displays with light Holothuroid populations. Fine zooplankton, epiphytic material, bacteria, and phytoplankton from a dynamic (surging) seagrass refugium, for example, may be a viable means for generating suitable prey for suspension feeders in closed aquarium systems. The weakly defined dietary needs of Sea Apples necessarily requires judicious experimentation and demands that casual and novice aquarists leave the keeping of such animals to the dedicated few.

Sea Apples will feed any time of day and are stimulated to do so solely by adequate water flow evidenced by the distension of feeding tentacles. Inadequate water movement will incite a Sea Apple to inflate and release itself from its mooring to drift about the aquarium. Be sure that if you observe a Sea Apple drifting about that it is likely dissatisfied with the water flow in its previous location. In aquaria, this may occur after some time of settled residence as pumps age or clog slowly, and the dynamic of water flow wanes. Aquarists have often made the mistake of shutting off water flow when target feeding with hopes of extending the feeding opportunities of the filter-feeders but in light of its overall needs, this is clearly counterintuitive. Sessile invertebrates on the whole are critically dependent on strong water flow to deliver food and remove waste products.

The presence of phytoplankton in the water tends to elicit a strong feeding response in filter-feeding Sea Apples. If using a prepared phyto-substitute, be sure to always use fresh product (no more than six months old) and make a suspension that has been whisked in an electric blender to reduce the particle size. Otherwise, the food matter may be too large to be digestible by suspension feeders. Some folks have found hand-held mini-blenders (for health food "shakes") to be very convenient for this purpose when a spouse does not favor the processing of stinky algae in the family's kitchen blender! Sea Apples feed at the same rate regardless of how much food (or how strong the concentration of prey) is in the water. In general, overfeeding is wasteful and ineffective. Instead, small frequent daily feedings are recommended. A food drip with a suspension may be the best strategy with target feedings of live rotifers, live phytoplankton, and like substitutes. Soaking prey with fatty supplements like Selcon may also be helpful for nourishing and conditioning Holothuroids. The potentially heavy and frequent feedings required for healthy suspension feeding sea cucumbers is just one of the many reasons for small, frequent water changes.

The Detritavores

Deposit feeding sea cucumbers (the detritavores) are typically much easier to feed. They can be quite useful for maintaining a clean and healthy deep sand bed but are limited

---

Some of the more common quirks of Sea Cucumbers include:

- ✓ If disturbed, some Cucumarid Sea Apples can inflate themselves to more than double their normal size and drift with the currents until they find a new spot to settle. In the wild, this is rare with most staying for months or even years in the same spot. Frequent movement in aquaria may indicate dissatisfaction with current or delivery/availability of food.

- ✓ The frequent "tentacle-sucking" of Cucumarid Sea Apples is not only to harvest trapped food particles from the filter-feeding appendages but to coat said tentacles with mucus. Interestingly, the feeding tentacles of sea cucumbers do not secrete mucus to trap food directly and need to be coated from a sac inside the mouth.

- ✓ Sloughing or shedding is a mucous response that cannot adequately identify the health or lack thereof in a sea cucumber. Both healthy and stressed sea cucumbers display the activity. In either case, it is not necessarily indicative of a severe problem, but the slough should be exported quickly if possible (manually, or with adequate water movement to whisk the product away for export by the protein skimmer).

- ✓ Some sea cucumber species have the ability to absorb their own innards in cooler waters and seasons… perhaps to stave off death by attrition.

Stichopus chloronotus, the Black Sea Cucumber: this one photographed in Roratonga, Cook Islands where it's contents are often consumed as "rori."

by that very medium. In some ways they can be a burden if overstocked. These detritavores are fairly non-selective (unlike their Cucumarid Sea Apple brethren) about the organic material they ingest. Substrate grain (or particle) size is the only required consideration for these creatures. Sea cucumbers can starve on coarse mediums like crushed coral and coarse dolomite. They mop their mucus covered feeding tentacles across the substrate and "suck" the randomly collected organic matter from their appendages. They also digest large amounts of organic matter off of ingested sand. A considerable amount of substrate is processed in this manner and the pelletized excrement is termed "**cast**." Because of the highly motile, albeit nocturnal, nature of the deposit feeding sea cucumbers, you will notice that they have markedly better-developed tube feet than filter-feeding Holothuroids.

Reproduction

Sea cucumbers are generally represented by separate sexes that are not dimorphic by outward appearance, although hermaphroditism has been documented in at least one deep-sea species. They have a single gonad unlike most echinoderms with paired gonads. Reproduction in Holothuroids occurs either sexually or asexually by fissionary split.

Acts of sexual reproduction in sea cucumbers are quite interesting. They may occur as an act of mass demersal spawning, larvae may be brood spawned, or in some cases there is coelomic development with the young rupturing through the parent's body wall!

Most sexual acts in sea cucumbers involve the discharge of sperm and eggs into the water upon environmental or chemical ques. Embryogeny is like Asteroids (Sea Stars) up through gastrulation. These juveniles grow quickly as plankton without feeding. In as little as three days they develop into an auricularia and then metamorphose into a barrel-shaped dololaria with five flagellated girdles. Very soon after, they develop tube feet and settle down for a benthic lifestyle.

Toxic Spawns

Events of sexual reproduction in captivity are of some concern to aquarists as the often-conspicuous green eggs of sea cucumbers are potentially toxic to fishes. It is likely that the eggs are simply noxious and distasteful rather than fatal to fishes (indeed, the successful ingestion of a palatable but poisonous egg otherwise would be illogical from an evolutionary perspective). Beyond the toxicity of shed sex cells there is also the matter of the sheer volume of product released. In some cases a cucumber may release nearly a third of its weight in sex cells! That can be a serious load in closed systems for biological faculties to handle suddenly.

Simple Fission

More commonly and with no ill-effects to the system, asexual reproduction by fission has been observed in some Cucumarids like the small yellow, *Colochirus robustus*. This small suspension feeding cucumber is one of the easiest Holothuroids to keep and reproduce in captivity.

In any case of captive reproduction in Holothuroids, the catalyst can be either stress induced or evidence of good health.

Compatibility

The compatibility of Sea Cucumbers in captivity is a relatively simple matter. They are limited by available food and they are "safe" with anything that won't try to eat them! It is a common mistake to overstock an aquarium system with too many Holothuroids. In aquariums with a mature and deep sand beds (over 3"/7.5 cm), deposit feeding common cucumbers may be stocked at a rate of one per 30 gallons approximately. Most Sea Cucumbers however are best suited for larger aquariums (100 gallons or more) that have been established for at least one year. Under the best circumstances, some sea cucumbers can be kept successfully for many years in closed aquarium systems.

Best of the Sea Cucumbers

Holothuria floridana (Atlantic/Caribbean): The Florida Sea Cucumber commonly occurs in sand and seagrass beds and as such may be slightly less suitable for heavily rockscaped reef aquariums. They are better off in systems with deep sand beds and refugiums. Grows to 10" (25 cm) with most specimens topping off at around 6" (15 cm) in captivity. Found in an array of colors from creamy tan to yellow and through brown to black.

Holothuria thomasi: The Tiger Tail Sea Cucumber is a reasonably good aquarium occupant despite its potentially large adult size of more than 4´ (120 cm). *H. hilla* from the Pacific is similar. Despite the rigors of import and inconsistent availability of organic sustenance in many aquariums, this species has proven to be durable and long-lived in captivity. Commonly occurs on the reef proper (hard substrates) and as such often assimilates in to reef aquariums nicely.

Pseudothyone, Neothyonidium and like species: The Hidden Sea Cucumbers to 1" (2.5 cm) are tiny suspension-feeding Holothuroids that are commonly acquired incidentally with live substrates. Placing few if any notable demands on the aquarist, this cucumber lives innocuously clinging to the rock in crevices, or burrowed in the sand. Moderate to strong water flow is necessary for success with all suspension feeding reef invertebrates.

Colochirus robustus (Eastern Indian Ocean and Western Pacific, Sri Lanka to the Philippines to Japan): The Yellow Sea Cucumber grows to 4" (10 cm) in length. A colorful and hardy aquarium specimen, this species is one of the best Dendrochirote Cucumarids for aquarium care. Like *Pentacta anceps*, the Pink and Yellow Sea Cucumber, they are suspension feeders that are best left undisturbed. Put them in an area of strong water flow and resist moving them if they choose to travel. They commonly reproduce in captivity and pose little threat to aquarium systems with reasonably good husbandry. Inline, fishless refugiums are highly recommended for success with these and like species for natural plankton culture.

Stichopus chloronotus (Indo-West Pacific; Hawai'i and the South Pacific): The Black Sea Cucumber is one of the many edible species (for human consumption). It can be a useful detritavore in reef aquariums when small. Commonly imported at 3-4" (7.5-10 cm).

Worst of Sea Cucumbers

Synapta sp., Euapta lappa, Opheodesoma sp, Synaptula sp. (circumtropical): The Synaptids (or Synaptulids) are commonly known as Sticky Sea Cucumbers or Medusa Worms. These highly specialized deposit feeders/detritavores are generally very difficult if not

The common Florida Sea Cucumber, *Holothuria floridana*, in an aquarium. (L. Gonzalez)

Another uncommon Sea cucumber, *Opheodesoma sp.*, the Alabaster Worm from the Indian Ocean. Only recorded from the Maldives. Nocturnal, hides by day. To eight inches in length (20 cm). A highly specialized detritavore.

impossible to keep long term in captivity. Most fail within months and few live beyond a year in captivity from their specialized association with sponges (Porifera) from which they are believed to obtain the staple of their diets. Occurring in a variety of colors ranging from white through brown to black, these fascinating sea cucumbers are easily damaged in transit and fall prey to the many hardware hazards common in aquariums (intakes and overflows). All Synaptids are best kept in species tanks by dedicated aquarists only.

Holothuria edulis (Indo-Pacific; Red Sea to Hawai'i): Commonly imported from Fiji, this edible species, as its name suggests, has a mixed reputation in captivity. Many of the problems associated with this sea cucumber stem from its marked sensitivity to the rigors of import and handling. Additionally it is known to grow somewhat large for aquariums (to 12"/30 cm). This species is not as hardy as many other Holothuroids.

Holothuria mexicana (Atlantic/Caribbean): As if the common name weren't offensive enough, the Donkey Dung Sea Cucumber is entirely inappropriate for average aquarium care. They grow to more than a foot long (30-50 cm) attaining a hefty girth and weight. Commonly occurring in grassy sand flats, their demand for food in closed systems is extraordinary, so much that few aquarists can realistically supply enough in time. It is definitely a voracious detritavore for the swimming pool sized reef aquarium.

Actinopyga agassizii (Tropical West Atlantic): the **Five-Toothed Sea Cucumber** is a hefty species (topping off at around 12"/30 cm) known to be potently toxic when expelling its sticky Cuvierian tubules. A sand flat dweller, this species faces the same difficult challenges at finding ample supplies of food as other large Holothuroids in this category.

Bohadschia argus: the **Ocellated Sea Cucumber** is widespread throughout the Tropical Pacific and Indian Ocean. This species is highly toxic and exceedingly large (length and weight) for captivity. It is not recommended for private aquarium life.

Pseudocolochirus species (Eastern Indian Ocean through the West Pacific): the amazingly beautiful Cucumarid **Sea Apples** must ultimately be included on the list of "worst" sea cucumbers. They are some of the most toxic Holothuroids and are more difficult to sustain in captivity due to obligate filter-feeding dependency and appreciable size at maturity. Few aquarium systems or

Holothuria edulis (Lesson 1830), the **Red and Black** or Edible Sea Cucumber lives in the open, on sand, seagrass beds, and under rocks. To one foot in length (30 cm). Aquarium image.

aquarists are well suited to keep this animal successfully. See the expanded detail of this group of sea cucumbers above.

With due consideration for the adult size of some species in this genus, many **Holothuria** can still be useful if not attractive invertebrates for aquariums with deep sand beds.

Right: Jack Melroy holding a *Thelenota anax*. One of the largest Holothuroids.

Below: Rare and unusual among this already unusual class of Echinoderms, this small *Synaptula* is always found in association with sponges (Brown Vase Sponge here). Congregations are often observed in residence and possibly feeding on the combined matter of organic films, microbial life and shed metabolites on the surface of Porifera.

Echinoderms: Cucumbers (Holothuroids)

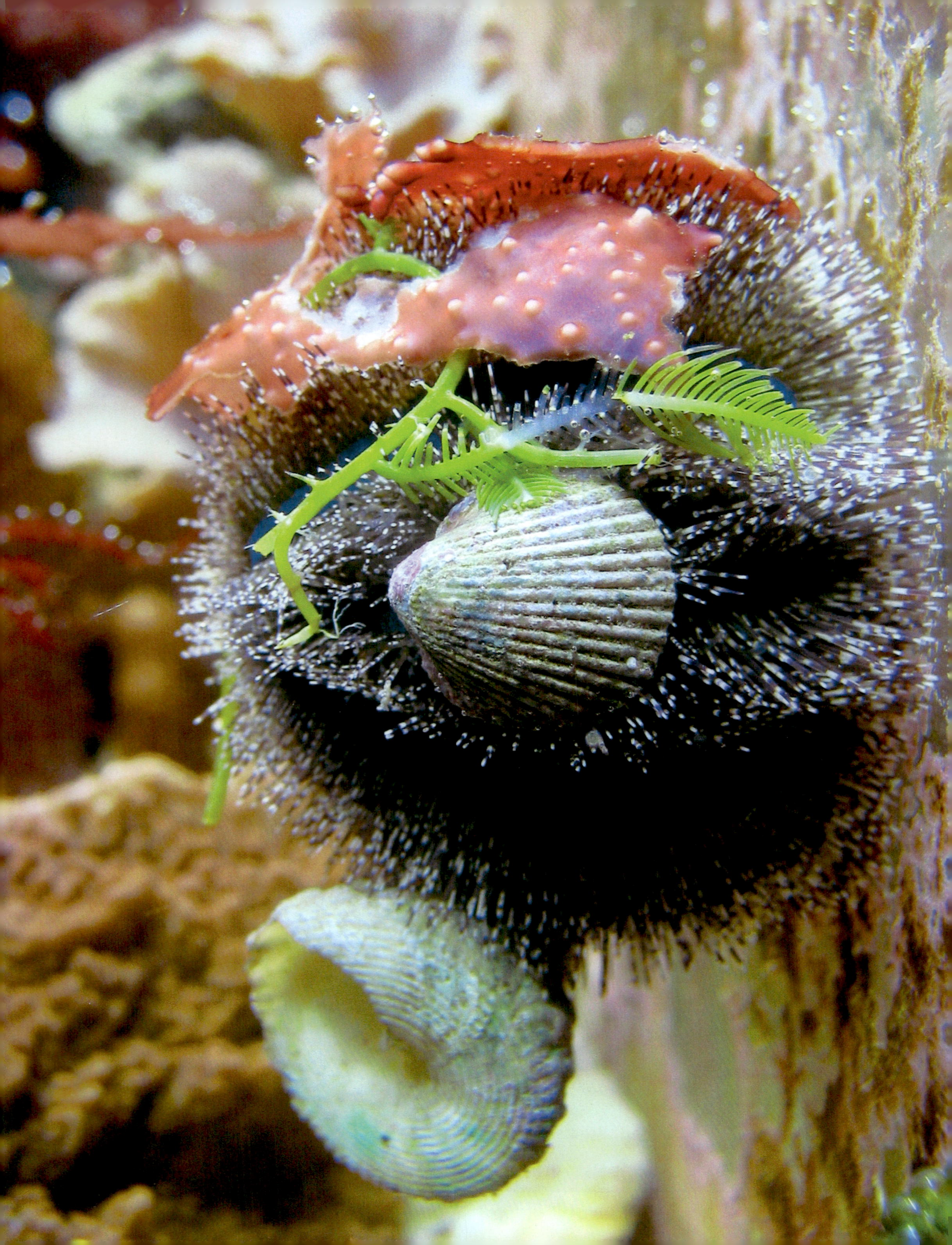

Urchins, Sand Dollars & kin
The Echinoids

The Clypeasteroids, Sand Dollars and Cake Urchins, are "reef safe" but difficult if not impossible to keep in home aquariums. They are mostly deposit feeders (with the exception of the suspension feeding *Dendraster*), requiring enormous tracts of deep live sand to forage in for survival. Few live beyond a year in home aquaria with most starving within months. *Clypeaster subdepressus* from the tropical West Atlantic pictured here.

Some of the most common questions asked of divers involve the danger of encounters with the likes of sharks, eels or barracuda. Most folks are surprised instead to discover that one of the most prevalent dangers to divers hides discreetly covered in spines: the humble Sea Urchin! In the home aquarium too, more aquarists have surely been punctured by the spines of a sea urchin than the formidable and notorious spines of a Lion- or Scorpionfish. Injuries to humans, by urchins at large, are rather ironic when you consider that their strategy is purely defensive and that a victim must literally impale themself upon the living weapon. Some myths and legends abound regarding the actual danger of contact with the spiny wonders. In most cases, envenomation is not an issue, but septic concerns need to be addressed… more about this below.

Sea Urchins, or *Wana* in Hawaiian, occur in a wide array of colors, shapes and sizes. Most mature to 3-6" in diameter (7.5-15 cm), but some Indo-Pacific species attain 18" (45 cm) in diameter. The spine length varies from modest nubbins to daunting spears over 15" in length (37.5 cm) as observed in the genus *Diadema* on tropical reefs.

One of the most fascinating symbiotic relationships on a reef occurs with some marine fishes and urchins. Several familiar groups of fishes commonly live within, or will retreat to hide in, the precarious spines of echinoids. Damselfish and Cardinalfish are quick to find cover in the spiny forest. Shrimpfishes spend a lifetime thereabouts. The pointed fortress offers remarkable protection for adult fishes and serves as a nursery for some piscine fry.

Left: *Mespilia globulus* - the Blue Tuxedo Urchin. It is known as a decorator urchin for collecting various living and non-living reef fragments to cloak its visage. (*L. Gonzalez*)

Echinoderms: Urchins & Kin

Heterocentrotus mammillatus (Linnaeus 1758), the Red Slate Pencil Urchin. Indo-Pacific; Red Sea to Hawai'i. Nocturnal, hiding in crevices by day in depths to thirty feet, emerging at night to rasp rocks. To one foot (30 cm) overall diameter. Hawai'i picture.

Geographically, Echinoids can be found over a wide range of habitats from very shallow to very deep water and from icy temperate seas through tropical waters. Almost 1000 species of Echinoid have been identified circumtropically. A truly cosmopolitan animal!

On Urchins Feeding and Feeding on Urchins...

Urchins are quite popular in the marine aquarium trade. They can be useful and interesting scavengers and many are effective algae eaters. Some eat meatier fare, others feed on suspended matter, but most are fairly indiscriminate deposit feeders on the substrate and to this end serve as helpful scavengers in marine aquariums. Algae constitutes a significant part of the diet for most Echinoids.

Aquarists and non-aquarists alike have found yet another use for sea urchins... *uni*. **Uni** is the fresh harvested roe (eggs) of sea urchin savored in the cuisine of Japan and other countries. Regarded as a delicacy, fresh *uni* has a sweet flavor and can be found at finer sushi bars near you! Some of the best *uni* in the world is collected off the coast of Southern California from sea urchins living in kelp beds. Only recently has the industry for *uni* developed, yet it has quickly become a multi-million dollar trade in the United States, mainly for export to Japan. Urchin species from the genus *Strongylocentrotus* are sought for this culinary delight and have become quite popular with humanoid diners. Fishes have been aware of this tasty meal as any savvy underwater photographer will tell you; one of the surest ways to attract fishes is to crack open a sea urchin! From just looking at these prickly monsters who'd have guessed so many creatures above and below the sea would hear the dinner bell?

Another fascinating use of Echinoids has been discovered with native peoples who have used the spines of Slate Pencil Urchins as... well... pencils to write on slate! Let us wonder no longer where the common name of these urchins comes from.

Taxonomy

This class of animals that we call Echinoids belongs to the phylum Echinodermata that includes many other familiar relatives like the sea stars and some perhaps unfamiliar species like the so-called Medusa worms, a Holothuroid "sea cucumber." We categorically break the Echinoid class down into "regular" and

Strongylocentrotus sanfriscanus is the larger, redder individual in the upper part of the frame. The urchins in the foreground are the more common, smaller *Strongylocentrotus purpuratus*. Both have their eggs used in sushi cuisine as *uni*.

Asthenosoma varium: at maturity, this urchin measures 6"/15 cm in test size and spines to 12"/30 cm. The common name "Fire" urchin has been ascribed for their excruciatingly painful sting. Their quick and jerky mobility has likewise earned them the name "Galloping" urchin. This toxic species is one of the most popular hosts for commensal animals (fishes, shrimp, and snails commonly)

"irregular" species. The aquarium trade is most familiar with "regular" species, which are so-named for their characteristically symmetrical and spherical shape. **Sea urchins** are "*regular*" echinoids and most have unmistakable spines to earn their family name; Echinoidea is Greek for "like a hedgehog." "*Irregular*" Echinoids encompass most species unfamiliar to the aquarium trade such as **Heart, Cake** and **Biscuit Urchins,** and **Sand Dollars**. This group is essentially comprised of sand-dwellers, most of which bury in the substrate for a significant part of their lives. All here are non-spherical ellipses, flattened discs or some irregular variation of an orb. Most irregular Echinoids are not popular in marine aquariums as they lack the durable jaw apparatus necessary to graze nuisance algae. Irregular Echinoids prefer to feed upon very fine organic particles in the sand and soft substrates. Few aquariums, even with deep sand bed methodologies, can supply enough food matter to sustain irregular Echinoids.

Like other Echinoderms, the Echinoids have evolved to use a water-vascular system (**ambulacral**) for their locomotory tube feet. The tube feet are called **podia**, which stick out from the **ossicles**. Ossicles are simply flattened skeletal plates that have fused together to make a solid **test** (skeleton). They have a true body cavity (**coelom**) supporting the calcareous internal skeleton, which is surprisingly fragile (image on page 334). All are covered by some form of spiny dress, apropos to be christened the "Spiny-Skinned Animals" (Echinoderms). The Echinoids also have specialized tube feet called **pedicellariae** between the spines used for cleaning and defense. Some of these are armed with venom glands and can inflict a painful sting. Human fatalities have been reported with some species.

> *Asthenosoma* "Fire" Urchins are well known and highly regarded by respectful divers and aquarists alike. This toxic species can inflict a very painful injury with venom-sacs on the tips of hollow spines. The spines of *Asthenosoma* can, in fact, inflict injury without penetrating the skin, but by mere contact. Beware!

Echinoderms: Urchins & Kin

Microcyphus rousseaui (Agassiz & Desor 1846), Rousseau's Sea Urchin can be found in the Western Indian Ocean; Red Sea to eastern Africa. Commonly in shallow water (0-5 meters) and mainly on upper reef slopes. This is another attractive urchin that should not be permitted into the reef aquarium. It casually feeds on algae and sessile invertebrates. An urchin of modest size at maturity: to three inches in diameter (7.5 cm). A Red Sea image.

Toxic Sea Urchins...

Some dangerous Echinoid sea urchins possess modified tube feet called **globeriferous pedicellariae,** which contain venom glands. Other toxic species have venomous spines, but as a rule, a dangerous species has one or the other means of defense but not both. Venomous sea urchins are principally tropical species, which fall into one of three categories:

• Long-spined with injected venom as with *Diadema* species or from broken, embedded hollow-tube spines (*Echinothrix* species).

• Short-spined with venom that can sting with a puncture (*Phormosoma* species) or from glands at the spine-tips (*Asthenosoma* and *Araeosoma* species)

• Venomous tube-footed species (*Toxopneustes* and *Tripneustes* species) with non-venomous spines

Of these three categories of toxic sea urchins, the Flower Urchins, *Toxopneustes* is reputed to be one of the most dangerous. The specialized venomous appendages (**pedicellaria**) may continue to envenomate even after being torn from the urchin's body. Fatalities by envenomation are unsubstantiated or anecdotal by most accounts, and any contact is likely to result at worst in severe discomfort. Ingestion of poisonous Echinoderms is an entirely different matter, with recorded fatalities.

Typical symptoms of injurious contact with toxic sea urchins include: severe burning sensation localized to the wound, hypersensitivity of the skin, vomiting/nausea and in severe cases paralysis and respiratory distress.

Beyond any concerns of envenomation, there are fundamental concerns of septic infections from puncture wounds by sea urchins. Long-spined urchins are some of the most commonly implicated offenders with a conspicuous pigment from fragmented spines leeching into the flesh (*Diadema* and other genera). Some of these spines have smaller recurves (like barbs) that serve to "ratchet" their way into a wound! Tetanus is but one concern from a puncture by a sea urchin. Competent medical attention and broad-spectrum antibiotics are likely in order for such injuries to prevent complications from the likes of *Vibrio*, *Aeronomas* and other pathogenic dangers.

OK… now that we have got the Echinoid *boogey-men* out of the way, on to selecting safe and hardy sea urchins for the marine aquarium…

Selection

Sea Urchins… Reef-Safe or Not?

With the popularity of reef aquariums, aquarists commonly want to know if sea urchins are safe to keep with their sessile invertebrates, like coral. Alas, it is a difficult question to answer categorically. Ultimately, species that most people would never guess to be nimble or possibly reef-safe may in fact be fine candidates (some long-spined *Diadema*), and other seemingly innocuous species are bulldozers (*Echinometra*) or indiscriminate opportunistic feeders (*Echinothrix*). There is no clear consensus on the suitability of sea urchins in reef aquaria. Most Echinoids in time grow too large or too hungry for the average garden reef display packed with precariously placed corals. There is some minor concern for their vigorous grazing abilities and the impact that it has on coralline algae species. To wit, such destruction of corallines may be noticed with little regard in systems where nuisance algae and its control by the urchins is a more important issue. Otherwise we might say that urchins are not reef-safe for the toll they may take on desirable corallines. There is a somewhat more legitimate concern with species that have short, less-tactile spines poking and tearing tissue of anemones and corals while moving about the tank. Some urchins simply dislodge corals and other invertebrates and cause damage incidentally (a fallen animal might be torn or burned by another cnidarian after being displaced). For this reason,

Tripneustes ventricosus (Lamarck 1816), the West Indian Sea Egg. This fascinating urchin hails from the Tropical East and West Atlantic coasts and is yet another collector or "decorator" species (like *Mespilia* and *Lytechinus* species) and is also (over) collected for its roe …*uni*. A larger urchin, this Sea Egg grows to eight inches in diameter (20 cm). Juveniles inhabit hard substrates while adults move out to sand and grass beds. Pictured above at night, in Cozumel, and below, a closeup in the Bahamas. The spines on this species are not venomous - but the tube-feet are!

Mespilia globulus, the Blue Tuxedo Urchin (Sphere Urchin of science). Eastern Indian Ocean to western Pacific... in shallows amongst algae it grazes on. To three inches (7.5 cm) in diameter. Needs hard substrates, shady areas. Can be kept solitarily or in small groups. Eats mainly algae, *including corallines*.

many aquarists seek the smallest specimens to enjoy for a longer term on the clean-up crew. "Reef-safe" with urchins is ultimately a matter of perspective and need. For most marine aquariums, however, sea urchins can be hardy, effective, inexpensive scavengers, and fascinating guests in their own right.

A wide array of Echinoid species are commonly imported small enough to go unnoticed on live rock. Most can be enjoyed for some months, or even a few years, before growing too large for comfort in the display. Some species are decidedly nocturnal or even boring (into the rock for life). Some of the boring species suffer the duress of import (dry shipping) and contribute significantly to a hard curing process for live rock if they die inside. Although some shallow species of sea urchin will tolerate exposure to air and are commonly shipped in this manner between sheets of wet newspaper, most are intolerant of exposure to air. As a rule, it is always best to capture, transport and move sea urchins under water at all times.

Another concern of Echinoid health in aquaria is the occasional occurrence of temperate species in the ornamental tropicals market. *Strongylocentrotus purpuratus* and *S. franciscanus* from the west coast of America are sometimes peddled as "tropicals." Regrettably, they will not survive more than a few weeks, or months at best, and may cause serious pollution in the system on their demise.

You might imagine, the criteria for evaluating an Echinoid candidate are as varied as their colors, feeding habits, spine shapes, and sizes. Take, for example, the attractive *Mespilia globulus* (AKA the Blue Tuxedo Urchin). This species is a fairly dedicated algae grazer that has been observed to eagerly consume nuisance filamentous algae. It grows to a manageable adult size and has a very handsome color pattern with short non-venomous spines. It rarely, if ever, eats coral and would seem to be a fine choice for most reef aquariums. However, it does have the somewhat irritating habit of being a decorator urchin. As such, small bits of debris from all about the tank are picked up and carried around the aquarium in adornment as a shield from the many stinging and non-stinging perils that abound. This activity will inevitably include the smallest and most expensive coral fragment or polyp in your reef collection! At length, the suitability of any given urchin for your aquarium must be determined on a case-by-case basis with regard for

Echinothrix calamaris (Pallas 1774), the Hatpin Urchin. Indo-Pacific; Red Sea to Hawai'i. It is unfortunate that this strikingly patterned urchin is so risky for aquarium use. An indiscriminate feeder, this species should be kept singly and may prey on cnidarian livestock. They need large spaces in rock to hide amongst by day and coarse substrates. Photographed in Cebu.

the possible risks and merits of the species.

The process of selecting a healthy sea urchin is rather akin to the way one would assess health in other popular Echinoderms like sea stars. The three fundamental criteria are: Appearance, Activity and Integrity (firmness)

Appearance: Does the specimen look healthy at a cursory glance? Individuals should have good color, appearing rich in pigment, without any hint of a dusky or pale appearance. Having researched a species in advance, consider how it compares with photographs that you have seen of wild specimens. There should not be any unnatural irregularities in color or form – ideally with no dull

Choosing the right Urchin species:

Which species best suits your aquarium will depend on the nature and needs of the display. These are some of the questions you might ask yourself:

• Does the system have enough algae growing to support an herbivorous species?

• Am I willing to target feed an omnivorous or carnivorous species?

• Are corals in the tank well-positioned and securely held to prevent tissue damage or dislodgement?

• Is there any concern for small pieces of rockscape tumbling from the leveraged movement of a stout urchin?

• What's that smell?!?! (OK… forget this last question. We were just checking to see if you were still paying attention.)

Eucidaris tribuloides, known by many names including: Mine, Pencil, and Club Urchin. This very common import is collected mostly in the Tropical West Atlantic. A solitary species, the Pencil urchin may feed on algae, bryozoans, sponges, and tunicates in the wild. It behaves like a carnivore most often, though, and is one of the least effective algae grazers for aquarium use. There are reports of pencil urchins using their spines to trap and kill shrimp, fishes and other macro-organisms. Frequent target feeding with meaty foods of marine origin will be necessary for long term success. (*L. Gonzalez*)

Echinoderms: Urchins & Kin

Echinometra mathaei, the Common Reef Urchin, is a smaller species growing to a maximum of about four inches overall diameter (10 cm) and found as deep as 60 feet. It mostly eats algae rasped from rocks by night. Some consideration must be given to the potential for such stiff spined species to pierce or dislodge sessile cnidarians in the aquarium. It is not reef safe! Urchins from this genus (*Echinometra*) are known as rock boring urchins; they prefer to hide in crevices, boring into rock if necessary, and burying in the soft substrates when found on sand.

patches or missing "spines." Be sure to scrutinize any prospective buy closely by having the seller rotate the specimen underwater in good light in front of you. The presence of healthy tube feet will be apparent as the animal moves about or when clinging to a hard substrate.

Activity: Perhaps the simplest way to make an initial assessment of an urchin's health is to have the animal moved to a less-desirable spot in the dealer's tank or by adding an attractant to the water (food or meaty juice). You should observe the urchin responding and moving deliberately, with vigorous activity in the tube feet. Sluggish or inactive specimens are likely stressed, damaged or even freshly deceased.

You might also fairly have the urchin placed upside down to see if it rights itself quickly. Most any healthy urchin will immediately (slowly but surely!) begin to manipulate its spines to flip itself right side up. Another healthy and peculiar behavior seen in urchins is a cloudy "eruption" from atop the test. In some species like the long-spined urchin from the Pacific, *Diadema savignyi*, a bright blue ring surrounds the anus atop the animal. This colorful spot is often mistaken for a "mouth" by the uninformed and the excrement spewed as some mysterious product. Well... there is no mystery of course: just a colorful anus and a healthy sea urchin proffering evidence of digestion. Gametes during reproductive events do in fact exit here

and have been commonly observed in captivity although few if any have yet been successfully reared.

Integrity: Evaluating the integrity of an urchin is perhaps easier said than done considering their spiked armor. One cannot test the *test* ("body" of this spiny-skinned animal) for soft spots as easily as other echinoderms like starfish. Nonetheless, one can look for drooping or missing spines and notice if any fall off during handling and inspection. Indeed, no spines should fall off easily on a healthy specimen.

Acclimation

Sea Urchins, like most invertebrates, are categorically sensitive to changes in water chemistry and quality, and

Echinometra mathaei - a common rock-boring species. This one an adult off Oahu, Hawaii.

Diadema setosum - the Long-spine, Blue-spotted Urchin. To about four inches in diameter (10 cm). Useful in coral bearing aquariums as these echinoids crawl nimbly among the rocks. A frequent "contaminant" on live rock imports. Fiji nighttime image.

Echinoderms: Urchins & Kin

Echinothrix calamaris, on aquarium glass.

may ship poorly under duress. They are so sensitive, in fact, that many public aquariums have used Echinoids as a beacon to degrading water quality in their display (looking for drooping or dropped spines as an indicator). Although there are exceptions among Echinoids that live in volatile, near-shore environments, most specimens seen in the hobby are not likely to tolerate abrupt changes. In a somewhat related anecdote, sea urchins are often shipped with ice packs; they seem to be able to weather a change from warm to cold much better than a sudden change from cold to warm. Please do consider this with regard for acclimation and incidences of disease and morbidity (mentioned below). Echinoids do not occur in freshwater habitats and should not be placed in freshwater for any reason.

Many urchins are quite sensitive to handling and transport and may even suffer from exposure to air. As such, it is best to move all sea urchins under water. Many wave-bearing species in fact have thick and tough tests ("skeletons") that may collapse in upon themselves if raised from the water! There are indeed some urchins commonly shipped from Florida that are transported "dry" with success. In such cases, the Echinoids are placed between layers of wet newspaper in an insulated box and shipped with ice packs to reduce activity and metabolism. The intent is largely to reduce the catastrophic fouling of shipping water with submerged multiple specimens. Dry shipping is by far the exception to the rule with all Echinoderms and can only be done with select species known to be tolerant. For all others, shipping individually and fully submerged is recommended.

On acclimation to a new system, a **simple drip method** for mixing new system water with shipping water is best. Thin airline tubing with a plastic needle valve can be used to siphon system water into the shipping container. A slow dilution over a 15 to 30 minute period will likely be fine for most, so long as the pack water is at least doubled. Ameliorating the shipping water with cups full of water from the destination system, at about five-minute intervals, would be equally effective. Be sure to insulate against falling water temperatures during the acclimation period as well. It is interesting to note that like some other familiar marine animals, including many seahorses and elasmobranches (sharks and rays) for example, sea urchins favor full strength seawater of a specific gravity of 1.025.

Care

Environment

Sea Urchins are adapted for living on rocks and other types of hard substrates with spines and podia for movement and secure placement to resist tidal and wave action. They need similar materials in their systems to live comfortably. Some study of your individual species' biotope is encouraged. Approximating the physical make-up, lighting, circulation, etc., will contribute measurably to your success. Such

Fragile as an eggshell, here's a denuded "Test" of *Stronglylocentrotus sanfranciscus*

The many faces of *Tripneustes gratilla* species, the Priest-Hat, Sea Egg or Collector Urchin… family Toxopneustidae. Indo-Pacific; Red Sea to Hawai'i. Venomous to the touch! This toxic sea urchin grows to about five inches in diameter (12.5 cm). It is mentioned here so that divers and hobbyists alike will avoid it. Shown here: specimens in the Red Sea at night, Andaman Sea and Hawai'i.

information is accessible through accurately researched periodical literature, hobby, trade and scientific publications available through the pet trade, public libraries and universities.

As a deposit feeder, the nature of the substrate is of critical importance for an urchin's long-term survival. Mature tanks are recommended with established organic material for improved feeding and nutrition opportunities. A healthy growth of green microalgae that has overcome brown diatom algae (algal succession) is a good bio-indicator. Be aware though that most Echinoids are active only at night. Not all are graceful, however, so please take heed regarding rock and coral placement. Even if your urchins look like they're able to negotiate the open spaces in the system by day, such may not be the case by night when they're active.

Urchins can impose remarkable leverage when moving between crevices, and rockslides can be quite damaging to many animals in the display - or to the tank itself!

Filtration

Above average filtration is required when keeping more than a few sea urchins. They can produce a remarkable amount of waste from their vigorous deposit feeding efforts. Ammonia is excreted through five pair of gills and solid feces exits, as mentioned before, through an aboral (top side) anus. As with all marine animals, but especially so with this voracious scavenger, adequate biological faculties of mineralization are necessary. A well-functioning protein skimmer should also be a part of most every captive marine system; with sea urchins it is perhaps

necessary.

Diseases of Sea Urchins

Although **epizootic** diseases of Echinoids have been studied by the scientific community, not much of practical worth is known to the hobby and science of aquariology in address of such pathogens. In some cases, storm activity has been linked to urchin disease as with the discovery of the marine amoeba, *Paramoeba invadens*, which causes **Paramoebiasis** in Echinoids. Two prominent theories cite the catalyst for this amoeba as 1) higher temperatures and 2) dissemination as an exotic from afar by virtue of the storm activity. The former is at least pertinent to aquarists as increased and prolonged water temperature is a recurring theme with various diseases in sea urchins. "**Bald Sea Urchin Disease**" is one of the

Echinoderms: Urchins & Kin 335

more prominent illnesses associated with Echinoids and has been linked to high temperatures. Also known as "**Red Spot Disease**," this condition is thought by some to be caused by bacteria; *Vibrio* and *Aeromonas* have generally been associated with the afflicted. Symptoms include darkly pigmented and swollen lesions and the loss of spines. There is no known cure for this condition presently and infected individuals must be quarantined in isolation. Judicious experimentation with various therapies including antibiotics may be in order. Be sure to report any efforts in such endeavors to the aquarium community!

As an interesting aside, some of the devastating diseases of coral in the Caribbean have been linked to diseases in sea urchins. Specifically, mass die-offs of *Diadema* Echinoids are believed to have afforded the aggressive growth of nuisance algae in direct and stressful competition for space with corals. "**Black Band Disease**" in coral has specifically been mentioned with regard for the loss of natural algae control by *Diadema* sea urchin species and the stress of increased encroachment of algae on corals.

Feeding and Nutrition

Types, Frequency, Amount, Wastes

As Ricky Ricardo might put it, "these thins will eat everythin' and anythin', bobbalooo!" They are nature's surest answer to "who's going to clean up?" To be specific, sea urchins are deposit feeders that forage for algae, sessile benthic animals and animal remains. In captivity, most are eager feeders on most everything and anything. Hide the family cat and keep your eyes on the urchins! Their healthy and indiscriminate appetite is the very thing that makes them such wonderful aquarium denizens. They can be employed to clean up serious nuisance growths on various substrates and then easily and satisfactorily fed alternate foods when the targeted "prey" wanes (eg. running out of nuisance algae and eating desirable coralline algaes).

All Sea Urchins have a celebrated chewing apparatus called Aristotle's Lantern, composed of five large calcareous plates, a number of smaller rod-like pies, and related special muscles. It is a real spiffy contraption but not very expedient by design; the consumption of a cluster of seaweed may take weeks. By species, echinoids run the gamut of committed herbivores through indiscriminate omnivores to deliberate predators and specialized feeders. Even the "best

Diadema savignyi, a Long-Spined Sea Urchin known from the Indo-Pacific; Africa to the South Pacific. The test measures about five inches maximum (12.5 cm) in diameter, with spines to sixteen inches long (40 cm). A blue ring around the anus is indicative of this species. Aquarists frequently discover this urchin as a contaminant with live rock imports. They scavenge hard substrates with agility and minimal disturbance to sessile invertebrates. (*L. Gonzalez*)

Astropyga radiata (Leske 1778), the Radiating Hatpin Urchin. Indo-Pacific; Africa to Hawai'i. An exceptionally colored species, this urchin can be found both day and night on sand to rubble substrates. It eats algae but will also consume invertebrates. Fares best in captivity if target fed meaty foods several times weekly. An uncommon import. This one photographed in North Sulawesi.

behaved" herbivorous urchins can turn cannibalistic if driven to hunger. It is not at all uncommon for urchin species to be intolerant of their own kind and kin. Unless you know an Echinoid species to be tolerant, it is best to keep them solitary.

Wilkens mentions an apparent need for supplying urchins with adequate sources of lime (calcium carbonate) for test & spine growth/strength. Crushed shells and other calcareous substrates may work well. He also alludes to the unquenchable appetite his specimen showed for *Caulerpa*. Again, they will eat anything and everything!

A captive diet for Echinoids in general should include a variety of dried and fresh frozen foods. Specifically, it should include a well-varied mix of plant matter and meaty foods of marine origin (shrimp, plankton, krill, fish, squid, etc). When feeding dead whole prey to urchins, provide bone and shell whenever possible (small whole silversides and unpeeled shrimp, for example). A homemade food recipe can be one of the best ways to deliver highly nutritious food to fishes and other macro-organisms like sea urchins. Various recipes can be made to each separately address specific dietary goals or needs for herbivores, carnivores and the like. You can find excellent recipes for making marine animal's food on the Internet and in Fenner's reference,

"The Conscientious Marine Aquarist."

Reproduction

All Echinoids are of separate sexes and reproduce by external fertilization. There is no readily apparent structural difference between the sexes (dimorphism). With urchins, the boys and girls basically look alike... just like the rock and roll "Hair Bands" of the 80's. In brooding species, the eggs are either retained in external body cavities or between spines. The feeding larvae are planktonic until their "skeleton" (test) begins to form, which begins when most species reach approximately one millimeter.

Sexual maturity is reached, as

Chondrocidaris gigantea. In Kona, heavily coated with encrusting algae. (*R. Fenner*)

mentioned above, between 1 and 3 years of age for many species. Urchins categorically produce very high numbers of eggs (around a million eggs is quite common). This is possible by the strategically hollow body cavity, which by design leaves considerable space for development of the gonads during breeding seasons. While the rearing of larvae by aquarists is not actively pursued, events of reproduction provide a very nutritious meal of gametes for cnidarian and filter-feeding invertebrates in closed systems. The production of gametes only happens within a specific range of temperature and salinity, however. Duress can even cause a sea urchin to reabsorb its gametes. Or, on the other side of the equation, stress of urchins can induce them to spawn *en masse* and die. In the confines of an aquarium, though, this can be a dangerous amount of matter for biological filtration to suddenly contend with. Furthermore, captive rearing is difficult because of the extended larval stage (30-60 days) and inherent difficulties maintaining pelagic larvae (plankton) in good health for the extended interval (water quality, avoiding pumps/filters, providing suitable first food, etc.).

Predator/Prey Relations

For the most part, due to prickly-ness (is this even English?!?) and toxicity (for some), sea urchins have very few predators under ordinary circumstances; but since when are aquariums ordinary? Some Triggers and various Puffers will nibble at urchins (and heaters, rocks, bubbles, and other inanimate objects!). Some predatory sea stars, large butterflies, parrotfishes, hogfish, wrasses, basses and crabs are also known to attack urchins on occasion. There

Heterocentrotus mammillatus Slate Pencil Urchin, in Hawai'i, with a small *Canthigaster sp.* Tobie (Sharpnose) Pufferfish.

Order Spatangoida: Heart Urchins and Sea Biscuits (AKA Sea Porcupines) include the genera *Meoma, Brissus, Maretia, Lovenia, Agasizzia, Metalia, Plagiobrissus*. These sand-dwelling animals are covered in fuzzy spines and spend an active life burrowing into richly organic substrates. It is doubtful that any but the largest aquariums have a chance at keeping this categorically short-lived echinoid alive for more than a few months captively. Although reef safe, aquarists are discouraged from keeping this urchin casually. Very large and deep sand beds are recommended if not necessary for any attempt in aquaria. They are commonly found in sandy flats and among seagrasses with heavy sediments. Photo, a test of a *Meoma ventricosa*, off Cozumel, Mexico.

Echinoderms: Urchins & Kin

Diadema antillarum, a beautiful long-spined species. This one photographed off Cozumel.

are even some wicked snails that savor echinoderms. Otherwise, most all macro-organisms will leave sea urchins alone and vice versa. Burrowing urchins don't seem to be too tempting to would-be predators (it's hard work extracting them from rock and sand!). Exposed forms are too crunchy, spikey, poisonous (to eat), venomous (to touch), or odd-shaped to consume.

Sea urchins are known to have at least several interesting, commensal relationships with other creatures exploiting the safe haven of their spiny forest. We commonly see snails and shrimp, like Coleman's shrimp (*Periclimenes colemani* family Palaemonidae) and Squat Urchin Shrimp (*Gnathophyllodes mineri*), nestled among Echinoid spines. Various cardinal and damsel fishes take refuge in and use the spiny-skinned urchins as a nursery for their fry. It is a truly magnificent sight to see an entire shoal of juvenile Kaudern's cardinalfish (AKA Banner or Banggai Cardinalfish) shifting and jockeying in position as the field of a long-spined urchin pitches and twitches. With some species, the relationship is quite strong and others it is somewhat more casual. On a much smaller scale, even the polychaetes worms don't want to be left out of the party; *Flabelligera affinis* has been found as a commensal among the spines of a sea urchin. And let's not forget the cnidarians - some urchin species have their spines colonized with zoanthid species! On closer examination with a trained eye or sight glass, you might be surprised to see who has hitchhiked along with your urchin.

Facing page: Juvenile *Pterapogon kauderni* (Banggai or Kaudern's Cardinalfish) take refuge in the sharp, defensive spines of a *Diadema* urchin. Aquarium Photo.

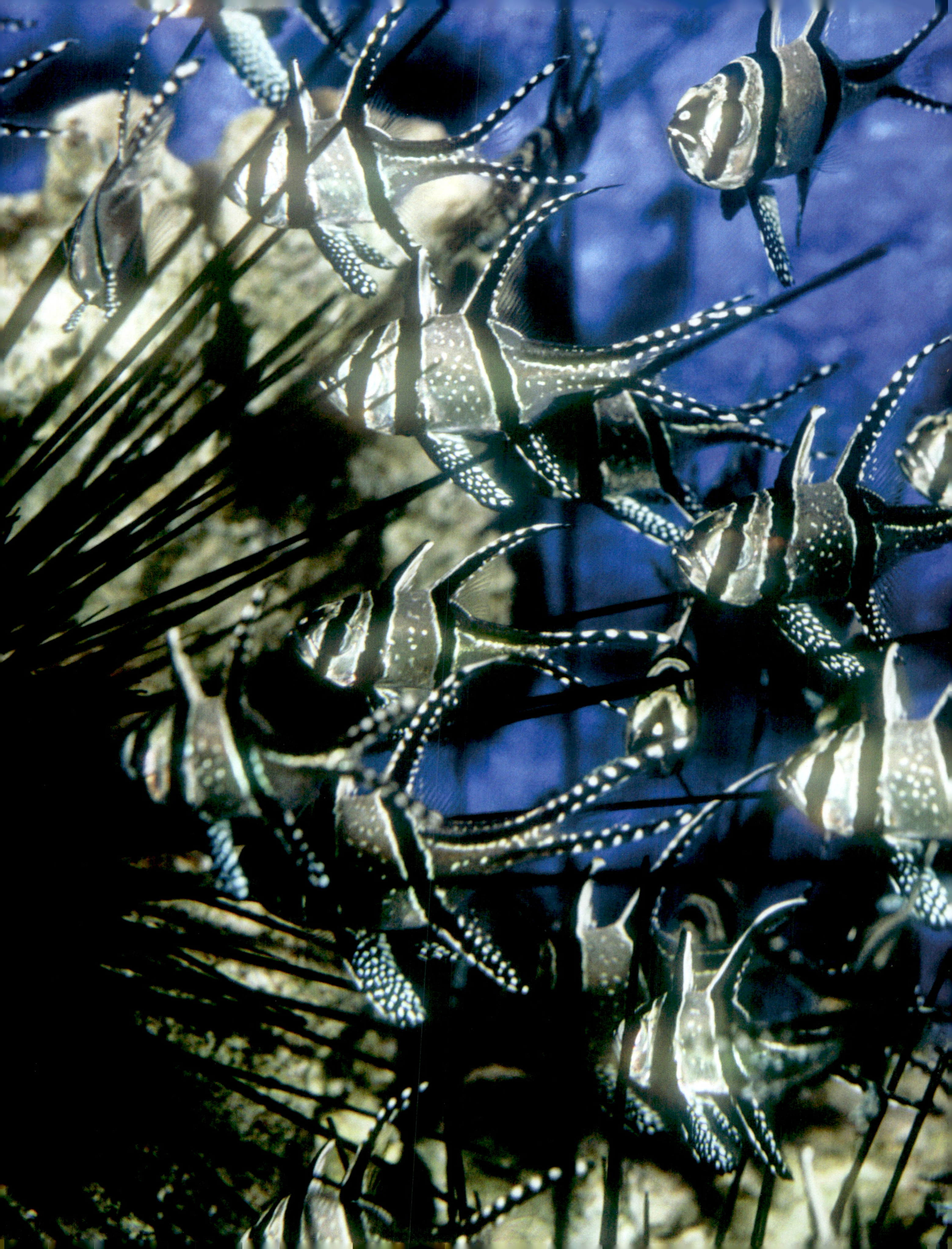

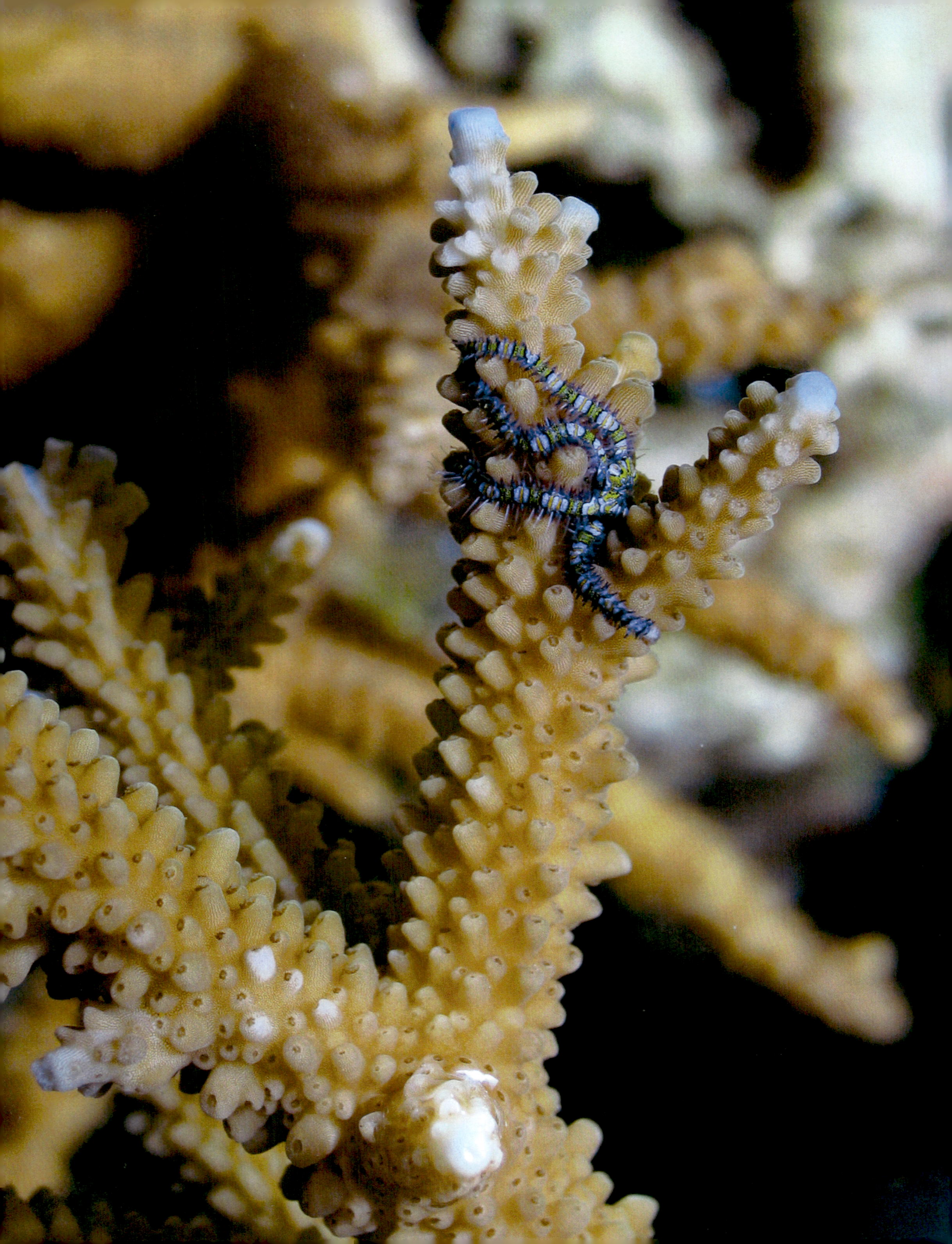

Sea Stars
Asteroids, Ophiuroids, & Crinoids

Fromia monilis in the Maldives.

The "spiny skinned" animals that we know as "starfish" are represented by three classes: **Asteroids** (sea stars), **Ophiuroids** (brittle and basket stars), and **Crinoids** (feather stars and sea lilies). All are found in the aquarium trade, many regrettably so. It may be tempting to call these animals starfish, but it is somewhat inappropriate. There is little relation with these or any echinoderms to the fishes. Most importantly, sea stars do not possess a higher-order brain or spine; however they are not complete dummies either. They do have a closely interacting nervous system that renders them tactile in a chemical and visual (light sensitive) sense with an acute awareness of their surroundings.

Echinoderms are also uniquely exclusive to saline habitats. They are an endless source of fascination for many people, and the desire to keep and study them in aquariums is little surprise. Many sea stars are found in nearshore habitats, occur in dramatic colors, and look so sturdy as to entice you to handle them. Looking closer you'll find that their endoskeletal elements (**ossicles**) give them a very solid and durable texture, bordering on rigidity. Alas, this group of animals is one of the most delicate in terms of acclimating, with a dismal record of survival in aquariums. There are species of sea stars that stand out in the trade as decidedly hardy and long-lived in aquaria, but even these individuals must run a taxing gauntlet upon import.

The primary reasons for most of the initial mortality of sea stars have to do with collection, holding and shipping practices commonly employed (exposure to air, shipped in too little water, varying and poor water quality). Typically, once collected,

Left: Unidentified brittlestar, perhaps an *Ophiothrix* species, entwined about the branches of an *Acropora* coral. (*L. Gonzalez*)

Protoreaster lincki, the Red African or Horned Sea Star: A beautiful, opportunistic omnivore of other invertebrates that can literally wipe out an aquarium of sedentary life! This seastar has even been known to eat sleeping fish. Photographed at a retailer. (*L. Gonzalez*)

Asteroids

Here is a group of animals that aren't what they appear. "Oh, a starfish, they're so beautiful, but such slow moving creatures." At a conference on "Excited Sea Star Behavior," a time-lapse film was presented depicting several hours in which the sea stars moved "really fast" - at a few feet per hour. Actually, there are some members of the class Asteroidea who clock in at a few feet per minute! From a layman's perspective, this group of sea stars includes the sturdy and stiff bodied species that you might recall seeing as dried curios, and are recognized by most as the classic sea star forms. Most members of this class are omnivores or dedicated predators. They categorically are not to be trusted with invertebrates, as many will eat desirable aquarium denizens such as corals, sponges, snails, and especially clams (most predatory sea stars, in fact, will make a bee-line directly for the most expensive *Tridacna* in your display). A somewhat reliable legend in aquarium science exists that the "thorny-backed" or "knobby" members of this class tend to be dangerous to keep with other invertebrates. Notorious they are roughly removed from their substrate (tearing the tube feet of sea stars and breaking the legs off of aptly named brittle stars), packed one on top another in very little water, kept in these poor conditions before shipping, and not fed during any of this time while navigating the chain of custody on import to your retailer or you. The process of transit from point of collection to an established display aquarium is often more than a week's time for imported animals.

It is ultimately ironic that so many of the most popular species of sea stars offered in the trade are outright poor candidates for captivity, due to their inherent sensitivity to shipping conditions and their historically poor rates of survival through the chain of custody. Surprising as it may seem, the most commonly encountered species (Blue *Linckia*, *Protoreaster* stars like the Red African and Chocolate Chip, and common Atlantic *Oreaster sp.*) are amongst the highest casualties at all levels. It is especially ironic when other durable and more appropriate species are overlooked or weakly endorsed. Due to this poor species selection, mishandling, and starvation, many of the sea stars kept by aquarists don't live a month after collection. They are frequently observed with **vacuolations** (missing spots) in their bodies, a cessation of feeding, and ultimately dissolving to death in the system unless noticed and removed. Sea stars, like most invertebrates, require even more consideration of handling and acclimation protocol than fishes.

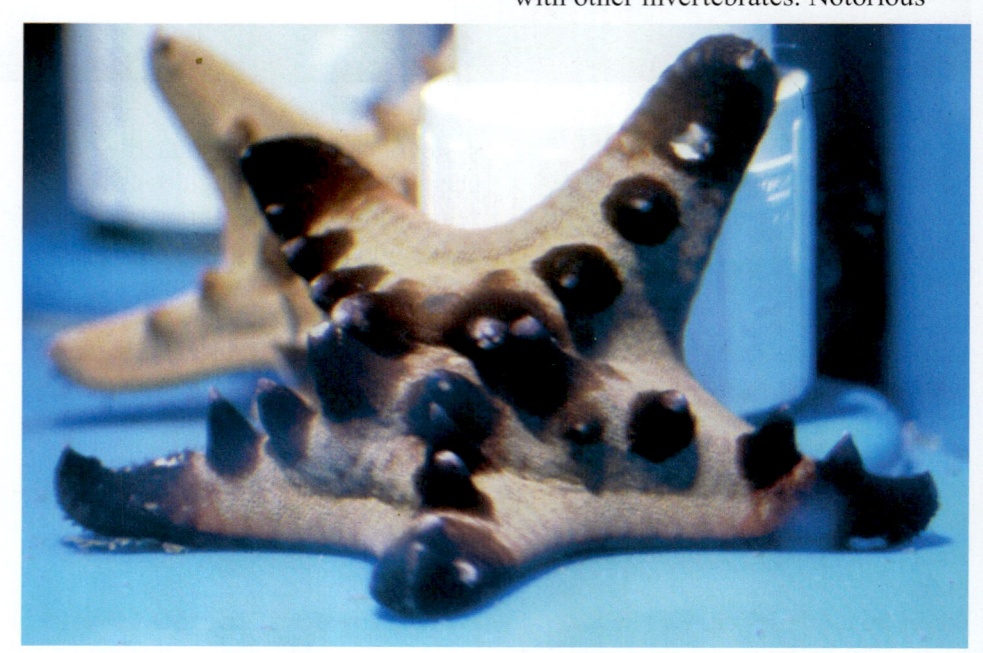

The Chocolate Chip Star, *Protoreaster nodosus*: An opportunistic omnivore on other invertebrates. Found widely throughout the tropical Indo-Pacific. Select for smaller 2-3 inch (50-75 mm) specimens and keep them well fed. Ultimately not reef aquarium safe. Wholesaler photo.

Ophioderma squamosissimum. The Red Caribbean Brittlestar. This beauty will predate on small motile invertebrates if not kept properly fed. This individual is a very unusual orange color morph. The inset photo shows more typical coloring. Aqarium photographs. (*L. Gonzalez*)

Fromia ghardaqana Ghardaqa Brittle Star. Endemic to the Red Sea. Grows only up to three inches (75 mm) in diameter.

carnivores include *Protoreaster* and *Pentaceraster* species. The not-quite-thorny, but bumpy, bulbous and bad biscuits of the genus *Culcita* are perhaps worse still for their large size, bulldozing carriage, and absolutely indiscriminate appetite. As referred to in common nomenclature as Sea Biscuit, Pin Cushion and Pillow star, *Culcita* species are known to eat coral tissue, mollusks, algae, bacterial slime and almost anything between it and the carbonate foundation upon which it sits. Needless to say, this creature is best suited for a species-specific display. Most of the Asteroid sea stars, as a rule, are best remanded to fish-only displays and are to be rarely, if ever, trusted with sessile invertebrates. There are, however, reef-safe members of this class, most of which include or resemble the doughy-armed *Linckia* species. The reef-safe genera of this class include: *Fromia, Linckia* and *Astropecten*.

Echinoderms: Sea Stars   345

Orange *Linckia sp.*, on glass at a retailer. (*L. Gonzalez*)

Ophiuroids

Ophiuroids (brittle stars) are the peaceful and lovable darlings of the aquarium trade in starfish. Once established, most can live for years upon years with very little maintenance, and only light feedings in well-established tanks. They serve as incredibly functional scavengers and detritavores with few demands. In heavily rockscaped aquariums they play an important role in keeping inaccessible areas cleaned up as they stir up and aerate the sand bed. Unlike their Asteroid brethren (many of which require a stocking density of one per one hundred gallons or more), brittle stars can be kept in small displays and in greater numbers, upwards of one per ten gallon densities, with regular feedings. Smooth armed specimens are collectively called "Serpent" stars while the spiny armed species are simply called "Brittle" stars.

The Brittle or Serpent Stars are grouped as the class Ophiuroidea are comprised of many families and are characterized by having highly mobile arms that can be used to assist in relatively rapid motion. These echinoderms are decidedly faster, and more delicate regarding handling than Asteroids. They are often mistaken by the uninformed at first as an "octopus" because of their similar agility and serpentine mobility. Their common name is derived from their sinuous, snake-like movements, and the fact that they are truly brittle and break away easily if attacked. In somewhat comical fashion, a frightened brittle star transforms from its usually supple form into a rigid, stiff-armed ornament (perhaps to increase the likelihood of a short and clean break away from an attacker). Industry professionals have long struggled with chagrin to this response while attempting to pack such stars in small shipping bags. It is a comical sight indeed with the echinoderm splayed arms stiff and wide out, as if to say, "You can't make me climb into that bag!" A technical distinction: the tube feet (**podia**) in this class are generally used as sensory organs, rather than for active feeding as with their kin, the Asteroid sea stars. There are more than 2,000 described species worldwide, and they are found congregating throughout shallow reef environments, hiding under

346   *Natural Marine Aquarium Volume I - Reef Invertebrates*

rocks, and within and between other living organisms (corals, sponges, gorgonians, etcetera). In and upon the benthic environment of a reef, such animals are exposed to a wide array of pathogenic and non-pathogenic organisms (some of which can *become* pathogenic in some circumstances). Ophiuroid sea stars have drawn some attention for their apparent resistance to disease and infection in the wild despite the formidable and harsh realities of reef life. Marine physiologists have observed that a thin layer of bacteria lives between layers of the brittle star's protective skeletal matrix. In a lab, these symbiotic bacteria have been demonstrated to be anti-microbial: preventing the growth of other bacteria responsible for pneumonia, staphylococcal skin infections, salmonella, and diphtheria, for example. Like so many other reef animals, there are great implications for study and discovery of useful elements of brittle stars, which could contribute to the fields of medicine and science. Beyond the science, however, as aquarists we may enjoy the benefits of this symbiotic relationship with faith that Ophiuroid stars are very durable aquarium inhabitants when provided with reasonably good husbandry.

Crinoids

Crinoids are construed to be the most ancient class of all Echinoderms, with fossil forms dating back to the Paleozoic era. Over 500 species are presently recognized. Although the size of extant (living) forms is a few inches to a couple of feet in dimension, some extinct crinoids were very large, such as *Extracrinus subangularis,* which had a stalk of nearly 21.5 meters in length. Any visit to your local or national natural history museum will impart an appreciation for how species-rich and dominant Crinoids were in reef communities of the geological past. They are further removed from the other living classes of Echinoderms

Asteroid sea stars encompass a wide range of shape, size and hardiness.

Another typical offering, the "Doughboy" Sea Star, *Choriaster granulatus*, is a big, bulky Indo-Pacific asteroid that scavenges in reef shallows. It should only be employed in systems of hundreds of gallons size and kept away from sessile invertebrates, as it is a coral eater.

Patiria miniata, the Bat Star

Pisaster ochraceus, the Ochre Seastar.

Echinoderms: Sea Stars

Asteroid tube feet up-close: the ambulacral system of a sea star in action. Very reactive to touch and the sensation of food in the water... a prompt response by the tube feet to stimulation indicates a potentially good specimen.

in having a mouth oriented away from the bottom surface, and having an early life history as attached juveniles. As young, their bodies are composed of an attached stalk; this condition persists for Sea Lilies, but Feather Stars break off, become stalk-less and lead free-living (mobile) lifestyles.

Crinoid Physiology and Function

The internal skeletal components (**ossicles**) of a Crinoid make their stalks appear jointed. Feather stars have **cirri** (anchoring appendages like feet or talons) on the end of the stalk to grasp the substrate. Growth of the stalk is from the crown where the arms come together. Five arms, with their jointed appearance, initially radiate from the periphery of the crown and in most, quickly branch into a sum total of many more arms. On each side of the sea star's arms are rows of jointed **pinnules** with (**ambulacral**) grooves on the oral surface. The arms' (branchlets) bare pinnules convey food by twitching their tube feet like an elaborate conveyor system. During feeding, the arms and pinnules are outstretched and their tube feet (**podia**) are erect. The **papillae** along the length of the podia secrete mucus and plankton becomes trapped in it. The podia then channel this food-embedded mucus into the ambulacral groove where cilia carry it to the mouth. The mouth, in turn, leads to a short esophagus before the intestines, which make one or more turns down around the underside of the animal before turning around to head back up to a short rectum and anus. The whole of this digestive tract is lined with undulating cilia. Waste is finally ejected as mucus balls from the anal cone.

NOTE: Crinoid Feather Stars and Sea Lilies are categorically tragic aquarium subjects from any perspective in the industry and throughout the chain of custody upon import. Both are very delicate to handle and ship, on top of which they are also extremely impractical, perhaps impossible, for the average aquarist to keep in an aquarium. The mortality of Crinoids as a rule is high when handled, and we wish to make it clear: the casual harvest and keeping of these animals is to be discouraged outside of species specific displays until the primary obstacles of collection and husbandry are finally surmounted. The Gorgonocephalid Basket Stars (an Ophiuroid) mirror the delicate nature and suffrage of Crinoids as aquarium subjects and all such admonitions apply here as well. General content presented here on starfish husbandry *does not* apply to Feather Stars, Sea Lilies or Basket Stars.

A Crinoid seastar unfolding for the night's feeding. Photographed in Fiji.

348 *Natural Marine Aquarium Volume I - Reef Invertebrates*

Pentaceraster tuberculatus, the Green Horned Seastar, seen here at a retailer, is an opportunistic carnivore and feeds on small molluscs and other invertebrates. (*L. Gonzalez*)

escape to safe harbor. As a fright-induced strategy, it is interesting to note that only the arms of the starfish are bio-luminescent while the more vulnerable central disk is not. It is an effective defensive strategy to direct a predator's attention to the easily reparable arms. Although Ophiuroids can indeed repair all parts of the central disk and even regenerate entire organs, an injury to the central disk is far more compromising than the loss of a brittle arm.

Aquarist-imposed fragmentary propagation is generally not recommended in sea stars or brittle

The Amazing Regenerative Powers of Starfish

Sea stars are pentaradially symmetrical. That is to say they are composed of five, alike-segmented bodies, although the actual number of arms can range tremendously. Each equal segment has redundant sets of vital organs. Due to this design, some have a remarkable ability to heal from severe traumas to the body and even form free-living clones from cleaved segments that are aborted or fall off with part of the central/oral disc attached. As an imposed or deliberate act, the event generates new individuals in a manner called **cometary fission.** For quite a while, the fragmented clone of a sea star looks like the celestial body of its naming (a comet). Some stars can even lose a significant part of the central disk and organs and still regenerate missing organs.

In Ophiuroids, the strategy of having brittle arms is equally impressive and highly effective. By offering a piece of an arm under attack to a predator, the brittle star can often escape and live to regenerate missing parts (**autotomy**). The sacrifice is taken to another level in some species that demonstrate bio-luminescence: fragmented arms can glow suddenly and brightly to shock and confuse a predator while the sea star gains additional time to

"Pinnules" of an unidentified Crinoid star, in the Andaman Sea.

Echinoderms: Sea Stars 349

Ophiuroids, Zoanthids, Poriferans - Oh, my! This close-up reveals just a hint of the marvelous complexity of life on a reef. Here we have a microscopic demonstration of commensal life between reef invertebrates including brittle starfish, stinging polyps and an encrusting sponge (*D. Fenner*)

stars, but should you find an "extra" arm lying about, don't be too quick to dispose of it. Instead, place it in the sump or refugium for quiet care and study. Asexual reproduction by fragmentation is not at all uncommon in many species of sea stars.

Reproduction

Cloning by fragmentation is only one form of reproduction for these unique reef creatures. Reproduction of sea stars may occur via sexual or asexual events. Unlike most echinoderms, many sea stars and brittle stars are monoecious (hermaphrodites). They reproduce mainly by releasing their eggs and sperm into the environment separately, induced by chemical and physical cues (they rarely self-fertilize and are weakly successful when it occurs). Spawned sea stars begin life as bilateral pelagic larvae, derived from the joining of gametes of separate parents. Pelagic sea stars swim or float about in the water column, hopefully landing in a suitable environment and developing through metamorphosis into miniature versions of their parents. Beyond asexual fragmentation and broadcast spawning (sexual), some starfish also brood their young (live-bearing) as self-fertilizing hermaphrodites. These tiny species, miniature serpent stars, have become quite popular with reef aquarists for their innocuous behavior, utility, and prolific mode of reproduction. A similar Ophiuroid, the striped micro star, enjoys a similar reputation for its useful participation as a hardy detritavore and due to its prolific mode of reproduction. The six-armed micro stars simply divide in half through asexual fission and produce two fully formed clones in short order. Summarily, the echinoderms that we call sea stars demonstrate some of their most fascinating behaviors through interesting reproductive strategies.

Selection

The proper selection of healthy sea stars revolves around three general criteria: Appearance, Activity, and Integrity (firmness).

Appearance: Does the specimen appear to be a healthy individual when compared with images you have studied of ones in the wild? Although there are many color varieties within some species, healthy specimens are bright in sheen (sharp, or brilliant color, so to speak), with no dull or dusky, discolored areas.

Sea stars can have a wide ranging number of arms... here's a "handy" species: *Pycnopodia helianthoides*, a Sun Star. Up to two feet across and an eating machine!

Fromia indica, the Indian Sea Star. Widely distributed across the Indo-Pacific. This one off of Queensland, Australia.

Echinoderms: Sea Stars 351

Inspect sea stars for discolored, sunken areas evidenced by looking at the dorsal side of the animal. This sea star is in trouble.

Be sure to inspect all the surfaces of a prospective buy by having the seller rotate the specimen underwater in good light in front of you. Also be alert for visible parasites or their presence by bumps on the surface, or evidence of missing tube-feet. Small parasites are almost usually best left on their host as more damage often results from attempts at physical removal. Given the choice, select specimens without small parasites, however incidental or harmless they might actually be.

Activity: is best determined by having the animal moved to a less-desirable spot (onto sand or off, depending on the species' preferences), or by offering an attractant (food/juice). With Asteroids, you should see the animal extend its tube-feet on every limb. For both sea stars and brittle stars, you should see them respond and move with clear and excited interest by food stimulation. Immobile specimens are likely to be stressed or damaged. A standard approach in sorting specimens is to place them on their "back" (upside down) for a chance to appraise the possible presence of parasites, breaks, weak spots in the underside **integument**, and to judge how well each turns itself back over - a good sign of vigor in all starfish. It also allows you to make sure the tube-feet are intact in Asteroid sea stars. Pillow and Pincushion Stars (*Culcita spp.*) will inflate themselves measurably when upturned: until their tube feet make contact with a hard surface so they may flip themselves right side up.

Integrity: Sea stars should have a solid integrity, their bodies and appendages should be firm to the touch. While handling the animal to determine the first criterion (Appearance), take the opportunity to feel the animal, if at all possible, and gently squeeze along its body (always underwater: avoid exposure to air for most). Infection or internal damage is often revealed in soft regions of the animal even when the sea star looks otherwise fine from outward appearance. Such injuries rarely heal successfully in captivity.

It is a good idea to let some time pass by in a dealer's aquarium before purchasing newly imported sea stars. Newly arrived batches fall into one of two wide categories: 1) mostly living and 2) almost all or mostly dead. Which group yours falls in will be revealed in time, a few days to a couple of weeks is a good period to wait while considering or holding a prospective purchase.

We would like to add a couple of comments regarding excess handling of these animals: if possible, please don't. With Asteroids, the stress and potential damage to their tube feet and bodies is considerable. With Ophiuroids, brusque handling will quickly illuminate why this group of animals is called "brittle" when

Culcita schmideliana (Retzius 1805), the Spiny Cushion Star. Indian Ocean; eastern Africa to Malaysia. To ten inches (25 cm) in diameter. This least likely candidate is one of the most acrobatic sea star species, inflating and deflating to navigate throughout the reef and easily flipping itself over when necessary.

Oreasteridae sp., attractively marked with an array of plates, spines, tubercles and warts.

pieces of arms fall off. If a sea star must be moved, you can either take the substrate with it, or if this is impractical, gently and slowly move the animal back and forth, side to side, until it releases itself. If it seems stuck to a substrate and unwilling to otherwise let go, utilize a stiff edged piece of plastic, like a credit card, to nudge the animal off the substrate. Lastly, as with so many other invertebrates, DO NOT lift these animals into the air. This is a major cause of their mortality and morbidity in captivity. Please bag them underwater to eliminate the risk of trapped air, which leads to infections.

Care

Unique Sea Star Physiology: The "shell" of sea star bodies that we see has a thin layer of external tissue overlaying the endoskeleton, which is composed principally of limestone elements (calcium carbonate), called **ossicles**, embedded in an organic and inorganic matrix body wall. The living parts are contained in a series of delicate, lined cavities in which are found all body organs and their principal locomotory, food-catching and manipulation structures/mechanisms, as well as the water-vascular or ambulacral system of tube-feet/valves. Another salient anatomical part of starfish is the sieve-like plate off center on their dorsal surface, the **madreporite**, where the animal draws in water. These structures are in intimate contact with the water about the animal, utilizing molecular pumps to maintain ionic integrity (to get about, feed) and they are extremely sensitive to changes in the physical make-up of their surrounding water.

Acclimation

Unlike many fishes with thick layers of tissue, echinoderms and most invertebrates are *very* sensitive to osmotic differences. A sudden change of one part per thousand of specific gravity in a short time can be fatal to a sea star and underscores the critical importance of slow and gentle acclimation of these animals to new aquarium water. The lesson here: pre-mix new seawater to as close as you can to current conditions (for water changes) and employ long-term acclimation protocols when moving these animals. *Special Note:* numerous aquarists have lost sea stars and urchins to this dynamic with a sudden change in the brand of synthetic sea salt normally employed, or with heavy metal additives to the water (inorganic supplements and medications). There should be a special emphasis here regarding the supply and maintenance of near-seawater conditions in captivity. Stability is crucial in an echinoderm's environment. Although it is true that many species of echinoderm are tidal and hail from dynamic nearshore environments,

Echinoderms: Sea Stars 353

A beautiful *Fromia sp.* star that is frequently available to hobbyists and a wonderful, reef-safe addition once established. This one photographed in Fiji.

most of the popular species observed in the aquarium trade do not. Select specimens from merchant systems that enjoy high and consistent water quality that mimics natural reefs, and strive to provide the same yourself. Appreciable differences should be made-up in quarantine over a matter of weeks in time.

Quarantine all Sea Stars:
It's best to place a new sea star in a sump or refugium if not a dedicated quarantine vessel, for several weeks to ascertain its overall health. Give it time to rest and acclimate to a new aquarium husbandry regime before transferring the specimen to the main system. In a smaller, less cluttered setting, like a sump, refugium, or quarantine tank, you'll be able to assess whether a sea star is feeding, moving at all, or in need of assistance. It can be a tragic mistake to place a new sea star directly into a fully rockscaped display with pitfalls and inaccessible regions where it may suffer or die unnoticed and pollute the system.

Captive Habitats

Essentially, all sea stars need aged and mature aquariums. Potential systems for sea stars should be running for more than six months (as a bare minimum) and will hopefully exceed 100 gallons in volume for most species. Think in terms of large and well established when endeavoring to keep echinoderms. Many popular species (such as *Linckia*) require large areas of otherwise non-competitive space in an aged aquarium to feed upon surface matter and detritus with their extensible stomachs. By *aged* we mean fully cycled, climax community established, mature aquariums, with no major incidents of nutrient accumulation or nuisance outbreaks. The reasons for this are many, not the least of which is a concern for osmotic shock if frequent or large water changes are necessary to control a nuisance algae or other such problem in a flawed aquarium system.

The rigidity of body form for most sea stars should also be a concern to aquarists in some different regards. Habitat make-up must allow for the animal's movements in and around the rockscape, seafloor, and other organisms. With the growing popularity of reef aquariums, tank mates for motile sea stars are a special concern as many sessile residents may be bumped off of their settings by a rampaging specimen (those tube-feet can open closed live shellfish), or conflicts may ensue with stinging animals (corals, anemones and the like) in the narrow confines of the display. Such unnaturally repetitive contact with stinging tank mates is yet another reason for utilizing larger aquariums with sea stars. There are exceptions to these guidelines, like the wonderfully flexible brittle (and serpent) sea stars. These reef-

Echinaster enchinophorus, the Thorny Sea Star. Photographed at a retailer. (*L. Gonzalez*)

354 *Natural Marine Aquarium Volume I - Reef Invertebrates*

Not All Aquarium Denizens are Starfish Safe!

Action (ouch!) photos of a big puffer (*Arothron stellatus*) with a matching appetite. This big boy was spotted out of the corner of an eye on a dive in Pulau Redang, Malaysia. What was really amazing was the *Linckia laevigata* he cleaved an arm off of shortly after taking the first photo! (*R. Fenner*)

safe specimens fare quite well in aquariums including small displays and refugiums and place very few demands on a system.

As slow, sightless animals it is no surprise that many sea stars in their natural habitat are cryptic or reclusive. Some notable species, however, are indeed quite conspicuous or even gaudy. Due consideration must be extended to all in captivity. Some live burrowed in the sand by day, many prefer to hide discreetly in crevices in live rock.

There are, in fact, many unique relationships among sea stars with both living and non-living substrates. This diverse group of echinoderms has evolved numerous specific relationships with sponges, zoanthids, plants, and corals. Some will spend their entire lives encased within a porous hard substrate, their filter-feeding arms stretching out from the safety of their holes in the rock or stony lair, while others have evolved in perfect camouflage with their commensal host or residence (Gorgonians, Neptheid branches, etc.).

Feeding

Sea stars are found in almost all benthic and sub-substrate (buried or embedded) habitats in the seas. Some are tiny, living between sand grains, while others are several feet across in diameter. None of them order take-out food to be delivered, so dinner is literally *on the reef.* In determining the food and feeding habits your sea stars, you'll see that these echinoderms are a conspicuous example of form following function.

Don't be put off on purchasing a Sea Star or Brittle Star with missing, or smaller limbs if the animal is otherwise healthy. (*R. Fenner*)

Echinoderms: Sea Stars

Gomophia watsoni, Watson's Brittlestar. Tropical Australia endemic. To four inches across (10 cm). Notably, this species of seastar is a grazer on detritus and algae.

The oral side of a sea star is underneath the body and reflects their bottom feeding habits. In Asteroids the anus is on the top side, while in Ophiuroids it is not separate from the oral/body cavity. Although a single recipe for feeding sea stars cannot be proffered for this very diverse group of animals, one can be assured that suitable food items are benthic organisms (and like substitutes) with a few rare exceptions including the surprisingly opportunistic **piscavore**, *O. incrassata*- the Green Brittle Star. Fodder for hungry starfish commonly includes: algae, mulm (organic film and bacterial slime), sponges, coral tissue, mollusks, bryozoans, tunicates (ascidians/sea squirts), plankton (for filter-feeders), other echinoderms, and other various micro-organisms. Please know that most sea stars will ultimately starve in aquariums, however slow it might be, if they are not deliberately fed. They are large animals that need larger amounts of food than the seemingly self-sufficient grazing gastropods (snails and the like), by comparison.

When target feeding sea stars, place food items near the arms but not under them, and let the animal crawl toward the food. Placing food directly under a starfish may elicit a flight response from the sudden handling. If the animal does not respond appropriately to food nearby it is quite possibly in dire straits. As mentioned before, sea stars are not as brainless as they might seem, and often can be trained to migrate daily to a feeding spot if the aquarist is diligent and timely in feeding schedule.

You will likely want to select specimens (as your best bets) for either generalized bottom feeding behaviors (detritivores) or omnivorous feeding behaviors that you can and will specifically target feed. Be careful though, if you keep other invertebrates in your aquarium, that you don't select a specimen that will prey on your other desirable species (which for some species can include eating most everything including other sea stars) Most sea stars are true omnivores. Species that feed upon bivalves demonstrate a most impressive display of patience and power when pulling apart the equally formidable opposing shells of a Tridacnid, for example. Even less conspicuous, but equally devastating in aquariums, are species that silently consume gastropods (snails) which can make a tremendous impact of the health and balance of a system. The Asteroids are close to, if not the highest, ranking group of predators on the floors of the world's seas. They are very important in determining the make-up and presence of the living seascape. As such, the living decor of a mature rockscape can be changed rather quickly by a hungry sea star.

It is commonly thought that all Asteroids feed via a cardiac stomach (eversing and covering hapless bivalves or corals, as in the case of the Crown of Thorns Starfish, *Acanthaster planci*). But many species have more or less specialized mechanisms and preferences. With perhaps more than 6,000 species of Echinoderms recognized, and numerous feeding adaptations within, aquarists will want to research the very specific needs of each specimen thoroughly before committing to its inclusion in an aquarium system. Use the guidelines here as a starting point for narrowing the field of candidates.

Asteroids: The most common error in selecting from this group is acquiring species that get too large or are ravenous predators. Not only will many species attack various types of favored reef invertebrates and fishes,

Pentaceraster cumingi, (formerly *Oreaster occidentalis*) the omnivorous Panamanian Horned Sea Star at a retailer.

but in aquariums they often cannot consume them in enough quantities to even survive. Unless you are willing to make a special effort to house and feed the larger, predatory species, it is best to stay with reef-safe choices, of which there are far fewer in this family compared to the Ophiuroids (brittle and serpent stars). Asteroids are generally omnivorous with a carnivorous tendency to eat desirable reef aquarium invertebrates. These spiny-skinned sea stars are best kept in mature, large (over 100 gallons) fish-only and species-specific displays because of their indiscriminate feeding habits. The generous inclusion of large portions of live rock (rotated with fresh pieces regularly) for them to graze upon is highly recommended. Staple food items can also include naturally occurring and prepared greenstuffs (processed spirulina and nori/seaweed). However, meaty foods of marine origin are the primary staple (mysis, gammurus, clam, shrimp, krill, and more).

Ophiuroids: are some of the easiest sea stars (excluding the Gorgonocephalid Basket Stars!) to keep and feed. They are successful and fairly indiscriminate detritavores that use their mucus-covered, serpentine arms to trap micro-organisms and plankton. Some employ a benthic feeding strategy, while others filter-feed almost exclusively by fishing with their sticky arms waving in the currents. Most species in this class are very reef-safe, and desirable additions for aquarists with mixed invertebrate aquarium systems. Most will accept supplemental feeding of sinking dry, prepared and thawed fresh meaty foods. Sinking greenstuffs are also accepted: algae, plant matter and similar substitutes of spirulina and seaweed, but limited terrestrial plant matter.

Crinoids and Gorgonocephalids: require little discussion within the scope and address of this text. These delicate creatures are **obligate** planktivores feeding on suspended matter from the water column. They use up to 20 arms (varying by group) adorned with numerous ancillary side branches (pinnules) to catch food at night. Difficulties with this species in captivity abound. Crinoids will not feed if currents are less than ideal, they only feed at night, and they require copious amounts of plankton that present enormous challenges to water quality for aquarists. As with most filter-feeders, particle size of prey is a very critical matter and the provision and verification of effective prey by casual aquarists is quite difficult. Even attempts to culture live baby brine shrimp or marine rotifers for daily live feedings fail to keep feather or basket stars alive, in part due to the severely limited diet (if not the tedious burden of food culture on the aquarist). Because of the often unsatisfied or unknown requirements of a given Crinoid species in captivity, unhappy specimens are often incited to relax their cirri and set off to swim away by flapping feathery "wings" (arms) in the current. A swimming Crinoid in a mixed invertebrate display is at serious risk of being carried into a cnidarian (stinging coral or anemone), or to meet a certain death at a pump intake in almost any aquarium. Again, we cannot begin to stress the inappropriate nature of collecting and keeping Crinoids for casual consumption by the ornamental pet trade. These magnificent creatures belong in the ocean, or rarely, in species-specific research systems.

Compatibility

As a rule, sea stars are compatible with most popular marine creatures that won't eat them. Toothy fishes

Ophiotheia danae, in the Red Sea.

The Basket Star: *Astrophyton muricatum*, in a typical daytime spot; curled up inside a brown sponge. This one in Belize.

Fromia indica, the Indian Brittle Star. Range includes the Indo-Pacific to the seas of Japan. Need mature aquariums with plenty of green algae. To nearly four inches (10 cm) in diameter.

and inquisitive reef bullies such as triggers, puffers, many eels, some elasmobranches, larger wrasses and Pomacentrid angels should all be avoided when keeping starfish. Other large or uncommon aquarium inhabitants should also be avoided, if for no other prudent reason than the inherent risks of great disparity in sizes. Small to medium popular community-type marine fishes are mostly compatible with sea stars. On the other side of the coin, a few interesting sea stars need to be respected and likely avoided for their piscavorous appetites. Biscuit or pillow seastars have a decidedly carnivorous appetite and the spiny Green Brittle Starfish (*Ophiarachna incrassata*) is a bold fish murderer! These crafty fishing sea stars arch and prop themselves up to make a hollow pocket in wait for weary, hapless fishes to crawl underneath for a nap.

Special Notes

One of the most fascinating aspects of keeping some types of sea stars is the chance to get a rare glimpse of a mysterious and luminary phenomenon: **bio-luminescence**. Bio-luminescence is a stunning emission of light that can be produced by many creatures large and small, although most are marine in origin. Fireflies are perhaps the most familiar exception and example of bio-luminescence to the masses. Other insects, fungus, dinoflagellates, bacteria, fishes, and some invertebrates are also capable of such illumination. Radiance is generated in most of these organisms chemically, although bacteria are prominently recognized for generating the luminous glow in light-organs found in squid and some fishes like the intriguing Anomalopid Flashlight fishes. Explanations for why this phenomenon occurs vary. In fishes it is thought to be employed to attract food or mates. Other organisms like sea stars seem to broadcast it suddenly in a fright-induced event, leaving us at a loss for a definitive categorical reason. Aquarists have reported the activity most often in small Ophiuroid brittle/serpent stars associated with live sand and rock (including but not limited to *Amphiura* and *Amphipholis* species). Tiny incidental stars found at the bottom of shipping bags and boxes (having abandoned live rock or other porous substrates with corals, sponges and the like) have been observed to flash suddenly and quite brightly with iridescent color when attempts at scooping them out of the vessel stimulate them. Industry professionals charged with handling and unpacking of late night shipments of such livestock are often treated to this show. Aquarists have also reported the activity on occasions when an aquarium is fed or disturbed late at night, lending to a magnificent

Possibly *Leiaster speciosus*, which is usually more red-colored than this aquarium specimen. Broad Indo-pacific distribution. This beautiful sea star feeds primarily on detritus and will usually accept prepared foods. Aquarium photo. (*L. Gonzalez*)

Ophiothrix suensonii. Like a few other brittlestars, this small Caribbean resident is a suspension feeder, capturing tiny organic particles. It is rarely seen in the hobby. Cozumel photo.

miniature display of living fireworks in a dark tank or in a dark room.

Stars on Parade: The Best and Worst

When looking for the *best bets* with sea stars, most aquarists should seek the **microphagous** species (smooth armed Asteroids and most all Ophiuroid brittle stars). They present the least risk to other tank inhabitants and will live peacefully for many years once established. Such microphagous species feed readily on detritus, algae and small invertebrate animals, and place few demands on a system. On the contrary, **macrophagous** sea stars (the knobby varieties) should be avoided or handled with caution. Most are consummate omnivores with occasion to behave like voracious predators. Also, please do not take the names commonly ascribed to newly imported sea stars for granted. Many sea stars are generically labeled as with the popular genus *Linckia*, which many aquarists recognize and trust by name. The same cannot be said for all known and less common genera such as *Fromia, Tamaria,* and *Leiaster.* In the interest of moving livestock expediently (if not out of an ignorance or unwillingness to properly identify said animals) many similar starfish are labeled as "*Linckias.*" In such situations, an *educated consumer* will make the difference between selecting an appropriate animal, or passing on an inappropriate one. *Caveat emptor!*

Best of Sea Stars

Amphipholis (circumtropical) and like species: **Live-bearing Mini Serpent stars** are self fertilizing hermaphrodites that prosper very well in captivity and have great benefits to marine aquarists. Because they brood their young, they reproduce quickly and quite successfully with little more than good water quality. They are effective detritavores that need little or no target feeding in systems with a moderate feeding schedule for fishes or corals; incidental food matter and inevitable detritus will sustain them well. Left unattended in a fishless refugium, mini stars are prolific and innocuous guests that contribute admirably to the balance of a healthy reef aquarium. Mini stars can be target fed and deliberately cultured, and aquarists are encouraged to share samples with each other for successful refugiums and displays. This desirable species has the wondrous attribute of being a bio-luminescent variety.

Ophiactis and like species: **Striped Micro Brittle stars** are popular reef denizens that are commonly observed hiding in live rock and other porous substrates, including living sponges and other sessile reef invertebrates. They are commonly imported with live rock from the Pacific and with clumps of algae from various locales. If a healthy aquarium is stocked slowly and without reef ravaging fishes, micro stars will populate and flourish. Evidence of their success and adaptability in aquariums can be seen as they adorn live rock and plant matter with a show of numerous arms peeking out from the substrate while filter-feeding in the current. Such species are generally quite small, with an adult size of less than 1" (2.5 cm). The arms are often but not always striped, and number five or six. Six armed varieties have been observed to reproduce by asexual fission by splitting in two and regenerating the missing half. These sea stars are quite prolific and utilitarian in reef refugiums and displays.

Common Ophiurids (circumtropical) Serpent and Brittle starfish are represented in the trade by numerous species from around the world. Adult sizes range from less than one inch (2.5 cm) to well over one foot (30 cm), however, 4-6" individuals are typical. They occur in a myriad of colors and patterns, but most shallow species are muted or mottled in color.

Linckia columbiae, the Fragile Seastar. Semitropical *Linckia* species found in the eastern Pacific. This one off of San Diego, California is in the process of sampling/eating a sea anemone.

Echinoderms: Sea Stars

Archaster typicus, "White (or burrowing) Sand Star." A popular and useful sand-sifter. (*L. Gonzalez*)

Uniquely patterned and popular varieties stand out, like the Harlequin sea star, *Ophioderma appressum*, from the Atlantic, which occurs in a striking pattern that is nearly black and white banded and reticulated. Equally dramatically colored and patterned species occur that display lovely shows of red, yellow, orange, cream, nearly white, striped, and spotted. Many arrive with regularity to the USA in imports from Tonga, Fiji, and Indonesia. But as a rule, burgundy, black, and light and dark brown colors dominate. Specimens labeled as "common" are often collected easily (nearshore, shallow access and/or in abundance) which, if handled properly, indicates a more likely suitable aquarium specimen. Such sea stars may be subject to shorter handling time and shipping routes, which predisposes them to make a smoother acclimation to captivity. Once established, Ophiuroid stars are rather hardy and long lived. They are some of the most useful reef scavengers for aquariums as indiscriminate omnivores and detritavores, while posing little threat to other desirable tank mates.

Ophionereis reticulata, the Reticulated Brittle Sand Sea Star: a common, hardy, and utilitarian species from the Atlantic. This delightful species is usually seen in an attractive range of light colors with short-spined arms. Commonly found in the sand and under rocks and coral, the Reticulated Brittle starfish is long-lived, low-maintenance and has even been known to reproduce in captivity. A fine detritavore for the marine aquarium.

Archaster typicus (Indonesia): The **White** (Burrowing) **Sand star** is a handsome, hardy, and utilitarian sand-sifting sea star. They are coveted by some reef aquarists for their extraordinary effectiveness as detritavores consuming algae, debris and micro-organisms from the sand bed surface. *Archaster* stars usually remain buried by day and rise at night to feed. They are very efficient at keeping the surface of a large sand bed nicely stirred and free of nuisance

Gomophia watsoni, Watson's Brittle star, is an Australian endemic (to 4"/10 cm) and highly variable in color.

360 *Natural Marine Aquarium Volume I - Reef Invertebrates*

Acanthaster "Crown of Thorns" stars should be handled only with a net and then only carefully. The spines evident in this photo are sharp and accompanied with a hemolytic (blood-cell splitting) toxin. This is a similar situation to that posed by the handling of many commercial sea urchins that many people handle haphazardly. Be careful! Here is a close up of a specimen in Pulau Redang, Malaysia... you get my point?

algae such as brown diatoms. Their efficient skills as detritavores can be a significant obstacle though, in smaller aquariums or very well "scrubbed" systems. A mature sand bed at some appreciable depth (over 4"/10 cm), in larger aquariums (100 gallons per specimen), is highly recommended. As with all burrowing animals, please make sure your rocky habitat is well founded on the bottom of the tank and not merely on the top of the substrate, for fear of rockslides.

Worst of Sea Stars

Pentaster and ***Protoreaster***: The thorny Pacific sea stars: Members of these genera are quite exemplary of the standard and stereotype for Asteroids. Most are voracious omnivores or decided predators. They categorically are not to be trusted with reef invertebrates and commonly eat corals, sponges, snails, and especially clams. *Pentaster* and *Protoreaster* tend to be imported large, grow large, require large aquaria, and need large filtrative faculties to manage their large appetites. Too many slowly starve in captivity over a period of 10-18 months. Although many species are quite hardy once established, these sea stars require large aquariums and dedicated displays to address their special needs (with live rock and like considerations for rockscape, substrate, water flow, and sufficiently high oxygen levels). Regular target feeding is a forgone conclusion and daily offerings may be necessary for some species. Some are suitable for beginners, but none are appropriate as unfed scavengers or ancillary decorations for the marine aquarium.

Asterina folium and like species: This tiny, asymmetrical sea star perhaps does not belong in either the best or the worst categories of sea stars, and is surely misunderstood. Nevertheless, it finds its way here into the less favorable category if for no other reason than for its inclination to be excessively prolific. *Asterina* has been accused of grazing some coral, which surely occurs in some cases. More often than not, though, they graze necrotic tissue (mistaken for killing an already dying coral) and algae, and even large populations can been maintained in aquaria with absolutely no harm to corals - effectively reef-safe. However, even when well behaved, the very likelihood of rapid proliferation makes this sea star a nuisance of possibly plague proportions, and amnesty from grazing live corals ultimately cannot be assured. It is unfortunate because they are by and large dedicated algae grazers and quite useful otherwise in aquariums. They may even be cultured as suitable prey for the Clown Harlequin Shrimp, *Hymenocera sp.* In refugiums and fish-only displays, this sea star can be a fascinating and effective scavenger. Monitor the population of this starfish in all aquariums.

Oreaster (Caribbean): Some members of this common genus roam the turtle grass beds of the tropical west Atlantic instead of hard substrates (the coral reef) looking for tasty sponges or mollusks. Because of its specialized diet, voracious appetite and large adult size (growing to 20 inches/50 cm in diameter), it is not recommended for casual aquarists. Like others in this fascinating class (Asteroid), they possesse a cleverly evolved arsenal of hydraulic tube feet connected to an elaborate water-vascular system that encircles the animals' mouth, and extends via five radial canals down the

A *Pisaster ochraceus*, the Ochre Seastar on the move. Those little tube-feet can carry this star at a speedy "many feet per hour." (R. Fenner)

Echinoderms: Sea Stars 361

center of each arm.

Ophiarachna incrassata (Pacific): Many words have been used describe the Green Brittle Starfish including "satanic" and "murderous." Altogether, this sea star actually does have many fine merits. It is remarkably hardy, beautiful, long-lived and behaviorally fascinating. It is also a very thorough and indiscriminate scavenger. Unfortunately, because of its large adult size (potentially spanning over 12"/30 cm) and exceptionally adept skill at catching and killing small fishes, a strong warning must be proffered to aquarists keeping this species. In hunting mode, *O. incrassata* will pose motionless in an arched back formation. The stiff posture creates an attractive hiding place for unsuspecting small fishes to fall prey to death from above when the starfish drops down upon them. When hungry, this brittle star has also been known to attack shrimps, bivalves (including expensive Tridacnids!) and quite a few other animals. The aggressively opportunistic and predatory nature of the Green Brittle Star makes it one of the only Ophiuroid brittle stars unsuitable for reef invertebrate aquariums. Larger,

A gorgeous Red Feather Star *Himerometra sp.*, a beautiful representative of this delicate group. Not recommended as a casual aquarium denizen.

Ophiarachna incrassata, the Green Brittle Starfish: This animal is a predatory fish eater that does indeed do a spiffy janitorial job when small - but grows quickly, and under darkness of night can/does eat aquarium fishes. This species has been documented to arch up in "sleeping caves" of captive fishes and drop down on unsuspecting meals. If you keep this species, watch it carefully, and keep a count on your piscine livestock.

reef-safe fishes, however, make fine tank mates for this frisky feeder. For fearless aquarists, enjoyment of keeping this animal can be multiplied by its intelligent behavior and habits as it responds quickly and favorably to training with food. Yes, it can even be hand-fed!

Ophioderma squamosissimum (tropical Atlantic): The Red Serpent Star (Bahama star) only makes this list of "worst" starfish by virtue of its exceptional sensitivity to the slightest changes in water quality (pH, bacterial count, low oxygen, etc.). Collection and shipping imposes great stress on this species and mortality is rather high. Once established in captivity, it can be a long-lived aquarium denizen. Nonetheless, the Red Serpent Star demands regular feedings and a very considerate diet with great variety of matter. The highest water quality and stability is necessary to have any success with this species. Large mature aquariums are strongly recommended. Aquarists should observe this species in a merchant's display for two or more weeks after import. Quarantine, extra feeding considerations, and observation for vacuolations or necrotic areas will be necessary for two to four weeks before moving any new specimen to a display aquarium. Never place a newly imported or otherwise freshly acquired specimen directly into a mixed reef garden display.

Gorgonocephalid Basket Star, **Crinoid** Feather Stars and Sea Lilies: Most aquarists are unable or unprepared to provide adequate husbandry for these filter-feeding sea stars. Raising children is arguably less work than keeping Feather and Basket Stars. Although some folks have reported reasonable success in their husbandry, these nocturnal animals are by no means easy to keep in captive conditions. They require very specialized care, large aquariums, and tedious diligence in their food preparation and administration. Species-specific tanks are necessary as many fish and invertebrates (some mollusks, many crustaceans, especially crabs, and other starfish) will harass or consume these vulnerable sea stars. Due to the vagaries of cost and current collection and transport techniques, the stalked and attached Sea Lilies specifically are rarely offered. When they do show up in the trade, they are generally already dead or nearly so from being thrashed about.

Advanced aquarists and researchers interested in attempting to work with such species should consider that plankton reactors and aged refugia (prolific and fishless) may need to be established inline for a year or more in addition to reliable live food culture to begin to have a chance keeping these stars successfully. Additional target feeding is a forgone

Linckia laevigata (Linnaeus 1758). Blue and greenish ones in Fiji. Also found in other colors, brown, tans, violet to burgundy, even mottled, and there are other species of the genus offered to the trade. This animal is very often doomed from the retailer to aquarists... having suffered too much damage and neglect in the process of collection, holding, shipping. Look for damage (seen in photo above) and avoid such obviously poor specimens. In the wild, this is an algae, bacteria, detritus feeder that needs space (hundreds of gallons) and mulm (muck, dirt, call it what you will), on the bottom of its system to survive.

Brooding and "live-bearing" miniature starfish have become delightfully commonplace in the refugiums and deep sand beds of modern marine aquariums. These tiny Ophiuroids (brittle starfish) are remarkably useful and peaceful scavengers with an adult size usually measured in milimeters. This "jumbo" individual (1 inch, tip to tip) was photographed climbing through *Caulerpa* in a seahorse display at a retailer. (*L. Gonzalez*)

Culcita novaeguineae, the Bun Starfish or Pincushion Star. Eastern Indian Ocean, Western Pacific. To 12 inches (30 cm) across. This animal requires large quarters with plenty of open space and feeding of bivalves, snails, fish meat, tablets and may eat your corals! Photographed in Indonesia.

conclusion (cultured rotifers, *Nauplii*, and other live plankton). They may need darkness or subdued light for successful feeding. Finely tunable apertures on multiple effluents of the water moving hardware of the system are recommended to adjust the path and power of the currents for these most demanding creatures. Most seem to be **positively rheotactic**: orienting themselves in the path of water flow to effectively filter-feed. Feeding attempts have included live cultured phytoplankton, rotifers, HUFA (highly unsaturated fatty acids) soaked brine *Nauplii* (too large for many species), *Cyclops*, diatoms and detritus suspensions from stirred substrates. Continuous food drips through the nocturnal photoperiod may be necessary and are certainly helpful. Superior water quality is essential to prevent unfavorable autotomy (self-breakage) in these sea stars. It is one of the most challenging obstacles for an aquarist to provide heavy or continuous feeding suspension (embattled against filter dynamics) while maintaining the highest water quality standards. Daily water changes are commonly employed in attempts to realize success with Crinoid and Gorgonocephalid stars.

Most Feather stars, Sea Lilies and Basket stars will not arrive at their final destination (retailer/consumer) due to traumas encountered in collection, transport and from starvation. These animals should not be touched or moved unnecessarily; it is possible to nudge them along or even move them by getting the specimen to perch on a wood dowel, net handle or like instrument to move them underwater into a sealed container. Categorically, none of the filter-feeding sea stars are recommended for casual aquarists under any circumstance. Advanced aquarists are advised to pursue the captive study of these creatures only with the utmost respect, preparatory work and prudence.

Acanthaster (Indian Ocean/Pacific): The many reasons for the Crown of Thorns Sea Star making the "worst of" list of starfish are painfully clear: literally and figuratively. The extraordinary spines are superficially a great barrier to predation in the wild and necessary handling in captivity. They should always and only be handled with a net (and even then only carefully). Furthermore, the wicked spines are quite sharp and packed with a **hemolytic** (blood-cell splitting) toxin. To top off the list of offenses, the Crown of Thorns sea star is a voracious predator that has become notorious for devastating reefs and negatively impacting entire localized ecosystems. Most aquarists are familiar with stories of the obligatory hand extraction of this scourge from reefs since higher predatory members of the food chain have been removed. Amazingly, specimens are still

Fromia sp. In the Red Sea.

Fromia elegans: A brown individual in Fiji.

imported into the aquarium trade. ***Choriaster*** is another too-common offering. *Choriaster granulatus*, the "Doughboy" Sea Star, is a big, bulky Indo-Pacific asteroid that scavenges in reef shallows. It should only be employed in systems of hundreds of gallons size and kept away from sessile invertebrates, as it is a coral eater. Realistically, few aquariums are large enough or prepared to support this sea star's special physical and dietary needs. For advanced aquarists and specialized displays.

Culcita, Pentaceraster and like species: Summarized at the top of this chapter, there's not much more to say in explanation of why this animal makes the list of "worst starfish" for captive care by most aquarists. This is truly an animal for a species-specific display where its needs are addressed acutely. Known as the Sea Biscuit, Pin Cushion and Pillow stars, these Asteroids are large, indiscriminate feeders that consume almost anything in their path. They need huge displays and powerful filtration systems to support their large appetites.

Linckia, Leiaster, Tamaria, Chaetaster, and ***Ophidiaster***: Much has been said in this chapter about the delicate nature of *Linckia* and similar species. Indeed, there are some species and individual specimens that are hardy and long-lived in aquaria. Most, however, are moderately to very difficult to keep successfully. Many obstacles befall these sea stars along the chain of custody, from collection to residence in captivity. Some have very specific and challenging diets to replicate (sponge, epiphytic matter, and bacterial slime). Many commonly are imported with parasites, and most all are rather sensitive to water quality in shipping and display. Any attempts at keeping these species must begin with a 2-4 week observation period in a dealer's display or better: a quarantine tank. Feed heavily when possible and relegate captive care and study to large, mature aquarium systems with plentiful supplies of live rock. These genera are best reserved for intermediate to advanced aquarists. *Fromia, Celerina, Neoferdina, Ferdina* and *Gomophia* species are somewhat like-bodied and occasionally mistaken for *Linckia* species. They are highly variable in suitability for captivity, with most being somewhat to very sensitive like the afore-mentioned *Linckia*. Approach captive care as you would with *Linckia* and take all precautions in handling and husbandry for your

Echinoderms: Sea Stars

Linckia laevigata, in an apparent state of repose on a bed of *Caulerpa racemosa* in captivity. *Linckia* are found in many colors like brown, tans, violet to burgundy and even mottled. Many other attractive species of this genus are also available to the hobbyist. (L. Gonzalez)

best chance at success here.

Nardoa (Indian to Central Pacific Oceans; Africa to the Philippines): Especially when started small, these species tend to get along well with most invertebrate livestock. Unfortunately, they are exceptionally difficult to import at present with high rates of morbidity and mortality. Large aquaria with a plentiful supply of live rock are highly recommended, if not necessary. Occasionally rotating live rock with fresh pieces and minimizing the number of competitive deposit feeders in the display is also recommended.

Echinaster (circumtropical): Several species in this genus from around the world are commonly seen in the trade and they may appear to be quite unrelated to each other in gross physiology. Although some species in this genus are hardy and long-lived in aquaria, more *Echinasters* are notably delicate to ship and feed. *Echinaster echinophorus*, a thorny red/orange species from the Atlantic, is all too commonly imported and popular for its beauty and typically modest price. Unfortunately, this animal lives in seagrass beds and eats sponge – an environment which is neither commonly nor easily replicated in the home aquarium. As such, most starve and die within months. *Echinaster* are also inclined to be predatory on clams and anemones or polyps. Few *Echinaster* can be recommended for the average marine aquarium.

Astropecten and **Iconaster**: Categorically, members of these genera are delicate. They require very mature aquariums, deep sand beds with rich infauna, mature live rock and a significant supply of food. Diet for these sea stars seems to be more narrowly defined and as such makes their care in captivity more challenging. With their sensitivity to the imposed rigors of shipping, and forgiving the gross generalizations here, we do not recommend these genera for casual aquarists when so many other beautiful and hardy species are available. Some *Astropecten* are collected in temperate waters off the Carolina coasts and are ill suited for captivity.

Patiria, Pisaster, Tosia, Pentagonaster, and more: (Temperate Sea Stars) A brief mention here of a despicable practice: the sale of coldwater organisms for tropical systems, as with collections off the Northern East and West Coasts (the Carolinas and California for example). Many popular animals fall into this category, including Garibaldi damsels, Leopard sharks, and Catalina gobies. In the case of sea stars, several are seen at least occasionally and need to be avoided. These animals rarely acclimate to warm water conditions, either falling apart within days or stress-starving to death in a few weeks. Avoid them unless you have a system designed (chilled) for their appropriate care. Most need to be kept consistently at temperatures under 69F (20C) in systems with a reliable chiller.

Fromia monilis, in Indonesia (D. Fenner). Right, an orange *Linckia sp.* at a retailer. (L. Gonzalez)

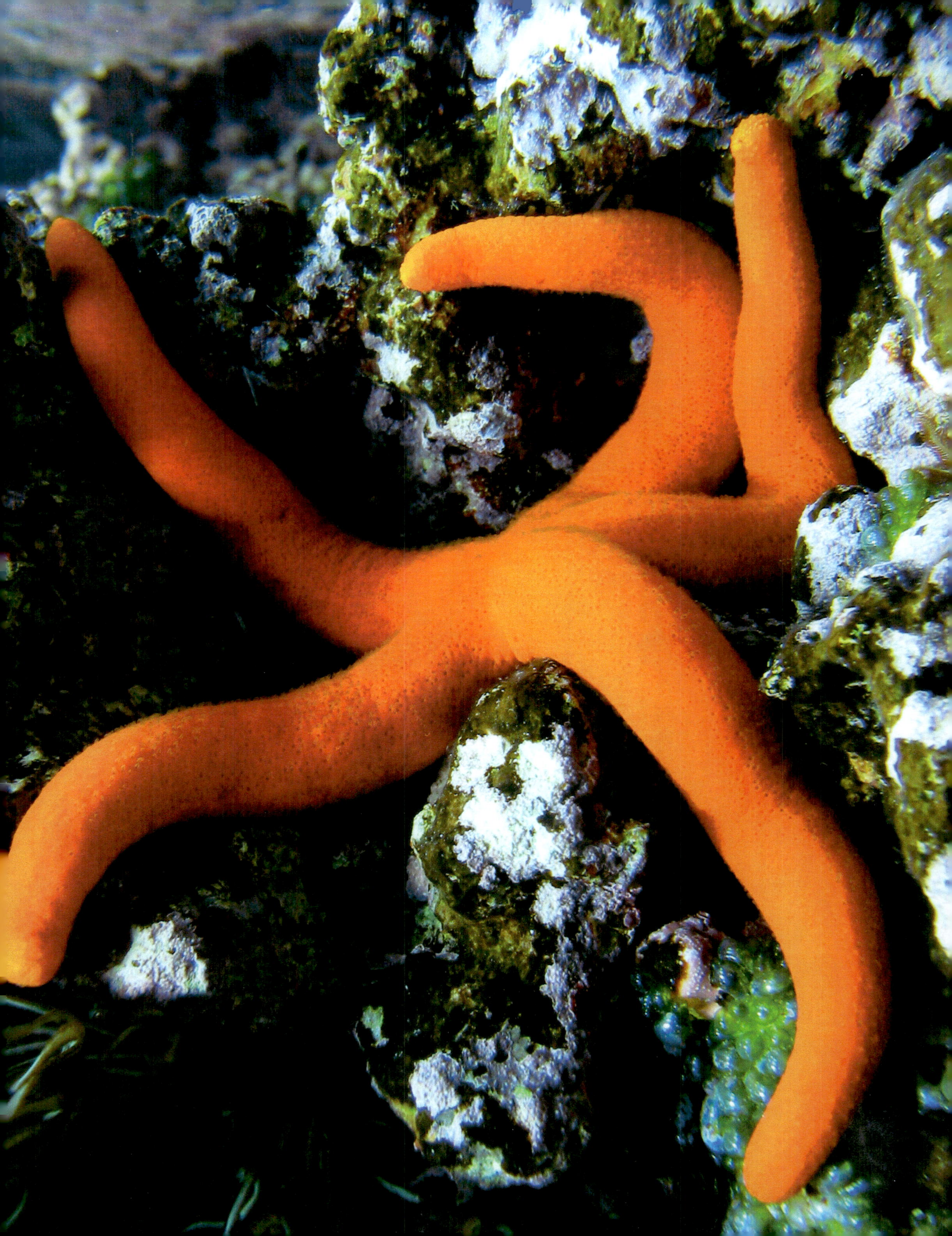

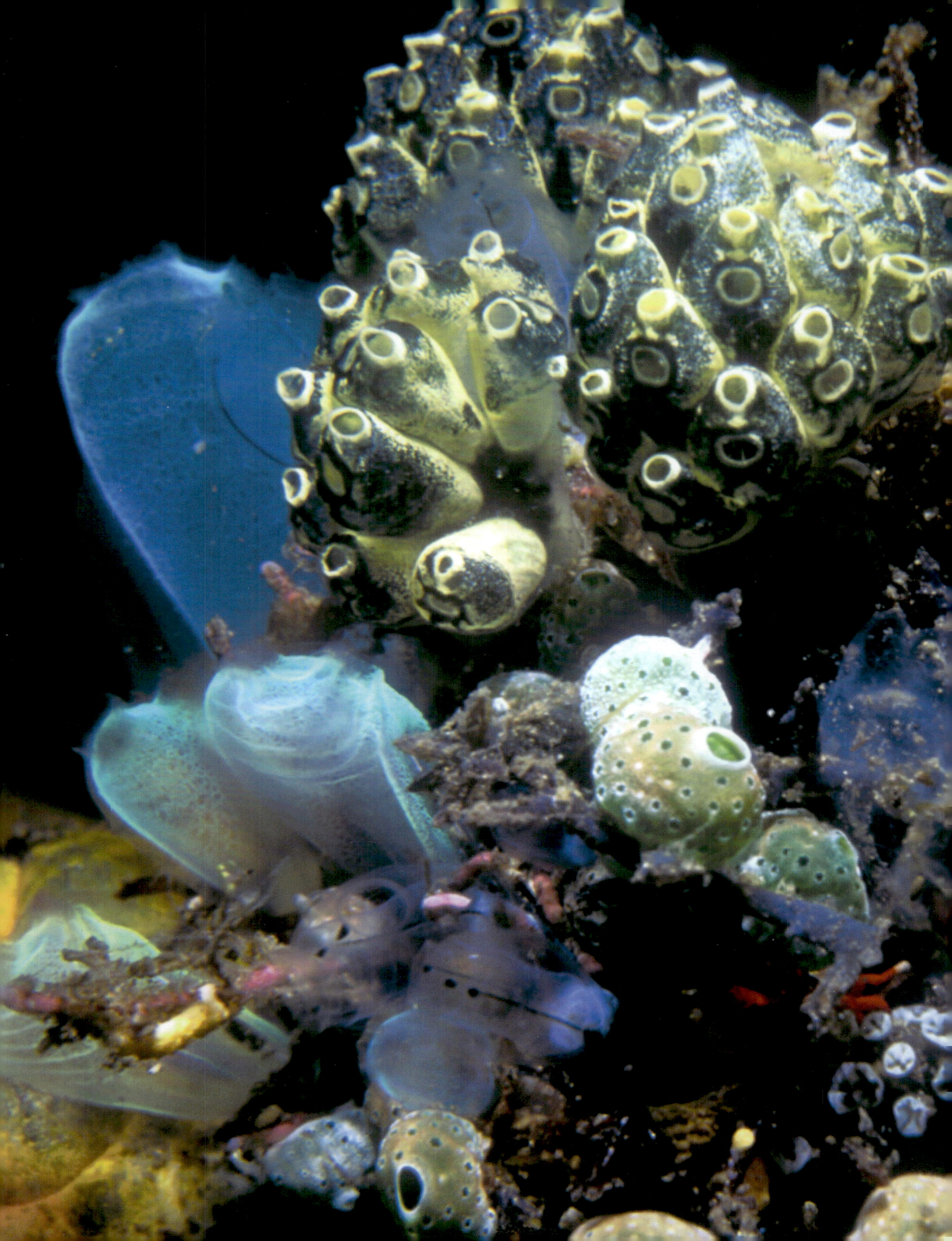

Tunicates, Sea Squirts
The Ascidians

Tunicates are some of the most fascinating and elusive marine invertebrates. Most every aspect of their life, behavior and physiology is unique and interesting, yet enjoyment of these animals in aquariums presents extraordinary obstacles. The simple truth of the matter is that casual aquarists are usually unable and should not attempt to keep most sea squirts in captivity. Sea squirts are best maintained in deliberate aquariums where care is directed toward their optimum study and health.

The fundamental challenges to keeping tunicates in aquariums are as follows:

❖ Many have a very brief natural lifespan of months with most less than 1 year; captive life spans are further abbreviated because good husbandry is undefined.

❖ They suffer high morbidity and mortality due to great sensitivity during shipping.

❖ Most filter-feed on bacteria which cannot be provided reliably or easily

❖ Because of the necessity for heavy and consistent suspensions of microscopic food, it can be challenging to maintain adequate water quality in mixed garden reef aquaria.

There are approximately two thousand described species of tunicates occurring worldwide from shallow to abyssal depths and from tropical to frigid polar waters. The largest, solitary sea squirts attain sizes in excess of one foot long (30 centimeters) and can live for several years. Most species however are quite small, at less than an inch or two in length, and have a natural lifespan of mere weeks to months.

As a striking taxonomic anomaly, tunicates are classified in the phylum Chordata, to which vertebrate mammals belong. Most of the roughly 40,000 species of vertebrate Chordates described are fishes, while the invertebrate Urochordates occupy three classes in a fraction of one percent of the phylum. Of these three classes, only the Ascidians (Ascidiacea) will ever be seen by aquarists, let alone most people (the other two classes are pelagic/planktonic).

These "lower" Chordates are called *tunicates* for the cellulose based "tunic" that comprises their body

Left: A diverse grouping of tunicates including the dazzlingly irridescent **Rhopalaea sp.**, an **Atriolum robustum** and another unidentified **Clavelina sp.**. Ascidians can often be found together with soft corals and other sessile benthic marine life forms. **Above right:** *A. robustum.*

Polycarpa aurata in Bunaken.

Sponge or Solitary Tunicate?

Most aquarists acquire their Ascidians incidentally, as explants on live rock or on hard substrates with other collected invertebrates. These hitchhikers are generally the hardiest species, particularly evident through their ability to travel the chain of custody on import and under duress with roughly handled live rock. Other sea squirts can be procured through businesses as deliberately collected specimens. These attractive species are often beautifully colored or patterned and yet usually the most difficult to support in captivity. Any aquarist considering the acquisition of a tunicate for aquarium life must give due consideration to the extraordinary challenges of keeping this group, and be prepared to address them in earnest.

Tunicate Physiology

The Urochordata don't look like fishes, birds, or mammals for they lack a backbone, but they do share four other important characteristics with the vertebrates. During larval development they possess a tail, a dorsal nerve cord, a dorsal (non-bone) stiffening structure (the notochord), and pharyngeal gill

covering, *sea squirts* for their capacity to forcibly eject water, and *Ascidians* from the Greek for "little body." They are distinguished as immobile, sessile creatures after they pass through their larval stage. That is to say, they spend the rest of their lives firmly attached to a hard substrate. As such, they are completely dependent on the surrounding environment for sustenance and life support, with currents to bring necessary nutrients and to carry waste away. Since aquarists know little about what these animals eat, let alone how to deliver it, at what size or in what portions, it stands to reason that deliberate culture of tunicates should be left to the dedicated few.

Tunicates live rather discreet and often obscure lives as sessile invertebrates. They are often disguised by incidental growths of living sponge, algae and bryozoan camouflage. But they are a major part of the "cryptic fauna" of live rock, often making up a sizable proportion of this product (second only to sponges). Although they may not be easily recognized, tunicates are among the most common and important marine invertebrates in reef systems serving a significant role in the sea *en masse* as tremendous filter-feeders.

Tunicates occur in both simple and compound formations: solitarily or colonial. The solitary and stalked species are commonly misidentified as coral (like the so-called "Lollipop coral": an *Oxycorynia/ Neptheis* tunicate) and colonial species are often mistaken for other encrusting reef growths like sponges.

Sponge or Solitary Tunicate?

Tunicate:

- ✓ Responds rapidly to being touched (closing apertures/openings)
- ✓ Two large openings (one atop and the other just slightly lower), or one large opening atop and many small holes around the body, systematically patterned
- ✓ Inward and mobile tentacles lining the apertures

Sponge:

- ✓ Slow or totally unresponsive to being touched (holes are not movable)
- ✓ Spiked outward projections around openings (spicules) but no inward projections
- ✓ Rough or porous texture (sponge) and not slimy/mucous or smooth (tunicates)

Is it a sponge, or a tunicate?

The Row Encrusting Tunicate (*Botrylloides*) at first glance looks quite similar to the Vase Sponge (*Mycale*) pictured opposite. Notice, however, the the smooth rim of this ascidian's large atrial (exhalent) openings in contrast to the sponge. To the touch, you will also feel and notice a difference as the tunicate is quick to respond by closing its apertures and has a slimy texture.

Mycale laxissima, the Strawberry Vase Sponge, can be readily distinguished from a tunicate by its thin crown of spicules at the excurrent openings (osculae). In contrast to the Ascidians, sponges are also very slow to respond to touch if at all.

clefts. A drastic metamorphosis occurs, however, between the larval and adult stages. The result is an adult form that is amazingly simple, compared to the juvenile.

Colonial tunicates are made up of individuals called zooids. These sessile individuals live fastened to the substrate by their cellulose tunics. Some of these clonal species are comprised of more-or-less separate individuals or "zooids," whereas in other species many hundreds or thousands of zooids share a common tunic. The tunic is the sheath-like exterior of the animal or colony and is secreted by the mantle. It is quite unique (in the animal kingdom) because the fundamental composition is a *plant* based cellulose.

The tunic is a noxious and effective barrier, filled with defensive elements that dissuade most predation. Alas, this very attribute is one of the many obstacles to captive husbandry with Ascidians, at their sudden death in an aquarium, and the subsequent degradation of the tunic can lead to a harmful if not fatal environment with the release of toxic compounds.

Solitary tunicates have two conspicuous siphons (incurrent & excurrent, or branchial & atrial) they use to pump water through the gill net in their bodies. In fact, their ability to contract their siphons is the one obvious trait that separates Ascidians from the similar-looking sponges (phylum Porifera), which have permanent openings (as described above).

Eusynstyela cf. misakiensis (Watanabe & Tokioka 1972). Indonesia, New Caledonia, Japan. Zooids about 1 cm. long placed next to each other with a conspicuous white mark between the siphons. Typically orange-red to purplish in color. Photographed in Pulau Redang, Malaysia. (*D. Fenner*)

Tunicates, Sea Squirts

Ascidians have a heart, unique among Chordates, capable of reversing the direction of blood flow! Their primitive heart is a simple tube-shaped organ, which beats a number of times in one direction, and then beats in the other direction to move blood through the necessary spaces (blood channels). It is a most peculiar circulatory system!

Although tunicates are regarded as wholly sessile invertebrates, some species are capable of limited mobility. The so-called locomotion of "motile" tunicates is little more than the strategic directional growth of a new tunic.

Selection

Like Sponges, Ascidians should never be lifted into the air. The pressure of gravity on the unsupported weight of their tissue and the likelihood of trapping air, is often fatal. Tunicates are simply not evolved to handle this kind of exposure. Purchased and transported animals should be handled under water at all times (both during bagging and releasing). Where possible, a close inspection of the tunicate should include a basic test of vigor and health through visual stimuli. Some tunicates can sense light and those specimens that are healthy will promptly retract, closing their siphonal openings. An aquarist will select tunicates much as one shops for cnidarians: looking for rich color, fast response to a stimulus, packing and handling under water whenever

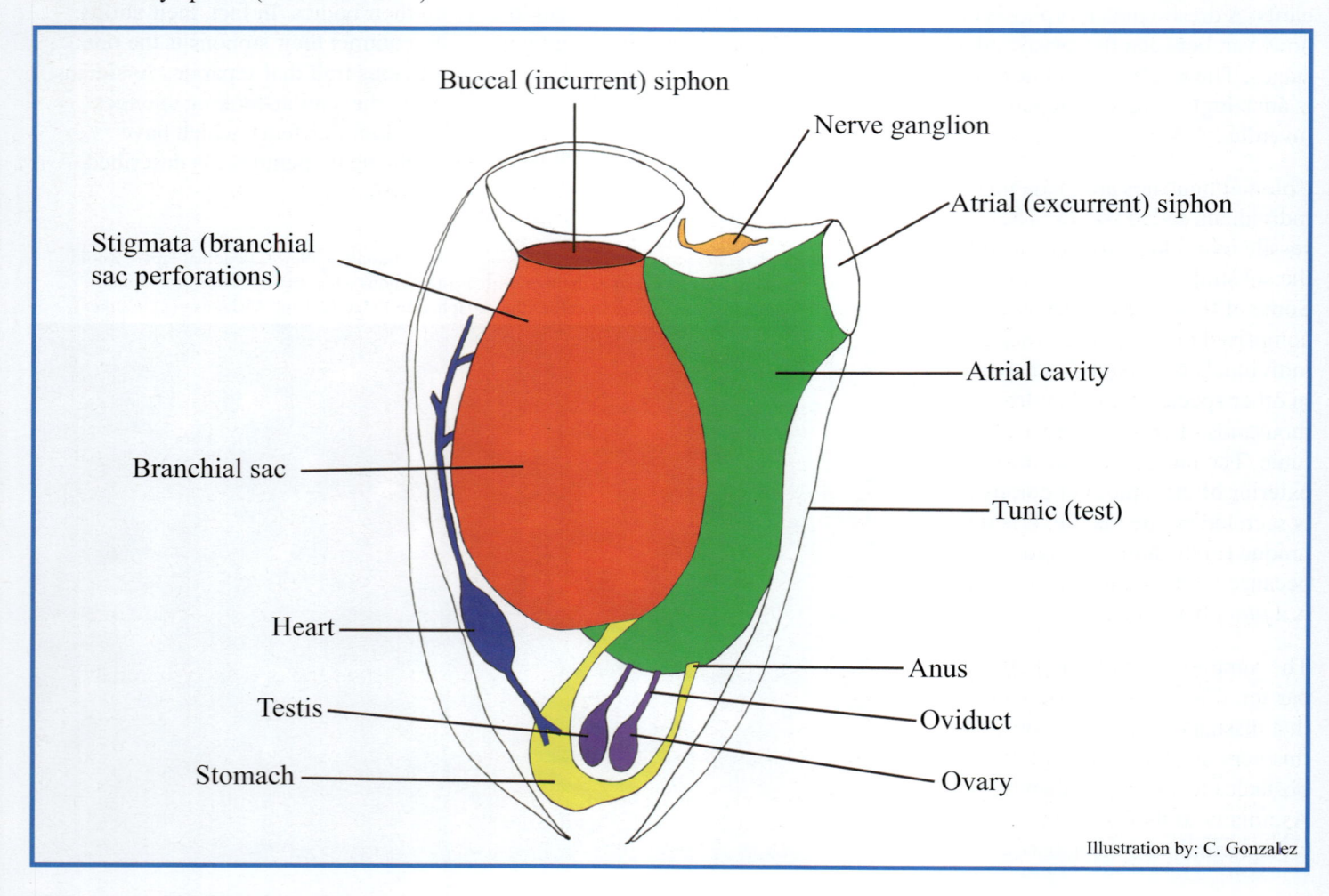

Illustration by: C. Gonzalez

372 *Natural Marine Aquarium Volume I - Reef Invertebrates*

Aplidium crateriferum, (Savigny 1816), Crater Sea Squirt. Tropical Indo-Pacific; Red Sea to Western Pacific. Usually grows under rocks (here a colony is doing well under a *Sarcophyton*). Can be grown in captivity (even reproduced). Photographed in Australia.

possible, and always slow and gentle acclimation to new light and water conditions.

A peculiar phenomenon of tunicates, commonly observed in captivity and not clearly explained, is the shedding of "heads" (on stalked animals) or the complete waxing and waning of colonial tunicates altogether. In these instances, animals have been observed to regenerate lost tissue as with stalks that give rise to new zooids. Although the cyclic occurrence seems to demonstrate their ability to survive, and may even be a natural strategy, it cannot be good in aquarium systems already incapable of meeting their basic nutritive requirements.

Anecdotally, these degenerative events have often been associated with symbiotic species that were poorly acclimated to new light intensities or have been receiving improper illumination altogether.

It is important to select small individuals, and only those that are firmly attached to hard substrates. Although loose tunicates can be glued to a new substrate, this action imposes unnecessary stress. Large individuals do not seem to survive the collection and transport process as well as small ones. Additionally, and from a practical point of view, larger specimens will need far more nutrition than is found in a typical aquarium. Feeding on bacteria is a biologically limiting strategy in captivity. In this regard, tunicates and sponges are very much alike. Typically, the success or failure with one group will bode likewise for the other. Any aquarist with an aquarium over six months old with little development of natural and incidental sponges should probably resist keeping tunicates.

Care of Tunicates

The three essential parameters for keeping tunicates are food, water flow

Tunicates, Sea Squirts

Atriolum robustum, (Kott 1983). South Africa to New Guinea, Philippines. Small urn-shaped zooids of about an inch in height. Large and few oral siphons. Most are rust-orange or greenish shades. Often confused with ***Didemnum molle***. The color is due to the endosymbiotic algae ***Prochloron***.

and light. Some Ascidians sponsor *endo*symbiotic species of algae within their tissue (like photosynthetic coral and anemones), but more are aposymbiotic, preferring to live in the dark or at least the shade. Careful observation of the other life on and around collected specimens will provide insight into what type of environment they favor. Some sponges or coralline algae species are easily recognizable as "shade-dwelling," while the presence of a symbiotic polyp or coral suggests something to the contrary. The species that are partly supported by endosymbiotic algae are easier to keep in captivity and numerous aquarists have already succeeded in growing and propagating some of these with typical reef aquarium care. Symbiotic species favor moderate to strong illumination but must be acclimated slowly, especially to very bright light (metal halide). VHO and PC fluorescent lighting systems may be ideal for keeping tunicates in less than 30" (75 cm) of water at depth.

Adequate water flow is perhaps as crucial as feeding because it is the key factor in food delivery and waste disposal. Moderate to strong water flow is highly recommended for all tunicates in aquariums (rather critical for stalked species). They also seem to favor, or at least tolerate, laminar flow where many cnidarians will not.

Feeding

Feeding tunicates is crucial to their successful care in captivity and is the subject of tremendous "myth-information" in the trade. Aggressive marketing of food products (like bottled "live" and "semi-live" phytoplankton) have been flawed attempts if not outright exercises in demagoguery, when directed at keepers of tunicates. While phytoplankton is a fundamental foodstuff in the reef ecosystem (and a benefit to gorgonians and some soft corals), tunicates apparently do not favor these algae. Most tunicates feed predominantly on dissolved organics and bacteria. Offering phytoplankton is unlikely to benefit tunicates directly and perhaps at best, serves only as organic fodder for decomposing

Polycarpa sp. Indo-Pacific; Australia, New Guinea, Indonesia, Philippines, the Micronesians. Bunaken photo.

bacteria, a tunicate's real food. The regular feeding of phytoplankton may be no more beneficial than feeding the aquarium a hamburger, as either in decay will increase the density of the microbial population and levels of dissolved organics in the system. At any rate, a healthy deep sand bed will metabolize available organic matter and help feed tunicates and other filter-feeders somewhat indirectly.

There are many similarities between tunicates and sponges regarding captive husbandry. Much like sponges, tunicates are strict filter-feeders requiring specific food particle size and specific type. They require food particles smaller than anything likely prepared or from a bottle, especially without mechanical manipulation (Note: always whisk bottled phytoplankton in a high speed electric blender to reduce particle size for any invertebrates experimentally fed with such matter). As well, both are quite "hungry" as reef invertebrates and may exceed an aquarium's ability to generate suitable natural bacteria and nanoplankton. They are truly extraordinary filter-feeders able to cycle tens to hundreds of gallons of water daily through their systems.

To capture desired nanoplankton (colloidal material, bacteria, etc.) tunicates utilize a very fine mucous net. Food trapped in the mucus is transferred by ciliary action to the digestive gland. Most tunicates do not appear to have any specialized excretory or osmoregulatory organs. Some waste may simply be passed out of the excurrent siphon. But they are also known to deposit nitrogenous wastes as solid crystals of guanine

Upper right: *Didemnum molle* - Usually 4 cm high and 2-4 cm in diameter but may reach diameters of 10 cm (4"). Photographed in Queensland.

Lower Right: *Didemnum cuculliferum* - Common on reef flats and slopes in the Philippines, New Guinea and here in Fiji. Membranous, spreading colonies that are pink to magenta in color, sometimes with light to white areas around their cloacal siphons.

or uric acid in the tunic (storage secretion). This action imparts a dusty look to the tunicate.

Hungry Ascidians benefit from the regular stirring of the sand bed in aquariums. Well-seasoned systems with refugiums, less-than vigorous protein skimming, and the healthy feeding of a strong bio-load (fishes and/or corals) are probably a boon to these efficient living filters. With reasonably good water quality otherwise, the provision of detritus and sedimentary matter seems likely to be useful if the polluted harbors and waterways well-populated by Ascidians are any indication! Most aquarists, however, will not want their display aquariums to even remotely resemble a polluted harbor and we do not have the benefit of an "infinite" ocean of water to dilute heavy feedings, so there are few places for tunicates in most modern systems. Specialized feeding is part of the specific care needed for systems success with Ascidians.

Reproduction

Tunicates are hermaphroditic animals, possessing both male and female structures that can self-fertilize, although cross-fertilization is the usual rule, with sperm from one animal triggering release of eggs by another. The Ascidians also reproduce asexually in a number of ways including vegetative/fissionary budding.

Swimming Ascidian larvae are non-feeding and resemble tadpoles after leaving the parent's body cavity (through the atrial siphon). The

Clavelina picta, the Painted Tunicate. Frequently found in clusters of hundreds, hanging on to gorgonians, black coral, and sponges. Photogaphed in the Bahamas on a Sea Fan.

Didemnum sp. off Queensland.

purpose of their short motile life is to find a suitable substrate for settlement. If successful, they undergo a dramatic change with their entire appearance and viscera twisting and metamorphosing. In a very short time, the swimming larvae settle as sessile organisms that we can easily recognize as sea squirts.

With some species, reproduction in captivity is likely and perhaps even inevitable. Asexual events are frequent and occur commonly in the colonial forms. Events of sexual reproduction in captivity may also be supported by the reality of a larval stage that in some species lasts mere minutes.

Compatibility

Despite their apparent defenselessness, tunicates are remarkably safe from

Rhopalaea sp. Translucent, cool colored groupings. Common throughout the tropical Indo-Pacific, reef flats and slopes, this group is predated by *Nembrotha* nudibranchs. The vertical whitish or brown lines are their sperm ducts. N. Sulawesi photograph.

predation. A few fishes, crustaceans, gastropods, echinoderms and flatworms do graze upon them, but by and large they live with little chance of attack. Like sponges, Ascidians are so potently laced with poisons and toxic elements as to be offensive or noxious to many would-be predators. Specifically, they contain and concentrate many different chemical compounds, including heavy metals and hydrocarbons that serve, in part at least, as anti-predation mechanisms. Researchers have used Ascidians as pollution indicators by studying concentrations of contaminants in their tissues. These elements are said to be sequestered in the tunic sheath to discourage predation or overgrowth by other species. Furthermore, the tunic sheaths manufacture secretions with an extraordinarily acidic value of 1.0! It is not surprising that some speculate about the health and safety of other animals because of toxic Ascidians in closed marine systems over time. Tunicates have been implicated in causing the deaths of other captive marine organisms when killed or molested themselves. However, there are whole groups of crustaceans, worms and other phyla that have evolved or adapted to live in association with, or even entirely within Ascidians. Some are mutualistic, while others are parasitic. On the living reef and in aquariums, when successful, Ascidians can be aggressive competitors for space and may out-compete their cnidarian neighbors.

378 *Natural Marine Aquarium Volume I - Reef Invertebrates*

Summary

Are they interesting? Absolutely! But for seemingly simple, sessile animals, the Ascidians are complex creatures with digestive, nervous, reproductive and circulatory systems (including a heart) that reveal their relationship with the "higher" chordates. Are tunicates useful? Again, the answer is yes. They are exceptionally efficient filter-feeders and nutrient sinks. If you have diverse healthy live rock, you assuredly have some sea squirts. These obscure animals are likely in residence, either too small or too camouflaged to be seen on, in or under your live rock. Tunicates will quietly reside, sieving the water many times over for nanoplankton and perhaps adding their own gametes occasionally to the mix of planktonic material.

In the end, despite their fascinating morphology, behavior and beauty, tunicates are still too difficult for most aquarists to keep deliberately. We advise new aquarists to avoid keeping tunicates altogether, advanced aquarists to study them judiciously, and anyone in between to avoid acquiring them unless their system is large and very well established (aged over one year, and preferably more than two years old). The nanoplankton that Ascidians need to survive is not something that can easily be cultured or prepared (bottled), they are fundamentally bacteriovores. This is one creature that is best admired from afar by the casual aquarist.

Upper left: *Aplidium tabascum*. Colonies are reddish to orange and opaque. Indo-West Pacific; common in northern Australia. This colony photographed in North Sulawesi.

Lower left: *Leptoclinides cf. reticulatus* photographed in Indonesia.

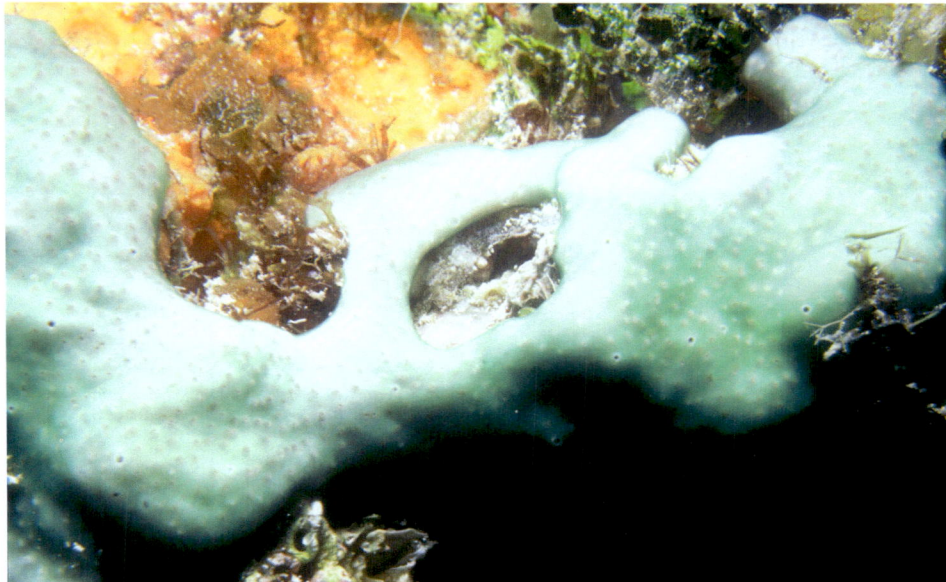

Top, an unidentified *Trididemnum* species. Middle, *Trididemnum solidum*. These beautiful encrusting Ascidians were photographed off of Cozumel, Mexico by Diana Fenner. Bottom is *Synoicum sp.* in Sulawesi.

Clavelina species occur in amazing textures and colors!

Left: *Clavelina robusta,* (Kott 1990). Western Pacific; Australia, Indonesia. Philippines, Japan, Solomons. Dense clusters of cylindrical zooids. Dark blue to gray in color with yellow, green or white rings about both siphons. Here in Bunaken, Sulawesi, Indo. **Right:** *Clavelina* sp. Bulb Tunicates. Grow in clusters. These ones common in Florida, Bahamas, the Caribbean. Here attached to a clam, which is in turn attached to a gorgonian.

Left: *Clavelina moluccensis* (Sluiter 1904). Australia, Indonesia, Singapore, Philippines. Colonies occur as many ovoid zooids on thick stalks. Color varies from dark to light blue to violet. A colony off Pulau Redang. **Right:** *Clavelina diminuta,* (Kott 1957). Western Pacific; Australia, Indonesia, Philippines. Colonies form masses of rounded heads of highly variable color. Kott (1990) moved this species to the genus *Pycnoclavella*. This batch off the north (Gilis) of Lombok, Indo.

Left: *Clavelina picta*, the Painted Tunicate. Florida, Bahamas, Caribbean. Translucent bodies of variable cool colors. Siphon rims reddish to dark purple. This one off St. Thomas. **Right:** an unidentified *Clavelina* species.

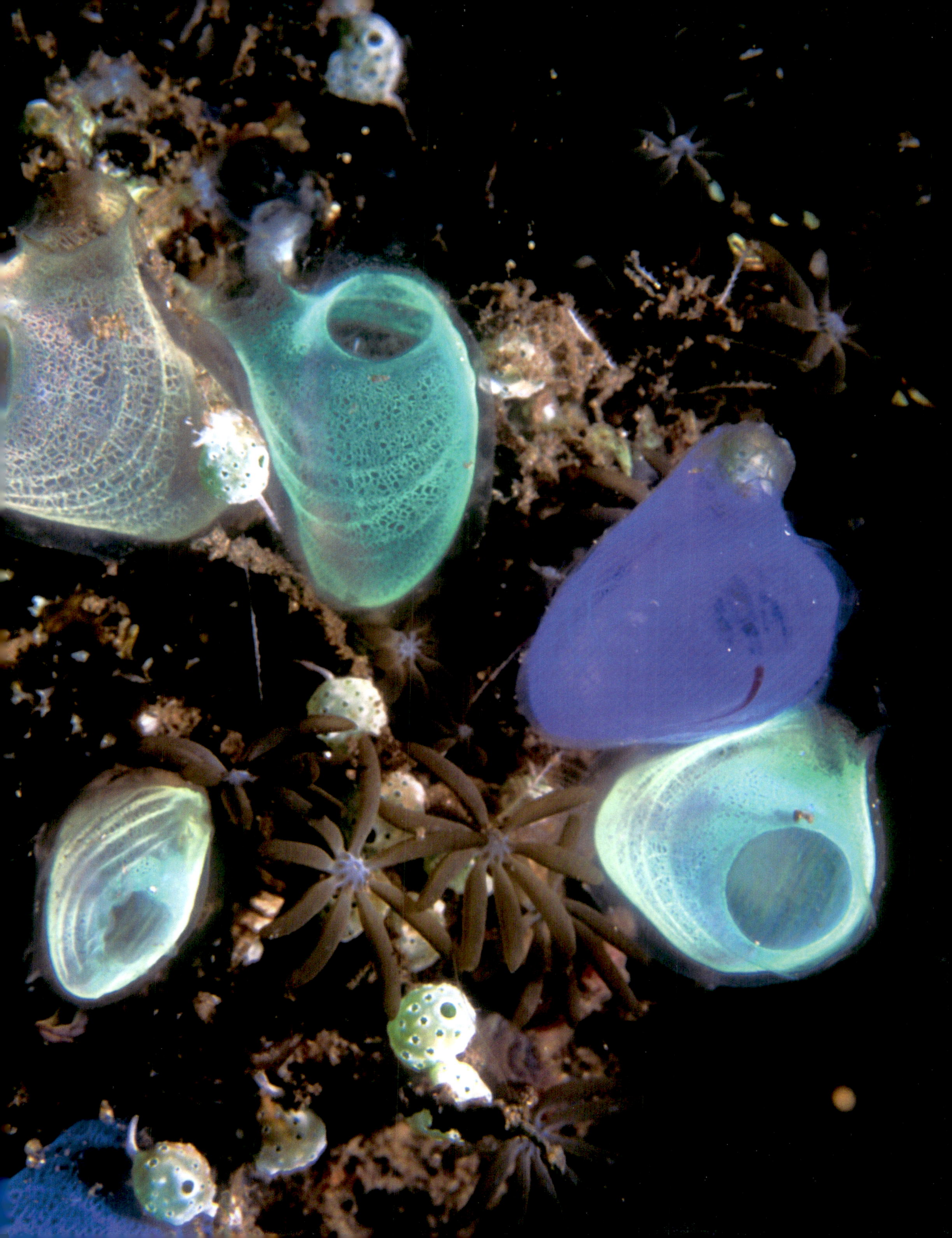

Serving Life and the Living:
Wisdoms proffered on responsible aquarium keeping...

Aquariums are much more than clear little boxes that hold water. Indeed, anyone who has endeavored to keep aquatic life will attest to their educational value and many important lessons they've come by from keeping them... many unwillingly, and some unwittingly. If you took high school level chemistry, physics, or earth science you likely were exposed to concepts like pH, alkalinity, hardness of water, etc. Likewise, in life science courses you were no doubt acquainted with the various phyla and taxonomy mentioned here. But think on how much more real these terms and organisms have become with your involvement in their care.

Us:

In the West we have finite, goal-oriented views that differ diametrically from the Orient with it's thoughtfully infinite, other-directed perspectives. Here, instead, some people think and feel that "might makes right," and "those that have, get what they want." We must be mindful, though, that the possessions of which we speak are living and dependent creatures and part of our environment collectively. Aquarium specimens are a responsibility and not merely a possession or color to compliment our living spaces. Being able to afford something is not the only or even the principal consideration for whether or not you should take an organism into your care. What's more, many of us are of the age or exposure as a generation that the world's resources seemed "limitless" or practically "inexhaustible." Some folks even have an ethic that we should get what we can while the getting's good. These attitudes and facts are false. There really is only so much that our planet can and will endure *vis a vis* our population and its utilization of living and non-living resources. Worldwide, coral reefs cover about 300,000 square kilometers... that's only a tenth of a single percent of the oceans surface area. From this small space come many human-used resources, and almost all reef invertebrates and fishes in our hobby.

Them:

There are people and organizations that push the rhetoric that keeping captive aquatics is wrong. They feel that the world's reefs would best or only be served by sequestering them from human touch altogether. Is this correct? That is to say, is it possible or practical? If so, who decides what the people who actually own the resource (coastal & island states and governments) are best off choosing to do? We have to wonder if the folks who would disallow the aquarium hobby have truly measured and duly considered the complimentary and beneficial activities of studying and keeping aquatic life. Aquarists inspire empathy and a curiosity if not passion for life and the life sciences in children and adults alike. As aquarists, we would argue that there are long-term and far-reaching benefits to responsible aquarium keeping. In fact, we are sure of it.

Reality:

We must consider how it is that we can best utilize the living resources of the coral reef that we admire so well. How best can we serve the greater good for all? Let us consider that there are at least three ways of looking at the use of a given resource base. The how-to, opportunity-cost, and null-hypothesis methods are how we will label these perspectives.

The how-to point of view simply poses the question individually, "how can I best utilize something." In the case of marine livestock, the attitude of some in the West including most of Western Europe is, "If you've got the money... buy it." This is the type of errant thinking that the "Them" groups, mentioned above, hope to restrict, tax (indirectly) or otherwise limit presumably for the environment's protection. Responsible aquaristics, however, considers how to limit dependence on living resources and even to support them by the captive propagation of reef life which, among other things, displaces the demand for wild-harvested product in a self-sustaining industry with a strong and remarkable precedent set by the freshwater aquatics trade.

The (business) opportunity cost perspective is one that addresses "what ways can we utilize our resources best or better?" The mindset is rather like self-employment, where one is faced with the daunting task of assigning scarce resource (budget monies) to direct and expand the business' chances for success. Here we ask how we, as consumers or folks engaged in the trade of marine livestock, can best apply our resources? "Highest and best use" is a theme common to the allocation of any resource, monies, real property, militaries, etc. What are the highest and best uses then of natural marine resources? Passive conservation? Pro-active conservation? Aquaculture? An interesting and necessary discussion of this matter entails equations of catch-per-unit effort and optimum (not maximum) sustainable yields. We need to determine what can we get for what we have now and in the future. That is to say: what is the greatest return we can get for the use of the resource now, such that its recovery in space and time is either unaffected (neutral impact on the environment) or that the resource is enhanced (stimulated by our activity)? In a tangible analogy that you can better understand, perhaps, we can use the example of new methodologies for rearing wild-harvested larvae. Some excellent researchers have been developing techniques for collecting wild-harvested larvae of reef organisms and rearing them in a laboratory facilities and farms. This strategy is much less expensive than traditional aquaculture where broodstock must be maintained and then stimulated to reproduce. Even when larvae can be produced with old techniques, culture through various stages of larval development is challenging, if even possible, for many species. Wild-collected larvae, however, can be strategically timed for harvest to spare all of the previous effort and expense at very small "cost" to the environment. We can truly say that the impact is so small because of known data on the naturally miniscule survival rate of broadcast spawned organisms (plankton!). Through such research, wild-caught, advanced or adult fishes will be displaced from demand and import for hardier (and higher survivability) lab-farmed larvae. The environment then benefits in turn by the continued activity of aquarists through education, understanding and desire to conserve by inspired children and peoples, discoveries for science, etc. In this manner there is almost no measurable impact on the environment. This is but one of the possibilities and explanations of a healthy "opportunity cost" perspective.

Then there is always the inevitable null-hypothesis perspective: a term used to describe the cost of doing nothing. In the real world, this option is as rare as hen's teeth. Almost all natural habitats are utilized by humans to some degree... some sustainably, others not. We exploit our environment for food extraction and cultivation, dumping wastes, run-off, transit (sailing the seas), recreation and more. Would they be better off without any human interaction, input? Perhaps and even likely in some cases (but not all... we do some good!). There's always the occasional meteorological or geological disaster: cyclonic storms, seismic waves, tsunamis and the like that can result in massive losses of aquatic life and species. But the events that are human-related... how detrimental are they? And should we allow them to exclude *Homo sapiens sapiens* at any cost? As stated, this is not a real question as these resources are almost all the domain or property of other indigenous peoples who use them for their own purposes. We truly have no place to tell the tropical island governments and peoples how to manage their resources. We can only participate within the realm of commerce and activity in the most responsible and useful way possible which includes encouraging desirable activities and discouraging undesirable practices by the very strongest means we have: our buying preferences (economics).

Consider this... what is the focus of pet-fisherfolk in these locales? Not simply to gather what they can in the here and now, but assuredly to guard against destruction of the base resource to allow for its further use in time. They are not merely fisherman, but farmers cultivating a resource. Corals, other invertebrates, algae, marine vascular plants and fishes are usually hand-caught with great human effort and low-technology means except in the case of poison collection. Compare this with destructive higher-technology activities of other industries: the extraction of "coral rubble" to make cement (sheer metric tons that dwarfs the aquarium trade's consumption of Scleractinia), explosives used for gathering fishes for human consumption, and the "incidental mortalities" that are a consequence of siltation, tourist activity, and transport by ships (recreational and commercial) through and onto reefs. Obviously, the impact of the pet-fish industry and hobby on the world's reefs is miniscule... even positive. With sustainable collection, the resource is cherished and preserved, while the locals are hired by hard-currency producing enterprise. Air freight is funded, and enough interest is generated such that local government steps in to regulate the resource and businesses using it. There is clear and significant evidence that the trade in aquatic life benefits many peoples directly and the environment itself.

A Conclusion:

The point of this mix of ideas and facts is that we, as aquarists, have a tremendous opportunity to experience, understand and share our broader awareness with other people. More importantly, we should make every effort to do so with faith that aquarium use of living resources has a miniscule negative impact on the world's reefs and a proven potential for a positive impact. In fact, as aquarists we are on the forefront in promoting awareness, enjoyment and conservation of these precious living resources.
In the end, what we perceive is not just ourselves. There is what is. This is not just "our" world, but "all of our world." We will only live quality lives for our brief times here on this planet by gently utilizing space, food, and other life... not by over-utilizing, polluting to extremes, and displacing others. Carry on gently in this world. Do consider this simple yet profound understanding in your endeavors, including aquarium hobby activities. We are making choices that influence many people and environments when we purchase and keep aquariums. Our buying preferences impact economies, the extraction of wild stocks, and the demand for captive-produced organisms. We must also consider electrical use, production and distribution, and the myriad of products it takes to set-up and maintain captive systems. To the extent that we can "afford" (in the broadest sense, not simply dollars and cents) to pledge our resources for their living resources, we are directing the use of the same, and should do our part to ensure not only the survival of the life in our care, but its sharing with others. Show off your systems to others, speak up regarding the enrichment the hobby brings you, and do consider how it is that you can add to the body of knowledge of aquaristics and science regarding the care and captive reproduction of your charges.

"When one man, for whatever reason, has the opportunity to lead an extraordinary life, he has no right to keep it to himself." - *Jacques Cousteau*

Resources

Bibliography:

A Homeowner's Guide to Trimming and Altering Mangroves, produced in cooperation with the Florida Sea Grant College Program, D.E.R. & D.N.R.

Abbott, Robert R.1990. "Retailing the giant clams". Pets Supplies Marketing. 10/95

Adey, W. H., and K. Loveland, 1991. *Dynamic Aquaria: Building Living Ecosystems*, Academic Press, San Diego, CA. 643pp

Allen, Dr. G. R., and R. Steene, 1994. *Indo-Pacific Coral Reef Field Guide*, Tropical Reef Research, Singapore. 378pp

Aw, Michael.1995. "Quick-draw Claws". Sport Diver.11/12, 95

Baensch, Hans & Helmut Debelius. 1994. Marine Atlas, v.1. MERGUS, Germany. 1215pp

Battisto, Cathy. 1999. "Live rock ala mode". Tropical Fish Hobbyist. 8/99

Baugh, Thomas M. 1991. "Dwellers of the sand (Mole Crabs)". FAMA 11/91

Baugh, Thomas M. 1991. "The Seaweed Cucumber". FAMA 4/91

Birkeland, Charles & John S. Lucas. 1990. "*Acanthaster planci*: Major Management Problem of Coral Reefs". CRC Press

Brach, Vince. 1996. "Aquarium "Gunmen" (alpheid shrimp)". FAMA 8/96

Brach, Vince. 2001. "The cleaner scene". FAMA 1/01

Branson, Helen Kitchen. 1999. "Natural live rock filters in saltwater aquariums (Hawaiian LR)". FAMA 4/99

Brockmann, Dieter. "Undated. Keeping Anemone Shrimps in the marine aquarium". Aquarium Digest International #53, p. 13

Bruce, A.J. Undated. Shrimps that live on tropical echinoderms. Underwater Magazine. #17

Bruckner, Andrew W. 1993. Tropical Shrimp. Social relationships in the aquarium. FAMA 2/93

Brusca, R.C., & G.J. Brusca, 1990. *Invertebrates*. Sinauer Associates, Inc. Sunderland, Mass. 922 pp

Burgess, Lourdes A. 1991. "*Nautilus*: the pearly chambered nautilus". TFH 5/91

Calfo, Anthony. 2001. *Book of Coral Propagation: Reef Gardening for Aquarists Volume 1*. Reading Trees, Monroeville, PA. 450 pp

Carlson, Bruce A. 1991. "Tridacna clams; true giants in their field". FAMA 4/91

Coletti, Ted. 1998. "Habitat tanks; like biotope tanks, but different." AFM 9/98

Colin, P.L., and C. Arneson, 1995. *Tropical Pacific Invertebrates*, Coral Reef Press, Irvine, CA. 296pp

Cranston, Bob & Cathy. 1992. "The sea tiger: a truly impressive predator". Discover Diving 3-4/92

Dakin, Nick. 1992. *The Book of The Marine Aquarium*, Salamander Books, Ltd.

Damone, Joe. 1999. "The Caribbean Basket Starfish (*Astrophyton muricatum*) in the Home Aquarium: Is it possible?". Aquarium Frontiers Online 2/99

Daum, Wolfgang. "Sea slugs- exotic additions to the marine aquarium". Aquarium Digest International #26 pp 27-30

Debelius, Helmut. 1999. *Crustacea of the World. Atlantic, Indian, Pacific Oceans*. IKAN, Germany 321pp

Delbeek, J. Charles. 2002. "Fighting Flatworms" AFM 7/02.

Delbeek, J.Charles., and J. Sprung, 1994. *The Reef Aquarium, Volume 1*, Ricordea Publishing, Coconut Grove, FL. 544pp

Delbeek, J. Charles. 1990. "Live rock algal succession in a reef system". FAMA 10/90

Dixon, Beverly A. 1991. "Fungus Among Us". AFM 4/91

Donovan, Paul. 1992. "Cuttlefish". FAMA 1/92

Duffy, J.E. 2002. "Biodiversity and ecosystem function: the consumer connection". Oikos 99:201-219

Duffy, J.E. and K.S. Macdonald. 1999. "Colony structure of the social snapping shrimp". Journal of Crustacean Biology 19:283-292

Duffy, J.E. 1998. "On the frequency of eusociality in snapping shrimps with description of a new eusocial species". Bulletin of Marine Science 62:387-400

Duffy, J.E. 1996. "Eusociality in a coral-reef shrimp". Nature 381:512-514

Duffy, J.E. 1996. "Species boundaries… of Alpheid shrimp". Biological Journal of the Linnaean Society 58:307-324

Duffy, J.E. 1996. "Resource-associated population subdivision in a symbiotic coral-reef shrimp". Evolution 50:360-373

Duffy, J.E. 1996. "*Synalpheus regalis*, new species, a sponge-dwelling shrimp from the Belize Barrier Reef". Journal of Crustacean Biology 16: 564-573

Duffy, J.E. and M.E. Hay. 1994. "Herbivore resistance to seaweed chemical defense". Ecology 75:1304-1319

Duffy, J.E. 1993. "Genetic population structure in two tropical sponge-dwelling shrimps". Marine Biology 90:127-138

Duffy, J.E. 1992. "Host use patterns and demography in a guild of tropical sponge-dwelling shrimps". Marine Ecology Progress Series 90: 127-138

Duffy, J.E. and V.J. Paul. 1992. "Prey nutritional quality and the effectiveness of chemical defenses against tropical reef fishes". Oecologia 90:333-339

Duffy, J.E. and M.E. Hay. 1991. "Host plants as food and shelter". Ecology 72: 1286-1298

Duffy, J.E. and M.E. Hay. 1991. "Amphipods are not all created equal". Ecology 72:354-358

Duffy, J.E. 1990. "Amphipods on seaweeds: partners or pests?". Oecologia 83:267-276

Duffy, J.E. and M.E. Hay. 1990. "Seaweed adaptations to herbivory". BioScience 40:368-376

Duffy, J.E. In press. "The ecology and evolution of eusociality in sponge-dwelling shrimp". In: T. Kikuchi (editor). Genes, behavior, and evolution in social insects. University of Hokkaido Press, Sapporo, Japan

Dybas, Cheryl Lyn. "The critters; tough but tender abalone stage a comeback". Sea Frontiers 5-6/94

Emmens, C.W. 1991. "Plankton, pts. I, II, III". FAMA 9, 10, 11/91

Erhardt, Harry & Horst Moosleitner. *Marine Atlas, Volume 2 & 3*. Mergus, Melle, Germany

Esterbauer, Hans. 1995. "The Crown-of-Thorns Starfish." TFH 11/95

Fatherree, James. 2002. "Ask the Reefer: The carbonates explained." TFH 10/02

Fatherree, James. 2002. A look at the Mollusks. TFH 6/2002

Fatherree, James. 1998. "A living fossil" in your aquarium". FAMA 5/98

Fenner, Robert., 1998. *The Conscientious Marine Aquarist*, Microcosm Ltd., 432pp

Fenner, Robert. 1996. "Nudibranchs. The beautiful naked-gilled sea slugs". TFH 2/96

Fenner, Robert. 1995. "Spiny-Skinned animals, phylum Echinodermata." FAMA 5/95

Fenner, Robert. 1995. "*Acanthaster*, Crown of Thorns Sea Stars." FAMA 1/95

Fenner, Robert. 1994. "A Diversity of Aquatic Life: Feather duster worms." FAMA 12/94

Fenner, Robert. 1993. "More Spiny-skinned Animals. Crinoids: Sea Lilies and Feather Stars." FAMA 10/93

Fossa, A. Svein & Alf Jacob Nilsen. 2002. *Modern Coral Reef Aquarium, Volume 3*. BSV Germany

Fox, Gregory A. 1993. "Buried Treasure: The Sand Dollar is sure to provide great commercial and aesthetic value for years to come." FAMA 3/93

Giwojna, Pete. 1991. "The acrobatic sex life of the Arrow Crab". FAMA 9/91

Glodek, Garrett. 1996. "A little bit about nudibranchs". FAMA 11/96

Glodek, Garrett. 1995. "The biology of nudibranchs". FAMA 5/95

Glodek, Garrett. 1993. "Plankton". FAMA 7/93

Goldstein, Robert J. 1993. "Rock wars. Opposing parties seek a lasting peace in the controversy over live rock". Pet Age 4/93

Gosliner, Terrence M, David W. Behrens and Gary C. Williams. 1996. Coral Reef Animals of the Indo-Pacific. Animal live from Africa to Hawai'i exclusive of the vertebrates. Sea Challengers, Monterey California. 314pp

Grosskkopf, Joachim. Undated. "Our family favorite: *Odontodactylus scyllarus*. the Mantis Shrimp". Aquarium Digest Intl. #53

Hahn, Kirk O., 1990. Annotated Bibliography of the Genus *Haliotis*. S.I.O. Reference

Harris, Linda K. 1994. "At the Division of Worms there are even worms to spare; Smithsonian has millions of them". Knight-Ridder Newspapers 11/30/94

Hashimoto Y, Fusetani N, Nozawa K. "Screening of the Toxic Algae on Coral Reefs". 569-572

Hay, M.E., J.E. Duffy, and W. Fenical. 1990. "Host-plant specialization decreases predation on a marine amphipod: an herbivore in plant's clothing". Ecology 71:733-743

Hay, M.E., J.E. Duffy, V.J. Paul, P.E. Renaud, and W. Fenical. 1990. "Specialist herbivores reduce their susceptibility to predation by feeding on the chemically defended seaweed *Avrainvillea longicaulis*". Limnology and Oceanography 35:1734-1743

Hopley D. et al (eds.) Proceedings of the sixth international coral reef symposium

Hoover, John P. 1998. "*Hawai'i's Sea Creatures; A Guide To Hawai'i's Marine Invertebrates*". Mutual Publishing, Honolulu, HI.

Hoover, John P. 1998. "Hawai'i's Lobsters". FAMA 10/98

Hoover, John P. 1997. "Hawaiian Hermit Crabs", parts I, II. FAMA 9,10/97

Humann, Paul., 1993. *Reef Coral Identification*, New World Publications, Inc., Jacksonville, FL. 239pp

Humann, Paul. 1992. *Reef Creature Identification*. New World Publications, Inc. Jacksonville, FL. 320pp

Hunziker, Ray. 1993. "The magic of tropical Shrimps". TFH 7/93

Jewell, Jack. 1997. "Breeding the Aiptasia eater". FAMA 12/97

Johnston, Elizabeth S. & John Forsythe. 1993. "An octopus in your house?". AFM 8/93

Jensen, Christopher. 1998. "Red Legged Hermit Crab". FAMA 4/98

Johnson, Don S. 1999. "Shrimp, Crabs and Lobsters in the marine aquarium". AFM 8/99

Kerstitch, Alex. 1994. "Farmers of the sea; the blue revolution, part 4, Scallops, Mussels, Abalone". FAMA 9/94

Kerstitch, Alex. 1993. "Thumb-Splitters". SeaScope v.10, Winter 93

Kerstitch, Alex. 1992. "Crabs in the aquarium. From Fiddlers to Pom-Poms". FAMA 2/92

Kerstitch, Alex. 1992. "Living jewels of the sea". FAMA 1/92

Kerstitch, Alex. 1991. "Living mollusks (giant clams)". FAMA 12/91

Kerstitch, Alex. 1991. "Living mollusks". FAMA 12/91

Kerstitch, Alex. 1990. "Molluscan touch". SeaScope Vol. 7, Summer 90

Kirkendoll, April. 2001. *How To Raise & Train Your Peppermint Shrimp*. Lysmata Publishing. 128 pp

Knaack, Joachim. 2000. "Collecting Mantis Shrimp". TFH 9/00

Knop, Daniel. 1996. *Giant Clams, A Comprehensive Guide to the Identification and Care of Tridacnid Clams*, Dahne Verlag GmbH, Ettlingen, Germany

Lamberton, Ken. 1993. "Some perfect marine gastropods". FAMA 11/93

Littler, Diane Scullion and Littler, Mark Masterson. 2000. Caribbean Reef Plants. Off Shore Graphics Inc, Washington, D.C. pp. 356-380

Lowrie, Jonathan & Eric Borneman. 1999. "A Survey of Marine Microbes". Aquarium Frontiers 5/99

Maier, Robert von, 1991. "Bristle Worms". Discover Diving. Sept./Oct. 1991

Mancini, Alessandro. 1992. "Fiddling with Fiddler Crabs". TFH 10/92

Mancini, Alessandro. 1991. "Starfishes in Tropical Marine Aquaria." TFH 9/91

Mancini, Alessandro, 1990. "Tropical Tubeworms in the Marine Aquarium." FAMA 8/90

Meinesz, Alexandre and Simberloff, Daniel. 1999. "Killer Algae". University of Chicago Press, Chicago. pp. 295-304

Mercier, Annie & Jean-Francois Hamel. 1998. "Tunicates: A crucial step in evolution." FAMA 3/98

Michael, Scott. "Tips for choosing your Shrimp Goby and Snapping Shrimp pair." Aquarium Fish Magazine. February 2002 pages 27-33

Michael, Scott. "Strange Burrow Fellow." Aquarium Fish Magazine. January 2002 pages 18-25

Michael, Scott. 1998. "Hermit danger. Some species of hermit crabs actually consider your fish their dinner". AFM 5/98

Michael, Scott. 1997. "Sea Cucumbers: These are not your garden variety of reef tank inhabitant, and can cause real problems." AFM 3/97

Michael, Scott. "Mantis Shrimp. Why do these two words strike fear in the hearts of marine tank aquarists?". AFM 2/97

Moe, M.A. Jr., 1992. *The Marine Aquarium Handbook: Beginner to Breeder,* Green Turtle Publications, Plantation, FL. 320pp

Nilsen, Alf & Sven Fossa. 2002. *The Modern Coral Reef Aquarium, v. 4.* BSV, Germany. 480pp.

Nilsen, Alf. 1991. "The successful coral reef aquarium. Pt. 5, The macro-life of the live rock". FAMA 1/91

Paletta, Michael. 1995. "Living with live rock; the "foundation" of successful reef setups". 6/95

Paul, Valerie J., M.E. Hay, J.E. Duffy, W. Fenical, and K. Gustafson. 1987. "Chemical defense in the seaweed *Ochtodes secundiramea* (Montagne) Howe (Rhodophyta): effects of its monoterpenoid components upon diverse coral-reef herbivores". Journal of Experimental Marine Biology and Ecology 114:249-260

Paul, Valerie J., and Fenical, William. 1986. "Chemical defense in tropical green algae, order Caulerpales". Marine Ecology Progress Series

Pennings, S.C., S.R. Pablo, V.J. Paul, and J.E. Duffy. 1994. "Effects of sponge secondary metabolites in different diets on feeding by three groups of consumers". Journal of Experimental Marine Biology and Ecology 180:737

Peters, Michael J. & Joyce Wilkerson. 1996. "Scarlet cleaner shrimp larval development". FAMA 3/96

Pro, Steven. 2002. personal communications

Reeves, Linda. 1994. "Florida lobster diving. Where to find and how to catch these wily crustaceans". Skin Diver 7/94

Riddle, D., "Opposite Ends of the Spectrum: Ultraviolet and Infrared Radiation in the Reef". Aquarium Frontiers, Vol.3, No.2

Riddle, D., 1995. *The Captive Reef,* Energy Savers Unlimited, Inc. 297pp

Ríos, R. and J.E. Duffy. 1999. "Description of *Synalpheus williamsi*, a new species of sponge-dwelling shrimp (Crustacea: Decapoda: Alpheidae), with remarks on its first larval stage". Proceedings of the Biological Society of Washington 112:541-552

Robles, Carlos. 1996. "Turf battles in the tidal zone (palinurid lobsters)". Natural History 7/96

Rohleder, P.G. Undated. "*Linckia* and *Fromia*- Two starfish for the reef aquarium". Aquarium Digest International #53

Sagasti, A., L.C. Schaffner, and J.E. Duffy. 2001. "Effects of periodic hypoxia on mortality, feeding and predation in an estuarine epifaunal community". Journal of Experimental Marine Biology and Ecology 258: 257-283

Sagasti, A., L.C. Schaffner and J.E. Duffy. 2000. "An epifaunal community thrives in an estuary with brief hypoxic episodes". Estuaries 23:474-487

Segars, Herb. 1992. "The North American Lobster". Discover Diving 1-2/92

Shimek, Ronald. 2001. "Bristlewormophobia". TFH 7/01

Shimek, Ronald. 2001. *Sand Bed Secrets: The Common Sense Way to Biological Filtration,* Marc Weiss Companies, Fort Lauderdale, FL.

Shimek, Ronald. 2001. "Dearest Mudder: The importance of deep sand." AFM 3/01

Shimek, Ronald. 1999. *The Coral Reef Aquarium,* Howell Book House, New York, NY.

Shimek, Ronald. 1998. " Reef Tank Plankton- A partial success." Aquarium Frontiers 3/98

Shimek, Ronald. 1998. "Crabby comments. Give proper conditions, crabs make happy, enjoyable and entertaining pets". AFM 2/98

Shimek, Ronald. 1996. "Nudibranchs: The bad nudes of the reef". Aquarium Frontiers 3/4, 96

Shimek, Ronald. 1996. "Gerkins or Dills: The cucumbers of the sea." Aquarium Frontiers 3:3/96

Sleeper, Jeanne Bear. 1992. "West coast lobster hunting". Skin Diver 10/92

Spalding, Mark D., Corinna Ravlious & Edmund P. Green. 2001. *World Atlas of Coral Reefs.* UNEP WCMC. 424 pp

Sprung, Julian., and J.C. Delbeek. 1997. *The Reef Aquarium, Volume 2,* Ricordea Publishing, Coconut Grove, FL. 546pp

Sprung, Julian. 2001. *Invertebrates: A Quick Reference Guide,* Ricordea Publishing, Coconut Grove, FL.

Sprung, Julian. 1995. " Magnificent Mangroves: You can actually grow mangroves in the comfort of your own aquarium." AFM 12/95

Stolzenburg, William. "The naturalist; the familiar stranger; in many ways, the alien octopuses act almost like humans". Sea Frontiers 7-8/93

Stratton, Dick. 1998. "Thumb Splitters". TFH 4/98

Sy, Leng, 2002. personal communications

Tavares, Iggy. 1998. "Beautiful slugs!?". FAMA 2/98

Toonen, Rob. 2002. "Sea Stars: *Linckia* spp.". Advanced Aquarist Online Magazine. 5/02

Toonen, Rob. 2001 MACNA on bottled phytoplankton products in review

Toonen, Rob. 2000. "Invert insights: Holothurians." TFH 7/00

Tullock, John H. 1999. "Crabs and their relatives". AFM 3/99

Tullock, John H. 1998. "The other mollusks. Meaning, the ones that aren't snails". AFM 7/98

Tullock, John H. 1997. "Little Marine Friends: Beneficial micro-organisms in the saltwater aquarium." AFM 3/97

Tullock, John H. 1997 Natural Marine Aquariums- Simplified Approaches To Creating Living Saltwater Microcosms Microcosm, Ltd., Shelburne, VT

Tullock, John H. & J.R. Shute. 1994. "Live rock", parts I, II TFH 10, 11/94

Tyree, Steve. 2000. *The Environmental Gradient, Cryptic Sponge and Sea Squirt Filtration Models,* 2d ed., Volume 1 of the Captive Maintenance Advanced Techniques CMAT Series

Volkart, Bill. 1990. "Sea Squirts: Tubular wonders." TFH 10/90

Volkart, Bill. 1990. "The cephalopods". TFH 9/90

Walls, Jerry G. 1995. "Crab watch: the Japanese Shore Crab, *Hemigrapsus sanguineus*". TFH 3/95

Walls, Jerry G. 1992. "Little White Worms." TFH 7/92

Weingarten, Robert A. 1991. "Tridacna- the giant clam". FAMA 2/91

Wilkens, Peter. 1998. "Death in a Colorful Package." Aquarium Frontiers Online. 5/98

Wilkens, Peter. 1991. "Keeping flame scallops". TFH 2/91

Wilkens, Peter. 1990. *Invertebrates, Stone and False Corals, Colonial Anemones,* Engelbert Pfriem Verlag, Wuppertal, Germany. 136pp

Wilkens, Peter, and J. Birkholz. 1986. *Invertebrates, Tube-, Soft-, and Branching Corals,* Engelbert Pfriem Verlag, Wuppertal, Germany. 134pp

Wilkerson, Joyce D. 1994. "Scarlet cleaner shrimp". FAMA 8/94

Wood, James B. 1994. "Don't fear the raptor; an octopus in the home aquarium". FAMA 4/94

Young, Forrest A. 1993. "Live rock aquaculture". FAMA 5/93

Internet References: (topic in brackets is not part of address)

http://www.breeders-registry.gen.ca.us [Breeder's Registry]
http://members.ozemail.com.au/~edrew/shed/halimeda/halmov/halmov.htm [*Halimeda*]
http://www.seaslugforum.net/elysviri.htm [Sea slugs]
http://people.ucsc.edu/~mcduck/nudifood.htm [Nudibranch foods]
http://www.animalnetwork.com/fish2/aqfm/1998/nov/wb/default.asp [Dinoflagellates]
http://www.projectlinks.org/dinoflagellates/ [Dinoflagellates]
http://www.typhoon8.net/reef/page_ochtodes.html [*Ochtodes* algae]
http://www.vims.edu/bio/mobee/duffypub.html [Alpheids]
http://www.vims.edu/bio/mobee [Seagrass]
http://www.mbari.org/~conn/botany/flora/mflora.htm [marine flora]
http://seaweed.ucg.ie/Algae/Ulva.html [*Ulva*]
http://www.anstaskforce.gov/SF4_01_meeting_summary.htm [nuisance algae]
http://web.uvic.ca/bmlp/news/news32.html [macroalgae culture]
http://www.ecosystemaquarium.com/index.html [mud filtration/marketing]
http://www.wetwebmedia.com [comprehensive aquatics]
www.dynamicecomorphology.com/depublish.htm [various]
http://lifesci.ucsb.edu/~biolum/ [bio-luminescent organisms]
http://biol.dgbm.unina.it:8080/ascidians/AscidianNews/an39.html [Ascidians]
http://www.animalnetwork.com/fish2/aqfm/1997/sep/wb/default.asp [Ascidians]
http://www.animalnetwork.com/fish2/aqfm/2000/jan/wb/default.asp [Sea Cucumbers]
http://www.emedicine.com/emerg/topic158.htm [Echinoderm envenomation]
http://www.reefs.org/library/article/r_toonen2.html [Sea Cucumber defenses]
http://www.nmnh.si.edu/iz/echinoderm/body_records.htm [Echinoderms]
http://arond-thomasonline.com/template.asp?articleid=554&categoryid=116 [holothurin]
http://chinesefood.about.com/library/blchineseing6.htm [Sea Cucumbers for humans]
http://www.spc.org.nc/coastfish/News/bdm/12/4.htm [Sea Cucumbers]
http://chinesefood.about.com/library/weekly/aa092200a.htm [Sea Cucumbers]
http://saltaquarium.about.com/blcucumberfam_holothuriidae.htm [Sea Cucumbers]
http://citd.scar.utoronto.ca/EESC04/SCMEDIA/INVPHYLO/Echinodermata/Table.html
http://www.catalinaop.com/uni.htm [Echinoids as food]
http://www.shop-maine.com/mill/mil_urc.asp [Echinoids as food]
http://underwaterphotos.com/article4.htm [Crustaceans]
http://www.seaweb.org/resources/13update/urchin.html [Echinoderm disease]
http://www.marlin.ac.uk/bio_pages/Bio_Eco_EIR.LhypR.htm [kelp beds]
http://coastal.er.usgs.gov/african_dust/events.html [reef diseases]
http://saltaquarium.about.com/library/weekly/aa052899c.htm [Echinoids]
http://www.marlin.ac.uk/index2.htm?demo/Echesc.htm [Echinoids as food]
http://www.blueboard.com/mantis/ [Stomatopods]
www.crustacea.net [Crustaceans]
http://www.amonline.net.au/invertebrates/cru/index.htm [Crustaceans]
http://www.nisk.k12.ny.us/nhs/bionet/Taxonomy%20links.htm [various]
http://www.oeb.harvard.edu/palumbi/people%20pages/paul_isopods.html [Isopods]
http://reefkeeping.com/issues/2002-05/rs/ [Isopods]
http://www.niwa.cri.nz/pubs/wa/09-3/isopod.htm [Isopods]
http://www.earthlife.net/chelicerata/pycnogonida.html [Sea Spiders]
http://slugsite.tierranet.com/ [Sea Slugs]
http://slugsite.us/bow/nudi_han.htm [Sea Slugs]
http://www.animalnetwork.com/fish2/aqfm/1999/may/wb [Sea Slugs]
http://www.divernet.com/biolog/nudib797.htm [Sea Slugs]
http://www.orsi.it/club/awarenes/mollusch.html [Mollusk taxa]
http://www.medslugs.de/Opi/Opisthobranchia.htm [Sea Slugs]
http://waquarium.mic.hawaii.edu/MLP/search/spiny_lobster.html [Lobster]
http://www.beach-net.com/horseshoe/Bayhorsecrab.html [Crab]
http://www.cephbase.utmb.edu/ [Cephalopods]
www.tonmo.com [Octopus]
http://www.advancedaquarist.com/issues/may2002/short.htm [Flatworm control]
http://www.angelfire.com/mo2/animals1/cephalopod/cephalopod.html [Cephalopods]
http://is.dal.ca/~ceph/TCP/octobite.html [Cephalopod bites]
http://www.dal.ca/~ceph/TCP/cuttle1.html [Cuttlefish]
http://ourworld.compuserve.com/homepages/BMLSS/cuttle2.htm [rearing Cuttlefish]
http://www.wikipedia.org/wiki/Nautilus [Nautilus]
http://animaldiversity.ummz.umich.edu/mollusca.html [Mollusks]
http://www.animalnetwork.com/fish2/aqfm/1999/nov/wb/default.asp [Snails]
http://www.nhm.ac.uk/zoology/taxinf/index2.html [Polychaetes]
http://www-biol.paisley.ac.uk/courses/Tatner/biomedia/units/anne1.htm [Annelids]
http://www.aquarium.net/0697/0697_2.shtml [Polychaetes]
http://www.aquarium.net/1096/1096_3.shtml [Polychaetes]
http://reefkeeping.com/issues/2002-06/rs/ [Polychaetes]
http://saltaquarium.about.com/cs/msubpestbworm/index.htm [Polychaetes]
http://www.alientravelguide.com/science/biology/life/animals/platyhel/turbella/acoela/ [Flatworms]
http://www.angelfire.com/va3/bryozoans/ [Bryozoans]
http://www.animalnetwork.com/fish2/aqfm/1997/dec/wb/default.asp [Bryozoans]
http://www.fishtech.com/speech.html [Abalone]
http://www.dohenystatebeach.org/ip-tidepoolprepack.htm [Arthropods]
http://biology.usgs.gov/s+t/SNT/noframe/ca166.htm [Abalone]
http://www.ma.org/classes/oceanography/fffox/home.html [plankton]
http://www.ncl.ac.uk/~nmscmweb/3rd_level/teaching/undergrad/lecture_notes/st2mb/msm225/lecture1.htm#def [plankton]
http://saltaquarium.about.com/gi/dynamic/offsite.htm?site=http://www.aquarium.net/0497/0497%5F1.shtml [Echinoids]

Other Resources:

American Conchologist, the official publication of the Conchologists of America, Inc. Quarterly. P.O. Box 1226, New Albany, In 47150

Florida Aqua Farms, Inc., 33418 Old Saint Joe Road, Dade City, FL 33525
phone 352.567.0226 web: http://www.florida-aqua-farms.com/ [aquaculture supplies... plankton starter cultures]

Inland Aquatics, 10 Ohio Street, Terre Haute, IN 47807-3417
phone 812.232.9000 web: http://www.inlandaquatics.com [aquaculture supplies... plankton starter cultures]

National Research Center for Cephalopods (NRCC)
Marine Biomedical Institute
301 University Blvd.
Galveston, TX 77555-1163
http://www.nrcc.utmb.edu

Glossary

A:

Aboral: opposite of the mouth
Adductor muscle: in bivalves- holds shell halves together
Aerobic: utilizes oxygen for life processes
Ahermatypic: usually non-photosynthetic and non-reef building species
Alkalinity: a measure of the buffering capacity of seawater; also used to describe the higher end of the pH scale
Allelopathy: chemical warfare between some plants and/or animals
Ambulatory: mobile
Animal filter: type of refugium dedicated to the fixing and/or export of nutrients via cultured animals (like corals, sponges or anemones)
Anaerobic: activity in an environment of little or no oxygen.
Anoxic: oxygen-poor state
Aposematic: warning coloration
Aposymbiotic: lacking a symbiont as with non-photosynthetic corals
Aragonite: crystalline calcium carbonate (dimorphic with calcite)
Asexual: having no dedicated gender; a form of reproduction that does not require representation by separate sexes
Autotomy: the sudden casting off a body part when stressed, injured or attacked (like brittle starfish shedding arms)
Autotrophic: an organism capable of synthesizing its own food using light or chemical energy.
Axenic: sterile
Azooxanthellate: lacking endosymbiotic algae

B:

Benthic: lifestyles and forms associated with the seafloor
Bioluminescent: the ability of an organism to produce its own light via chemical or biological processes
Biomass: a sum of living matter within a given area
Bio-minerals: elements used in metabolism and processes like calcification
Budding: an asexual event of reproduction (clonal)

C:

Calcite: crystalline form of calcium carbonate (dimorphic with aragonite)
Capitulum: the top or crown
Carapace: the hard, chitinous exoskeleton of a crustacean
Cerata: dorsal aspects of Aeolid nudibranchs that perform functions of respiration, digestion and defense
Chitinous: a substance that is formed of proteinaceous chitin
Choanocytes: the flagellated cells in sponges, which are responsible for moving currents through the colony
Cilia: small, hair-like structures.
Circumtropical: found in all tropical or equatorial areas around the world
Cirri: a tendril or like-aspect
CITES: an acronym for Convention on International Trade in Endangered Species
Chelipeds: the crustacean appendages that bear claws
Chromatophores: highly specialized cells filled with liquid pigment that can be manipulated in size and form to dramatically change the appearance of an organism
Cloaca: the cavity into which the intestinal, genital, and urinary tracts open in some animals
Clonal: a cell, group of cells, or whole organism that is genetically identical to a common ancestral donor
Coelom: a body cavity
Commensal: a symbiotic relationship in which one species benefits while the other is unaffected
Cometary fission: the natural (or imposed) fragmentation of a sea star, which causes the regenerative arm to grow to resemble its namesake
Crepuscular: active during twilight and predawn hours
Cryptic: visage of obscurity in an attempt to blend into the environment
Ctenida: gills found in bivalves.

D:

Demersal: occurring at or near the bottom of the sea as with reef invertebrates that deposit their spawn on the seafloor (hard or soft substrates)
Deposit Feeder: feeding on the substrate/seafloor
Detritus: settled organic particles of debris
Diatom: microscopic algae comprised largely of silica (commonly observed as "brown slime algae")
Digitate: having fingerlike projections.
Dinoflagellate: protozoans (typically characterized by having two flagella and a cellulose covering)
Diurnal: active during daylight hours
DOC: Dissolved Organic Compounds.

E:

Endoperoxide synthetase: an enzyme that can be activated with Hydrogen Peroxide to induce mollusks to spawn
Endosymbiotic: incused protection of a symbiont within a host, as with zooxanthellae in cnidarian tissue
Epiphytes: an organism that grows upon another but does not prey upon it
Epipodium: one of the (lateral) lobes of the foot in certain gastropods
Errantiate: motile
Extant: still living (not extinct)

F:

Fallow: in aquaristic terms- to allow an aquarium to run "empty" without a viable living host for the purpose of controlling or limiting an unwanted organism, like a parasite
Family: a taxonomic category of related organisms ranking below an order and above a genus
Fileclams: bivalves including the so-called "flame scallops"
Filter-feeder: an organism whose feeding strategy is wholly dependent on nutrition carried to it from the water around it
Fission: an asexual reproductive strategy of division like budding or splitting
Flagella: a whip-like extension of certain cells that facilitates movement
Filtrants: chemical or physical materials used to remove matter selectively
Frags: in reef keeping, refers to asexual fragments taken from an organism for the purpose of propagating the colony

G:

Gametes: sex cells
Ganglia: a brain-like assemblage of nerve cells
Gemmules: asexually produced bud or packet of "essential cells"
Genus: a taxonomic category ranking below a family and above a species
Glitter lines: beautiful dappled light, refracted from the disturbed surface of water. Also known as "God-beams"
Gonads: sexual organs
Greenwater: unicellular, free-floating algae
Gut-loaded: a reference to whole food/prey items that are "full-bellied" with other nutritious foods/vitamins

H:

Hermaphrodite: possessing both male and female sexual organs
Hermatypic: usually possessing endosymbiotic algae and regarded as reef building
Heterotrophic: unable to synthesize its own food and must actively feed
Holdfast: attachment aspects of macroalgae
HUFA: Highly Unsaturated Fatty Acids
Hyponome: a modified "tentacle" found within the mantles of cephalopods used for propulsion

I:

Infauna: organisms that live in the substrate
Intertidal: existence and exposure between high and low tide

J:

K:

Kalkwasser: calcium hydroxide
Kelvin: a temperature scale used in aquaristics to rate the color of light

L:

Laminar: uni-directional, planar
Live Rock: a non-living mineral matrix (generally carbonate in nature) infused with living micro- and macro-organisms
LPS: **L**arge **P**olyp **S**tony coral

M:

MACNA: **M**arine **A**quarium **C**onference of **N**orth **A**merica
Mantle: the protective tissue of mollusks that envelops vital organs and secretes the shell
Maxilliped: paired appendages of crustaceans used in feeding
Meiofauna: tiny organisms living on marine substrates
Meroplankton: organisms that spend only part of their life cycle as plankton before settling out
Mesocosm: an intermediate sized model of the (reef environment)… usually refers to multi-tank systems and larger/public displays
Metabolites: a product of, or necessary for, metabolic life processes (food, waste, etc)
Microcosm: a small model of the (reef) environment
Microcrustaceans: in reef keeping, a collective term used to describe organisms like amphipods, copepods and mysid "shrimp"
Midden: "trash" pile, as can be seen outside of octopuses' dens
Mixotrophic: strategy of both seeking and manufacturing own food
Morbidity: incidence of sickness
Morphology: the form and structure of organisms
Mortality: incidence of death
Motile: mobile, movable
Mutualism: relationship between two different species in which both members benefit

N:

Nanoplankton: very fine plankton measuring 2 to 20 um
Necrosis: tissue die-off
Nekton: matter, unlike plankton, that can move freely in water and includes sizes ranging from microscopic to whales (all fishes are nekton, e.g.)
NSW: Near/Natural Sea Water (in regards to water quality standards)
Nori: dried seaweed

O:

Obligate: only able to exist or survive by assuming a specific role
Ocellated: having patterned eyespots- a form of mimicry

Octocoral: cnidarians characterized by eightfold radial symmetry
Oolitic: fine, round grains of limestone composed of calcium carbonate
Operculum: plate-like covering (gills in vertebrate fishes, or as found in snails and tube worms)
Order: a taxonomic category of organisms ranking above a family and below a class
Organic dyes: in aquaristic terms, agents used as medicants (like Malachite Green and Methylene Blue) which are helpful with many fishes, but can be harmful or fatal to most invertebrate life
Organismal feeding: consuming whole prey- may be microscopic as with some filter-feeders
Oscula: an excurrent opening of Porifera (sponge)
Osmoregulation: maintenance of osmotic pressure in the body of a living organism
Osmosis: diffusion of fluid(s) through a semi-permeable membrane until there is an equal concentration on either side
Osmotic shock: stress of an organism due to a solute/solvent change in the environment
Ossicles: small calcareous structures used to help compose the skeleton of echinoderms
Ostia: incurrent openings of Porifera (sponge)
Ovary: the (usually) paired female or hermaphroditic reproductive organ that produces ova (eggs)
Oviparous: egg-laying

P:
Papillae: tiny bumps
P.A.R.: Photosynthetically **A**ctive **R**adiation- a useful measure of the quality of light for photosynthetic organisms
Parthenogenesis: meaning "virgin birth" from the Greek, it is a form of reproduction in which an egg develops without fertilization
Pathogen: an agent that causes disease
Pelagic: living in open oceans or seas
Photic zone: the region of water where light is visible
Photoadaptation: ability of photosynthetic organisms to adapt to changes in light
Photoinhibition: the cessation of photosynthesis from excess illumination
Photosynthetic: using the products of synthesis from light energy
Phylum: a primary division of a kingdom (above a class in size)
Plankton: pelagic organisms (heterotrophs and autotrophs) mostly found near the surface of the water
Plankton reactor: a vessel, like a refugium, for culturing live plankters
Planulae: the larvae of cnidarians
Plenum: the water-filled space under a sand-bed
Pleopods: abdominal, paired, leg-like swimmerets on crustaceans

Podia: foot-like aspect
Polymorphic: the occurrence of an organism in different forms, stages, or types
Porocytes: incurrent aspects of a sponge forming ostia
Rheotactic: the movement of an organism in response to a current (as with filter-feeders)
Propagule: seed
Protein skimming: (AKA foam fractionation) a form of chemical and particulate filtration designed to exploit the attraction of various dissolved compounds in water to the air-water interface of fine bubbles

R:
Radula: a hard, tongue-like feeding aspect found in most mollusks
Random turbulent: non-linear, multi-directional water flow
Redox: the reduction-oxidation potential of water: a measure of water quality kept in favor around 400mv
Refugia/Refugium: place(s) of safe-haven for the protection and/or propagation of targeted organisms
Rhinophores: tentacle-like olfactory organs on the back of the head of a sea slug

S:
Saturation point: the point at which no additional light can increase the productivity of photosynthesis
Scute: a hard, external plate or scale, like the ridges of a Tridacnid clamshell
Sedentary: living on or attached to the substrate
Senescence: the state of growing old; having a defined lifespan
Sepia: exuded by squid or octopus in flight or fear, commonly called "ink"- it's a solution of the pigment melanin and mucus, and tyrosinase
Septa: partitions of a shell or corallum
Sessile: not free-moving, living/fixed upon the substrate
Setae: bristle- or hair-like aspects
Siliceous: containing or resembling silica
Siphon: a tubular aspect of many reef invertebrates through which water is moved
Siphuncle: the tube that passes through all chambers of the shell of *Nautilus*, which is used to finesse buoyancy
Slough: to shed (mucus typically)
Spicules: small, needlelike structures composed of silicate or calcium carbonate and used in skeletal support and defense
Spongin: a horny, sulfur-containing protein (related to keratin) that contributes to the skeletal structure of Porifera (sponges)
SPS: Small **P**olyp **S**tony (coral)
Stridulation: sound made by lobsters with the rubbing of antennae against ridges on their head
Subchelate: in reference to the (smaller) claw-like paired legs of crustaceans

Subclass: a taxonomic category of related organisms ranking between a class and an order
Substrate: the surface or body of media upon which an organisms grows on and in
Subtidal: the aquatic region that does not experience periods of tidal exposure
Sump: the lowest region/reservoir in a plumbed aquarium system
Symbiotic: a close relationship between two or more different organisms that may or may not benefit mutually

T:
Taxon: a taxonomic category or group
Trochophore: the ciliated larva of various invertebrates like mollusks and annelid worms
Trophic level: a given place in the food chain
Turbulent flow: voluminous random movement of water.

U:
Unicellular: single-celled organism, which can be smaller than "greenwater" (phytoplankton) or as large as *Caulerpa*

V:
Vascular: pertaining to vessels that carry or circulate fluids through an organism
Vegetable filter: type of refugium dedicated to the fixing and/or export of nutrients via cultured plants or algae
Veliger: a larval stage of mollusks
Velum: a ciliated organ that develops in certain larval stages of most marine gastropods- used for swimming
Vermiform: worm-like
Viviparous: live-bearing

Z:
Zooplankton: includes animal plankters of which more than half of all in the ocean are copepods
Zooxanthellae: various algae that live symbiotically within the cells of other organisms, such as corals, clams and other reef organisms

Read about cnidarians like corals, anemones and jellyfish in a future installment of the Natural Marine Aquarium series. (*L. Gonzalez*)

Index

A

abalone 136, 186, 187, 189
Acanthaster 356, 361, 364
Acanthurus 84, 114, 120
Acervochalina 161
Acetabularia 91
Acmaea 202
Acoel 177, 178, 179, 181, 211
Acropora 244, 273, 280, 343
Actinopyga 307, 322
Actinotrichia 91
Aeolid 208, 209, 219
Agelas 157
Aiptasia 70, 117, 264-265, 268
algal scrubber 74, 109
alkalinity 14, 17, 24, 26, 33, 43, 73-78, 87, 91, 96, 98, 101, 107-108, 122-123, 241
allelopathy 161
Allgalathea 293
Allonautilus 253
alpheid 266, 274, 275, 276
Alpheus 274
Amblyeleotris 274
Amblygobius phalaena 84
Amphinomid 165, 172, 173
Amphipholis 358, 359
amphipod 300, 301
Amphiroa 91, 111, 112
Amphiura 358
Anamobaea 171
Anampses 180, 181
anemone shrimp. *See Periclimenes*
Anemonia 55, 70, 117
Aniculus 279, 283
Annelid 32, 135, 165, 172, 174
Anoplodactylus 281
Aplidium 373, 379
Aplysia 211
Aplysilla 213
Aplysina 144, 146, 152
aposymbiotic 42, 63, 78, 151, 169, 227, 374
aragonite 15, 33, 36, 62
Arca 224
Archaster 30, 32, 360
Architeuthis 248
Arctides 299
Arothron 355
arrow crab. *See Stenorhynchus*
Artemia 47
arthropod 61, 85, 169, 170, 172, 175, 185, 249, 251, 255, 261, 281, 279, 282, 290, 291, 295
Ascidian 13, 19, 29, 50, 136, 140, 179, 197, 204, 212, 217, 228, 356, 369-379
Aspidochirote 308
Asterina 64, 272, 361
Asteroid 85, 345, 346, 347, 352, 365
Asthenosoma 327, 328
Astraea 47, 81, 83, 98, 137, 186, 190, 191, 204
Astraea phoebia 191
Astropecten 345, 366
Astrophyton 357
Astropyga 286, 337
Atergia 146
Atrina 228
Atriolum 369, 374
Avicenia germinans 104

B

banded coral shrimp. *See Stenopus*
banggai cardinalfish. *See Pterapogon kauderni*
Baseodiscus 182
Bathynomus 302
Berghia 208, 210
BGA - Blue Green Algae. *See* cyanobacteria
Bispira 168, 169, 171
Bivalvia 136, 185, 222-229
black-spot disease (of fishes). *See Paravortex*
black frilly limpet. *See Scutus*
blood shrimp. *See Lysmata debelius*
Blue Green Algae (BGA). *See* cyanobacteria
blue leg hermit. *See Clibanarius tricolor*
blue Linckia starfish. *See Linckia laevigata*
blue ring octopus. *See Hapalochlaena lunulata*
blue tuxedo urchin. *See Mespilia globulus*
Boergesenia 92
Bohadschia 270, 314, 322
Boodlea 91
Bornella calcarata 212, 221
Bossiella 91, 111, 112
Botryocladia 71, 92
boxer shrimp. *See Stenopus*
box snail. *See Heliacus*
bristleworm 27, 135, 164-165, 172-177, 211, 270, 289
brittle starfish. *See* Ophiuroid
Bryopsis 92, 101, 189
bryozoan 20, 136, 177, 209, 212, 217, 229, 356, 370
bubble algae. *See Valonia*
Bulla 211
bumble bee snail. *See Engina (Pusiostoma) mendicaria*
byssal 226, 229, 234, 236, 237

C

cake urchin. *See* Clypeasteroid
Calappa 279, 283, 287, 290
Calcinus 285, 289
Calliactis 283
Callyspongia 143, 147, 153, 162
camel shrimp. *See Rhynchocinetes*
Camposcia 281, 290
Cancer 155, 158, 287
Cassiopeia 47
Catalaphyllia 34, 39
Caulerpa 48, 55, 69, 72, 77, 87, 88, 92-95, 189, 211, 241, 363, 366
Centropyge 84
Cephalopod 136-137, 185, 246
cephalotoxin 250
cerata 210, 219
Ceratosoma 213
cerith 33, 50, 85, 186, 194, 283
Cerithium 194
Chaetaster 365
Chaetomorpha 63, 65, 75, 88, 95, 101, 130
Chelidonura 178, 208, 211
chiton. *See* Polyplacophorid
Chloeia 173
Chlorodesmis 95, 96, 112
Chondrocidaris 338
Choriaster 270, 347, 365
christmas-tree worm. *See Spirobranchus*
Chromodoris 118, 206, 209, 212-214
Cirolanid. *See* Isopod
cladiella 100
Cladophora 85, 96, 211
Clathrina 148
Clavelina 369, 376, 380
cleaner shrimp. *See Lysmata*
Clibanarius tricolor 282-283, 285, 289
Cliona 146, 159
clown shrimp. *See Hymenocera elegans*
Clypeaster 325
Clypeasteroid 325
Codium 85, 95-96
Colochirus 310, 316, 321
Colpophyllia 119
conch. *See Strombus*
cone snail. *See Conus*
Conus 200
Convoluta 178
Convolutriloba 177-178
copepod 300, 301, 302
Corallina 91
coralline algae 12, 14, 20, 26, 27, 96-97, 103, 106-108, 234, 329, 336, 374
Coriocella 188
cowry. *See Cypraea*
Cribrochalina 154
Crinoid 270, 274, 276, 291, 293, 348-349, 357, 363-364

crown conch. *See Melongena corona*
Cryptocentrus 274
Ctenochaetus 84, 98, 111
Ctenogobiops 274
Cucumaria 309
Cucumarid 305, 307-310, 319-320, 322
Culcita 270, 345, 352, 364-365
cuttlefish. *See* Cephalopod
Cyamus 300
cyanobacteria 58, 70, 73, 81, 91, 97-98, 151
Cyphoma 188, 199
Cypraea 197, 198, 204

D

Dardanus 283, 284, 287, 291
decapod 138
decorator Crab 279-282, 289-290
decorator urchin 325, 330
Dendronotid 212
Dendostrea (oyster) 228
denitrification 30-31, 34, 36-37, 42, 62
Dentiovula 199
Derbesia 89, 102
detritivore 34, 42, 44, 50, 60, 61, 85, 128, 164-166, 172, 173, 182, 202, 289, 356
Diadema 47, 52, 83, 85, 92, 103, 106, 107, 114, 128, 132, 274, 305, 325, 328, 329, 332, 333, 336, 340
diatom 14, 28, 33, 36, 58, 60, 65, 69-74, 78, 81, 90-91, 98, 116, 128, 182, 189-190, 194, 203, 211, 335, 361, 364
Dictyopteris 99
Dictyosphaeria 92, 97, 99
Didemnum 374, 375, 376, 379
dinoflagellate 58, 70, 7381, 83, 91, 98-99, 182
Diodora 202
Dorid 208, 212, 219
Dromid 282
Drupella 200

E

Echinaster 354, 366
echinoderm 13, 59, 64, 93, 107, 139, 185, 209, 229, 263, 271, 272, 305, 307, 313, 317, 320, 323, 326, 328, 331,-334, 340, 343, 346, 350, 353, 354, 356, 377
Echinoid 20, 50, 85, 92, 97, 139, 161, 228, 256, 271, 298, 305, 325-341
Echinometra 329, 332, 333
Echinothrix 328, 329-330, 334
eel grass. *See Zostera*
egg cowry. *See Ovula*
Elacatinus 116, 181
Elysia 85, 96, 208, 211, 212

emerald crab. *See Mithraculus*
Engina (*Pusiostoma*) *mendicaria* 201
Enoplometopus 294, 296, 297
Enteromorpha 100-102, 211
epitoky 135
Epitonium 200
Eucidaris 331
Euglena 99
Eunice 174
Eupolymnia 167
Eurythoe 166, 172, 173
Eurythoe complanata 166
Eusynstyela 371
Euxiphipops 119

F

fanworm. *See* Sabellid or Serpulid
Fasciolaria tulipa 201
feather duster. *See* Sabellid or Serpulid
feather starfish. *See* Crinoid
fiddler crab. *See Uca*
fighting conch. *See Strombus alatus*
Fiji 13
fileclam. *See Lima*
Filograna 166
Filogranella 166
fireworm. *See Eurythoe*;
    See also Hermodice
fire shrimp. *See Lysmata debelius*
fire urchin. *See Asthenosoma*
Flabelligera affinis 340
Flabellina 220
Flabellinopsis 221
flagella 134, 148, 153
flame "scallop"… a fileclam.
    See Lima scabra
flamingo tongue. *See Cyphoma*
flatworm 47, 99, 135, 165-166, 177-181, 208-211, 234, 377
foam fractionation. *See* protein skimmer
foraminiferan 50, 99, 209, 211
Fromia 345, 351, 354, 358, 359, 364-366

G

Galaxaura 91, 100
Gambierdiscus 73, 99
Gammurus 300
gastropod 42, 50, 92, 101, 136, 185-189, 194, 202, 229, 249, 251, 290, 356, 377, 385, 388
Gelloides 157
Glossodoris 213, 214
Glyptoxanthus 280
Gnathophyllid 271
Gomophia 270, 356, 360, 365
Gonodactylaceus 260
Gonodactylus 257
gorgonian 58, 131, 199, 200, 228, 268, 281, 347, 355, 374, 376, 380

Gracilaria 51, 65, 66, 74, 88, 100, 101, 130
Grayella 157
greenwater 57-59, 69, 72, 79, 153, 240-241, 389, 391
Gymnodoris 208

H

hair algae 47, 57, 75, 84, 89, 91-92, 95-96, 98-102, 107-108, 221. *See also Bryopsis*; *Chaetomorpha*; *Derbesia*; *Enteromorpha*
hair worms. *See* Spionid
Halgerda 215-216
Halichoeres 234
Haliclona 143, 146, 150, 152
Halimeda 34, 52, 83, 86, 100, 102, 108, 111, 112, 387
Halophila 108-110
Halymenia 99, 101-103
Hapalochlaena lunulata 250
harlequin shrimp 263, 271, 272, 361. *See also Hymenocera picta*
heart urchin. 339 *See also* Spatangoid
Heliacus 188, 189
hermit crab 42, 60, 63, 83, 85, 138, 194, 250, 279, 282-289, 291, 385-386
Hermodice 165, 166, 172, 173
Heterocentrotus 326, 339
Hexabranchus sanguineus 209, 215
Hippolytid 269
Hippopus 231, 233, 235, 242, 243, 244
Holothuria 32, 308-309, 312, 318, 321-323, 386
Holothuria thomasi 321
Holothurid 307, 308, 310, 317
Holothuroid 32, 50, 139-140, 162, 270, 305-313, 316-322, 326
Homarus 296
Homotrema rubrum 50
Hoplophrys 279
horseshoe crab 255, 279, 291-292. *See also Limulus*
Hyastenus 279
Hydrolithon 97
Hymedesmia 157
Hymenocera elegans 272
Hymenocera picta 271, 272
Hyotissa 225
Hypnea 103
Hypselodoris 214

I

Ianthella 157
Iconaster 366
Ircinia 144, 157, 162
Isochrysis 59
Isognomon 224
isopod 300-303

Istigobius 32, 274

J

Jania 103, 111

K

kalkwasser 26, 45, 78, 98, 389
kelp 69, 72, 91, 100, 103, 107, 108, 204, 326, 387. *See also Macrocystis*; *Sargassum*
keyhole limpet 204 *See also Acmaea*

L

Laguncularia racemosa 104
Lambis 194, 195
Latrunculia 155
Lauriea 281, 371
Leiaster 358, 359, 365
Leucetta 150, 151
Leuconia 157
Limaria 226, 228
Lima scabra 223
limpet 50, 85, 97, 136, 186-187, 202-204
Limulus 291, 292
Linckia 139, 189, 270, 272, 344-345, 354, 359, 363, 365-366, 386
Linckia laevigata 272, 355, 363, 366
Lissocarcinus 280, 293
Lithophyllum 97, 101, 103
live rock 8, 9, 11-29, 31, 39, 40, 44, 46, 47, 48, 57, 67-70, 80, 82, 86, 91-92, 102, 106-108, 115-119, 128, 130, 133, 135, 140, 144, 148, 150, 185, 194, 201, 202, 224, 256, 257, 261, 274, 279, 280-285, 289, 307, 317, 330, 333, 336, 355-359, 361, 365-366, 370, 379, 384-389
live sand 9, 11, 14, 30-45, 130, 144, 202, 273, 282, 289, 325, 358
Lobophora 103, 113
lobster 119, 138, 146, 161, 255, 279, 281, 291, 294, 295-299, 385-387, 390
Loimia 50
Lopha 229
Lucapina 202
Lybia 290
Lysiosquilla 259
Lysmata 138, 242, 264-270, 276, 385
Lysmata debelius 139, 266, 268
Lysmata wurdemanni 264, 265, 268
Lytechinus 329

M

Macrocystis 103, 108, 204
macrophytes 113
Majid 85
manatee grass 107, 110. Also see *Syringodium*
mangrove 34, 49, 52, 55, 60, 62, 66, 104-106, 108, 384, 386
mangrove conch. *See Melongena corona*
mantis shrimp 138, 242, 255-261, 264, 272-273, 285, 385-386. *See also* Stomatopod
mantle 136-137, 185, 189, 198, 202-203, 207-209, 211-212, 215, 219, 223, 225, 227, 229, 232-234, 236, 243, 246, 371, 389
Manucomplanus 283. *See also Polypagurus*
Margarites 190
medusa worm (errantiate). *See* Synaptid
medusa worm (sedentary). *See* Terebellid, *Timarete, Loimia*
Megathura 204
Melongena corona 194
Meristiella 69
mermaid's cup 91
mermaid's cup/wineglass. *See Acetabularia*
Mesophyllum 97
Mespilia globulus 325, 330
Metasepia 246, 251
Microcyphus 328
mimic octopus 248
Mithraculus 85, 92, 280, 287, 289, 293
mithrax. *See Mithraculus*
mollusk 11, 13, 14, 20, 23, 32, 59, 81, 85, 97, 117, 119, 136, 161, 185-186, 197, 200, 206, 209, 222-223, 229, 234, 241, 251, 257, 261, 282, 284, 290, 297, 298, 345, 356, 361, 363, 385-388, 390
Monanchora 143, 157
mud snail 201, 202. *See also* Nassarius
murex snail. *See* Muricid
Muricid 200
Mycale 159, 371
mysid/*Mysis* shrimp 32, 50, 57, 60, 95, 266, 269, 389
Mysidopsis 301

N

Nardoa 366
Naso 18, 84, 92, 109, 114, 120
Nassarius 33, 50, 98, 141, 202
Nautilus 136, 185, 246, 247, 251-253, 384, 387, 390
Nemastoma 104, 106
Nembrotha 217, 377
Neomeris 104, 106
Neopetrolisthes 281, 284, 286, 289, 291
Nerita 193
Nerites 85, 186, 193
Neritina 193
Nitophyllum 99, 107
Notodoris 212
nudibranch 85, 119, 136, 185-188, 206, 209-213, 221, 271, 280, 290, 377, 385-386

O

Ochtodes 71, 88, 386-387
octopus 35, 119, 136-137, 146, 148, 185, 246-252, 261, 295, 346, 385-390
Odontodactylus scyllarus 256, 258, 385
Ophiactis 359
Ophiarachna incrassata 358, 362
Ophidiaster 365
Ophioderma 32, 345, 360, 362
Ophioderma squamosissimum 345
Ophiotheia 357
Ophiothrix 343, 359
Ophiuroid 50, 139, 305, 343, 346, 347, 348, 349, 350, 352, 356, 357, 358, 359, 360, 362, 363
Opisthobranch 136, 186, 189, 206-211, 387
Oreaster 344, 356, 361
Ostreobium 104, 107
Ovula 199
Ovulid 199
oyster 36, 37, 136, 200, 222, 224, 228-229, 315

P

Padina 82, 103, 106, 107
Paguristes cadenati 283, 285, 289
Paguroid 279
Palaemonids 271
Panulirus 296-299
paper shell snail. *See Stomatella*
Paravortex 181
Parribacus 298, 299
Patiria 347, 366
peacock mantis. *See Odontodactylus scyllarus*
peanut worm. *See* Sipunculid
pearlfish 315, 317
Pedum (Scallop) 227
Pelecypods 222
Penicillus 34, 107, 108, 112
Pentacta 310, 311, 313, 316, 321
Pentagonaster 366
pen shell. *See Atrina*
peppermint shrimp. *See Lysmata wurdemanni*
Percnon gibbesi 85, 289, 290
Periclimenes 263, 267, 269, 270, 340
Perna 224, 226
Petrochirus 283, 285, 291
Petrolisthes 284, 288-289, 291
Peyssonnelia 96-97, 107-108
Phenacovolva 200
Phestilla 221
Phormosoma 328
Phyllidia 215, 216

Phyllidiella 216
Phyllidiopsis 216
Phyllodesmium 219
Phyllospongia 143, 158
phytoplankton 14, 33, 54, 56-59, 69, 72, 127,-129, 131, 153, 189, 227, 240, 291, 310, 318-319, 364, 374-375, 386, 391
phytoplankton reactor 227, 240
pillow seastar. See *Culcita*
pistol shrimp. See Alpheid
Plagusia 280
planaria. See Acoel (flatworms)
plankton 11, 14, 37, 39, 47-49, 52, 53, 56-60, 62, 67, 70, 73, 85, 95, 96, 98, 101, 106, 108, 117, 126, 130, 136, 137-138, 143, 153, 189, 193, 210, 222, 224, 227, 229, 238, 240-241, 249, 256, 271, 276, 289, 291, 307-308, 318, 320, 321, 337, 338, 348
plenum 8, 40-41, 390
Pleurobranchus 212
Polycarpa 370, 374
polychaete 14, 22, 60, 135, 209, 234, 261, 270, 289, 340, 387
Polyclad. See Flatworm
Polypagurus 283
Polyplacophorid (Chiton) 187, 203
Polysiphonia 102, 107, 108
pom-pom crab. See *Lybia*
porcelain crab. See *Petrolisthes*
Porcellanella 291
Porifera 134, 143-163, 204, 206, 212, 215, 274, 322-323, 350, 371, 390
Porites 159, 227
Porphyra 113
Portunus 283
Prochloron 374
propagule 105
Prosobranchs 186, 211
protein skimmer 25, 31, 59, 66, 67, 79, 81, 82, 98, 121, 123, 125, 319, 335
Protoreaster 344, 361
Protoreaster lincki 344
Pseudanthias 115
Pseudaxinella 151
Pseudobalistes 115
Pseudobiceros 178, 179
Pseudoceros 177, 179
Pseudocheilinus 176, 234
Pseudochromis 177
Pseudocolochirus 307, 309-313, 322
Pteraeolidia 220
Pterapogon kauderni 114, 340
Pteria 228
Ptilocaulis 151
Puperita 193
Pusiostoma 201. See also *Engina (Pusiostoma) mendicaria*
Pycnopodia 351

Pyramidellid 189, 232, 234
pyram snail. See Pyramidellid

Q

queen conch 194-196. See also *Strombus gigas*

R

rabbitfish 32, 81, 84, 109
RDP 53, 57, 70, 74
red-footed snail 204
red/rust-brown "planaria." See Acoel (flatworms)
redox 14, 55, 75, 77, 78, 87, 98, 169, 210, 390
red (scarlet) leg hermit. See *Paguristes cadenati*
refugium 8, 9, 11, 14, 25, 28-29, 31, 33-39, 42, 44, 46-67, 69, 73-74, 76, 82, 85, 88, 90, 92-98, 101, 105, 106, 108, 109, 112, 117, 118, 126, 128-130, 133, 140, 143-144, 150, 153-155, 157, 159, 164, 166, 167, 170, 182, 185, 193-194, 201, 222, 241, 269, 276, 281, 285, 289-291, 295, 307-308, 318, 319, 321, 350, 354, 359, 361, 363, 375, 388, 390
Reticulidia 216
Rhipocephalus 108
Rhizophora mangle 34, 80, 103-104, 108, 193
Rhodymenia 108
Rhopalaea 369, 377
Rhynchocinetes 263, 264, 268
Roboastra 217
rotifers 33, 133, 154, 289, 319, 357, 364

S

Sabellastarte 167, 168
Sabellid 50, 63, 135, 165-171
Saccoglossa 208, 211
Salarias 84
sally lightfoot (spray) crab. See *Percnon gibbesi*
sand dollar 139, 305, 325, 327, 385
Sargassum 71, 72, 74, 101, 103, 107, 108, 111
Saron 266, 269
scallop 222-224, 226-228, 385, 387, 389
See also fileclam
Schizophrys 281, 282, 288
scud. See Amphipod
Scutus 202, 203
Scyllarides 298, 299
seagrass. See *Syringodium*; See also *Thalassia*
sea star. See Asteroid
sea apple. See Cucumarid
sea biscuit 339, 345, 365

sea cucumber. See Holothurid
sea hare. See *Aplysia*
sea lettuce. See *Ulva*
sea slug. See Opisthobranch
sea spider 279, 281, 283, 290, 387
sea squirt. See Ascidian
sea urchin. See Echinoid
Sepia 248, 250-253, 390
serpent starfish. See *Ophioderma*; See also Ophiuroid
Serpulid 50, 165-169, 171
setae 137, 139, 165-166, 172, 390
sexy shrimp. See *Thor amboinensis*
shrimp goby 273-274, 386
Siphonochalina 161
Siphonodictyon 155, 157
Sipunculid 50, 182-183
slipper lobster. See *Scyllarides*
snapping shrimp. See Alpheid
spaghetti worm. See Terebellid
spanish dancer. See *Hexabranchus sanguineus*
Spatangoid 339
sphere urchin. See *Mespilia globulus*
spider snail. See *Lambis*
Spionid 135, 165-166, 172
Spirastrella 157, 162
Spirobranchus 166-168, 170
Spirorbid 165-167
spirulina 249, 295, 357
Spondylus 222, 224, 229
sponge crab 281, 282
Sporolithon 97
spray crab. See *Percnon gibbesi*
squat lobster 281, 291
Squilloid 257
star turbo snail. See *Astraea phoebia*
Stenopodid 139, 176, 269
Stenopus 139, 176, 264, 266-270
Stenorhynchus 175-176, 264, 266, 289, 292
Stichopus 270, 307, 309, 321
Stomatella 33, 50, 85, 186, 202
stomatopod 27, 138, 175, 248, 255-261, 273, 285, 387
Stonogobiops 274
Strombus 33, 50, 85, 98, 189, 194-196, 201, 204
Strombus alatus 85, 194-195
Strombus gigas 196
Strongylocentrotus 326, 330
Stylopodium 99, 111
Stylotella 149
Sycon 50
Synalpheus 274, 384, 386
Synapta 270, 310-311, 321
Synaptid 165, 308, 310
Synchiropus 180-181
Synoicum 379
Syringodium 34, 87, 107, 110-111

T

Tamaria 359, 365
Tambja 217, 218, 219
Tedania 153
Tegula 190
Terebellid 165, 166, 167, 172, 183
Terebris 200
textile snail. *See Conus*
Thalassia 34, 41, 87, 89, 107, 109-111
Thelenota 308, 323
Theonella 154
thorny oyster. *See Spondylus*
Thor amboinensis 266
Thycra 189
tiger tail sea cucumber. *See Holothuria thomasi*
Timarete 170
Titanoderma 97
Tosia 366
Toxopneustes 328
Trapezid 273, 280
Tridachia. See Elysia
Tridacnid 50, 80, 99, 117, 120, 136, 166, 178, 189, 222-224, 226-229, 231-245, 285, 356, 362, 384-387, 390
Trididemnum 379
Tripneustes 328-329, 335
Tritoniopsis 208
Trizopagurus 283
Trochus 85, 141, 186, 190-191, 204
Tubastrea 63, 120, 200, 221
tulip snail. *See Fasciolaria tulipa*
tunicate. *See* Ascidian
turban 190, 204
Turbellarian 177
Turbinaria 108, 111
Turbo 81, 85, 98, 141, 186, 189-190, 192, 194, 201, 204
turbo snail. *See Astraea; Trochus; Turbo*
turkey wing clam. *See Arca*
Turris 200
turtle grass. *See Thalassia*
Tydemania 112

U

Uca 289, 290
Udotea 34, 52, 111, 112
Ulva 112-113, 211, 387
Ulvaria 113
upside-down jellyfish. *See Cassiopeia*
Urochordates. *See* Ascidian
Urosalpinx 200

V

Valenciennea 32, 180, 181
Valonia 74, 89, 92, 96, 113, 234, 289
veliger 136, 210, 229, 391
Ventricaria 92, 113

Vermetid 50, 167, 189, 197

W

Waminoa 178, 182
whelk 201
wonderpus 247, 248, 249

X

Xanthid 273, 280
Xenocarcinus 281, 282
Xestospongia 160

Z

Zebrasoma 84, 92, 114, 119, 181, 276
Zebrida 286
zooplankton 32, 34, 35, 37, 39, 46, 47, 56-59, 62, 66, 74, 98, 116, 128, 129, 131, 154, 157, 176, 255, 290-291, 310, 319, 391
Zostera 34, 108, 109, 110, 113

Facing: a *Mithraculus* crab cleaning *Pentaceraster cumingi* sea star in an aquarium. (*L. Gonzalez*)

Photography & Illustration Credits

In alphabetical order:

Skip Attix

Scott Boyer

Anthony R. Calfo - www.readingtrees.com

Kevin Carroll

Jason Chodakowski - www.hyperworx.com

Jamie Cross

Diana Fenner - www.disaquatics.com

Robert M. Fenner - the great majority of the photographs are from Bob's personal archives. Photos that do not have a credit explicitly indicated in the caption are from his collection. Please visit www.WetWebMedia.com for more information.

Christina Gonzalez - www.aufsteigen.net

Lorenzo Gonzalez - www.LGonzalez.net

Ken Gosinski - www.KensReef.homestead.com

Richard Hilgers - www.theculturedreef.com

Daniel Knop - www.KnopProducts.com

Ed Kruzel

Denis Lebrun

Jim Nastulski

Barry Neigut - www.ClamsDirect.com

Paul Ponder - http://home.mchsi.com/~epponder/homereef.htm

Greg Rothschild - www.GregRothschild.com

Henry C. Schultz III - www.saltyendeavors.cimaonline.us

James Troeger

All photographs herein are copyrighted to their respective owners/photographers. Reuse or reproduction in any form must be negotiated with the photographer directly.
Some photographs are used in a decorative or graphic spread without specified photographer credit. In these cases, the photographs appear elsewhere within the book, with a photographer credit as appropriate.

About the authors

Anthony Calfo was born in Hawaii and lives in Pennsylvania. He is a lifelong aquarist and an aquarium industry professional that has worked the better part of the last decade as a commercial coral farmer and wholesaler, producing cultured reef invertebrates in a greenhouse environment for the ornamental and zoological trade. He has authored the reef aquarium title, "Book of Coral Propagation, Volume 1" (ReadingTrees.com) and articles for print and electronic journals at large. Anthony travels frequently to visit organizations and clubs to present information on the aquatic sciences. He co-founded the Pittsburgh Marine Aquarist Society with the inimitable Bob Dolan. Schooled at Carnegie-Mellon University and the University of Pittsburgh, he has a BA in English Literature. When not at work generating content for Wet Web Media and Reading Trees publications, he is out seeking gainful employment in a position that involves good beer, fine wine or educating people to the wonders of the natural world. He will have reached nirvana when commissioned to do all three simultaneously.

Jason Chodakowski (tirelessly scanned hundreds of the photos herein) and pal Bob Fenner (right) are shark food at the Scripps-Birch Aquarium in San Diego

Robert (Bob) Fenner has lived the science, hobby and business of aquatics in the Philippines, Japan and United States in all phases- collection, wholesale, jobbing, retail, design, construction and maintenance. He has accomplished the yeoman's chore in doing so at all levels- manager, owner, hatchery worker, retail clerk, and technician. He has worked nearly his entire life in the field of ornamental aquatics. Academic accomplishments includes eleven years of college, a couple of life science degrees and a teaching credential for chemistry, physics and biology. Published works include several studies on aquatic biological and chemical questions, and an extensive publishing (books, articles, etc) and photographic background in the industry and hobby of aquatics. Bob has also taught high school sciences and marine science and aquariology courses at the University of California. He has been an avid aquarist since childhood and is active in hobbyist and scientific organizations. He has served on numerous boards, judged shows and given many presentations and helped form and operate (President) the employee-owned corporation, Nature Etc., Inc. in San Diego- started in 1973. It was a unique turn-key operation in the field of ornamental aquatics, designing and building ponds, lakes, fountains and waterfalls (Aquatic Environments), designing, installing custom aquarium systems and maintenance (Aquatic Life Services), and operating retail outlets (Wet Pets). Currently he does consulting and content provision to the trade, sciences and hobby of aquaristics.

Christina and **Lorenzo Gonzalez**, with the help of son **Kieran** (master chef of peanutbutter/banana/nutella on fine seven-grain bread) had the honor of assembling this book. The creativity flows in the Gonzalez household. They are artists in numerous mediums including, but never limited to, photography, music, juggling(!?), and studio arts.

Look for The Natural Marine Aquarium Series volume 2 - Reef Fishes